Molecular Biology Biochemistry and Biophysics

8

Protein Sequence Determination

A Sourcebook of Methods and Techniques

Edited by Saul B. Needleman

Second Revised and Enlarged Edition

With 80 Figures

Springer-Verlag New York · Heidelberg · Berlin 1975

Saul B. Needleman, Ph. D.

Clinical Regulatory Specialist
Scientific Affairs, Diagnostics Division
Abbott Laboratories
North Chicago, Illinois 60064, USA
Departments of Biochemistry and Neurology
Northwestern University Medical School
Chicago, Illinois, USA

Distribution in the British Commonwealth Market
by Chapman and Hall London

ISBN 0-387-07256-X 2nd edition Springer-Verlag New York Heidelberg Berlin
ISBN 3-540-07256-X 2. Auflage Springer-Verlag Berlin Heidelberg New York
ISBN 0-387-05210-0 1st edition Springer-Verlag New York Heidelberg Berlin
ISBN 3-540-05210-0 1. Auflage Springer-Verlag Berlin Heidelberg New York

Library of Congress Cataloging in Publication Data. Needleman, Saul Ben, 1927— . Protein sequence determination. Molecular biology, biochemistry and biophysics; 8) Bibliography: p. Includes index. 1. Proteins-Analysis. I. Title. II. Series. QD431.N43. 1975. 547'.75. 75-19025.

Typesetting and printing: Carl Ritter & Co., Wiesbaden. Bookbinding: Brühlsche Universitätsdruckerei, Gießen.

For Sondra

and

Marty, Arthur, Beth, Heidi

Preface to the Second Edition

In the few years since the first edition of this book made its appearance, undertakings in the area of protein sequence determination have increased at an almost logarithmic pace. The logic governing such efforts has, predictably, retreated from a simple sense of curiosity in achieving something that had not previously been done, to be replaced by an absolute requirement for sequence information as the best means for understanding heretofore unattainable aspects of chemical, and structural and functional mechanistic phenomena. Witness, for example, recent volumes of the Annual Review of Biochemistry and other review journals, which treat the genetic control of mitochondrial proteins, mechanisms of enzyme action, the immunochemistry of collagen, and the role of lysosomal enzymes in the lipid storage diseases — all on the basis of known protein sequence information. One must note, too, the appearance of related works in the area of nucleic acid sequencing.

The technologic advances in chemical sequencing procedures have proliferated at a remarkable rate. It is in the area of instrumentation, however, that the greatest advances have taken place. As a consequence of this, each chapter (with one single exception) in this book has been extensively revised and updated. Entire chapters have been replaced in several cases, without minimizing the value of their original content. In addition, a second volume will appear soon, dealing with subjects not covered in the present book.

Again, it represents a prodigious undertaking to thank all the people who have contributed to the success of this enterprise. To these, I can only express my sincerest gratitude for their sustained encouragement and criticism . . . and friendship.

Chicago, May 1975 | SAUL B. NEEDLEMAN

Preface to the First Edition

"All the King's horses and all the King's men couldn't put Humpty Dumpty together again."

It is entirely possible that the difficulty facing "all the King's men" was principally the lack of a sufficient guide to the techniques of reassembling from a series of small components, the original entity. It is the sincere hope of the editor of the present work and of each of the contributing authors that the modern researcher will not face a similar predicament in his endeavours to reconstruct the complete primary sequence of a protein from the array of component amino acids. Rather, it is the intent that, with this volume, he may proceed untimorously if not with outright confidence toward achieving his purpose.

To the newcomer in protein sequencing, compelled by necessity, or fascination, to investigate the exact order of amino acids in proteins, the question of "where to begin" — or "how to do it" is urgent. To those more skilled, a ready source of additional techniques should nevertheless be of value. This volume attempts to present in a single source a discussion of the methods and techniques useful to the determination of the primary structures of proteins and peptides. Hopefully, this book will tell the reader "how to do it".

Theoretical considerations, the derivation of mathematical concepts, discussions of active sites and of mechanisms of action are held to a minimum. While such topics are of considerable interest to many, they would burden the newcomer with details beyond his immediate needs. Ample references to the literature have been included to provide an extensive bibliography for the advanced worker and to satisfy the curiosity of the graduate student. Although not intended primarily for this purpose, the diversity of topics included make this volume a conceivable choice as a text for a senior or graduate course in biochemical techniques and methodology.

The contributors to this volume have been carefully selected to insure a balance of ideas from established leaders in the field and from "second-generation" workers who are even now in the process of making their primary scientific contributions. The international origin of the authors reflects the widespread interest in primary structure and conformational studies of proteins. The selection of authors and of subject matter was made solely by the editor and any reproach resulting from these selections must be directed towards him.

As is usual in such cases, credits must be given to many persons who directly or indirectly have contributed to the creation of this work. Perhaps the greatest vote of thanks must be given to every worker in the field who has come before, for it is they who made possible, who provided the means for that which appears in this volume, to aid those who will follow. In a more immediate sense, however, the editor must specifically mention with thanks and gratitude Dr. ALFRED GOLDSTONE, who was more a friend and advisor than co-author on this undertaking, and Mr. THEODOR K. MAYER who shared a load not really his. Finally, but not in the least, how does one thank SONDRA, my wife, who spent the many hours listening patiently to words beyond her ken but who gave them meaning.

Highland Park, Illinois
July 1970

SAUL B. NEEDLEMAN

Contents

Contributors

EDMAN, PEHR	Max Planck-Institut für Biochemie, Martinsried/München, W. Germany
FONTANA, ANGELO	Istituto di Chimica Organica, Universita di Padova, Padova, Italy
GISH, DUANE	Institute for Creation Research, San Diego, California, USA
GOLDSTONE, ALFRED D.	Neurology Department, Northwestern University Medical School and Veterans Research Hospital, Chicago, Illinois, USA
HARE, P. E.	Geophysical Laboratory, Smithsonian Institution, Washington, D. C., USA
HENSCHEN, AGNES	Max Planck-Institut für Biochemie, Martinsried/München, W. Germany
KASPER, CHARLES B.	McArdle Laboratories, University of Wisconsin, Madison, Wisconsin, USA
MATSUO, HISAYUKI	Institute for Protein Research, University of Osaka, Osaka, Japan
NAKAJIMA, TERUMI	School of Medicine, Hiroshima University, Hiroshima, Japan
NARITA, KOZO	Institute for Protein Research, University of Osaka, Osaka, Japan
NEEDLEMAN, SAUL B.	Scientific Affairs, Diagnostic Division, Abbott Laboratories, North Chicago, Illinois 60064, USA
PISANO, JOHN J.	National Institutes of Health, National Heart Institute, Bethesda, Maryland 20014, USA
SCOFFONE, ERNESTO †	Formerly: Istituto di Chimica Organica, Universita di Padova, Padova, Italy
VAN HOLDE, KEN E.	Department of Biochemistry and Biophysics, Oregon State University, Corvallis, Oregon, USA

Chapter 1

General Considerations

Saul B. Needleman

Protein sequence determinations have reached new levels of interest in recent years such that these studies no longer are confined to a few major protein laboratories. Now, researchers in widely diverse areas of biochemistry and the medical sciences face the requirements for knowing the primary sequence of a protein. Indeed, the time is approaching, perhaps, when the true essence of many research programs will necessitate the complete characterization of the amino acid sequence of the protein of central interest. The reasons for wanting to know the sequence of a protein probably are as great and as complex as the number of research problems dealing with proteins.

Unfortunately, wanting to know the sequence — or even requiring that information — does not accomplish it. Perhaps even more unfortunate is the prevailing idea that sequencing is difficult to do. It must be said that this is only partially the case. From the fundamental techniques of the early workers have evolved a diversity of approaches to sequence studies which have become standard procedure in the protein laboratory. In the 30 years since Sanger's revolutionary work on the structure of insulin, it is safe to say that development of new techniques or modifications of the old ones have increased in direct proportion to the number of workers in the field. While this might appear to add to the difficulties or complexity attending a detailed examination of the amino acid sequence of a protein, this diversity in available techniques has, instead, made the basic resolution of the problem attainable in the moderately equipped laboratory — when provided with sufficient literature sources to guide the way.

What does this book intend to accomplish? It is the intention of the contents of this volume to provide a quick but detailed discussion of the techniques involved in sequencing a protein — to provide a handy and fairly complete reference to the literature on this subject — to tell you "how to do it".

It is essential — in fact, an absolute requirement — that the purity and homogeneity of the protein whose sequence is to be determined be well defined. There is nothing more disconcerting, at the final stages of reconstruction of the completed sequence from the collection of short peptides, than to have segments "left over". It is puzzling to find one or two peptides, apparently obtained through rigorous experimental procedure, which do not seem to fit into an otherwise logical primary protein structure. More than once has the presence of a second "isomorphic" structure been confirmed on the basis of small bits remaining from a tedious study inadvertently initiated on an impure protein. One might imagine that the presence of even small amounts of contaminating material would create considerable confusion, even with the use of computer programming. Towards this end, in Chapter 2, is presented a discussion of analytical procedures used in establishing the purity and homogeneity of a protein. The physical characterization of the protein molecule is examined according to a series of convenient parameters — the mass, charge and molecular size of the molecule. Thus the application of sedimentation methods in the analytical ultracentrifuge provides one means for establishing the homogeneity of a protein. Electrophoresis differentiates members of a series according to charge and gel filtration is discussed as a technique for separation according to molecular size. The problems of purification are approached in terms of the criteria to be applied to the selection of the most reasonable procedures for the protein of interest and for establishing that purification has proceeded to the point of readiness for further analysis. Included in this chapter is a discussion of the special problems introduced by the presence of subunits in the protein (see below).

Once the protein has been obtained in the requisite state of purity, it becomes of interest to determine whether or not subunits are present in the total structure. Multi-subunit structures can be reduced to the individual subunits by appropriate concentrations of various salts or by changes in pH — or may require disulfide bond reduction followed by alkylation. Further separation and purification of the subunits must then follow.

The presence of subunit structures may be confirmed further by determining the nature and number of amino acid residues which occupy the amino-terminal and carboxyl-terminal positions in the protein. Nevertheless, knowledge of the terminal residues is of value as a guide in the further sequencing process and the techniques used in obtaining this information are described in Chapters 3 and 4. The three most commonly used procedures for N-terminal analysis are presented in detail together with information on several other more specialized methods which have appeared. Special attention is given to the case of the masked amino-terminus and to proteins having unusual configurations. Several procedures for carboxyl-terminal analysis are described in Chapter 3 and special attention is devoted to the chemistry of the tritiation procedure in Chapter 4.

With few exceptions, the protein, as isolated from its native source, is too large, too complex for direct application of sequencing techniques. While it might be possible to establish the order of the first half dozen or so amino acids in the protein with little difficulty, further progress would be impeded by the increased resistance with each succeeding step to undergo reaction. Thus, at each step, incomplete release of amino acids would lead to the accumulation of proteins in varying stages of stepwise degradation. Interpretation at each additional step would become increasingly difficult and meaningless. A much simpler approach to sequence determination involves cleavage of the original protein in a known manner into a group of short peptides. These can be fractionated into the component peptides each of which in turn can be subjected to detailed sequence analysis with considerably less difficulty than could the original intact protein. In Chapter 5 are presented two chemical procedures for fragmenting a protein into short peptides for sequence studies. Several enzymic methods are discussed, these including procedures using enzymes having highly limited specificity as well as those with broad specificity. In addition to the fragmentation methods, there is presented an additional discussion of techniques for the separation of the peptide mixtures obtained through these methods. This supplements the procedures outlined in the second chapter.

The analysis of each small peptide is aided by knowing whether specific amino acids are present in the peptide. Special color reactions are known for several of the amino acids and the necessary reagents and details for performing the color analysis are given in Chapter 6. Similar color reactions may be used on the intact protein to guide in the selection of proper non-enzymic procedures for the chemical cleavage of the protein to smaller units prior to enzymic hydrolysis to the peptide level.

The chapter on amino acid composition analysis has been completely revised and as such represents a departure from — but a supplement to — the information contained in the corresponding chapters in the first edition of this book. The earlier work dealt largely with the conventional analytical procedures for composition analysis; the present chapter discusses in some detail the parameters governing the several aspects of analysis permitting one to select special analytical conditions for specific analytical requirements and thereby to depart from the more commercial equipment which is available. Thus one is encouraged to design instrumental approaches tailored to one's own particular requirements.

The chapter devoted to the chemical release of single amino acids from the N-terminus of a peptide or protein has been extensively revised. The details of the Edman procedure presented in the first edition constituted the only complete discussion of the theoretical basis for this technique theretofore available. The chemical foundation for the reactions involved has been re-evaluated in terms of later information and the section on the application of the method to automatic instrumentation has been greatly elaborated beyond the scope presented earlier.

Chapter 9 represents another fresh contribution to this second edition. While the original material dealt with the application of gas chromatography to the analysis of total amino acid composition — and, as such, retains its utility in terms of theory and methodology of the gas chromatographic technique, the present chapter deals exclusively and extensively with the application of gas liquid chromatography and thin layer chromatography to the characterization and identification of the phenylthiohydantoin derivatives of amino acids obtained through the Edman degradation procedure and with certain closely related substances. Considerable attention is given to the preparation of columns and to the parameters governing their use.

The chapter on the re-alignment of known peptide fragments into the complete protein sequence has been up-dated but, alas, no computer-based procedure has yet emerged which can be applied with reasonable simplicity and accuracy towards this reconstruction of the total sequence of a protein from data generated through precise chemical or enzymic procedures. Where the computer technique exists, the input is the result of inprecise procedure or is of such complex nature as to make computer application to the problem essentially useless in comparison to the manual technique. The addition of several highly specialized enzymes of unique specificity to the armatorium of the protein biochemist has greatly simplified the sequencing process and has obviated some of the tedium heretofore encountered. Such methodology is illustratively applied to determining the sequence of an artificial representative peptide.

The final chapter remains unchanged from the first edition. The extensive description of the variety of chemical blocking and unblocking procedures remains as current and as useful now as when it was first presented. The major advances in this field have come in the application of these techniques to the Merrifield Solid Phase procedure. This area was treated only minimally in the present chapter but becomes the subject of a separate and major contribution to the second volume of this book.

As the matter of sequencing a protein has grown in complexity, the procedures for such studies have increased in sophistication. This is in line with the increased numbers of scientists applying their skills to larger and larger molecules. Nevertheless, in this edition more attention has been devoted to the less fully equipped laboratory; the emphasis is less towards commercial instrumentation and more towards the "do-it-yourself" approach. The theoretical parameters for each phase in the sequencing process are described in sufficient detail to permit a personalized approach within the available limitations of any reasonably stocked biochemical laboratory. Where details prove to be insufficient in the text, recourse can be had to the almost 1500 references to the original literature.

Each chapter stands alone as a primary reference to one aspect of the overall problem, that of sequencing a protein. Yet, the whole can only be the sum and continuity of its parts. Thus this volume represents in fact an integrated presentation. This book serves as a road map through a challenging and thought-provoking adventure.

Humpty Dumpty sits on the wall. The egg is not yet broken. It is time to begin!

Chapter 2

Physical Characterization of the Protein Molecule

Ken E. Van Holde

I. Introduction to the Problem

Before any protein can be considered for sequence analysis, there are certain fundamental questions about it that must be answered. These include, in a reasonable order of inquiry:

1. Is the Protein Preparation Pure? There is no sense in expending the time and effort required by sequence analysis on proteins that contain appreciable amounts of other, contaminating proteins. This is a minimal requirement.

2. About How Large Is the Active Protein Molecule? The kind of information required here is a semiquantitative estimate. This information is primarily of importance (to the sequencer) with respect to the next questions.

3. How Many Polypeptide Chains Does the Functional Protein Contain? How Large Are They? Are They Identical? These questions are of greatest importance, for together with [4], they define the magnitude of the sequencing task. If one is fortunate, and there is only one kind of polypeptide chain in the protein, it is still important to have a very good idea of the chain weight and, hence, the approximate number of residues to be expected.

4. If the Chains Are Not Identical, How Many Kinds Are There, and How Can They Be Separated? If it is found that the active protein consists of two or more distinct kinds of polypeptide chains, it will be necessary to find ways of effecting dissociation and preparative separation. Each then becomes a separate sequencing problem.

It is in answering these fundamental questions that the techniques of physical biochemistry can be of aid to those involved in sequence determination. It is important that the principles and limitations of these methods be thoroughly understood by the researcher, for such studies lie at the very root of his analysis. Substantial errors here will result in much misdirection of later research. The techniques that will be discussed in this chapter have many other uses. In each kind of application the requirements as to precision, and the dangers to be avoided are different. In this chapter, it will be presumed that the aims of the physical characterization are to answer the questions listed above, in preparation for sequence analysis.

In the following sections, a number of major techniques used for physical characterization and separation of protein molecules will be briefly described. In the concluding section, the way in which these might be used to answer the questions listed above for a typical protein will be illustrated.

II. The Accessible Physical Parameters

At the present time, there are a limited number of physical parameters which may be readily used to characterize macromolecules in solution, or which may be employed in separation. These include *molecular weight*, *density*, *charge*, and in a somewhat less precise way, *molecular dimensions*. A macromolecular preparation can be heterogeneous or homogeneous in any one, or all, of these parameters. As an example, let us consider molecular weight. Three different situations may be distinguished within two categories:

a) Homogeneity: All of the macromolecules in the sample have exactly the same *molecular weight*.

b) Paucidispersity: The sample contains a wide range of molecular species. This kind of system is frequently found with synthetic (and some natural) polymers. Proteins that are to be the subject of sequencing studies are very unlikely to be polydisperse, so we shall not deal with such systems in any detail.

Table 2-1. Average molecular weights

Designation	Defining equation	Measured by
Number average	$M_n = \Sigma W_i / \Sigma (W_i / M_i)$	Osmometry, vapor pressure, sedimentation equilibrium
Weight average	$M_w = \Sigma W_i M_i / \Sigma W_i$	Sedimentation equilibrium, light scattering
Z Average	$M_z = \Sigma W_i M_i^2 / \Sigma W_i M_i$	Sedimentation equilibrium
Z + 1 Average	$M_{z+1} = \Sigma M_i M_i^3 / \Sigma W_i M_i^2$	Sedimentation equilibrium
etc.		

A complete description of a paucidisperse system would list the molecular weights and amounts of all species present. However, many physical methods do not yield such detailed information, but give instead *average* quantities. In considering molecular weight, for example, averages may be defined with respect to the weight fraction (W_i) of the various species, as shown in Table 2-1. It can be easily shown that for a molecularly homogeneous substance, $M_n = M_w = M_z = M_{z+1} = \ldots$, whereas for a heterogeneous material, $M_n < M_w < M_z < M_{z+1} < \ldots$. The identity of several properly measured molecular weights is one critical test for homogeneity.

Heterogeneity in molecular weight is by no means the only kind. In every measurable physical parameter, a sample may or may not be homogeneous. For example, preparations may be obtained which are roughly homogeneous in molecular weight, but heterogeneous in density. Two polypeptide chains may have almost exactly the same weight, but differ in charge, because of amino acid substitution. Finally, it is possible, at least in principle, that a molecularly homogeneous protein might exhibit conformational heterogeneity. It is because of these possibilities that more than one physical technique is usually needed to establish the purity of a protein preparation.

III. Sedimentation Methods

A. Sedimentation Velocity

The moving-boundary sedimentation velocity technique is probably the best known of all sedimentation methods, and often the first to be employed in the physical investigation of a protein.

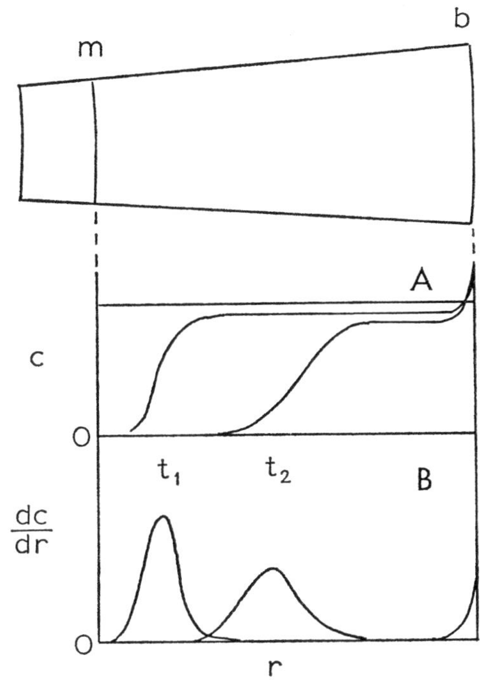

Fig. 2-1 A and B. Concentration (A) and concentration derivative (B) in a typical sedimentation velocity experiment with a high molecular weight solute. Patterns at zero time and for successively later times (t_1 and t_2) are shown. Sedimentation is from left to right; the meniscus is marked *m*, the cell bottom *b*. A schematic view of the ultracentrifuge cell is shown at the top. Note that the concentration difference across the boundary decreases with time, because of the radial dilution effect, and that the boundary spreads because of diffusion

As shown in Fig. 2-1, a macromolecule solution, initially at uniform concentration in an ultracentrifuge cell, is subjected to a high, constant centrifugal field. If the macromolecules are sufficiently large, a boundary is formed, which sediments down the cell. It can be followed by either measuring the light absorption in the cell, as in "scanner" optical systems, or by measuring the refractive index gradient (as in the "schlieren" system). The former will yield a trace proportional to the protein concentration (Fig. 2-1 A), the latter a pattern proportional to concentration gradient (Fig. 2-1 B). For details of technique, the reader is referred to a number of books and reviews [252, 1079, 1306].

The sedimentation coefficient, s is defined as

$$s = \frac{dr_b/dt}{\omega^2 r_b} \tag{1}$$

where r_b is the boundary position, ω the angular velocity of the rotor (radians/sec) and t the time. Since the boundaries are always broadened by diffusion, the point r_b needs definition. It is usually taken as the point of maximum gradient. (But see Ref. [1306], p. 103, for a more detailed discussion). Calculations are usually done from the integrated form of Eq. (1):

$$\ln r_b = \omega^2 st + C \tag{2}$$

where C is a constant of integration. Since s can, and often does depend upon concentration, a series of experiments at different concentrations should be done and extrapolated to infinite dilution.

The sedimentation coefficient, as such, is not a fundamental molecular parameter, for it depends not only on the molecular weight, but also on the partial specific volume ($\bar{v}$) and the shape and hydration, as reflected in the frictional coefficient f. We have

$$s = \frac{M(1 - \bar{v}\varrho)}{Nf} . \tag{3}$$

Here ϱ is the solution density and N is Avogadro's number.

While the complicated dependence of f on both size and shape of protein molecules makes it impossible to provide an exact, general relationship between s and M, there do exist approximate relationships which, if used with caution, can provide a rough value of the molecular weight from s values alone. If the protein molecule were a sphere, one could say that $f = 6\pi\eta R$, where R is the molecular radius and η the solution viscosity. Further, if hydration is negligible, $R = (3M\bar{v}/4\pi N)^{1/3}$, so that

$$s = \frac{M^{2/3}(1 - \bar{v}\varrho)}{6\pi\eta\,(3v/4\pi N)^{1/3}} . \tag{4}$$

This implies that $s\bar{v}^{1/3}/(1 - \bar{v}\varrho)$ should be a universal function of M for spherical, unhydrated proteins. Actually, real globular proteins appear to differ from this approximation in a reasonably consistent way, so that a graph of $s\bar{v}^{1/3}/(1 - \bar{v}\varrho)$ *vs* $M^{2/3}$ is in fact nearly a straight line (Fig. 2-2). What this means is that if one is willing to gamble that the protein in question does not have an anomalous shape (like myosin or fibrinogen) a reasonable estimate of M can be found from s and $\bar{v}$. Since $\bar{v}$ can be estimated from amino acid analysis [239], only s is needed.

The relationship shown in Fig. 2-2 is of the same order of reliability as estimates of molecular weight obtained from such techniques as gel permeation chromatography. Since it is at least as easy to do a sedimentation velocity experiment as to calibrate and run a gel column, this method should be considered for similar purposes. All such techniques, it must be emphasized, suffer from the danger that wildly erroneous results can sometimes be obtained. Note the error that would occur from using the $s\bar{v}^{1/3}/(1 - \bar{v}\varrho)$ value of myosin in this fashion! Such estimates should never be attempted on proteins suspected of being fibrous or denatured.

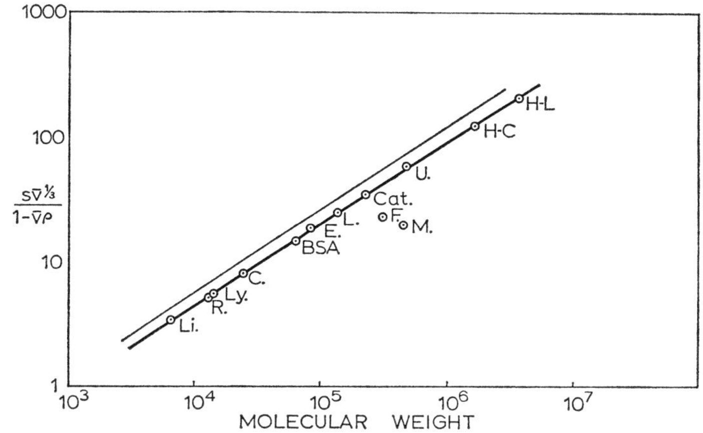

Fig. 2-2. The quantity $s\bar{v}^{1/3}/(1-\bar{v}\varrho)$ *vs* M on a log-log plot. Data are shown for a number of proteins, the molecular weights of which are known with precision. All sedimentation coefficients are expressed in Svedberg units (1 Svedberg = 10^{-13} sec) and are corrected to water at 20° C, and extrapolated to $c = 0$. The heavy solid line through the points has the equation $s\bar{v}^{1/3}/(1-\bar{v}\varrho) = 1.12 \times 10^{-2}\, M^{0.65}$. The light line is the theoretical line for unhydrated, spherical molecules [Eq. (4)]. Code for proteins: *Li* = lipase; *R* = ribonuclease; *Ly* = lysozyme, *C* = chymotrypsinogen *A*; *BSA* = bovine serum albumin, *E* = enolase; *L* = lactic dehydrogenase; *Cat* = catalase, *U* = urease, *H-C* = hemocyanin from *Callianassa*, *H-L* = hemocyanin from *Loligo*, *F* = fibrinogen, *M* = myosin

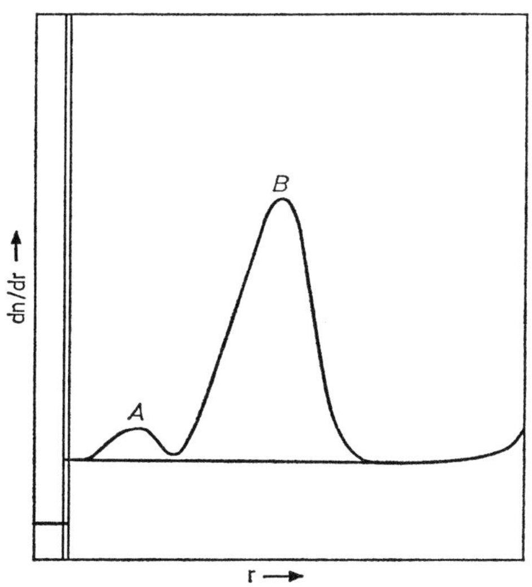

Fig. 2-3. A schlieren pattern from sedimentation of a hypothetical protein. The main component (*B*) is contaminated with a small amount (ca. 10%) of a more slowly sedimenting impurity. The main component may or may not be homogeneous; the slight skewing of the boundary could result from concentration dependence of *s*

In addition to its use in giving rough estimates of protein molecular weights, sedimentation velocity analysis is frequently employed to check for sample homogeneity. The analysis is all too often incomplete; in some cases only a single experiment at a single rotor speed and concentration is reported. A typical result is shown in Fig. 2-3. In this example, two boundaries are observed.

Following such an analysis, the statement will frequently be made that "with the exception of a small amount of slowly sedimenting material, the protein appeared to be homogeneous in the ultracentrifuge". A statement of this kind is almost devoid of meaning. In the first place, what is meant by "a small amount" of slowly sedimenting material? To give a more precise statement, one might determine the fraction of a small component by measurement of the areas of the two peaks. This will give (approximately, see below) the weight fractions of the component. (More exactly, it will give the refractometric fraction, probably about the same thing.) How is this to be determined? It must be remembered [1079] that the area under each peak will diminish with time because of the radial dilution effect. Precisely, the apparent concentration (c_p) will vary as

$$c_p = c_0^{(-2s\omega^2 t)} \tag{5}$$

where the s value is for the boundary in question, and c_0 the initial concentration. Boundaries with different velocities will vary differently in apparent area with time; the area of each should be extrapolated back to zero time. However, this may still not yield an accurate representation of the relative amounts of the two components. If the sedimentation coefficients of these materials vary with protein concentration, as is usually the case, an irksome complication called the "Johnson-Ogston" effect will be involved. The slow component will sediment more rapidly in the region behind the fast boundary, where it is not interacting with the fast component. It will therefore "pile up" behind the fast boundary, leading to an overestimation of the amount of slow component and an underestimation of the amount of fast component. This effect can be corrected for (see [1079]) but the correction requires data not usually available at this level of analysis.

In many cases (especially if the boundaries differ greatly in s) the Johnson-Ogston effect is not too serious. In this event, correction or extrapolation for the effect of radial dilution leads to a reasonable analysis. Suppose the weight fraction of slowly sedimenting material is small (say 10%). Is this significant? The answer depends upon what kind of studies are to be made with the mixture. Suppose that end groups are to be determined, and that molecular weight of the slow component is only $^1/_4$ that of the fast. Then about 31% of the molecules (number fraction) will be of the smaller species. Such contamination might be disastrous for studies of this kind. It would be essential to subject the sample to some kind of fractionation procedure (gel permeation chromatography, for example — see 2. III.) before continuing the study. A "minor" contaminant may not be inconsequential!

Let us analyze the hypothetical statement about Fig. 2-3 further. It is implied that the major component is homogeneous, simply because a single boundary is observed. There are many ways in which a heterogeneous substance can yield a single, fairly sharp boundary. To list a few:

1. The substance may be pauci- or polydisperse, with a number of components of very similar sedimentation coefficient, not resolved because of diffusion broadening

of the boundary. This danger is particularly acute with low molecular weight ($M < 20000$) substances, for diffusion will be rapid and sedimentation slow. Such mixtures will not give boundaries of the same shape as would be expected for a single component. However, there is no easy test to apply, for the shape of the boundary for even a homogeneous material is not strictly a Gaussian diffusion boundary [400]. Furthermore, even the sum of two Gaussian curves looks Gaussian if the peaks are not resolved. Problems of boundary resolution are considered in some detail in SVEDBERG and PEDERSEN [1248].

2. Because of concentration-dependent sedimentation, the boundary may be sharpened. A word of explanation will suffice here: in general, s will decrease with increasing solute concentration. This means that molecules on the low concentration edge of a boundary will tend to catch up with molecules on the high concentration side. In extreme cases, this can cause a very heterogeneous substance to appear homogeneous if the concentration used to load the cell is high.

3. Substances which are in rapidly reversible equilibrium may not be resolved in sedimentation. For example, a substance in rapid monomer-dimer equilibrium will yield a single boundary with s varying between that of monomer and dimer as the concentration is increased. For a detailed discussion of the complexities of sedimentation of associating systems, see the book by CANN [204].

The pitfalls described above could be largely avoided by more experiments. To test the homogeneity of a low molecular weight substance critically, the highest rotor speeds and lowest practical concentrations should be employed. Variation of the concentration over wide limits will usually detect concentration-dependent boundary sharpening and rapid equilibria. For these reasons, the frequent admonitions that sedimentation velocity data be extrapolated to zero concentration are not nit-picking.

Even after the best has been done, the sedimentation velocity method does not easily give assurance of homogeneity. The very best would be this: to use a known diffusion coefficient and concentration dependence of s to predict the boundary shape at different times [400], and compare with observation. This is almost never done. However, intelligent use of the sedimentation velocity method, utilizing variation of both rotor speed and concentration, can at least provide evidence against gross paucidispersity. Together with this, it can give an estimate of the molecular weight of the protein under well defined solution conditions.

B. Sedimentation Equilibrium

For critical tests of homogeneity in mass, the sedimentation equilibrium method is more powerful than the velocity techniques. This is because it allows (with proper care) the unambiguous determination of several molecular weight averages.

Let us consider what is observed in such an experiment. We shall not be concerned here with either the theory of sedimentation equilibrium or details of experimental procedure. For these, the reader is referred to SCHACHMAN [1079], FUJITA [400], or to a number of review articles (WILLIAMS et al. [1379], VAN HOLDE [1304], TRAUTMAN [1290].) Instead, we shall emphasize the problems and potentialities in the analysis of the sedimentation equilibrium results.

Suppose that the ultracentrifuge has been operated, at a relatively slow speed, for a time sufficient for equilibrium to be attained. Then for a homogeneous solute, or for each component in a mixture, the concentration will be distributed exponentially in the cell. At a point r cm from the center of rotation the concentration will be:

$$c\,(r) = c\,(a) \exp\{M\,(1 - \bar{v}\varrho)\,\omega^2\,(r^2 - a^2)/2\,RT\}\,. \tag{6}$$

Here a is the distance of the meniscus from the center of rotation, $\bar{v}$ is the partial specific volume of the solute, ϱ the solution density, R the gas constant, and T the absolute temperature. This equation assumes that the solution behaves ideally. In practice ideal behavior can only be approached, (although often very closely) in experiments at finite concentration. The use of Eq. (6) requires that the concentration, or some quantity proportional to the concentration, be measured at different points in the ultracentrifuge cell. This will be easy to do if a scanning absorption optical system is employed; it may even be possible with such an optical system to measure the concentrations of different components independently. More commonly, how-

$$\ln c(r) = \text{constant} + M(1 - \bar{v}\varrho)\,\omega^2 r^2/2\,RT\,.$$

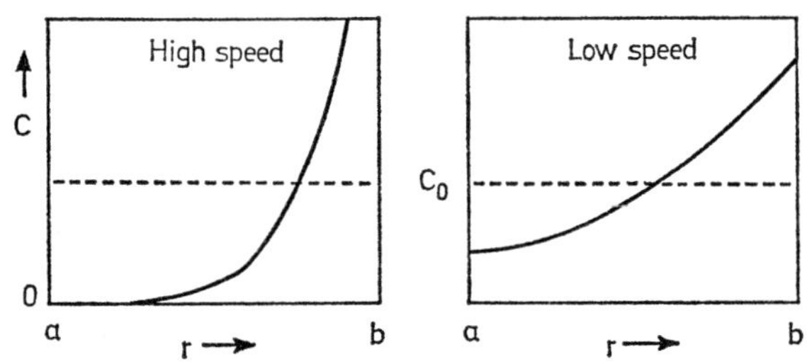

Fig. 2-4. The difference between the "low speed" and "high speed" sedimentation equilibrium methods. (See text)

ever, one uses the schlieren or Rayleigh optical systems. The former yields the refractive index gradient, the latter the refractive index difference between any two points. Obviously, the schlieren patterns can be integrated to yield concentration differences between points as well. Note, however, that neither yields the absolute concentration needed for Eq. (6). To cope with this problem, two different lines of approach have been used:

1. In the "high speed" method (also called "Yphantis method" or "meniscus depletion method" [1304, 1413], the rotor velocity is chosen sufficiently high that the concentration near the meniscus becomes negligibly small (see Fig. 2-4A). Then concentration differences between point a and any other point (r) yield the total concentration at r.

2. In the "low speed" method [1304, 1305], (Fig. 2-4B) the concentration is not zero at the meniscus; ideally the concentration ratio between the ends of the cell is about 3:1. To obtain the meniscus concentration (and thus the concentration at other points) one makes use of the conservation of mass, which yields

$$c\,(a) = c_0 - \frac{1}{b^2 - a^2} \int_a^b (b^2 - r^2)\,\frac{dc}{dr}\,dr \tag{7}$$

where b is the bottom of the solution column, and c_0 is the initial concentration. The latter is frequently determined in terms of area on a schlieren diagram, or number of interference fringes, in a "synthetic boundary" experiment in which solvent is layered over solution.

With $c(r)$ known as a function of r, a substance may be tested for homogeneity by use of a form of Eq. (6). Taking natural logarithms, Eq. (6) converts to the form

$$\ln c(r) = \text{constant} + M(1 - \bar{v}\varrho)\,\omega^2 r^2/2RT. \quad (8)$$

Thus, for a homogeneous substance, a graph of $\ln c$ *vs* r^2 should be a straight line. If the substance is not homogeneous in molecular weight, the total concentration (which is what will be determined by most optical methods) will be a sum of exponential functions, and $\ln c(r)$ *vs* r^2 will show upward curvature. The observation of such curvature is direct evidence for heterogeneity but appreciably heterogeneous substances may, under some circumstances, give nearly straight lines. The low speed and high speed methods differ somewhat in sensitivity to heterogeneity.

Methods for treating the data so as to most easily detect heterogeneity are discussed by Fujita and Williams [400 A]. In general, the high speed technique, which produces greater resolution of components in the cell, will be more sensitive. In Fig. 2-5 are shown the kinds of $\ln c$ *vs* r^2 curves to be expected for several mix-

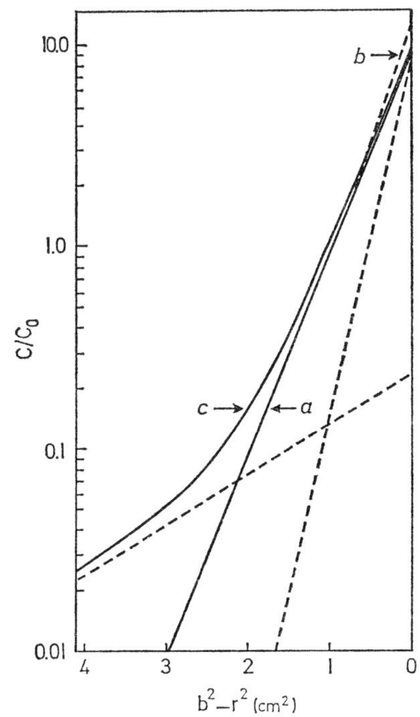

Fig. 2-5. Approximate graphs of $\ln c$ *vs* r^2 for hypothetical mixtures. In each case the high speed method has been used. Curve (*a*) represents a homogeneous protein. In curve (*b*), 10% of this material has irreversibly dimerized; in curve (*c*) 10% has irreversibly dissociated to $^1/_4$ molecules. The dotted lines show expected slopes for dimers and $^1/_4$ molecules. For (*b*), $M_z/M_w = 1.18$, for (*c*) $M_z/M_w = 1.06$

tures. Note that the high speed technique is capable of revealing a fairly small amount of low-molecular weight contaminant. With high molecular weight impurities it does not fare quite so well, since it may not be possible to resolve data all of the way to the bottom of the cell. This can be remedied by carrying out experiments at very low initial concentrations, but under these conditions the ultracentrifuge must be operating very smoothly to avoid convection [1413]. The use of $\ln c$ *vs* r^2 graphs as a critical test for homogeneity is not to be recommended. A better method is to obtain average molecular weights at different points in the ultracentrifuge cell. For example, the weight average molecular weight at each point can be obtained by numerically differentiating the $\ln c$ *vs* r^2 curve. For a heterogeneous mixture it can be shown (see, for example YPHANTIS [1413])

$$\frac{d \ln c}{dr^2} = \frac{M_{wr}(1-\bar{v}\varrho)\,\omega^2}{2RT} \tag{9}$$

here M_{wr} is the weight average molecular weight of the mixture existing at the point r, where the derivative was taken. Methods exist for calculating M_{nr} and M_{zr} as well [1304, 1413].

A rather different way of detecting heterogeneity involves the computation of different average molecular weights of the *entire sample*. In this calculation the low-speed method has considerable advantages, since the entire solution column can be examined. For example, the weight average can be easily measured from the total concentration difference across the cell

$$M_w = \frac{2RT}{\omega^2(1-\bar{v}\varrho)(b^2-a^2)} \frac{c(b)-c(a)}{c_0} \tag{10}$$

and the Z-average from

$$M_z = \frac{2RT}{\omega^2(1-\bar{v}\varrho)} \frac{\frac{1}{b}\left(\frac{dc}{dr}\right)_b - \frac{1}{a}\left(\frac{dc}{dr}\right)_a}{c(b)-c(a)}\,. \tag{11}$$

If this calculation yields $M_z > M_w$, the sample must be heterogeneous. Values of M_z/M_w for the mixtures shown in Fig. 2-5 are given in the legend. Since, with careful work, M_w can be determined to better than 1%, and M_z to a few percent, relatively small heterogeneity can be detected. Under some circumstances, other averages may be obtained to confirm the analysis.

In a single experiment, neither identity of two averages, nor apparent lack of curvature in $\ln c$ *vs* r^2 graphs prove homogeneity. This is because macromolecular solutions may be non-ideal. In general, non-ideality will tend to give apparent molecular weight values which decrease with increasing concentration. This will tend to produce downward curvature in the $\ln c$ *vs* r^2 graph; such downward curvature may nearly compensate for the upward curvature due to heterogeneity (see Fig. 2-6). This is one reason why one must be very critical of statements that linear $\ln c$ *vs* r^2 graphs indicate homogeneity. Similarly, non-ideality will usually depress the apparent Z-average molecular weight more than the weight average. It is possible, then, to have $M_z^{\text{app}} = M_w^{\text{app}}$ for a non-ideal, heterogeneous substance at some finite concentration. Such pitfalls can usually be avoided by carrying out experiments at several concentrations, and extrapolating apparent molecular weights to zero concentration (see 2. V., Fig. 2-14).

Usually, aqueous solutions of globular proteins are not too non-ideal, so the above dangers are not too severe. Exceptions are:

1. Highly asymmetric proteins.
2. Proteins dissolved in denaturing solvents, such as guanidine hydrochloride. An example of such a system is given in 2. V.
3. Proteins carrying a high net charge, especially when dissolved in buffer of low ionic strength. It is a good general rule to have the ionic strength at least 0.1, unless the protein is isoelectric.

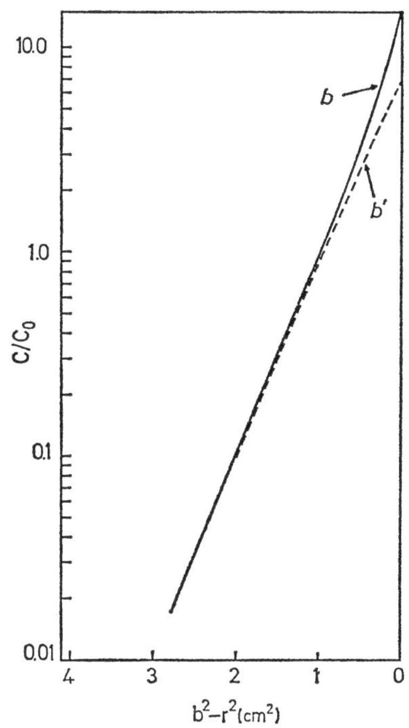

Fig. 2-6. A schematic representation of the effects of non-ideality. Curve (*b*) represents the same mixture as curve (*b*) in Fig. 4; (*b'*) shows the effect of appreciable non-ideality on this graph

C. Density Gradient Techniques

The development of density gradient methods has been of enormous utility to the investigation of crude and partially purified protein preparations. They have been less used, however, for the kind of problem we consider here — the proof of homogeneity in materials to be subjected to detailed chemical analysis. We shall consider the reasons for this, as well as some ways in which gradient methods can be useful.

Of the two kinds of density gradient methods — sedimentation equilibrium in a density gradient [858] and density gradient sedimentation velocity [176, 818], the former has been little employed with proteins. This is probably because the breadth of the band which forms at the isodensity point is inversely proportional to $\sqrt{M}$;

most proteins, being fairly low molecular weight macromolecules, will form very broad bands, making resolution difficult. Furthermore, many proteins are insoluble in the concentrated salt solutions needed for such studies.

It is quite possible, however, that this technique will become of greater importance in the study of certain kinds of conjugated proteins, such as lipoproteins and glycoproteins. Heterogeneity in such substances may reside largely in the amount of bound lipid or carbohydrate, with an accompanying heterogeneity in specific volume. Already, this technique has been used with success in investigations of lipoprotein preparations (see, for example [1376]).

Sedimentation velocity studies in density gradients (often sucrose gradients) are customarily used for rather different purposes. By the use of isotopic labeling, or by enzymatic assay, it is frequently possible to study the sedimentation of protein in very dilute solution or in grossly contaminated mixtures. For the kind of material which is considered in this volume — available in moderate quantities and in at least fair purity, the density gradient technique has little to recommend it over the more conventional sedimentation velocity measurement. In general, s values cannot be obtained as precisely by this method, nor is the resolution markedly superior. However, one combination of techniques would seem to provide excellent evidence for homogeneity. If both enzymatic activity and total protein concentration can be determined, the demonstration of uniform specific activity across a sedimenting band should be a quite convincing demonstration of purity. The criterion seems to have been little used to date.

The application of such methods to preparative problems has had some popularity, for the degree of separation that can be achieved in sucrose gradient centrifugation, for example, is much greater than in conventional sedimentation methods. However, many are finding gel permeation chromatography a simpler and more efficient way of separating molecules of different sizes.

IV. Gel Permeation Methods

Recent years have seen the rapid development of a new class of methods for the determination of heterogeneity, the rough measurement of molecular weights, and the separation and purification of macromolecules. These are the so-called "gel permeation" or "gel filtration" techniques. Basically, the idea is simple. Small cross-linked particles of a substance like dextran or agarose will be swollen by solvent and made permeable to small molecules. But their permeability to large molecules, such as proteins, will depend upon the "mesh" of the cross-linking. If such particles are packed into a column, we may distinguish two volumes, one accessible to all molecules (the so-called *void volume*, V_0, the volume between the particles) and the other accessible only to molecules able to penetrate completely into the gel particles. This latter *internal volume*, V_i, is determined by the dry weight of gel and its solvent uptake per gram.

In operation in the most usual manner (the zonal technique) a layer of the protein solution is placed atop the column, and then eluted through with the same buffer with which the column has been equilibrated. Different solute molecules, which can penetrate to varying degrees into the gel particles, will have different volumes available to them in a column made up of such particles. Then in elution from such a column, the elution volume (V_e) will be given by

$$V_e = V_0 + K_d V_i \tag{12}$$

where K_d is a distribution coefficient ranging from zero for very large solute molecules to unity for very small ones.

There have been two general points of view as to the mechanism which accounts for the values of K_d. On the one hand, many have assumed that equilibrium is maintained in the column, and that K_d represents that fraction of the internal volume available to a given type of molecule. The opposing view holds that the rate of diffusion of molecules into the gel particles is the controlling factor. The experiments of ACKERS [5] indicate that, at least to some extent, diffusion must play a role. While equilibrium and column determination of K_d were in good agreement for tightly cross-linked gels, for the coarser gels and larger proteins considerable discrepancies were found. Furthermore, calculations based on the Stokes' radius, as given by diffusion coefficient measurements, led to good agreement with the data. The Stokes' radius (a) is an empirical measure of effective molecular size, defined by the Stokes-Einstein diffusion equation

$$D = \frac{kT}{6\pi\eta a} \tag{13}$$

where D is the diffusion coefficient, k is the Boltzman constant, T, the absolute temperature, and η, the solvent viscosity.

Thus, from these researchs, it appears that gel filtration measures some quantity closer to a diffusion coefficient than to the molecular weight itself. It should be emphasized strongly that the Stokes' radius is neither an actual molecular radius nor a direct measure of molecular weight. For example, an elongated molecule will have a Stokes' radius (as determined from D) which does not correspond to any particular dimension of the molecule itself, and which is different from the value of a for a spherical molecule of the same weight. Furthermore, it is quite possible that K_d will not measure the same value of a as does D, in the case of a very asymmetric particle. This means that the use of gel permeation as a measure of molecular weight is no more straightforward than the use of either s or D alone for this purpose. The correlations which have been presented between K_d and M (see Fig. 2-7) derive in part from the fact that series of globular proteins, of roughly comparable hydration and asymmetry, have been used. They are, however, no better than the correlation shown in Fig. 2-2, and are subject to even more exceptions. For example, either glycoproteins or proteins that interact strongly with polysaccharides (like lysozyme) are apt to behave anomolously.

While these remarks are meant to caution against uncritical reliance on gel permeation as a method for accurate molecular weight determination, they are not meant to detract from the importance of the technique. As a preparative method, as a way for testing for size heterogeneity, and as a rapid means for rough M determination, the method is superb.

Two general techniques have been employed. In the column method [1385], the gel beads are equilibrated with buffer and packed in a chromatographic column.

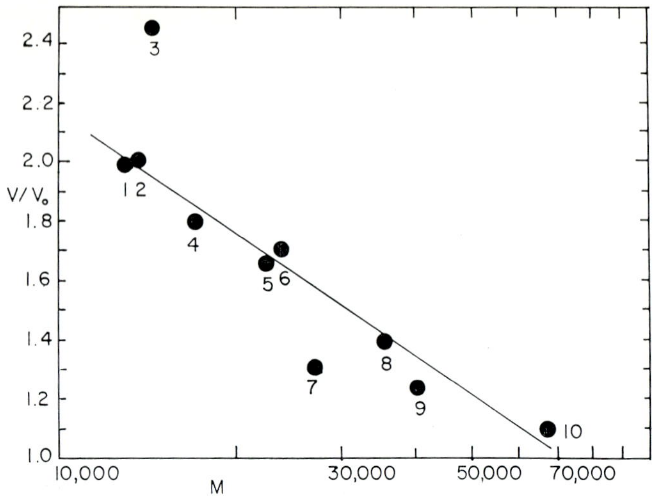

Fig. 2-7. Elution volume in gel filtration as a function of molecular weight. Conditions: 1.2 × 184 cm column of G-75 Sephadex, 25° C to 27° C, pH = 6, ionic strength = 0.494. Proteins are identified as follows: 1, cytochrome c, 2, ribonuclease, 3, lysozyme, 4, myoglobin, 5, α-chymotrypsin, 6, trypsin, 7, ovomucoid, 8, pepsin, 9, peroxidase, 10, serum albumin. The two which behave most anomalously are the glycoprotein, ovomucoid, and lysozyme, which has a carbohydrate substrate. The data are taken from WHITAKER [1364]

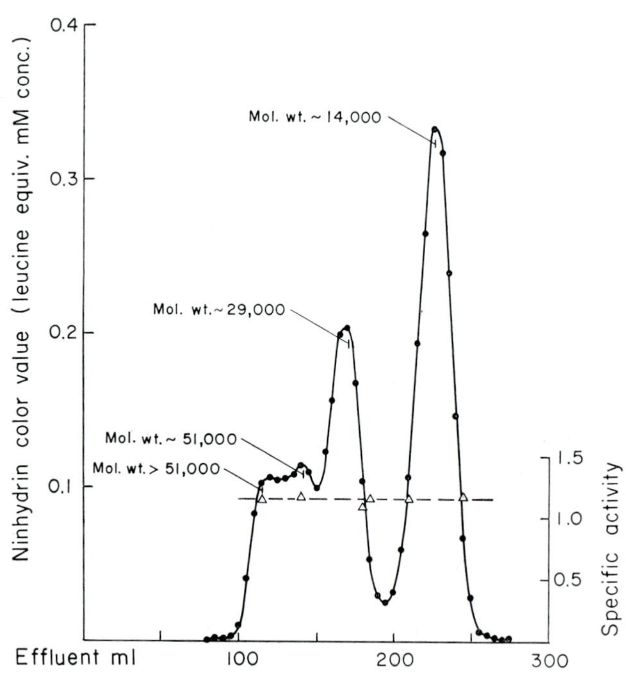

Fig. 2-8. Resolution of ribonuclease A and its aggregates on Sephadex G-75. Monomer, dimer, trimer and higher aggregates can be detected. The specific activity against cyclic CMP is also shown. Data of CRESTFIELD et al. [255]

A band of the protein solution is applied to the top, and eluted with the same solvent. Marker substances, such as low molecular weight dye compounds and large colloids (india ink or "blue dextran") are often added to define elution volumes corresponding to $K_d = 1$ and 0, respectively. In some applications, recycling is employed, to increase resolution [1385]. Typical results, showing the heterogeneity of a ribonuclease sample caused by irreversible aggregation, are shown in Fig. 2-8.

Ideally, the band eluted from a column loaded with a very narrow band will be nearly Gaussian in shape [1385]. Any marked deviation from such form may be indicative of heterogeneity, association, or dissociation of the protein. The concentration dependence of the elution volume is also of significance. A small increase in V_e with c is found for many homogeneous proteins, whereas a decrease with increasing concentration will usually signify association.

In the second method, a layer of the gel is laid down (by use of a special applicator) to uniform thickness on a glass plate. The solution to be tested is placed on a spot and elution carried out along ths length of the plate. Staining then reveals the migration of the protein solutes. This method requires, of course, only minute quantities of the sample. Its resolution does not appear to be as good as that of the column methods, especially since recycling can be employed in the latter. This method has little use for preparative purposes but is economical of material in initial investigations of samples.

There are now available a wide variety of gel types for such studies, ranging in the molecular weight exclusion limit from a few thousand to many million. While polysaccharide gels of various types have been most widely used, some investigators are beginning to utilize porous glass beads for such purposes. These have the advantages of incompressibility, which allows high flow rates, and chemical inertness. Whatever material is used, it is imperative that the columns be packed as uniformly as possible and that the gel be first washed, then rinsed by elution with the buffer to be employed. Large changes in pressure should be avoided and flow rates should be low, both to avoid compressing organic gels and to allow equilibration through the column. These requirements are especially strong with the gels designed for separation of very high molecular weight materials.

V. Electrophoresis

All proteins are polyampholytes and carry an electric charge, determined by the amino acid composition, pH, ionic strength, and the nature of the buffer ions. Since a homogeneous protein should have, under given conditions, a definite value for this charge, methods which can effect separation on the basis of charge are powerful tools in characterization. The simplest way in which such separation can be effected is on the basis of the migration in an electric field: electrophoresis.

A rigorous analysis of the process of electrophoresis is beset with difficulties [1261, 1306]. If we consider the simplest possible situation, an isolated charged particle in a non-conducting medium, the velocity of migration can be calculated by equating the coulombic force on the particle to the force of viscous resistance to motion through the solution.

$$fv = ZeE\,. \tag{14}$$

Here v is the velocity, f, the frictional coefficient, Z, the number of electron charges (of magnitude e each) and E, the electric field. The electrophoretic mobility is defined as the ratio of velocity to field:

$$U = \frac{v}{E} = \frac{Ze}{f} . \tag{15}$$

For spherical particles of radius a, $f = 6\pi\eta a$ (Stokes' law), so

$$U = \frac{Ze}{6\pi\eta a} \tag{16}$$

where a is the molecular radius and η, the solvent viscosity. These equations embody the most important statement about electrophoresis: to a first approximation, the mobility is proportional to the charge.

Unfortunately, such a treatment misses many important complications involved in the study of electrophoresis of proteins in aqueous solutions. A protein is not an isolated particle. It is surrounded by an environment of other small ions; in general these must form an ion atmosphere, predominantly of opposite charge, which will interact in a number of ways influence the motion of the particle. For these reasons, Eqs. (14) and (15) are rough approximations, and it is not a straightforward calculation to obtain the net charge Z from mobility data.

However, for purposes of separation or the determination of homogeneity, such complications are not of serious consequence. The fact remains that two protein molecules, identical except for a difference in a single charged group, often may be separated and distinguished by electrophoresis. Thus, the method is remarkably sensitive to the parameter for which it is designed.

A wide variety of techniques has been used for electrophoresis. These differ greatly in their sensitivity, the amount of protein required, and the conditions under which each may be employed. A few of the more widely used methods are described below.

A. Moving Boundary Electrophoresis

This is the classical method developed by Tiselius [1282]. While it is still the most accurate method for the determination of mobilities and quantitative analysis of mixtures, it is little used today, for it requires large amounts of material and is very inefficient in separation and resolution. Since the method is described in detail in the older literature (see, for example, the review by Alberty [20]) we shall not discuss it further. A recent discussion of the subject with special emphasis on interacting systems is to be found in the review by Cann [204]. Rather, we shall concentrate on the multitude of new techniques that have evolved in recent years. These may be largely classed under the general title of "Zonal electrophoresis". While such techniques are sometimes performed in solution with a sucrose gradient for stabilization, it is much more common to utilize some kind of solid supporting medium.

B. Zonal Electrophoresis on Supporting Media

The development of suitable solid, permeable supporting media has allowed the exploitation of the possibilities of zonal techniques in electrophoresis. The general difference between these techniques and moving boundary electrophoresis is illustrated in Fig. 2-9. Whereas in the Tiselius technique only portions of the fastest and slowest components can be isolated, the electrophoresis of a zone through a solid medium which prevents convection can lead to a "spectral" analysis of the material into individual components. These bands may be revealed by staining, by autoradiography, by fluorescence, or by elution of strips of the medium followed by assay. Most recently, the use of direct methods for scanning the UV absorption of the zones has become very popular. This has the advantages of allowing non-destructive selection of individual bands for excision and promises to become the most accurate quantitative method for analysis of concentrations. The latter technique allows at least partial recovery of the individual components. Furthermore, the methods are often rapid and may require only minute quantities of protein.

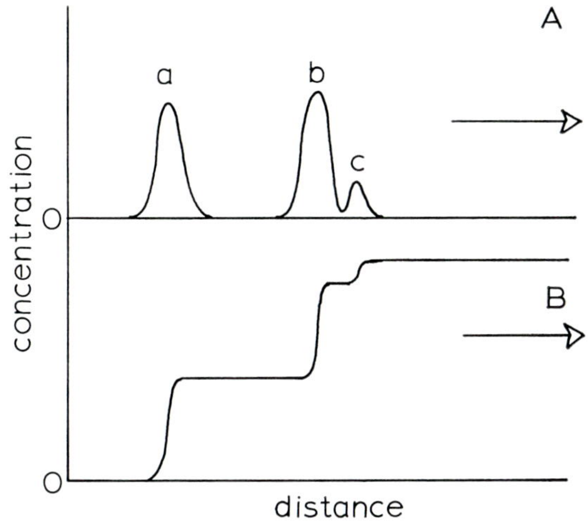

Fig. 2-9 A and B. A comparison of zonal (A) and moving boundary (B) electrophoresis. The same mixture is presumed to be present in each case, and migration is to the left. Note that in (B) only a small portion of component *a* can be obtained in pure form from this composite boundary, whereas in (A) complete separation of all three components can be effected

With all of these advantages, it is small wonder that zonal techniques have largely superseded the old moving boundary method. However, some price is paid in the use of solid media. In the first place, the rate of migration through such a substance as starch, polyacrylamide, or paper is even hardeı to account for theoretically than the electrophoretic mobility in aqueous solution. The viscosity appropriate to Eq. (16) is difficult to define, the effective cross-section of the cell is ambiguous, and interaction between the solute and the supporting media (absorption, gel filtration effects, etc.) enormously complicates the picture. For these reasons, analysis of zonal electrophoresis on solid media is almost always carried out empirically, by comparison of the distance of migration of the substance being investigated with that of standard substances.

A further complication, peculiar to solid media, may or may not be of importance depending upon the chemical nature of the medium. This is endosmosis, the migration of solution through a charged solid under the influence of a field. These effects depend upon the nature and magnitude of the charge on the supporting medium, and the nature and ionic strength of the buffer. In unfavorable cases, endosmosis may be so great as to more than compensate for the electrophoretic mobility of the protein solute.

Table 2-2. Some solid media for zonal electrophoresis

Medium	Description	Advantages	Disadvantages	Reference
Paper	Filter paper, moistened with buffer, held between electrode chambers.	Fast, simple requires only small amounts of protein.	Endosmotic effects quite severe, evaporation may be a problem.	[54]
Cellulose acetate strip	Strips of thin cellulose acetate sheet, used much like paper.	Can use exceedingly small quantities of protein.	Somewhat more difficult to use than paper.	[708]
Starch block	A slurry of starch placed in a trough connected to electrode chambers.	Simple to prepare, can be used for fairly large scale separations. Elution of products fairly easy.	Endosmosis, impurities in starch, gel filtration effects.	[739]
Starch gel	A gel is prepared from partially hydrolysed starch, in trough or between vertical plates.	High resolving power, great versatility.	Some impediment of motion of large proteins through the gel, removal of samples more difficult.	[1180]
Agar gel	Like starch gel, but more dilute gel can be employed.	Open gel structure gives conditions approaching free electrophoresis. Little absorption.	Endosmosis can be severe in alkaline solutions.	[429]
Polyacrylamide gel	Acrylamide is polymerized in small tubes, or slabs, to form gel. These are placed between electrode chambers.	Easy to work with, clear, controllable in porosity. No contaminants, endosmosis negligible.	Generally restricted to small scale since the gel is of low electrical resistance.	[1016] [1189]

A very wide variety of supporting media has been employed. We shall not here attempt to describe the various experimental procedures, which differ considerably in detail. Instead, general information, together with specific references, are given in Table 2-2. This is only a partial list; an enormous number of substances has been tried. Not listed here are a number of other media (cellulose powder, glass beads, etc.) most often used in preparative work in column apparatus. Choice of a technique for a particular problem may require considerable investigation. However, for rapid, convenient, and sensitive studies of homogeneity, the acrylamide gel technique has

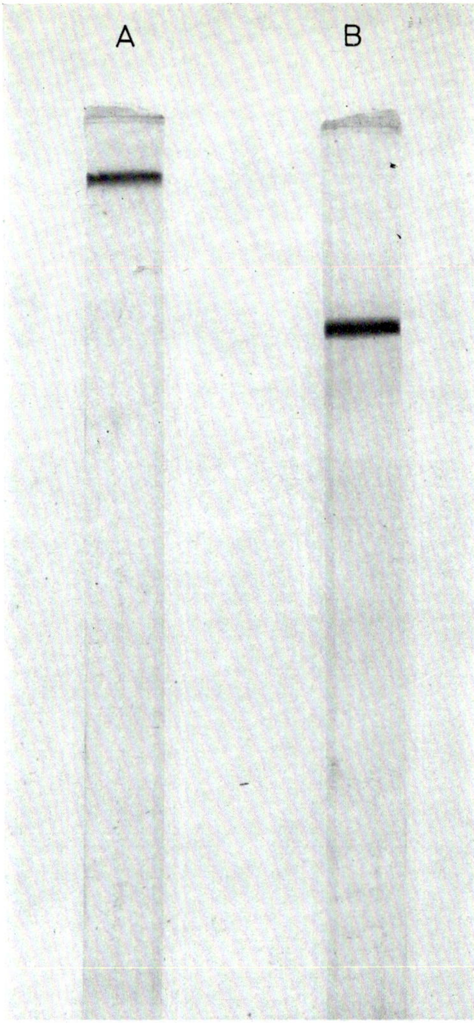

Fig. 2-10 A and B. Polyacrylamide disk gel electrophoresis of (A) bovine skeletal muscle and (B) bovine liver pyruvate kinases. The two purified enzymes have been studied at pH 9.5. (Courtesy of Dr. J. Cardenas and Dr. R. Dyson)

a great deal to recommend it, and it has become most widely used. Since polyacrylamide gels can be readily prepared over a wide range of acrylamide and/or cross-linked concentrations, it is possible to produce a great variety of conditions for the electrophoresis. An example is shown in Fig. 2-10. Polyacylamide gels are particularly well suited to UV scanning procedures.

Other techniques that appear to retain wide acceptance among the many that have been tried include paper and cellulose acetate strip electrophoresis. Fig. 2-11 shows cellulose acetate patterns of an isozyme system. A typical arrangement for this type of electrophoresis is shown in Fig. 2-12.

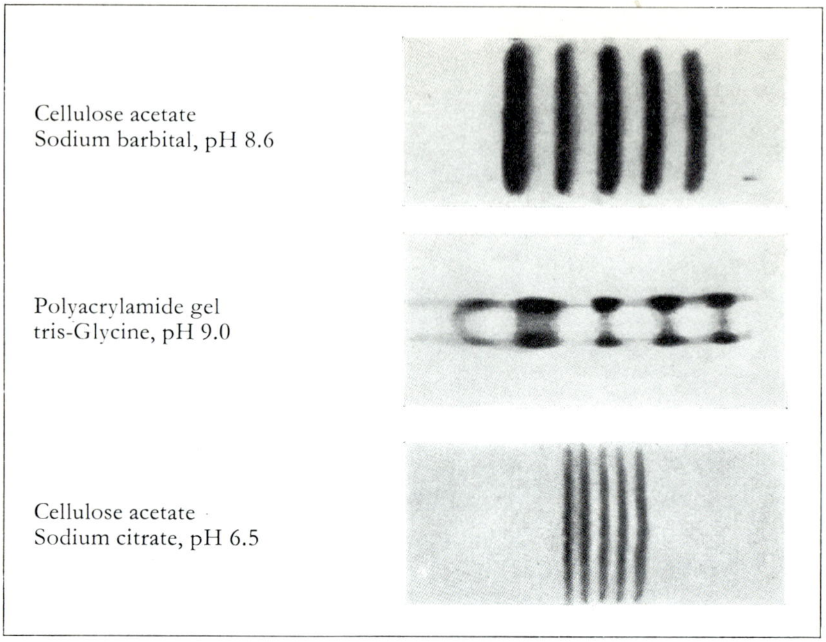

Fig. 2-11. Cellulose acetate electrophoresis pattern for the isozymes of aldolase A and C. From PENHOET et al. [967]

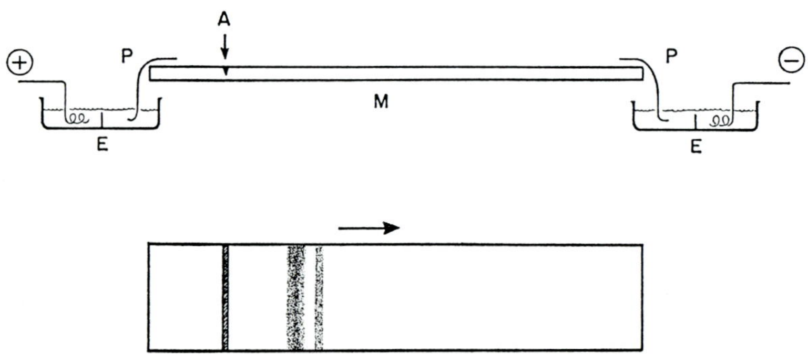

Fig. 2-12. Simplified drawing of apparatus for zonal electrophoresis on supporting media like paper, or cellulose acetate. The block or strip of medium (*M*) is connected to the electrode vessel (*E*) by strips of paper (*P*). The solution is applied at a slit (*A*). Below is shown a top view of the pattern after electrophoresis has proceeded

C. Isoelectric Focusing

Recent years have seen the development of an exceedingly sensitive method for the separation of molecules purely on the basis of electrical charge. This is *isoelectric focusing* [473]. In the most common form of the method, a stabilized pH gradient is set up in a liquid column that has been stabilized against convection by a pre-formed sucrose gradient. In such a column, each protein will migrate to the position corresponding to its own isoelectric pH and concentrate there. The final result will be a quasi-equilibrium, in which the tendency of the protein to diffuse out of the band will be balanced by the electrophoretic force pulling the protein back into the band.

Since the pH gradient can be made very shallow, exceedingly high resolution in terms of isoelectric point is readily attained. The method cannot be applied with equal ease to all proteins, however. It is necessary to have appropriate low molecular weight amphoteric substances to buffer the pH gradient, and this may be difficult in some pH regions, or these may interact with the proteins. Further, some proteins become quite insoluble at the isoelectric point. Nevertheless, when it can be used, this method provides probably the most sensitive test for charge heterogeneity available to date. It can also be used with considerable efficiency as a preparative technique.

D. SDS Gel Electrophoresis

A variant of the polyacrylamide gel electrophoresis technique has proved to be an exceedingly powerful method for the resolution and enumeration of the individual polypeptide chains in a protein. In some respects, as will be demonstrated later, the technique is superior in resolving power to any other currently available method.

The basic idea is simple. A gel electrophoresis experiment is carried out in the presence of low concentrations (often 0.1 to 0.2%) of sodium dodecyl sulfate (SDS). The protein is usually first treated at elevated temperatures with a more concentrated

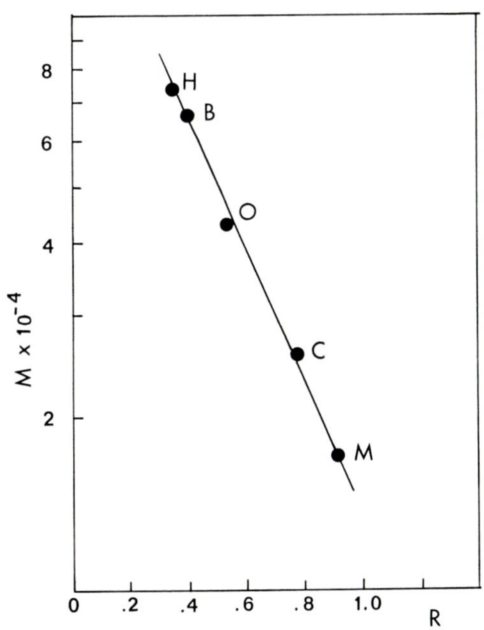

Fig. 2-13. Molecular weight (on a log scale) versus relative mobility (R) for a number of proteins in 8% polyacrylamide gel containing 0.1% SDS. Protein code: H = *callianassa* hemocyanin, B = bovine serum albumin, O = ovalbumin, C = chymotrypinogen A, M = myoglobin

SDS solution, in the presence of a reducing agent like mercaptoethanol or dithiothreitol (see Chapter 5.III and Chapter 6.IV). The reductant will cleave disulfide bonds and the SDS denatures the protein and binds to it strongly.

Remarkably, the migration rate of the protein in the acrylamide gel is quite insensitive to the net charge or amino acid composition of the native protein, but depends only on the molecular weight of the polypeptide chain. Even quite complex multi-subunit proteins will, under these circumstances, be separated into individual chains, unfolded, and separated on the gel, if chains of different *size* are present.

While the physical basis for the technique cannot be said to be fully understood as yet, experiments by Reynolds and Tanford [1022] have provided very strong clues. These workers find that reduced proteins, in dilute SDS solutions, bind sufficient SDS to form micellar structures. The diameter of these micelles appears to be roughly constant and the length is proportional to the polypeptide chain length, and hence to the molecular weight. This observation can explain, at least qualitatively, the behavior of proteins of different weight on the gel. Fig. 2-13 shows an empirical calibration obtained in a typical SDS gel experiment.

The techniques used, except for the addition of the detergent, SDS, are very similar to those described above for the conventional polyacrylamide gel experiment. Either tubular or slab-type gels can be used, and the detergent is simply included in the polymerizing mixture as well as in the buffers used in the electrode vessels.

A number of articles have appeared that detail the techniques for SDS gel electrophoresis and should be consulted for further details [296, 1338].

VI. Protocol for the Characterization of a Typical Protein, in Preparation for Sequencing

In this concluding section an idealized example will be considered. We shall suppose that the investigator has purified a protein and is considering sequence determination. For simplicity, it will be assumed that this is a "globular" protein and that the molecular weight is large enough ($>$ 15000, say) that he may reasonably suspect it to contain a number of subunits. He might proceed as follows.

A. Tests for Homogeneity of the Native Protein

A necessary (but by no means sufficient!) criterion is that the protein yield only one symmetrical peak in sedimentation velocity. If more than one component is revealed, the investigator should test, by varying the protein concentration, whether or not these components are in association equilibrium. If not, it will probably be possible to separate them by gel chromatography. If an association equilibrium is involved, conditions should be sought that yield the smallest unit (the protomer).

While an estimate of the molecular weight of the single component or protomer can be obtained from sedimentation velocity, a sedimentation equilibrium determination is to be preferred. The high speed method is simplest for such routine measurements and is sensitive to impurities.

Finally, zonal electrophoresis of the purified protein, under a variety of conditions, *may* yield evidence for the existence of more than one kind of polypeptide chain.

For example, isozymes are frequently nearly identical in molecular weight and will appear to be homogeneous in the ultracentrifuge, but may (but need not!) yield multiple bands in zonal electrophoresis. Isoelectric focusing, if applicable, is exceedingly sensitive at this point.

B. Detection and Enumeration of Subunits

Virtually every protein should be suspected of being made up of subunit chains; if the molecular weight is greater than 50000, the likelihood is very great. Before any sequencing can be considered, the number of kinds of such chains must be deduced, and they must be separated.

The non-covalently linked multi-chain structures are generally held together by relatively weak forces (hydrogen bonds, hydrophobic bonds, electrostatic interactions, etc.). For this reason even mild perturbation of the solution environment may cause dissociation. Some of the methods which have proved successful are:

1. Change of pH

This is the simplest of all procedures and is successful in a considerable number of cases. The critical pH ranges appear to be around pH 3 to 4 (the carboxyl titration region) and 9 to 10 (the lysine-tyrosine titration region). While simply titrating a protein to pH below 3 or above 10 to cause dissociation appears ideally simple, there are a number of complications which should be considered. In the first place, it is desirable to avoid extremes of pH, lest especially labile bonds be hydrolyzed, even at moderately low temperatures. On the other hand, at moderate pH values dissociation may not be complete. This can lead to serious errors in the determination of the weights, and hence numbers of the subunits. Furthermore, physical characterization of highly charged subunits is complicated, in every case, by the existence of charge effects. These will appear, in the most part, as strong solution non-ideality. Most of these problems can be dealt with, but caution and a thorough program of experimental work are required. A case in point is to be found in some of the early studies of aldolase [269]. Dissociation was effected in pH 2.9 citrate buffer. The subunits appeared, from sedimentation equilibrium experiments, to be nearly homogeneous, with a molecular weight of about 50000. Since the weight of native aldolase was then believed to be 152000, three equal subunits were postulated. Of course, this answer is now known to be incorrect; there are clearly four units [674, 967].

Apparently, the error resulted from a combination of circumstances: (a) an incorrect weight for the native protein; it is now believed to be about 160000. (b) incomplete dissociation, and (c) non-ideality, which masked the heterogeneity of the "subunit" solution. More experiments should have revealed these problems.

2. Chemical Modification of the Charge

In some cases, multi-subunit proteins which do not dissociate in a practicable pH range may do so if positively or negatively charged groups are added by chemical modification. The classic example of this type is the use of succinylation to add negative charges [716].

Since such chemical modification is sometimes difficult, and complicates further chemical studies, it should perhaps only be considered as a next step if some milder methods fail. Of course, the comments made above concerning incomplete dissociation and charge effects carry equal force here. On the other hand, peptide bond cleavage is less likely to be a complication.

3. Strong Denaturants (Guanidine Hydrochloride, Urea)

It has been known for a long time that concentrated solutions of materials like urea or guanidine hydrochloride [Gdn · HCl] denature most proteins. More recently, it has been realized that this denaturation is almost invariably accomplished by the dissociation of a protein into its constituent polypeptide chains. This is most clearly pointed out in the recent studies by TANFORD and co-workers [1262].

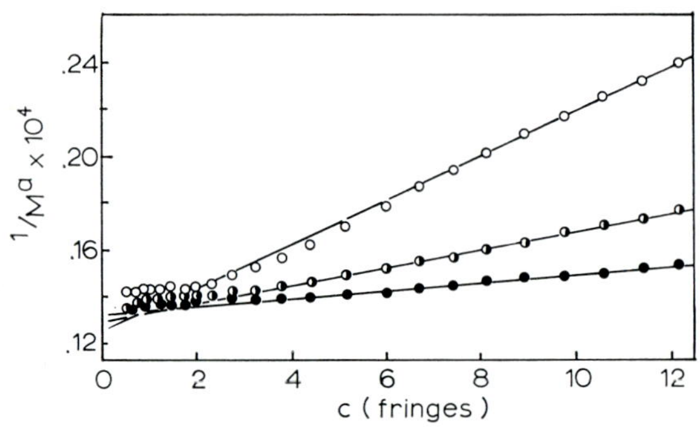

Fig. 2-14. Sedimentation equilibrium of the hemocyanin of *Cancer magister* in 6.0 M Gdn · HCl, 0.1 M mercaptoethanol. Reciprocals of M_z^a (○), M_w^a (◑) and M_n^a (●) are plotted versus concentration in the cell. These data are from a single high speed experiment. The high concentration data all extrapolate to a molecular weight of about 80000, but the deviations at low C may suggest slight heterogeneity (see text)

The principal disadvantages in employing such agents result from the high concentrations required. Solutions containing 6 M guanidine hydrochloride or 8 M urea are clearly unsuitable for many chemical studies and physical investigations are seriously complicated. Removal of the denaturant often leads to re-association and/or random aggregation. Sedimentation velocity, sedimentation equilibrium and light scattering experiments are all complicated by selective binding of either the denaturant or water by the protein. Such effects can lead to serious errors in molecular weight. To use the example of aldolase again, two competent groups have arrived at different results for the number of subunits [361, 1080]. A study of these papers reveals the complexities encountered with concentrated denaturants.

Nevertheless, such denaturants have begun to be widely used, and with proper care useful information can be obtained. Usually, Gdn · HCl or urea are used in conjunction with a reducing agent such as mercaptoethanol or dithiothreitol. Molecular weights can be estimated by gel permeation chromatography in a column loaded with the denaturant [211], or by intrinsic viscosity measurements [674].

More accurate results can be obtained by sedimentation equilibrium, but necessary precautions must be taken, for difficulties will be encountered. Preferential interaction of the denaturant with protein changes the effective partial specific volume [211] and measurement of the effective $\bar{v}$ in such solutions is difficult. Often one can assume, as an approximation, that the appropriate value of the apparent partial specific volume is 0.01 to 0.02 ml/gm less than the value measured in dilute buffer. Secondly, the solutions are very non-ideal. Fig. 2-14 shows the results of careful studies of this kind on subunits of a crab hemocyanin [207]. Each of the apparent molecular weights varies severely with the variation of concentration across the solution column. While the substance behaves *approximately* as if it were homogeneous, with a molecular weight of about 80000, one cannot be sure. Subtle heterogeneity might be hidden by the strong concentration dependence. As will be seen in the next section, this is in fact the case.

4. SDS-Polyacrylamide Gel Electrophoresis

Sodium dodecyl sulfate is a powerful dissociating and denaturing agent, and electrophoresis in such gels has remarkable resolving power. For example, the crab hemocyanin described above is revealed, in SDS gel electrophoresis, as having *two* kinds of subunits, in roughly equal proportions [207, 789]. These have molecular weights of 76000 and 84000. Resolution of so closely spaced components is not possible with any other technique known to the author. Thus, it would seem that at least at the present time the best technique for investigating subunit homogeneity is electrophoresis in gels containing SDS, as described in V. D.

Even if a protein yields only one subunit band on SDS electrophoresis, it cannot be assumed that only one kind of subunit is present. Subunit types may differ only slightly in amino acid composition or sequence; in fact, this is a common situation. If conditions can be found where dissociation to subunits is effected in the absence of strong denaturants, one can investigate the behavior in zonal electrophoresis at a number of pH values. Caution is required in two respects here: (1) many amino acid substitutions will not lead to detectable electrophoretic differences and may be missed; (2) artifactual band splitting can occur as a consequence of interaction with certain kinds of buffers [203]. In such cases, either heterogeneity should be shown in several buffer systems, or individual bands should be eluted and re-run. If they split again, artifact may be suspected.

The result of the above analysis may be the demonstration that the purified protein actually contains only one kind of polypeptide chain. If so, efforts at sequencing may begin. If not, extra effort will have to be expended to separate chains.

C. Preparative Chain Separation

Most of the techniques described above for analysis of chain heterogeneity can be scaled up for preparative purposes. In some cases, chromatography on ion-exchange resins will also be successful. Unfortunately, almost every case is unique and the details for effective separation must be worked out empirically in each case.

Chapter 3

End Group Determination

Kozo Narita, Hisayuki Matsuo, and Terumi Nakajima

I. Introduction

Most polypeptides and proteins occurring in nature have been shown to consist of a single polypeptide chain, but several proteins, insulin, chymotrypsin, immunoglobulin, hemoglobin and lactate dehydrogenase, for example, have been shown to be composed of more than a single chain. In protein molecules consisting of two or more polypeptide chains, the chains are held together either by covalent interpeptide-chain disulfide bonds of cystine residues (insulin, α-chymotrypsin, immunoglobulin etc.) or by the cumulative forces of ionic interaction, hydrogen bond and hydrophobic interaction between side chain groups of amino acid residues belonging to the neighboring different polypeptide chains (hemoglobin, lactate dehydrogenase etc.). Each polypeptide chain usually possesses an unsubstituted α-amino group at one end of the chain (amino terminal group or N-terminal group) and an unsubstituted α-carboxyl group at the other end (carboxyl terminal group or C-terminal group).

Quantitative analysis of both end groups provides valuable information concerning the purity of a protein in question (also see Chapter 2), the number of polypeptide chains in one molecule of a protein, the molecular weight of which has previously been determined by physicochemical techniques (Chapter 2), and branching or ramification of a polypeptide chain. Careful quantitative end group analysis on a protein whose molecular weight was not previously known allows the calculation of a minimum molecular weight.

Each analytical method for the terminal groups, to be described later in this chapter, has certain characteristic merits on one hand but also intrinsic disadvantages based upon the chemical principle involved. The use of at least two methods, which are governed by entirely different principles, is recommended in order to draw a valid conclusion based upon the results of end group analyses.

When a protein is shown to contain cystine or cysteine residues by amino acid analysis (Chapter 7), it is desirable to reduce disulfide bonds with β-mercaptoethanol or dithiothreitol in the presence of 8 to 10 M urea or 4 to 7 M guanidine hydrochloride (Chapter 6), and EDTA followed by alkylation of the resultant sulfhydryl groups prior to the end group determinations. A low concentration of the denaturants is not enough to cleave reductively all of the disulfide bonds with the thiol compound. For instance, a part of the disulfide bonds in immunoglobulin was reported to be resistant to β-mercaptoethanol in 8 M urea but was cleaved in 10 M urea or 6 M guanidine hydrochloride [186]. The chelating agent EDTA is required to minimize the unfavorable contamination by heavy metal ions which can catalyze re-oxidation of the resultant sulfhydryl groups. The sulfhydryl groups usually are alkylated with monoiodoacetic acid [254] (or its amide) or ethyleneimine [1004, 1005]. The former reagent converts the sulfhydryl group to an anionic carboxymethylcysteine residue while the latter converts it to a cationic aminoethylcysteine residue, which has a side chain length similar to that of the lysine residue and which becomes susceptible to trypsin at the carboxyl side in the peptide linkage. S-Aminoethylcysteine residues in peptides are more resistant to the air-oxidation than S-carboxymethylcysteines during fractionation of the peptides. For these reasons the aminoethylation reaction currently finds much favor. Alternatively, the performic acid oxidation of a cystine residue to two residues of cysteic acid has been used extensively since its introduction by Sanger in his study of insulin [1067]. As an undesirable side reaction, the oxidation of cystine residues is accompanied by the oxidation of methionine to methionine sulfone and of tryptophan, partly to N'-formylkynurenine and partly to unknown substances. For a small peptide, devoid of methionine and tryptophan resides, for which the primary structure already has been elucidated the performic acid oxidation procedure is effectively used to determine the location of the disulfide bonds in the polypeptide chain.

Some difficulties are to be expected if reduction and subsequent alkylation of a cystine-containing protein is not performed prior to end group analyses. The N-terminal group of chymotrypsinogen, for example, is a cysteine residue which is linked to a second cysteine located in a position much further along the polypeptide chain [103]. In such a case, when the dinitrofluorobenzene method of Sanger [1065] is used, monodinitrophenyl (DNP)-cysteine will first be formed by acid hydrolysis but two molecules of the monosubstituted cystine can reversibly interchange cysteine residues as shown in reaction (1). Thus, half the expected amount of bis-DNP-cystine will at best be determined (DNP-cystine is rather labile to acid hydrolysis) [1052]. If the phenylisothiocyanate method of Edman [310, 311] were used (see Chapter 8) no N-terminal phenylthiohydantoin derivative would be formed.

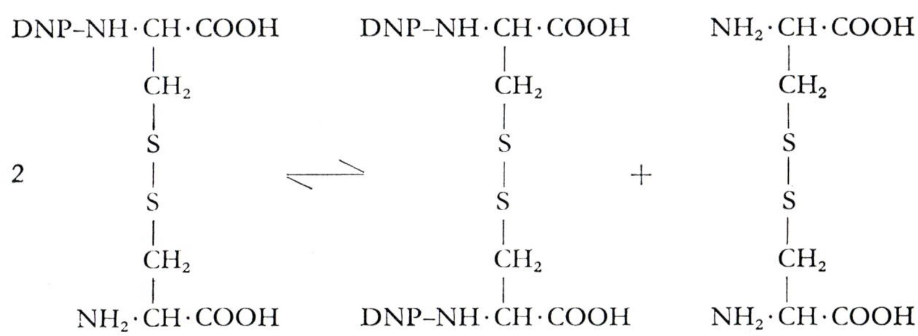

(1)

When a single linear polypeptide chain containing no disulfide bridges is being examined, the difficulties described above are not encountered during the determination of the primary structure of proteins. Methods that cleave disulfide bonds in proteins are described by Kasper (Chapter 5) and by Scoffone and Fontana

(Chapter 6). In attempting to elucidate the structure-function relationship of a protein, however, the native protein with intact disulfide bonds must of course be used.

Principles and methods of the various end group analyses have been outlined and discussed by BAILEY [54], and a detailed description of the techniques of some of the typical end group analyses was made by HIRS et al. [545, 551].

II. N-Terminal Group Determination

The general principle for amino end group determination is based upon the introduction of a marker group (colored, fluorescent, ultraviolet absorbing etc.) onto the amino function followed by the quantitative isolation and characterization of the derivatized terminal amino acid. Forty-two methods for N-terminal end group determination were reviewed recently by ROSMUS and DEYL [1039].

Although all of the terminal analysis methods described below can be used for both proteins and peptides, the cyanate method is especially suitable for proteins and large peptides while the dansyl method is preferred for oligopeptides. The phenylisothiocyanate method is described in detail by EDMAN in Chapter 8.

A. Dinitrofluorobenzene Method

In 1945 SANGER [1065] reported that 2,4-dinitro-1-fluorobenzene (FDNB) reacted quantitatively with amino groups of insulin under mild conditions (2) and that one mole of dinitrophenyl (DNP)-glycine, DNP-phenylalanine and ε-DNP-lysine could

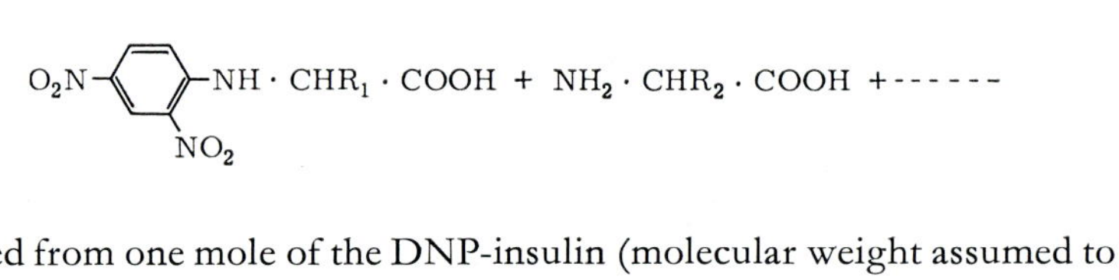

(2)

be isolated from one mole of the DNP-insulin (molecular weight assumed to be 6000) on acid hydrolysis after suitable corrections. The reagent FDNB is reactive not only with α-amino (and α-imino) and ε-amino groups but also with the sulfhydryl groups of cysteine, the phenolic hydroxyl group of tyrosine and the imidazole group of histidine residues in polypeptide chains. When lysine, cysteine, tyrosine residues are located at the N-terminal positions, they are recovered as their di-DNP derivatives from DNP-proteins. One merit of the Sanger method is that α-DNP-amino acids are rather resistant to acid hydrolysis in contrast to acyl amino acids and can be recovered with yields ranging from 25 to 90% after boiling in azeotropic hydrochloric acid (5.7 N HCl) or heating with 12 N HCl at 105° C for 12 hrs. Possible exceptions are DNP-glycine, DNP-cystine, DNP-proline and DNP-hydroxyproline. Stability of DNP-amino acids in hydrochloric acid was studied by PORTER and SANGER [991] whose results are listed in Table 3-1.

Table 3-1. Approximate breakdown of DNP-amino acids[a]

	Hydrolysis in boiling 5.7 N HCl		Hydrolysis for 16 hrs in 12 N HCl at 105° C
	Time of hydrolysis (hours)	Amount unchanged (%)	Amount unchanged (%)
DNP-Alanine	12	80	75
DNP-Arginine	12	90	75
DNP-Aspartic acid	24	60	75
Bis-DNP-Cystine	12	25	0
DNP-Glutamic acid	12	75	75
DNP-Glycine	8	40	50
DNP-Hydroxyproline	4	40	50
DNP-Isoleucine	12	80	75
DNP-Leucine	12	80	75
Di-DNP-Lysine	8	95	75
ε-DNP-Lysine	12	95	75
DNP-Methionine	12	75	75
DNP-Phenylalanine	12	70	50
DNP-Proline	2	10	50
DNP-Serine	12	90	75
DNP-Threonine	24	90	75
DNP-Tryptophan	12	90	0
Di-DNP-Tyrosine	12	75	50
DNP-Valine	12	80	75

[a] Data from Ref. [991], by courtesy of the Biochemical Society.

Although the decomposition reaction of DNP-amino acids with acid has not been studied extensively, the breakdown products seem to be the regenerated parental amino acid, 2,4-dinitrophenol (DNP-OH), 2,4-dinitroaniline (DNP-NH_2) and uncharacterized materials [1018]. As a result of heating DNP-proline with hydrochloric acid, δ-chloro-α-DNP-aminovaleric acid and α-chloro-δ-DNP-aminovaleric acid are produced in addition to regenerated proline [1078]. The low recovery of DNP-proline is mainly attributed to this pyrolidine ring-opening reaction (3), which also may apply to the lability of DNP-hydroxyproline.

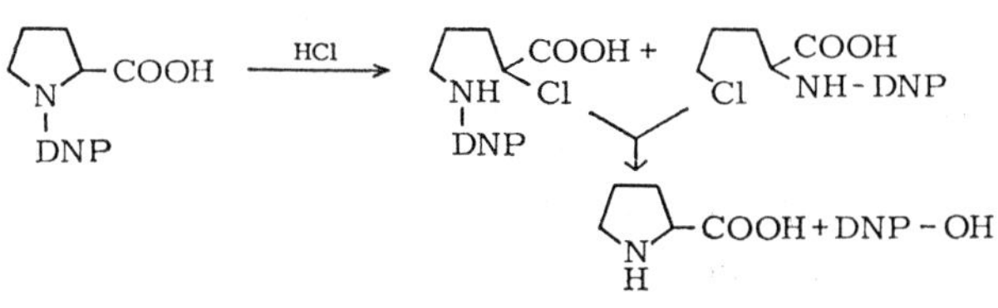

(3)

The extent of decomposition of the terminal DNP-amino acid during acid hydrolysis of a DNP-protein is not identical with that of the free DNP-amino acid alone and differs from one DNP-protein to another, depending upon the nature and amounts of constituent amino acids and upon the nature of prosthetic groups. For example, the terminal DNP-amino acid is destroyed more extensively in DNP-glycoprotein than in DNP-simple proteins. When an accurate estimate is required for a particular terminal DNP-amino acid, a recovery factor must be obtained by heating the particular DNP-amino acid with an equimolar amount of protein at a scheduled period, usually 6 to 15 h, with redistilled azeotropic hydrochloric acid in an evacuated

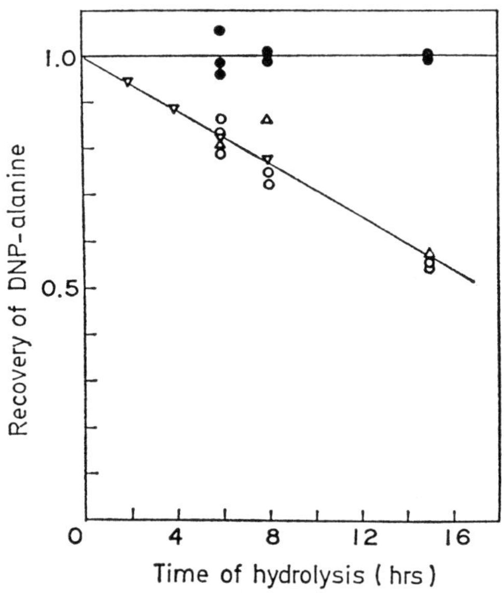

Fig. 3-1. Decomposition of DNP-alanine by heating with 6 N hydrochloric acid at 105 to 110° C. ▽, DNP-alanine alone; △, DNP-alanine plus equimolar amount of Taka-amylase; ○, DNP-Taka-amylase, of which terminus was alanine; ●, DNP-alanine recovered from DNP-Taka-amylase after correction for decomposition during acid hydrolysis. (The figure is drawn from the data in Ref. [10])

sealed tube at 105° C. In Fig. 3-1, the decomposition rate of DNP-alanine with 6 N HCl at 105 to 110° C as a function of time, in the presence and absence of α-amylase from *Aspergillus oryzae* is shown. The N-terminal residue of the amylase is alanine and the corrected values of DNP-alanine released at several periods of the acid hydrolysis from DNP-amylase are also plotted. When the amounts of released DNP-amino acid at various hydrolysis times are extrapolated to zero time, the amount of the terminal group can be estimated without obtaining a correction factor during the acid hydrolysis.

Besides the decomposition reaction of DNP-amino acids with acid, attention should be given to the photodecomposition of DNP-amino acids [11]. DNP-Amino acids are quite sensitive to light, expecially to visible light. Therefore when small amount of DNP-derivatives are treated in a dilute solution, the coupling reaction, hydrolysis and subsequent chromatographic separation must be performed in the absence of light. With the exception of DNP-tryptophan, α-DNP-amino acids yield 4-nitro-2-nitrosoaniline, the corresponding aldehydes and carbon dioxide by photolysis in 1% sodium bicarbonate [200].

Generally, α-DNP-amino acid derivatives are acidic and can easily be extracted from the acid hydrolysate mixture into ether. However, DNP-cysteic acid, DNP-arginine and a part of di-DNP-histidine remain in the water phase after ether extraction together with ε-DNP-lysine, O-DNP-tyrosine and imidazole-DNP-histidine located originally at non-terminal positions in peptide chains. Generally, *imid*-DNP-

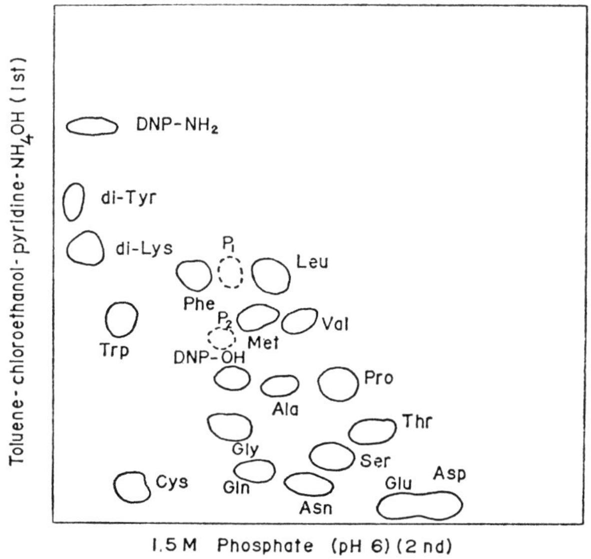

Fig. 3-2. Two-dimensional paper chromatogram of a synthetic mixture of DNP-amino acids according to the method of Levy [768]. Development was first carried out with "toluene" method of Biserte and Osteux [112] and then with phosphate shown in Table 3-2. P_1 and P_2, respectively, denote the position of α-DNP-amino-δ-chlorovaleric acid and δ-DNP-amino-α-chlorovaleric acid

histidine is very labile toward acid at high temperature and this derivative cannot be identified in the acid hydrolysate of DNP-proteins. Water-soluble DNP-amino acids are extracted with *n*- or *sec*-butanol.

Separation and characterization of DNP-amino acids were originally carried out by chromatography on silica gel columns by Sanger [991, 1065]. Modification of this procedure by using silicic acid-Celite columns was made by Green and Kay [440]. The use of an Amberlite CG-50 column was recommended by Seki [1148]. The mechanism of separation of DNP-amino acids with ion exchange resins seems to involve adsorption rather than an ion exchange reaction. Paper chromatography or thin layer chromatography is more convenient than column chromatography and has been widely used. Many systems for paper chromatography have been proposed, but the two-dimensional method recommended by Levy [768] appears to be most useful. Development in the first direction is made *via* the toluene-chloroethanol-pyridine-aqueous ammonia solvent reported by BISERTE and OSTEUX [112] and in the second by 1.5 M phosphate buffer. The locations of DNP-amino acids by this system are shown in Fig. 3-2. The separation of DNP-aspartic acid from DNP-glutamic acid is less satisfactory. Levy recommended the use of 2.0 to 2.5 M phosphate buffer instead of 1.5 M buffer, but at this concentration the other DNP-amino acids are separated less effectively [381]. As the first solvent, *n*-butanol saturated with diluted ammonia [706] can be used in place of Biserte's "toluene" system. The pattern of distribution of the DNP-amino acids is quite similar to that obtained with the "toluene" system, but separation of di-DNP-lysine from di-DNP-tyrosine, and

Table 3-2. Composition of solvents for paper chromatography of DNP-amino acids

Solvent	Composition	Remarks
Toluene-pyridine-2-chloroethanol-0.8 N NH_4OH	50 ml of toluene, 18 ml of pyridine, 30 ml of 2-chloroethanol and 30 ml of 0.8 N NH_4OH	Use the organic phase. Prepare freshly each time.
n-Butanol saturated with NH_4OH	*n*-Butanol saturated with 0.1% NH_4OH. Much concentrated NH_4OH, for instance 2 N, can be used without appreciably different *Rf* values.	Use the organic phase. Prepare freshly each time.
tert-Amyl alcohol saturated with pH 6 phthalate	*tert*-Amyl alcohol shaken with excess pH 6 phthalate buffer, which is prepared by mixing 50 ml 0.1 M potassium hydrogen phthalate in 45.5 ml 0.1 N NaOH and made up to 100 ml.	Papers are first dipped in the buffer and dried at room temperature.
1.5 M Phosphate buffer	138 g (1 M) $NaH_2PO_4 \cdot H_2O$ and 71 g (0.5 M) Na_2HPO_4 are dissolved in 1 liter H_2O	Chromatogram should be developed in a warm room (about 20° C) in winter, otherwise crystals will appear in the solvent during development.
2.5 M Phosphate buffer	229 g $NaH_2PO_4 \cdot H_2O$ and 118 g Na_2HPO_4 are dissolved in 1 liter H_2O	

DNP-valine from DNP-leucine is less satisfactory. Nevertheless, the *n*-butanol-NH_4OH system is useful since it is easy to prepare and requires no equilibration of papers with the solvent in a developing chamber, in contrast to the "toluene" system. Alternatively, the *tert*-amyl alcohol-pH 6 phthalate system of BLACKBURN and LOWTHER [114 A] has also been used. This system is limited either to one dimension or to the second solvent in two-dimensional chromatography, since a paper sheet must previously be buffered with 0.1 M phthalate at pH 6. Composition of solvents and *Rf* values of the ether soluble DNP-amino acids are listed in Tables 3-2 and 3-3. *Rf* values of the acid soluble DNP-amino acids are shown in Table 3-4.

Two-dimensional thin layer chromatography on silica gel G plates has also been used routinely with solvent systems reported by BRENNER et al. [170]. In Figs. 3-3 and 3-4 are shown typical two-dimensional chromatograms. MUNIER and SARRAZIN [892] separated all of the ether-soluble DNP-amino acids by two-dimensional thin layer chromatography on powdered cellulose without binders. The Biserte "toluene"

Table 3-3. *Rf* values of ether extractable DNP-amino acids[a]

	Toluene-pyridine-2-chloro-ethanol-NH_4OH[b]	*n*-Butanol saturated with 0.1% NH_4OH[c]	*tert*-Amyl alcohol saturated with pH 6 phthalate[c]	1.5 M Phosphate[c]	2.5 M Phosphate[c]	*n*-Butanol saturated with water[d]
DNP-Aspartic acid	0.08	0.08	0.03	0.62	0.44	0.12
DNP-Glutamic acid	0.09	0.10	0.06	0.56	0.37	0.14
DNP-Glycine	0.35	0.36	0.14	0.31	0.18	0.36
DNP-Serine	0.30	0.29	0.13	0.46	0.31	0.32
DNP-Threonine	0.36	0.41	0.17	0.54	0.33	0.43
DNP-Alanine	0.45	0.52	0.25	0.41	0.22	0.50
Di-DNP-Histidine	0.70	0.54	0.32	0.13	0.09	0.35
Dinitrophenol	0.43	0.62	0.45	0.31	0.20	0.56
Di-DNP-Lysine	0.72	0.65	0.49	0.11	0.06	0.72
Di-DNP-Tyrosine	0.74	0.75	0.55	0.02	0.01	0.78
DNP-Methionine	0.59	0.63	0.50	0.41	0.25	0.65
DNP-Phenylalanine	0.66	0.76	0.58	0.28	0.15	0.71
DNP-Methionine sulfoxide	0.35	—	—	—	—	—
DNP-Methionine sulfone	0.48	—	—	—	—	—
DNP-Valine	0.56	0.69	0.68	0.47	0.30	0.68
DNP-Leucine	0.64	0.78	0.70	0.40	0.30	0.74
DNP-Proline	0.43	0.53	0.50	0.52	0.41	0.48
DNP-Tryptophan	0.62	0.75	0.69	0.39	0.22	0.70
Bis-DNP-Cystine	0.32	0.24	0.28	0.16	0.10	—

[a] Dinitroaniline travels very fast in the organic solvents and very slowly in phosphate buffers. The *Rf* values of the DNP-amino acids are variable and the values listed should be considered as an indication of position.
[b] From Ref. [112]. [c] From Ref. [864]. [d] From Ref. [977].

Table 3-4. *Rf* values of acid soluble DNP-amino acids[a]

	Rf[b]	Color of spot	Ninhydrin reaction[c]
DNP-Arginine	0.64	Yellow	No reaction
DNP-Cysteic acid	0.25	Yellow	No reaction
ε-DNP-Lysine	0.55	Yellow	Brown-purple
O-DNP-Tyrosine	0.70	Colorless	Purple
imid-DNP-Histidine	0.33	Colorless[b] (or faint yellow[c])	Brown

[a] Solvent: *n*-Butanol-acetic acid-water (12:3:5, v/v). Prepare freshly each time.
[b] From Ref. [864]. [c] From Ref. [1071].

system and saturated ammonium sulfate solution-water-sodium dodecylsulfate (100 ml:700 ml:0.576 g) are used, respectively, for the first and the second dimensions. Several solvents were proposed by WANG and WANG [1327, 1328] for two-dimensional thin layer chromatography of DNP-amino acids on polyamide sheets. ROSMUS and DEYL reviewed comprehensively various chromatographic separations of DNP-amino acids [1040].

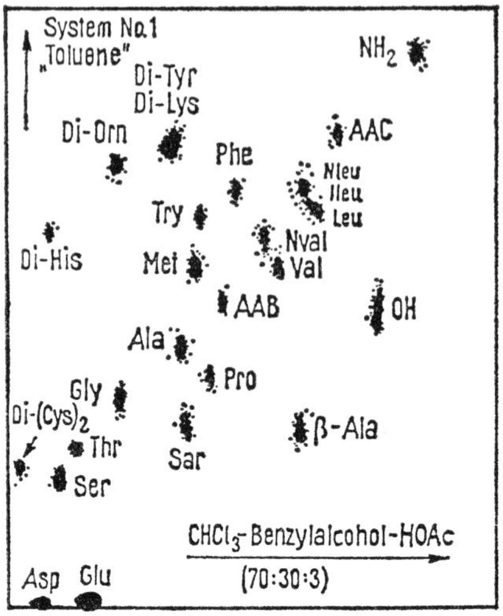

Fig. 3-3. Two-dimensional thin layer chromatogram of a synthetic mixture of DNP-amino acids (0.2 μg each). Abbreviations: OH, dinitrophenol; NH_2, dinitroaniline; AAB, α-amino-butyric acid; AAC, α-amino-caprylic acid. (Reprinted from Ref. [170], by courtesy of Birkhäuser Verlag)

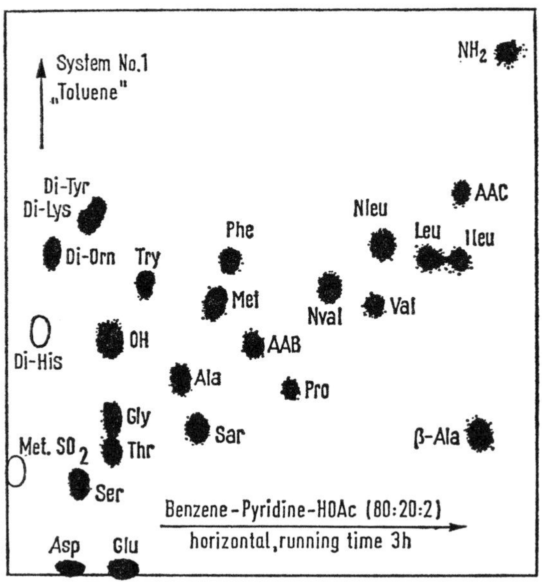

Fig. 3-4. Two-dimensional thin layer chromatogram of a synthetic mixture of DNP-amino acids (1 μg each). Abbreviations as in Fig. 3-3. (Reprinted from Ref. [170], by courtesy of Birkhäuser Verlag)

Most DNP-amino acids are yellow, except for O-DNP-tyrosine and *imid*-DNP-histidine which are colorless. Absorption maxima are near 260 nm, but those of DNP-proline and -hydroxyproline are at 385 nm in 1% $NaHCO_3$, the usual solvent for the extraction and estimation of DNP-amino acids from paper chromatograms.

In 1% $NaHCO_3$ solution, the ratio of optical density of most DNP-amino acids at 390 nm to that at 360 nm is 0.55 to 0.65, but that of DNP-proline and DNP-hydroxyproline is 1.1 to 1.4 and that of dinitrophenol is approximately 0.8. Dinitrophenol and artifacts produced can easily be distinguished from DNP-amino acids by measuring the above ratio of optical densities at the two wave lengths. The best method is to obtain the absorption spectra between 250 nm and 500 nm of the chromatographed yellow substances. Absorption of DNP-amino acids in an alkaline medium differs from that in an acidic medium.

The identified DNP-amino acids on chromatograms can be confirmed by regenerating free amino acids from the corresponding DNP-derivatives after hydrolysis with saturated barium hydroxide [795] in a sealed tube at 100° C for 1 hr or with concentrated ammonium hydroxide [865] at 100° C for 2 hrs.

Methods for the preparation of the standard DNP-amino acids are not described here since these are available commercially.

1. Dinitrophenylation of Proteins and Peptides

A weighed amount of a protein (1 to 2 μmole, minimum 0.2 μmole) and an equal weight of $NaHCO_3$ are dissolved or suspended in 10 volumes of water and an additional 2 volumes of 5% ethanolic solution of 1-fluoro-2,4-dinitrobenzene (FDNB) are added. The reaction mixture is shaken for 2 hrs at room temperature. The reaction mixture is then acidified with a few drops of concentrated HCl. The usual DNP-proteins are insoluble in water. The precipitated DNP-protein is collected by centrifugation and washed successively with ether and ethanol to remove excess FDNB and dinitrophenol produced from the reagent, and further washed successively with water, acetone and ether. The DNP-protein thus obtained is dried in an evacuated dessicator on $CaCl_2$ or P_2O_5.

When DNP-derivatives of proteins and oligopeptides are soluble in water, the reaction mixture is diluted with two volumes of water to decrease the alcohol concentration and excess FDNB is removed several times with ether. The reaction mixture is then acidified with HCl to pH 1 and the DNP-derivatives are extracted

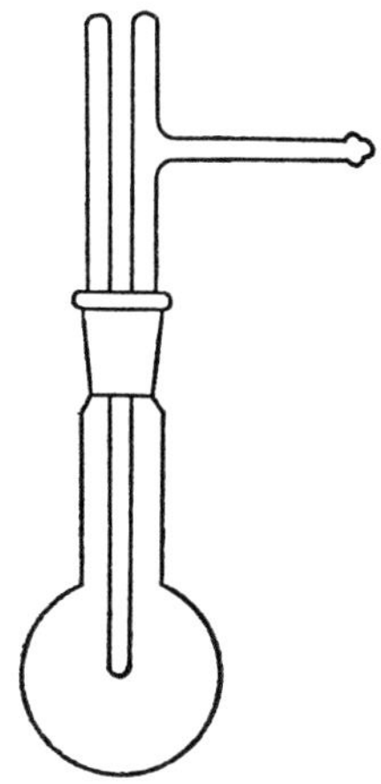

Fig. 3-5. Vacuum sublimation apparatus according to MILLS [866]

with ether, ethyl acetate, or *sec*-butanol. To prevent conversion of DNP-methionine to its sulfoxide, solvents used for the extraction of DNP-derivatives must be freed from peroxides by shaking with concentrated ferrous sulfate solution. After the organic extracts are washed with water, the DNP-derivatives are recovered on evaporation *in vacuo*. Dinitrophenol produced is removed by vacuum sublimation at around 60° C using an apparatus suggested by MILLS [866] as shown in Fig. 3-5. Into the inner tube is placed crushed dry ice. Dinitrophenol that has sublimed on the outer surface of the inner tube after 15 min is washed away with acetone using a polyethylene flush bottle. The surface of the reaction mixture covering the bottom of the outer tube of the sublimation apparatus is renewed by dissolving the mixture in acetone and again the content is dried *in vacuo*. Sublimation is repeated three to four times until no more dinitrophenol is sublimed. The end point of sublimation can easily be recognized by observing that no yellow color appears in water, into which the acetone wahings are poured.

The DNP-peptide remaining in the aqueous phase of the above extraction can be recovered by adsorbing it on the top of a talc column (1 × 5 cm), which has previously been washed with 1 N HCl. The column is washed with 1 N HCl (about 20 ml) to elute free peptides which escaped reaction and the adsorbed DNP-peptide is eluted with 1 N HCl-ethanol mixture (1:4, v/v).

When a small amount of peptide (0.2 μmole, for instance) extracted from a paper chromatogram is treated, the peptide is dissolved in 0.1 ml of 1% triethylamine, and 0.2 ml of 5% FDNB alcohol solution is added [1069]. After 2 hrs at room temperature, a few drops of water and triethylamine are added and the excess FDNB is extracted 3 times with ether. Salt-free DNP-peptide can be obtained on evaporation of the aqueous phase.

2. Hydrolysis of DNP-Proteins

The dried DNP-protein is weighed (minimum 0.2 μmole) into a small glass test tube and constant boiling HCl (5.7 N HCl), redistilled with a glass apparatus (to remove interfering metal ions) is added such that the DNP-protein is sufficiently covered with the acid. The test tube is then evacuated with a vacuum pump while cooling the contents with a dry ice bath for 5 to 10 min and sealed. In general, 80% of the weight of the DNP-protein roughly corresponds to the original protein. The molecular weight of the DNP-protein can be calculated by adding a factor, 166 X (where X is the number of DNP-groups introduced), to the molecular weight of the original protein, assuming that all of the constituent lysine, tyrosine, histidine, cysteine and the terminal amino groups have been quantitatively substituted [381]. SANGER [1065] originally estimated the protein content of DNP-insulin by analyzing its amide groups since amide groups of asparagine and glutamine are known not to be substituted with nor hydrolyzed by FDNB during the reaction under the alkaline conditions used.

The optimum time of hydrolysis at 105° C for the end group analysis differs from one DNP-protein to another and if a sufficient amount of protein is available, the best method is to follow the released amount of the terminal DNP-amino acid as a function of hydrolysis time as discussed previously. If the sample is not large enough, hydrolysis should be carried on for 6 to 8 hrs. This period is not enough to release DNP-amino acids quantitatively from the protein derivative but is sufficient to identify unstable DNP-amino acids during acid hydrolysis except for DNP-proline. Based on the result of a preliminary analysis, further experiments should be performed with reference to the stability of each DNP-amino acid as shown in Table 3-1.

3. Characterization and Estimation of the DNP-Amino Acid

The hydrolysate is diluted to about five volumes with water to decrease the concentration of HCl to 1 N, and the N-terminal DNP-amino acid is extracted with 5 ml of ether (should be freed from peroxide) until no more yellow color is extracted into the ether layer. This usually requires 3 to 5 extractions. The ether extracts are washed with small portions of water and the washings are returned to the aqueous phase. DNP-Arginine, DNP-cysteic acid and di-DNP-histidine cannot be completely extracted into the ether layer since these DNP-amino acids partition between the two phases. The use of *n*- or *sec*-butanol is recommended for their extraction, but ε-DNP-lysine, O-DNP-tyrosine and DNP-derivatives of oligopeptides are also extractable with these solvents.

The above ether extracts are evaporated *in vacuo* and the residue is dissolved in 0.1 ml of acetone. Fifty μl of the solution is transferred quantitatively with the aid of a micropipette to a sheet of filter paper (18 × 22.5 inch, Whatman No. 1) to be used for two-dimensional chromatography. Two paper chromatograms can therefore be prepared from a single sample hydrolysate. The first development is usually carried out with Biserte's "toluene" system in an ascending manner. The paper sheet is curled into a cylinder and placed in a dish of suitable size (e.g. 7.75 inches in diameter). The paper is equilibrated for 4 h with 200 ml of 0.8 N ammonia in a small dish placed concentrically inside the larger dish in which the chromatographic solvent is added later. The entire assembly is placed in a cylindrical glass jar with a well fitting plate glass lid (e.g. 8.75 inches in diameter and 18 inches in height). An air-tight seal of the rim of the outer cylinder is maintained with silicone grease and by placing a weight on top of the lid. Biserte's "toluene" solvent is prepared according to the composition shown in Table 3-2. The mixture is allowed to stand for 4 hrs (duration of the equilibration period) and the lower aqueous layer is withdrawn. The volume of the upper layer is enough to develop two paper sheets. The organic layer is added through a hole in a glass lid to the annular space between the inner and outer dishes through a funnel fitted with a long stem. Development of DNP-amino acids is then started and continued overnight (about 15 to 16 hrs) in the dark; during this period the solvent rises to within about one inch of the top of the paper cylinder. The paper is then removed and dried in an air oven at 40° C for at least 3 to 4 hrs; insufficient removal of the first solvent causes trailing when chromatography in the second dimension is performed.

In place of the "toluene" system, *n*-butanol saturated with 0.1% NH_4OH can be used as discussed previously.

The dried paper cylinder is then unfolded and suitable controls (5 μl of 0.01 M solutions of DNP-amino acids in acetone) including dinitrophenol are then spotted on either side of the unknown yellow spot. The second development is then carried out overnight in the dark with 1.5 M phosphate buffer of pH 6 listed in Table 3-2 either by the ascending or descending technique in a chromatographic apparatus containing a tray of water at the bottom. In the second development with phosphate buffer, no equilibration is necessary, so no air-tight chamber is required. When ascending development with phosphate is performed, care must be taken to prevent weakening of the filter paper due to absorption of the dense buffer solution.

After the paper chromatogram is air-dried, yellow spots (ultraviolet-absorbing spots under ultraviolet lamp) are marked with a pencil about 0.1 inch beyond their visible peripheries. Characterizations of the yellow spots are made from the positions in relation to the dinitrophenol spot comparing with a standard map prepared from a mixture of the authentic DNP-amino acids and dinitrophenol (Fig. 3-2). The marked spots are cut out, placed in test tubes, and 4 ml of 1% $NaHCO_3$ is added. The tubes are placed in a water bath at 55 to 60° C for 15 min to allow elution of the yellow color. A few blanks cut from each sheet are treated similarly. The extinction values of all of the solutions are read at 360 nm and 390 nm using a set of quartz cuvettes against the extract of the blank paper (the actual optical density of the blank is 0.001 to 0.002 per cm^2 against distilled water). The ratio of the optical density at 390 nm to that at 360 nm ranges from 0.55 to 0.65 with the common DNP-amino acids while those of DNP-proline and dinitrophenol are 1.4 and 0.8, respectively, as mentioned above. When the ratio of the optical densities is largely different from the above, complete absorption curves should be measured. Millimoler extinction coefficients of usual DNP-amino acids are 17 to 18. Therefore the approximate micromoles of the end group of the protein in question is calculated by multiplying by the optical density reading 17 and by the recovery factor and by dividing by 4 (volume of the extracting solvent). In the cases of bis-DNP-cystine and di-DNP-lysine, it is necessary to multiply further by the factor 0.6 due to the presence of two substituents. The recovery factor of each amino acid consists of the chromatographic recovery (about 90%) and the hydrolytic recovery (varies depending on the amino acid and on the hydrolysis time). It is recommended that recovery factors of all DNP-amino acids after chromatography and after hydrolysis-chromatography, respectively, be individually obtained.

Another convenient characterization system for DNP-amino acids is the *tert*-amyl alcohol-phthalate method of Blackburn and Lowther [114A]. *tert*-Amyl alcohol saturated with the buffer is used for development either in an ascending or descending manner and the buffer saturated with *tert*-amyl alcohol is placed in a chromatographic chamber. DNP-Glutamic acid moves ahead of DNP-aspartic acid which moves only slightly ahead of unmovable DNP-carboxymethylcysteine in this system in contrast to the "toluene" system. Mutual separation is enhanced considerably by developing with pH 5 phthalate buffer (50 ml of 0.1 M potassium hydrogen phthalate + 23.9 ml of 0.1 N NaOH made up to 100 ml) for long periods by the descending technique.

B. Dansyl Chloride Method

Hartley and Massey [516] reported in 1956 that non-fluorescent dansyl chloride (1-dimethylaminonaphthalene-5-sulfonyl chloride, sometimes abbreviated as DNS-Cl) gives a strong yellow fluorescence upon sulfonamide formation with amino group of amino acids and of peptides as shown in reaction (4). This reaction has been extended to the end group determination of peptides and proteins along similar lines to the FDNB method of Sanger [1065]. Its main advantage over the Sanger method is the 100-fold increase in sensitivity making it possible to handle less than 1 nmole of sample, and the increased stability to acid hydrolysis of the DNS-amino acids over the corresponding DNP-amino acids.

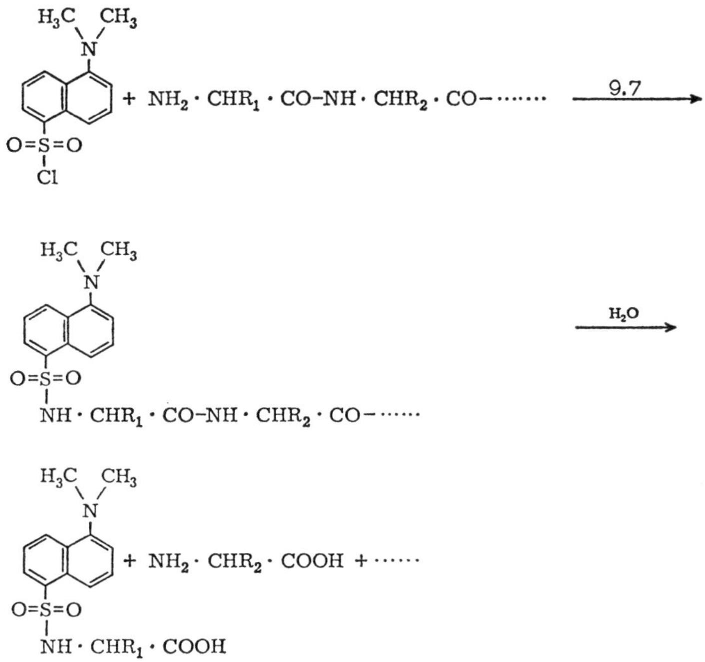

(4)

The reactivity of functional groups of amino acids and peptides toward dansyl chloride in aqueous solution depends mainly upon their strength as bases. In order to determine optimal conditions for labeling, GRAY and HARTLEY measured the velocity constants for the reaction of dansyl chloride with water or with a large concentration excess of each amino acid as shown in Table 3-5, and they found that the reaction with primary amines follows a Brønsted dependence:

$$\log K = 0.52\, pKa - 4.0\,.$$

They calculated the degree of reaction of various amino acids with dansyl chloride and demonstrated that the reactive groups having pKa of 9 to 11 could be completely labeled at pH 9.5 to 10.5 with sufficient concentrations of dansyl chloride and amino acids. Optimum labeling conditions for most amino acids require the use of a several-fold excess of the reagent in 5 mM concentration over amino acid (1 mM) at pH 9.7 at 37° C for 1 hr [517]. Since the reagent is hydrolyzed to DNS-OH (1-dimethylaminonaphthalene-5-sulfonic acid) by water and by hydroxyl ion, amino acids are labeled as a function of their ability to compete for limited amount of the reagent under the above conditions. Above pH 10.5 the reagent is hydrolyzed too rapidly, and at a pH lower than 9.5 the unreactive protonated form of amino acids is present in too high a concentration. Addition of acetone supresses the ionization of the amino group and solubilizes the reagent, and it was shown that 50% acetone with bicarbonate buffer gave optimal results.

Table 3-5. Reactivity of various bases toward dansyl chloride

Base	*pKa*	*k*[a]	% Labeled[b]
H_2O	—	4.1×10^{-5}	—
OH^-	—	15	—
NH_3	9.26	0.5	39
Alanine	9.87	10	99
Arginine (α)	9.04	6.4	99
Asparagine	8.80	3.1	97
Aspartic acid	9.82	4.4	93
Cysteic acid	9.0	5.7	99
Glutamine	9.13	4.8	99
Glutamic acid	9.67	5.0	97
Glycine	9.78	13	99
Histidine[c] (α)	9.18	12	99
Histidine[c] (imid.)	6.1	0.5	53
Isoleucine	9.76	14.5	99
Leucine	9.74	13.5	99
Lysine[c] (α)	8.95	3.6	98
Lysine[c] (ε)	10.50	42	99
Methionine sulfone	8.6	3.2	98
Phenylalanine	9.24	10	99
Proline	10.60	360	99
Serine	9.15	4.6	99
Threonine	9.12	4.9	99
Tryptophan	9.39	35	99
Tyrosine[c, d] (α)	9.11	(10)	99
Tyrosine[c] (ionized phenolic group)	10.07	280	99
Valine	9.72	13	99
Diglycine	8.17	3.5	99
Triglycine	7.91	3.0	97
Leucylglycine	8.2	2.4	94

[a] *k* is the second-order rate constant for the reaction between DNS-Cl and the appropriate base. Measurements were made in water at 22° C.
[b] Calculated percentage labeling of amino acids at pH 9.8, using 5 mM DNS-Cl.
[c] Value of *k* is that for the corresponding group of the free amino acids.
[d] The value for *k* is only approximate.
Data from Ref. [438].

Dansyl chloride reacts not only with amino groups, but also with imino, phenolic hydroxyl, thiol, and imidazole groups, but the stabilities of the products differ. The DNS-derivatives of thiol and imidazole compounds are very labile in both acid and alkali and are broken down completely under the conditions used to hydrolyze labeled peptides [1259]. Furthermore, when glutamyl and glutaminyl peptides are subjected to the reaction, especially in highly dilute solutions, they are partly converted into DNS-pyrrolidone carboxylyl peptides, the orange fluorescence of which is very weak compared with that of α-DNS-amino acids [1258]. Dansylation of peptides containing amino acid residues side chain groups of which are unreactive to dansyl chloride generally proceeds smoothly, but peptides containing lysyl, tyrosyl and histidyl residues and the N-terminal glutamyl peptides often give multiple products due to varying reactivities of the above side chain groups and to varying stabilities of the dansylated functional groups. When the dried dansylated product is treated with 10% formic acid for 1 to 2 hrs at room temperature, the unstable imidazole dansyl group of the histidine residue can be removed. By this formic acid treatment, single DNS-products could be obtained from glycylhistidine, histidyltyrosine and angiotensin, respectively.

During the reaction of dansyl chloride with amino acids, side reactions are known to occur. When excess reagent is present in the reaction mixture, DNS-NH_2 (1-dimethylaminonaphthalene-5-sulfonamide) is produced probably by reaction (5) [905]. The extent of the side reaction is dependent upon the nature of the side chains of amino acids as shown in Table 3-6.

$$\text{DNS-NH} \cdot \text{CHR} \cdot \text{COO}^{\ominus} + \text{DNS-Cl} \rightarrow \text{DNS-NH} \cdot \text{CHR} \cdot \text{COO-DNS} \xrightarrow{\text{OH}^{\ominus}} \text{DNS-NH}_2 + \text{RCHO} + \text{CO}_2 + \text{DNS-OH} \quad (5)$$

As mentioned, it is necessary to use the reagent in a concentration of 5 mM for effective labeling of amino groups. When very small amounts of a sample are treated, the volume of the reaction mixture can be correspondingly reduced to keep the concentration of the sample at about 1 mM. At the end of the reaction, the mixture contains several times more DNS-OH produced through hydrolysis than the DNS-peptide formed. It is best to remove the by-product, DNS-OH, prior to acid hydrolysis of DNS-peptides and the subsequent characterization of the DNS-amino acids in the hydrolysate by paper chromatography, thin layer chromatography, or paper electrophoresis. However, the removal of DNS-OH is not necessary when dansylation is carried out in the nanomole range.

Table 3-6. Percent formation of DNS-NH_2 during the reaction of dansyl chloride with amino acids[a]

Amino acids	% Formation of DNS-NH_2
Glycine	2
Alanine	11
Valine	12
Leucine	15
Proline	1

[a] The reaction mixture consists of 0.1 mmole of amino acid, 0.11 mmole of dansyl chloride, 0.5 mmole of $KHCO_3$, 4.5 ml of acetone and 3.5 ml of water. Data from Ref. [905].

For the removal of DNS-OH and DNS-NH_2 from the reaction mixture, Tamura and Nakajima [1258] recommend the use of a small column of DEAE-Sephadex (acetate form, 1 to 2 ml) prior to hydrolysis of DNS-peptide. The reaction mixture is placed on the small DEAE-Sephadex column and unadsorbed DNS-NH_2 is first washed out with 50% methanol. The adsorbed DNS-peptide is then eluted with 50% methanol containing 1 to 5% acetic acid and the yellow fluorescent bands eluted are collected. Under these conditions, DNS-OH is retarded and moves more slowly than the DNS-peptides.

The recoveries of various DNS-amino acids under the standard conditions of acid hydrolysis of DNS-peptides (5.7 N HCl at 105° C for 18 hrs) are listed in Table 3-7. Upon prolonged hydrolysis with acid di-DNS-lysine yields both the α- and ε-DNS-lysine monoderivatives in addition to the di-DNS-derivative. DNS-Tryptophan is completely destroyed under these conditions.

When much milder conditions are used, recoveries of the acid labile DNS-amino acids increase, but DNS-valyl or DNS-isoleucyl peptides fail to undergo complete hydrolysis [1258, 1259], since under the above standard conditions, some peptide linkages adjacent to DNS-valyl or DNS-isoleucyl residues resist acid hydrolysis. Such DNS-dipeptides can sometimes be identified in addition to DNS-valine or DNS-isoleucine. The DNS-derivatives of glutamic acid and glutamine are partly converted into DNS-pyrrolidone carboxylic acid under conditions of acid hydrolysis [1258]. DNS-Carboxylmethylcysteine undergoes similar changes, partially to the corresponding 6-member ring compound [1258] and partially to DNS-cysteic acid [438]. Upon ammonia treatment, these ring compounds yield their corresponding amides.

Table 3-7. Stability of DNS-amino acids to acid hydrolysis[a]

Derivative	% Remaining	Fluorescent decomposition products
DNS-Proline	23	DNS-OH[b]
DNS-Serine	65	$DNS\text{-}NH_2$[c] (25%), DNS-OH (10%)
DNS-Glycine	82	DNS-OH
DNS-Alanine	93	DNS-OH
DNS-Valine	100	no detectable breakdown
DNS-Leucine	100	no detectable breakdown
DNS-Isoleucine	100	no detectable breakdown
DNS-Tryptophan	0	$DNS\text{-}NH_2$, DNS-OH (traces of DNS-Gly, DNS-Ser, DNS-Ala)

[a] Percentage remaining as unchanged DNS-amino acid after 18 hrs of hydrolysis with constant boiling 5.7 N HCl at 105° C. Data from Ref. [438], by courtesy of Academic Press.
[b] 1-Dimethylaminonaphthalene-5-sulfonic acid.
[c] 1-Dimethylaminonaphthalene-5-sulfonamide.

Pronase digestion is effective on peptides containing the acid resistant peptide bonds such as DNS-valyl or DNS-isoleucyl linkage. In pronase digestion, ε-DNS-lysine and O-DNS-tyrosine are recovered with the exceptions of those bonding with proline residue [1259].

DNS-Amino acids should be viewed under an ultraviolet lamp. Both long-wave (365 nm) and short-wave (254 nm) mercury lamps are effective in exciting fluorescence. However, the former is preferred because short-wave radiation promotes photo-oxidation of DNS-amino acids to some extent, though this is not serious for short exposures with the exceptions of O-DNS-tyrosine and *imid*-DNS-histidine. These DNS-derivatives are readily decomposed into bluish-green fluorescent compounds by a few minutes exposure to ultraviolet light, both long-wave and short-wave or the sun light.

The fluorescence can be visually detected with as little as 10^{-2} nmole of material per spot. The fluorescence intensity of the DNS-amino acids or peptides is much higher in the basic than acidic solvent systems to be used for chromatography on paper and thin layer plates. DNS-amino acids and -peptides are usually unstable on thin layer plates, especially on silica, developed with either system. After one day, the fluorescent spots disappear and they are converted into non-fluorescent yellow-orange spots. Storage of the chromatographed plates in the dark does not prevent such decomposition. Addition of a few drops of β-mercaptoethanol to the solvents to be used for development in chromatography and storage in a place saturated with *n*-butanol vapor are somewhat beneficial [1259]. However, such destruction of DNS-derivatives is largely reduced on polyamide sheets and this is a significant advantage of polyamide sheets for thin layer chromatography of DNS-amino acids.

For characterization of DNS-amino acids can be carried out by paper chromatography, paper electrophoresis or thin layer chromatography. Boulton and Bush [750]

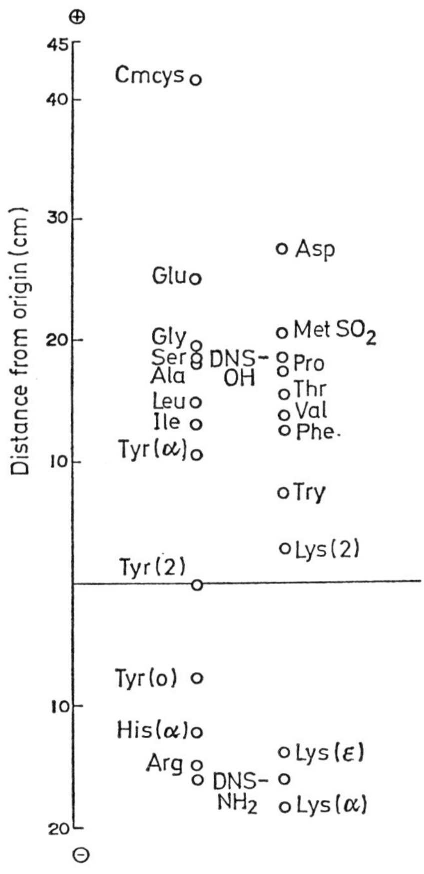

Fig. 3-6. Electrophoretic mobilities of various DNS-amino acids at pH 4.38 (80 V/cm, 2.5 hrs, 15° C). Spots are drawn approximately to scale, to give an indication of the resolution attained. Abbreviations: Tyr(α), α-DNS-Tyr; Tyr(2), N,O-di-DNS-Tyr, etc. (Reproduced from Ref. [438], by courtesy of Academic Press)

reported several useful systems for the paper chromatography of DNS-amino acids. The DNS-derivatives of glycine, alanine, phenylalanine, proline, valine, leucine and isoleucine can be separated with a mixture of light petroleum-acetic acid-water (100:90:10, v/v) and DNS-derivatives of asparagine, methionine sulfone, serine aspartic acid, glutamic acid, tyrosine, and glutamine can be separated with benzene-diisopropyl ether-acetic acid-water (1:1:1:1, v/v).

Gray [438] has shown that most DNS-amino acids can be identified by high voltage paper electrophoresis at pH 4.4, pH 1.9 and sometimes at pH 12.7, using an apparatus of the flat-bed type [449]. Apparatus of the Michl type [859] cannot be used since DNS-amino acids are extracted into the coolant. With electrophoresis at about pH 4.4 (80 v/cm, 2.5 hrs at 15° C), DNS-derivatives of cysteic acid, carboxymethylcysteine, aspartic acid, glutamic acid, methionine sulfone, tyrosine, lysine, histidine and arginine can be identified with certainty, and others with a reasonable degree of confidence (DNS-alanine, -threonine, -leucine, -valine, -isoleucine, and -phenylalanine) as shown in Fig. 3-6. DNS-Derivatives of glycine, serine and, sometimes, alanine are obscured in most cases by the strong blue fluorescent band of

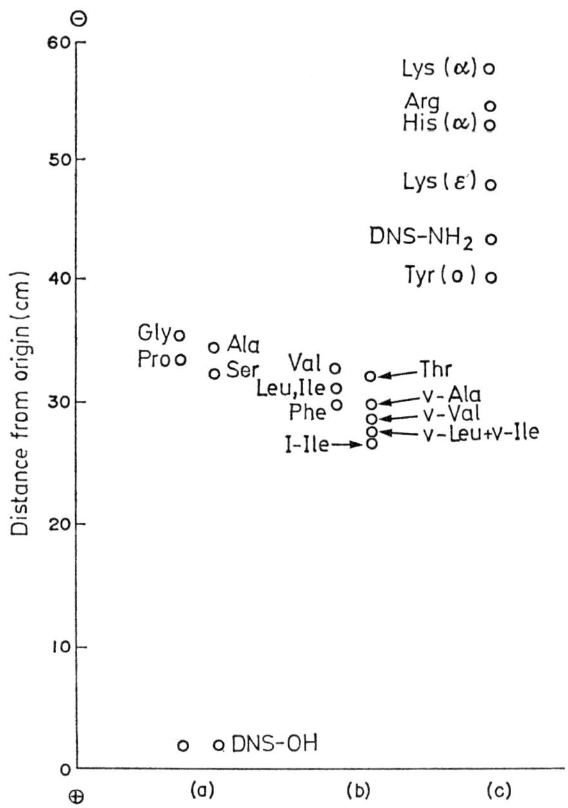

Fig. 3-7. Electrophoretic mobilities of DNS-amino acids at pH 1.9 (50 V/cm, 2 hrs, 20° C). *(a)* DNS-amino acids which run near DNS-OH at pH 4.4 (Fig. 3-6). *(b)* Certain other DNS-amino acids and peptides. Abbreviations: I-Ile, Ile-Ile; V-Ala, Val-Ala; etc. *(c)* Basic DNS-amino acids. (Reproduced from Ref. [438], by courtesy of Academic Press)

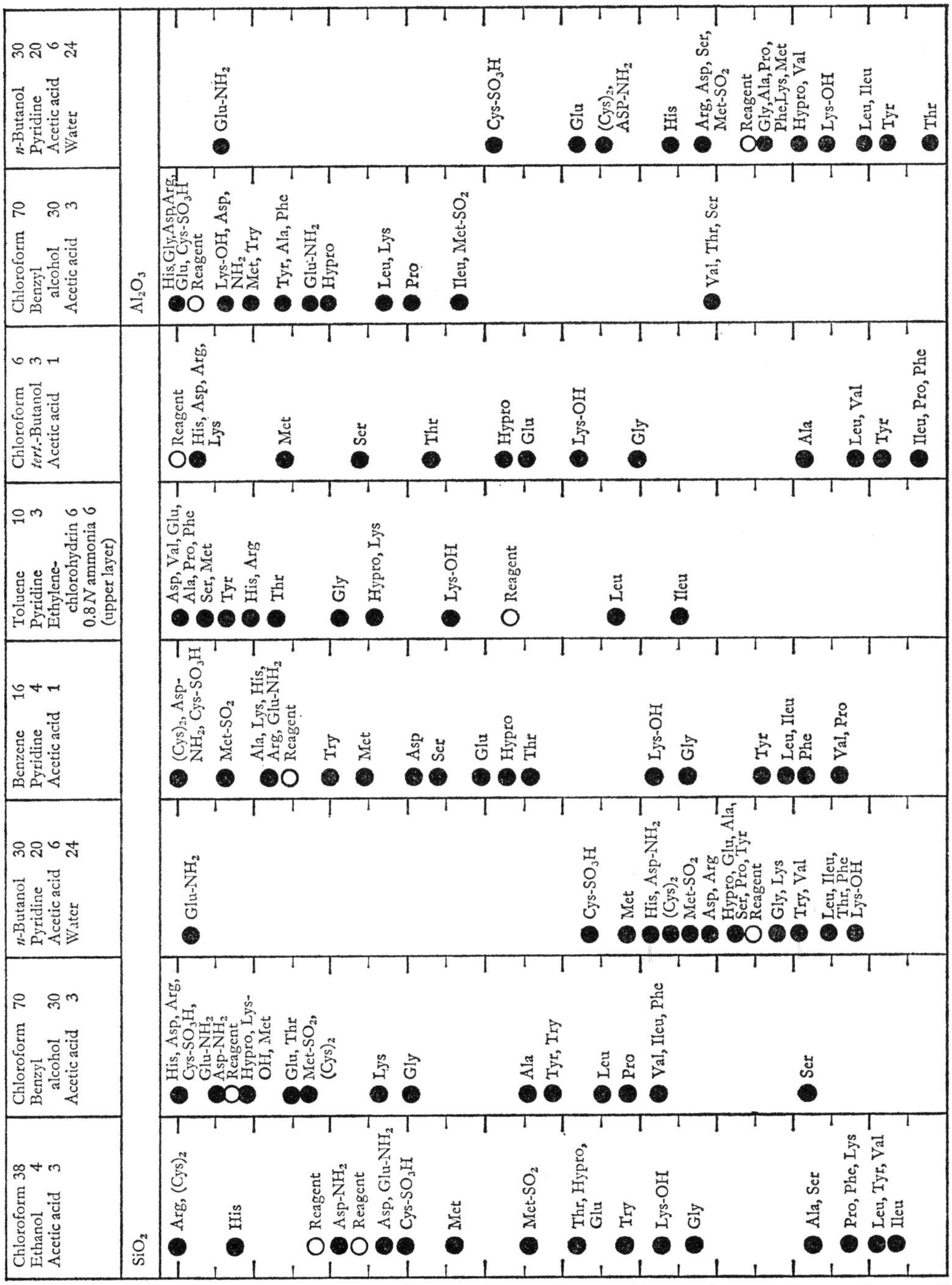

Fig. 3-8. *Rf* values of DNS-amino acids in thin layer chromatograms with different solvent systems and on different carriers. (Reprinted from Ref. [282], by courtesy of Elsevier Publishing Co)

DNS-OH. These DNS-amino acids can clearly be identified by electrophoresis at pH 1.9 as shown in Fig. 3-7. Sometimes electrophoresis at pH 12.7 is also used to identify α-DNS-tyrosine, O-DNS-tyrosine, DNS-proline, DNS-threonine, DNS-leucine, and DNS-arginine.

The identification of DNS-amino acids also can be made by thin layer chromatography on plates of silica gel G (or silica gel H, without binders). The extent of

Table 3-8. *Rf* values of DNS-amino acids on silica gel G[a]

	Benzene-pyridine-acetic acid (80:20:2)	Chloroform-*tert*-amyl alcohol-acetic acid (70:30:0.5)	Chloroform-*tert*-amyl alcohol-acetic acid (70:30:3)
DNS-Isoleucine	0.92	0.89	1.0
DNS-Leucine	0.89	0.63	1.0
DNS-Valine	0.85	0.75	1.0
DNS-Proline	0.79	0.50	0.93
DNS-Phenylalanine	0.69	0.42	0.94
DNS-Methionine	0.65	0.33	0.92
DNS-Alanine	0.61	0.29	0.89
DNS-Tyrosine	0.55	0.23	0.82
Di-DNS-Lysine	0.48	0.33	0.93
DNS-Glycine	0.38	0.12	0.65
DNS-Tryptophan	0.34	0.25	0.87
DNS-Histidine	0.27	0.09	0.59
DNS-Glutamic acid	0.18	0.08	0.42
DNS-Threonine	0.18	0.02	0.25
DNS-Serine	0.15	0.02	0.20
DNS-Aspartic acid	0.06	0	0.02
DNS-Carboxymethylcysteine	0.05	0	0.02
DNS-Glutamine	0.03	0	0.07
DNS-Asparagine	0.02	0.03	0.02
ε-DNS-Lysine	0.02	0.04	0.08
DNS-Arginine	0	0	0

[a] *Rf* represents the migration relative to $DNS\text{-}NH_2$. (From Ref. [887], by courtesy of Academic Press).

migration relative to $DNS\text{-}NH_2$ of the standard DNS-amino acids with several solvent systems is listed in Table 3-8. In Fig. 3-8, representative patterns of one-dimensional separation of DNS-amino acids on thin layer plates of silica gel G developed with several systems are shown [282]. In most solvent systems $DNS\text{-}NH_2$ runs near the solvent front. The presence of DNS-OH sometimes interferes with the separation of DNS-amino acids, but the blue fluorescent band of this substance is useful as an internal standard due to its easy differentiation from the yellow fluorescence of DNS-amino acids.

During the development of DNS-amino acids, the addition of β-mercaptoethanol (0.05%) is recommended to eliminate destruction of DNS-amino acids on silica, and particularly to prevent the oxidation of DNS-methionine to its sulfone [1258]. In thin layer chromatography of DNS-amino acids, the *Rf* values are quite variable depending upon the concentrations of salts present in the sample. The unknown sample should be mixed with a series of standard DNS-amino acids and these applied at the origin as shown in Fig. 3-9 [1258]. Although the *Rf* values of the sample and standard DNS-amino acid may be influenced by the presence of salt and appear to be the same, separation of the two can be effected and the relative order maintained when the two compounds are chemically different.

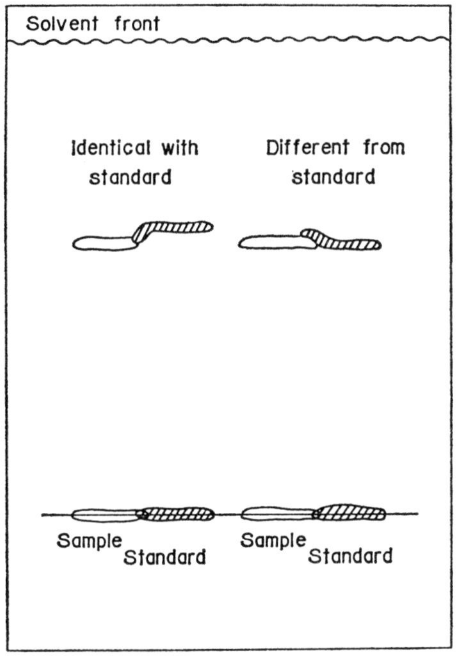

Fig. 3-9. Application of samples and standards of DNS-derivatives on a thin layer plate. (Reproduced from Ref. [1258])

Two dimensional thin layer chromatography on polyamide sheets introduced by WOODS and WANG [1391] gives extremely fine resolution of DNS-amino acids. HARTLEY and collaborators [517] replaced the technique of paper electrophoretic separation of DNS-amino acids with this simple chromatographic procedure using polyamide sheets, since 0.01 nmole of DNS-amino acid can easily be detected on a 5 × 5 cm sheet. Polyamide sheets can be prepared by coating a polyester film with polyamide prepared from ε-caprolactam in formic acid [1326], but this is commercially available (Cheng-Chin Trading Co. Ltd., No. 75, Sec. 1, Hankow St., Taipei, Taiwan). Two dimensional separation patterns of DNS-amino acids on polyamide sheets are shown in Fig. 3-10. The first development is carried out with solvent I and the second with solvent II. Separation of DNS-aspartic acid from DNS-glutamic acid, DNS-serine from DNS-threonine, and DNS-arginine, DNS-histidine from ε-DNS-lysine is not complete and sometimes DNS-alanine overlaps with $DNS\text{-}NH_2$. For the mutual separation of these, the third development (with solvent III) is performed after inspection under an ultraviolet lamp in the same direction as that with solvent II. By the triple development procedure, DNS-arginine still cannot be resolved from ε-DNS-lysine and a fourth development with solvent IV is required in the same direction for the separation of the two [1344]. When DNS-cysteic acid is present, it lies below the large blue fluorescent DNS-OH spot, but can be resolved by using 1 M ammonia-ethanol (1:1, v/v) in the direction of solvent II [92].

This two dimensional thin layer chromatography on polyamide sheets seems to be the best among separation systems discussed above.

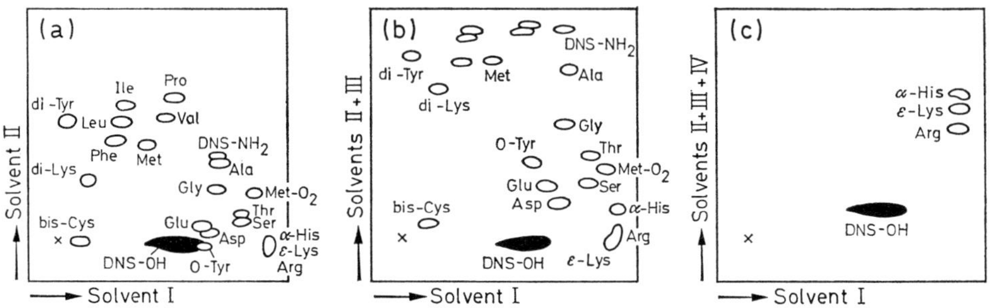

Fig. 3-10a—c. Two dimensional chromatograms of DNS-amino acids on polyamide sheets. (a) After the first and second developments with solvents I (1.5% formic acid) and II (benzene-acetic acid, 9:1, v/v); (b) after the third development with solvent III (ethyl acetate-acetic acid-methanol, 20:1:1, v/v); (c) after the fourth development with solvent IV (0.05 M Na_3PO_4 in 25% aqueous ethanol). α-DNS-ε-phenylthiocarbamyl-lysine runs with leucine in the first two developments, and with methionine in the third development

Solvent I 1.5% formic acid;
Solvent II benzene-acetic acid (9:1, v/v);
Solvent III ethyl acetate-acetic acid-methanol (20:1:1, v/v);
Solvent IV 0.05 M Na_3PO_4 in 25% aqueous ethanol.

A number of chromatographic systems for separation of DNS-amino acids are reviewed by Rosmus and Deyl [1039].

Due to its high sensitivity, the present dansyl chloride method, in combination with Edman's stepwise degradation method, furnishes a powerful tool for sequence analysis of peptides. After removal of the N-terminal residue of the peptide by the Edman method, the newly released terminal amino acid may be characterized by the present method using only very small amount of the sample. Using this technique, Gray and Hartley [439] elucidated the sequence of a hexapeptide derived from the chymotryptic digestion of *Pseudomonas* cytochrome c_{551} as well as other peptides from different proteins [515, 1172]. Bruton and Hartley [182] determined the N-terminal tetrapeptide and an octapeptide sequence containing a lysine residue of the catalytic site of methionyl-tRNA synthetase from *E. coli* using only 1 to 2 mg of the enzyme by this dansyl-Edman technique. Weiner et al. [1344] also elucidated the N-terminal hexadecapeptide sequence of *lac* repressor by this method on a nanomole scale.

The reaction of dansyl chloride and its application to biochemical analysis have been reviewed recently by Seiler [1144].

1. Reaction of Dansyl Chloride with Peptides and Proteins

A solution of peptide (0.1 to 1 nmole) is transferred into a small test tube (4 × 30 mm). After drying *in vacuo* the peptide is redissolved in 10 to 15 μl of 0.2 M $NaHCO_3$ and the solution is centrifuged, if necessary, and again evaporated *in vacuo* to remove traces of ammonia. The dried residue is redissolved in 15 μl of deionized water and an equal volume of dansyl chloride solution (2.5 mg/ml in acetone) is added to give a final concentration of 5 mM DNS-chloride, 1 mM peptide and 50% acetone. The pH of the solution must be 8.5 to 9.8 at this time. If the pH is too low, more base must be added. The final concentration of the reagent is about 5 mM. The tube is then covered with Parafilm and the reaction is allowed to proceed for 1 h at 37° C. The reaction mixture is then dried *in vacuo*.

TAMURA et al. [1259] recommended the following procedure for quantitative labeling of a peptide with dansyl chloride. A peptide (0.1 to 1 nmole) is dissolved in 10 μl of 0.1 M triethylamine in the cold and to the solution is added 10 μl of cold 0.5% dansyl chloride in acetone. The mixture is kept in a refrigerator for 16 hrs and the solvents are evaporated to dryness under a stream of nitrogen. 5 μl of 90% formic acid are added to the residue followed by 50 μl of water in order to hydrolyze the unstable dansyl derivative of the histidine imidazole residues. The solution is kept for 1 to 2 hrs at room temperature and then dried. The pH of the dansylation mixture under the above conditions is more basic and this promotes the quantitative dansylation of the tyrosine hydroxyl group and prevents the conversion of DNS-glutamic acid into DNS-pyroglutamic acid. By this technique the dansylated product of glutamylalanine gave a single spot on a thin layer chromatogram.

For the labeling of proteins, GRAY [438] recommended the following conditions: about 50 to 250 μg of performic acid-oxidized protein is placed in a small test tube (4 × 50 mm), and 50 μl of 1% SDS is added. The protein is then dissolved by heating the mixture in a boiling water bath for 2 to 5 min to ensure unfolding of the peptide chain. After the solution has cooled, 50 μl of N-ethylmorpholine (sequencing grade) is added and mixed thoroughly. To the solution is added 75 μl of dansyl chloride solution (25 mg/ml) in anhydrous dimethylformamide (specially purified for silylation reactions; the reagent solution must be freshly prepared since the solvent rapidly absorbs moisture, the reagent hydrolyzes) and the solution is mixed thoroughly. Dansylation is allowed to proceed for 1 hr or longer at room temperature. In order to precipitate the labeled protein, about 0.5 ml of acetone is added to the reaction mixture and the tube is inverted several times after covering with Parafilm to mix the solution thoroughly. The protein is pelleted by centrifugation and the sediments are washed once with 500 μl of 80% acetone and dried.

2. Hydrolysis

In most cases it is not necessary to remove DNS-OH from the reaction mixture. To the dried residue, 50 μl (0.5 ml for DNS-proteins) of redistilled constant-boiling HCl (5.7 N) is added and the tube is evacuated and sealed. Hydrolysis is carried out at 105° C for 16 to 18 h [438]. When DNS-valyl-, DNS-isoleucyl- or DNS-leucyl-dipeptides are present, the hydrolysis time must be longer. The hydrolysate is then dried under a nitrogen stream. The DNS-amino acids are dissolved in 10 μl of 50% pyridine (dissolves very insoluble di-DNS-tyrosine and -lysine) for application to the polyamide sheet for chromatography.

3. Identification of DNS-Amino Acid by Two-dimensional Thin Layer Chromatography on Polyamide Sheets [517, 1344]

Commercially available polyamide sheets are coated on both sides (15 × 15 cm) and are cut to 5 × 5 cm with a scissors. The hydrolysis residue is dissolved in 10 μl of 50% aqueous pyridine (v/v). Using a λ pipette with finely drawn tip, about 1 μl or less of unknown sample solution is repeatedly spotted onto a corner of one side of the sheet by touching the tip of the pipette to the polyamide sheet. This does not damage the sheet, which is tough and virtually unaffected by small scratches or compression. The application spot should not exceed 2 mm in diameter. The sheet is dried for a few seconds between applications in the hot air stream of a hair drier. Only 10% of the sample is applied, since the dansyl spots will trail severely if the sheet is overloaded.

A standard mixture containing about 0.05 nmole of each DNS-amino acid is spotted on the back of the sheet opposite the unknown. The sheet is dried by a hot air stream. Incomplete evaporation of the pyridine will produce trailing in the first dimension, due to local neutralization of the 1.5% formic acid. Ascending chromatography is first carried out for 5 min with solvent 1 (1.5% formic acid) in a glass bottle with wide mouth of an appropriate size or in a 150 ml beaker covered with a stretched sheet of Parafilm. The lower edges of the sheet must not touch the walls of the glass bottle or the beaker, otherwise capillarity may distort the solvent front, especially in the development with solvents II and III. After the first development, the sheet is dried for 10 min and then the second development with solvent II (benzene-acetic acid, 9:1, v/v) is performed for 5 min in the perpendicular direction to that with solvent I. After 5 min drying, the identification of the unknown DNS-amino acid is carried out by referring the locations of the standard DNS-amino acids on opposite side under an ultraviolet lamp.

Identification of DNS-derivatives of aspartic acid, glutamic acid, serine, threonine and histidine should be carried out after the third development with solvent III (ethyl acetate-acetic acid-methanol, 20:1:1, v/v) for 5 min in the same direction as solvent II. DNS-Arginine separates from ε-DNS-lysine after the fourth development with solvent IV for 15 min in the same direction as solvent III.

Used polyamide sheets can be re-used at least five times after washing with either acetone-29% NH_4OH (9:1, v/v) or acetone-90% formic acid (9:1, v/v) by soaking for 6 hrs followed by washing with purified methanol several times [1321].

A standard mixture of DNS-amino acids is prepared in the following way [1344]: 100 μl of an amino acid standard mixture to be used for calibration of an automatic amino acid analyzer (2.5 mM in each amino acid in 0.01 N HCl) is evaporated and redissolved in 0.5 ml of 0.1 M $NaHCO_3$, and dansylated by adding 0.25 ml of freshly prepared dansyl chloride solution at 5 mg/ml in acetone. The tube is covered with Parafilm and incubated for 30 min at 37° C. The reaction is terminated by adding 25 μl of 88% formic acid. The dansylated mixture is used directly for spotting (0.2 μl contains 0.05 nmole of each). This dansylated mixture contains bis-DNS-histidine, which is converted to α-DNS-histidine by acid hydrolysis.

4. Dansyl-Edman Method for Sequence Determination of Peptides and Proteins

For sequence analysis of peptides by the dansyl-Edman technique, Hartley [517] recommended the following procedure: a solution containing 20 to 100 nmole of a peptide is placed in a small stoppered tube and the solvent is evaporated. 150 μl of water is added and 10 μl is withdrawn for the N-terminal analysis by the dansyl chloride method described above. To the remaining solution is added 150 μl of 5% phenylisothiocyanate in pyridine (w/v) and the atmosphere in the tube is replaced with nitrogen by flushing. The stoppered tube is incubated for 1.5 hrs at 45° C and then the solvents and excess of the reagent are removed *in vacuo* at 1 mm Hg at 60° C for about 30 min or more over conc. $H_2SO_4 + P_2O_5 +$ NaOH. To the dried residue is added 200 μl of trifluoroacetic acid and the tube is flushed with nitrogen. The mixture is incubated for 30 min at 45° C and evaporated over NaOH for 10 min. The residue is dissolved in 200 μl of water and the anilinothiazolinone derivative of the N-terminus formed is extracted twice with 1.5 ml of *n*-butyl acetate. The aqueous layer is evaporated and redissolved in 150 μl of water. 10 μl of the solution is subjected to the N-terminal determination by the dansyl chloride method in order to identify the newly released terminus.

The above procedures can be scaled down so as to allow sequence analysis of 1 nmole of peptide. The procedures are exactly the same as those described above except that all volumes are reduced to one-tenth.

WEINER et al. [1344] modified the dansyl-Edman method suitable for sequence analysis of proteins on a nanomole scale. The sample protein (5 to 10 nmoles) is dissolved in 200 μl of coupling buffer (0.5 M $NaHCO_3$ adjusted to pH 9.8 with NaOH) and to the solution is added no less than 20 μl of 10% SDS solution (w/v). For efficient detergent action, SDS must be present in at least 1.4-fold weight excess over protein. Then 10 μl of phenylisothiocyanate is added to the tube, which is immediately warmed to 50° C in a water bath, flushed with nitrogen, and sealed with several layers of Parafilm and mixed thoroughly. The reaction proceeds for 30 min at 50° C with occasional vortexing every 5 or 10 min to keep the coupling buffer saturated with the reagent.

In order to precipitate the phenylthiocarbamylated protein, 2 ml of acetone is added to the reaction tube and after thorough mixing, the resulting flocculant precipitate is centrifuged for 2 min. The yellowish supernatant is withdrawn with a Pasteur pipette and the precipitate is washed with 2 ml of acetone. The colorless supernatant after centrifugation is discarded. The precipitate is spread around the walls of the tube by vortexing, and the tube is placed in the heated vacuum desiccator in an horizontal position where the damp slurry is dried under vacuum for 20 min. Rapid evaporation of acetone under reduced pressure will cause lumps on the bottom to explode and be blown out the mouth of the tube.

For cyclization of the N-terminal residue into the anilinothiazolinone derivative with concomitant cleavage of the first peptide bond, the tube is allowed to cool to room temperature, and 0.2 ml of trifluoroacetic acid is added. The tube is flushed with nitrogen and the mixture is incubated at 50° C for 5 min. Then about 1 mg of dried crystalline SDS is added to the tube and the tube is vortexed to dissolve the added SDS while still warm. The trifluoroacetic acid is blown away by a nitrogen stream, so that a residue forms on the walls of the tube. The tube is placed in the heated vacuum desiccator and dried *in vacuo* for 20 min. The residue is then dissolved in 0.2 ml of coupling buffer and, to the solution, is added at least 20 μl of 10% SDS solution. An aliquot (0.25 to 0.5 nmole in 10 μl) is withdrawn for dansylation. The Edman degradation may be continued by adding 10 μl of phenylisothiocyanate to the remaining solution and the procedures described above are repeated.

C. Cyanate Method

Stark et al. [1214] observed that contaminating cyanate ions in urea solution react with α- and ε-amino groups in proteins to yield carbamyl derivatives. On the basis of this observation, Stark and Smyth [1215] extended the carbamylation reaction with potassium cyanate to the N-terminal analysis of proteins. This method is suitable only for the determination of the N-terminal residue of proteins and large peptides. Potassium cyanate reacts quantitatively with amino groups of proteins at neutral or slightly alkaline pH in the presence of denaturants to yield carbamylated derivatives. Upon heating the carbamyl protein in strong acid, the N-terminal carbamyl residue cyclizes to the corresponding hydantoin, with accompanying cleavage of the peptide bond according reaction (6).

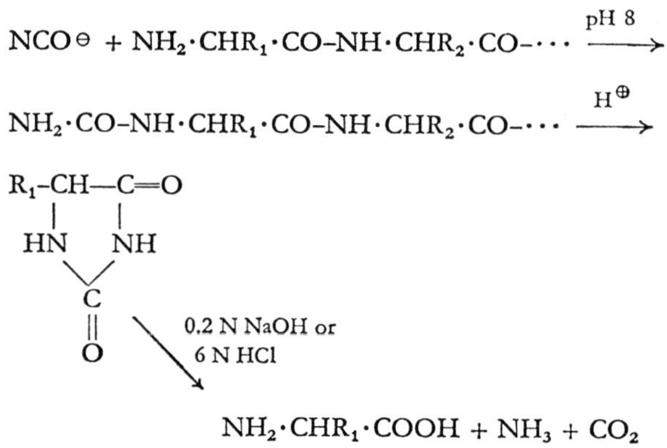

(6)

In contrast to the phenylthiocarbamyl group, the carbamyl moiety requires more drastic conditions for cyclization to the hydantoin of the terminal residue resulting in partial hydrolysis of the residual protein fragment during the cyclization reaction. Thus the cyanate method cannot be used repeatedly as is Edman's stepwise degradation method. The ε-amino group of lysine residues also undergoes carbamylation to yield homocitrulline residues with cyanate at a slow rate (half life is 2 to 3 hrs at 50° C) in comparison with α-amino groups (half life is 3 to 4 min) [1215], but the homocitrulline residues do not change under the cyclization conditions. An N-terminal lysine is converted into a carbamyl homocitrulline residue which is then cyclized to the hydantoin of homocitrulline upon heating with strong acid. Sulfhydryl groups also react rapidly with cyanate [1214]. In general, reduced and S-alkylated proteins or performic acid-oxidized samples are used in end group determinations of cystine- or cysteine-containing proteins. S-Carbamylcysteine is labile to acid hydrolysis; the products formed are predominantly ammonia and cysteine, plus a small quantity of cystine [1214].

Hydantoins are separated from the cyclization mixture by chromatography on Dowex 50 (H^+) as will be described later. When an oxidized peptide containing cysteine or cystine is hydrolyzed during the cyclization, cysteic acid is produced. The cysteic acid overlaps with hydantoins derived from the N-terminal carbamylated residues upon elution with water. For this reason, reduction of sulfhydryl groups is preferred to oxidation when the present cyanate methods are employed. However, the N-terminal carboxymethylcysteine does not yield its hydantoin by strong acid treatment and the carbamyl group attached to the terminal nitrogen atom seems to split off (private communication by P. E. WILCOX).

Cyclization of carbamyl amino acid residues into their hydantoins is carried out in a solution of equal volumes of 50% acetic acid and concentrated hydrochloric acid because of the enhanced solubility of carbamyl proteins in 50% acetic acid. Most hydantoins formed from the corresponding carbamyl amino acid residues are quite stable in the mixed acid solvent at 100° C for 1 hr; thus, with the exception of proline (21% yield), less than 3% of any amino acids are formed from the corresponding hydantoins. If only hydrochloric acid is used, approximately twice as much free

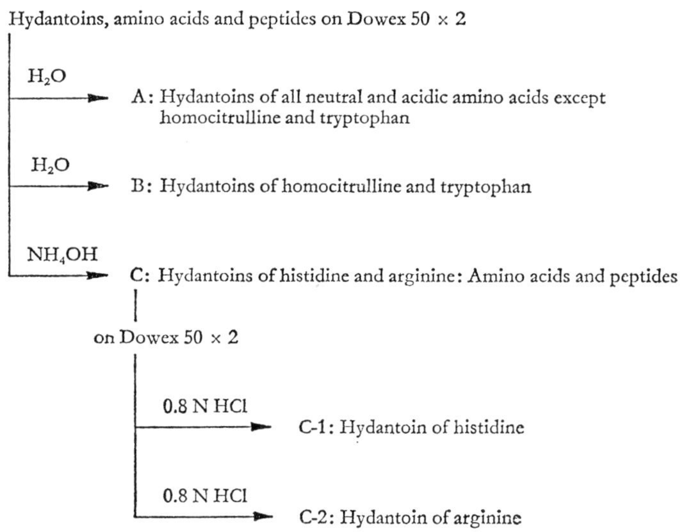

Fig. 3-11. Separation scheme of hydantoins from the products of partial acid hydrolysis according to STARK and SMYTH [1215], by courtesy of the American Society of Biological Chemists

proline is released. During the cyclization reaction, the thioether sulfur atom in methionine and S-alkylcysteine is partly oxidized in the presence of air. Therefore cyclization must be performed in an evacuated sealed tube.

Hydantoins of acidic and neutral amino acids are, respectively, acidic and neutral. These hydantoins can easily be separated from other amino acids and peptides in the cyclization mixture by passing the mixture through a column of the hydrogen form of a strongly acidic ion exchanger such as Dowex 50. By this treatment, the hydantoins appear at the breakthrough point, but amino acids, peptides and hydantoins of basic amino acids (histidine and arginine) are retained on the resin column. The basic hydantoins are eluted with 1 N NH_4OH together with amino acids and peptides. The evaporated eluate is repeatedly applied to a Dowex 50 column which has been equilibrated with 0.8 N HCl, and the basic hydantoins are separated by eluting with the same acid as used for the equilibration. Isolation of various hydantoins described above is summarized in Fig. 3-11 [1215]. When performic acid-oxidized proteins are used, cysteic acid and peptides containing cysteic acid are contained in the water eluate from the Dowex 50 column together with the hydantoins of neutral and acidic amino acids. In order to separate cysteic acid and its peptides from the hydantoins, the water eluate from Dowex 50 column used previously must be subjected

Table 3-9. Recoveries of amino acids after end group determination[a]

Amino acid	Recovery (%)
Lysine	84
Histidine	89[b]
Arginine	76[b]
Tryptophan	77[c]
S-Carboxymethylcysteine	68[d]
Aspartic acid	99
Threonine	30
Serine	20
Glutamic acid	60 (97[b])
Proline	73
Glycine	98[e]
Alanine	98[e]
Valine	96
Methionine	99[f]
Isoleucine	93
Leucine	97
Tyrosine	96
Phenylalanine	99

[a] All hydrolyses were performed for 20 hrs in 0.2 N NaOH at 110° C, except as otherwise indicated. Data from Ref. [1215], by courtesy of the American Society of Biological Chemists.
[b] Hydrolysis for 96 hrs in 6 N HCl at 110° C.
[c] Determination was performed in the presence of serotonin (see text).
[d] Hydrolysis for 48 hrs in 6 N HCl at 110° C. Hydrolysis in alkali gives no carboxymethylcysteine.
[e] Glycine and alanine were determined separately, not in the presence of other amino acids.
[f] Methionine plus methionine sulfoxides.

to further chromatography on a Dowex 1-X8 (acetate form) column (0.9 × 6 cm). The hydantoins of neutral and basic amino acids (except tyrosine hydantoin) appear at the breakthrough position by washing the Dowex 1 column with 0.5 N acetic acid with a flow rate of 60 ml/h. The hydantoins of glutamic acid (eluted within 50 ml of effluent) and aspartic acid (100 ml) are somewhat retarded, but strongly acidic cysteic acid and cysteic acid peptides are retained on the column. Tyrosine hydantoin overlaps the positions of the glutamic acid and aspartic acid derivatives. Direct characterization and estimation method of hydantoins by chromatography on

AG 1-X8 columns have been reported [410], but its application to proteins has not appeared yet. In general, amino acids regenerated from their hydantoins are analyzed.

Heating hydantoins with 6 N HCl regenerates the parental amino acids in nearly quantitative yields with the exception of serine, threonine, homocitrulline and tryptophan. Recoveries of serine and threonine are very low due to their destruction under acid conditions and tryptophan is completely decomposed. Homocitrulline hydantoin yields lysine (about 25%) plus homocitrulline upon acid hydrolysis. In the case of alkaline hydrolysis with 0.2 N NaOH, tryptophan is recovered and the yields of the two β-hydroxy-α-amino acids increase, but arginine is converted into ornithine, which overlaps with lysine during amino acid analysis under standard conditions. With alkaline hydrolysis, homocitrulline hydantoin regenerates lysine quantitatively, but recovery of histidine decreases and isoleucine partly epimerizes to alloisoleucine. Therefore the preferred hydrolytic agent for hydantoins changes according to the nature of the hydantoins fractionated. In Table 3-9 are listed the recoveries of amino acids after the end group determinations. Except for serine and threonine, these recovery factors may be used routinely. Serine and threonine hydantoins decompose somewhat during the cyclization, and the extents of their decompositions are variable; each produces some glycine upon hydrolysis with 0.2 N NaOH, especially threonine hydantoin. For these amino acids, it is therefore advisable that each investigator determine the recoveries under his own conditions rather than to use the recoveries listed in the table.

The regenerated terminal amino acids are usually characterized and estimated by an amino acid analyzer. The sensitivity of this method is therefore dependent upon the sensitivity of the analyzer used. Usually small amounts of amino acids which are not derived from the terminus appear in the amino acid analysis. It is, therefore, better to subtract the blank value from the quantities of amino acids obtained from the carbamylated sample. A blank determination is carried out on a sample protein after the entire procedure but without exposure to cyanate. The procedures using the cyanate method is described in detail by STARK [1213].

1. Carbamylation

Denaturants such as 8 M urea or 6 M guanidine chloride must be employed to obtain carbamylation in homogeneous solutions. Two ml each of N-ethyl morpholine (colorless redistilled preparation, stored in the cold) and water are mixed and the pH of the mixture is adjusted to 8 with glacial acid. To the above buffer is added 2.4 g of solid urea and the final volume is adjusted to 5 ml with water. To a solution of about 50 mg of protein in 2.5 ml of 8 M urea, is added 2.5 ml of the above buffer followed by 250 mg of KCNO. The carbamylation reaction is allowed to proceed overnight at 50° C.

Five ml of glacial acetic acid is added slowly with stirring to the reaction mixture to decompose the excess cyanate into carbon dioxide and ammonia. If the protein precipitates at this time, it is collected by centrifugation. The precipitate, often a clear gel, should be washed thoroughly, first with distilled water, then with acetone, and finally dried in the air. If no precipitate was formed on addition of acetic acid, the carbamyl protein is dialyzed against water. In many cases the precipitate forms during dialysis. The contents of the dialysis sac are lyophilized. Alternatively, gel filtration on Sephadex or Bio-gel can be used in place of dialysis.

2. Cyclization to Hydantoins

A half mole of carbamyl protein in 1 to 3 ml of 50% acetic acid is placed in a heavy-walled test tube (Pyrex, 1.8 × 15 cm) and an equal volume of concentrated HCl is added. The tube is then evacuated and sealed. After cyclization at 100° C for 1 hr, the contents are evaporated to dryness and 1 ml of water is added to the residue. If the protein contains tryptophan, 50 mg of serotonin-creatine sulfate should be present during the cyclization, to prevent destruction of N-terminal tryptophan.

3. Separation of Hydantoins from the Mixture

A column of Dowex 50-X2 (H^+ form, 200 to 400 mesh) is prepared in a 0.9 × 20 cm tube, fitted with a coarse sintered glass disk, such that the column height is 10 cm. The solution of hydantoins is placed onto the column with the aid of three 1 ml rinses of water. The column is developed with water at 20 to 25° C at up to 100 ml/h. The first 30 ml, including the water displaced by the sample, is collected as fraction A (contains acidic and neutral hydantoins) and the next 40 ml as fraction B (homocitrulline and tryptophan hydantoins). These solutions are evaporated to dryness. The column is washed with 40 ml of 1 N NH_4OH, retained as fraction C, and this solution is evaporated to dryness. The residues from fraction C are dissolved in 2.0 ml of water and 1.8 ml of the solution is placed on another column of Dowex 50-X2 (H^+ form, 0.9 × 10 cm), which has been equilibrated with 0.8 N HCl. The effluent from the column with 0.8 N HCl at a flow rate of about 60 ml/h at room temperature is collected in two fractions, 20 through 35 ml (fraction C-1, contains histidine hydantoin), and 35 through 70 ml (fraction C-2, contains arginine hydantoin). These solutions are evaporated to dryness.

Fraction A contains pyrrolidone carboxylic acid and pyrrolidone carboxylyl peptides, if present, derived from glutamine and glutaminyl peptides by cyclization during the cyclization of carbamyl amino acid residues. Hydrolysis of the pyrrolidone groups to glutamic acid or glutamyl peptides is effected with 5 ml of 3 N HCl as the solution is heated at 100° C for 30 min. After evaporation of the acid, the residues are dissolved in about 2 ml of water and transferred quantitatively onto a Dowex 50-X2 column (H^+ form, 0.4 × 5 cm) with an additional 1 ml of water. The column is washed with 6 ml of water and the effluent is concentrated to dryness. In this author's experience, small amounts of non-terminal aspartic and glutamic acids were always found in fraction A. Therefore, the partial acid hydrolysis of this fraction and retreatment of the hydrolysate with a small Dowex 50 column as described above is recommended.

4. Hydrolysis of Hydantoins

a) Fraction A

Two ml of freshly prepared 0.2 N NaOH is added and 1.8 ml is transferred to a hydrolysis tube (1.8 × 5 cm). The tube is evacuated and sealed, then heated at 110° C for 20 hrs. After hydrolysis, the NaOH is neutralized with 1 ml of 1 N HCl and the solution is evaporated to dryness in the tube. Two ml of citrate buffer of pH 2.2 is added and the solution is centrifuged to remove any precipitate. The supernatant (1.8 ml) is subjected to amino acid analysis for determination of the neutral and acidic amino acids. If the analysis reveals N-terminal glutamic acid, it may be desirable to perform a 96 hrs acid hydrolysis of the A fraction, since an excellent yield of this amino acid is obtained as shown in Table 9.

b) Fraction B

The residues from fraction B are dissolved in 2.0 ml of 0.2 N NaOH and hydrolyzed exactly as fraction A. The hydrolysate is analyzed with the short column (usually 15 cm) of an amino acid analyzer through the position of lysine.

c) Fraction C

The residues from fractions C-1 and C-2 are dissolved separately in 2.0 ml of 6 N HCl, and 1.8 ml of each solution is hydrolyzed in an evacuated sealed tube at 110° C for 96 hrs. The HCl is removed by evaporation and the residues are dissolved in 1 ml of 0.2 N NaOH. The solution is evaporated nearly to dryness to remove excess ammonia. After addition of 1 ml of 1 N HCl, the solution is evaporated to dryness. The residues are dissolved in 2.0 ml of pH 2.2 buffer and 1.8 ml is chromatographed on the 15 cm column, following the positions of histidine for C-1 and arginine for C-2. It is essential that the first 10 ml of the effluent should not be mixed with ninhydrin. Large quantities of amino acids derived from neutral peptides are present in this volume which could, if allowed to react with ninhydrin, precipitate in the reaction coil of the analyzer.

D. Other Methods

1. Chemical Methods

Three chemical methods widely used for the analysis of the N-terminal group have so far been described. In addition, a number of other chemical methods have been proposed but most of them seem to be inadequate for general application to proteins with a few exceptions and are seldom used at present. Some of these methods are briefly outlined below. About forty methods for amino end group determination are reviewed by Rosmus and Deyl [1039], especially on chromatographic systems for separation of derivatized amino acids. A discussion of the phenylisothiocyanate method, undoubtedly one of the most invaluable methods is reserved for the description by Edman in Chapter 8 of this book.

Aryl sulfonyl groups attached to amino groups, in general, are more resistant toward acid hydrolysis than peptide bonds or acyl amino linkages. If proper hydrolysis conditions, under which peptide bonds are cleaved but an aryl sulfonyl amino group is intact, can be found, aryl sulfonyl halides can be used as reagents for N-terminal analysis. Dansyl chloride, described in the previous section is a typical example. In addition to dansyl chloride, other aryl sulfonyl chlorides such as benzene sulfonyl chloride [460], β-naphthalene sulfonyl chloride [1253] and *p*-iodobenzene sulfonyl (pipsyl) chloride [1299, 1316] have been examined. The aryl sulfonyl amino acids derived from proteins after the reaction and acid hydrolysis are difficult to identify in small amounts, since no characteristic marker groups are involved in these reagents. Pipsyl chloride containing a radioactive I^{131} or S^{35} atom, however, has been shown to be of value for terminal analysis. Udenfriend and Velick [1299,

1316] used this method for the N-terminal analysis of several proteins. Detailed procedures of pipsyl chloride method are described by LEVY [770] and FLECHER [362]. The half life of the radioactive isotope of iodine or sulfur is relatively short, however, compared with the time necessary for terminal determination. Other aryl sulfonyl chloride reagents proposed for N-terminal analysis are *p*-azobenzene sulfonyl chloride [677], sulfoindonyl chloride [623] and N-methyl-2-anilino-6-naphthalene-sulfonyl chloride (mansyl chloride) [249]. Mansyl amino acids are fluorescent like dansyl amino acids but a separation system for mansyl amino acids has not yet been established, though this reagent is reported to be more sensitive than dansyl chloride.

2,4-Dinitrobenzene sulfonate [325] reacts only with amino groups and releases bisulfate in alkaline media (pH 10.5 to 11.5) yielding the same DNP-derivatives as with Sanger's reagent. Dinitrobenzene sulfonate is sometimes used for the modification of amino groups of proteins, but there seems to exist no advantage as an end group analysis method over the Sanger method. 2,4,6-Trinitrobenzene sulfonate [942, 1074, 1076] gives the trinitrophenyl amino group which is quite labile to hydrolysis and cannot be used for end group analysis, although the reagent is of value for the modification of proteins. The weakest point in Sanger's DNP-method is the destruction of the N-terminal DNP-amino acid under hydrolysis conditions of DNP-protein with hydrochloric acid. In order to overcome such disadvantage, various aryl halides have been reported for N-terminal analysis. 3,5-Dinitro-2-chloropyridine is the most interesting; the reagent reacts with amino groups under mild conditions (pH 8.5, room temperature, 2 hrs) in a manner similar to Sanger's reagent and dinitropyridyl amino acids including glycine and proline are released nearly quantitatively from the protein derivatives with 6 N hydrochloric acid containing 30% formic acid at 60° C for 8 to 10 hrs [1162]. The protonated nitrogen atom of the pyridine ring facilitates the hydrolytic cleavage of the first peptide bond. This catalytic effect due to the nitrogen atom is enhanced by the nitro group at the ortho position [1161] as shown in scheme (7).

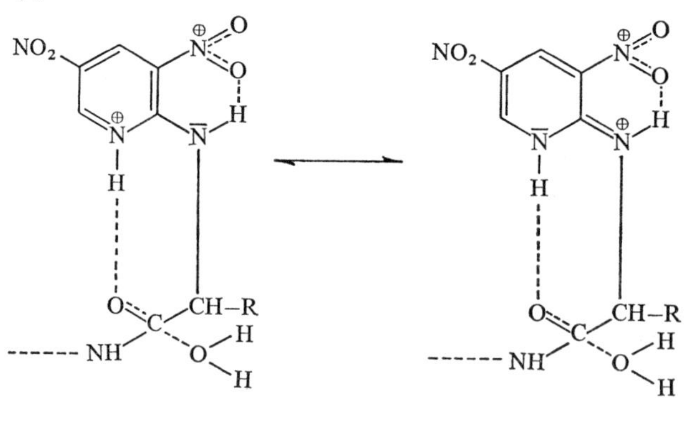

(7)

Characterization and estimation of dinitropyridyl amino acids can be carried out by the aid of chromatography in ways similar to those for DNP-amino acids [12]. This dinitropyridyl method seems to be superior to the DNP-method.

2-Chloro-5-nitropyrimidine [1162] is reported to have similar characteristics to dinitrochloropyridine described above. 2-Fluoro-3-nitropyridine and 2-fluoro-5-nitropyridine [1159] also have been proposed and 3-nitro-2-pyridyl amino acids are released from the modified proteins 100 times as easy as 5-nitro-2-pyridyl derivative. This phenomenon can be explained by scheme (7). 3-Fluoro-4-nitropyridine N-oxide [1163], 2,4-dinitro-5-fluoroaniline [89, 1014], nitro-2-methanesulfonyl-4-fluorobenzene [1142] and 7-chloro-4-nitrobenzene-2-oxo-1,3-diazole [412] are reported to be useful for the end group determination. There seems to exist no advantage over Sanger's DNP-method or dinitropyridyl method described above solely for the end group analysis, although some of these are reported to be useful as stepwise degradation procedures.

Holley and Holley [575] introduced the 2-nitro-4-carbomethoxyphenyl group onto the amino group of peptides, reduced the nitro group catalytically on platinum black to an amino group, and split the terminal peptide bond by heating in neutral aqueous solution with concomitant formation of a lactam ring as shown in the reaction of scheme (8).

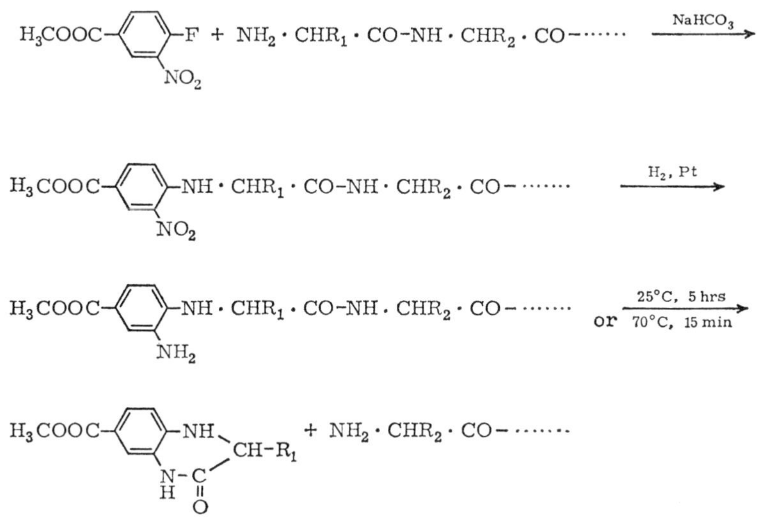

(8)

Ingram [608] and Jutisz and Ritschard [653] applied the principle of this reaction to DNP-peptides. 2,4-Diaminophenyl peptides produced by catalytic reduction were unstable to atmospheric oxidation and the lactam formation was carried out under a hydrogen atmosphere. Scoffone et al. [1141] carried out selective reduction of DNP-peptides with ammonium sulfide at 45 to 50° C in an ethanolic solution to yield 2-amino-4-nitrophenyl peptides, avoiding the formation of the sensitive 4-amino compound. The reduced product was cyclized at pH 2 to 3 by heating at 70° C for 15 min. The same principle was previously used by Akabori and Narita [12]. They reduced DNP-Taka-amylase with hydrogen sulfide in 70% pyridine and heated the resulting mononitroderivative at 100° C for 1 hr to liberate the terminal residue as the corresponding quinoxaline derivative. This reaction was used as a stepwise degradation method. They characterized the newly released end group as a DNP-amino acid according to the Sanger method and succeeded in determining the terminal dipeptide sequence of the amylase as H-Ala-Gly-. However, they observed several DNP-amino acids at the third step of the reaction. The Holley principle was also used for the cleavage of the first peptide bond at 35° C and pH 8 for 3 hrs of

6-amino-2,4-dinitrophenyl peptide which had been derived from N-*tert*-butyl-6-amino-2,4-dinitrophenyl peptide by the action of trifluoroacetic acid for 15 min. Yields of the quinoxaline derivatives of the N-terminal amino acids are 40 to 70% per cycle if stepwise degradation was repeatedly carried out [696]. A possible side reaction is the formation of benzimidazolone derivative which reduces the yield of the quinoxaline formation. The same principle was applied to 2-nitromethane-sulfonyl peptide after reduction of the nitro group at the ortho position with Pt/alumina catalyst at room temperature in 0.5% $NaHCO_3$ solution. Cyclization to the corresponding quinoxaline derivative of the terminal aminomethane-sulfonyl amino acid residue with concomitant spliting of the first peptide bond was carried out at pH 6 and room temperature for 15 hrs [272].

Another procedure for stepwise degradation was proposed by Signor and Bordignon [1160]. They reduced dinitropyridyl peptide to the corresponding dinitrotetrahydropyridyl peptide with $NaBH_4$ in 2% $NaHCO_3$ in an ice-bath for 5 min. Selective reduction of the aromatic nucleus enhances the basicity of the nitrogen atom and the protonated nitrogen at pH 5 to 6 provide the catalytic effect needed to split the first peptide bond as shown in reaction scheme (9).

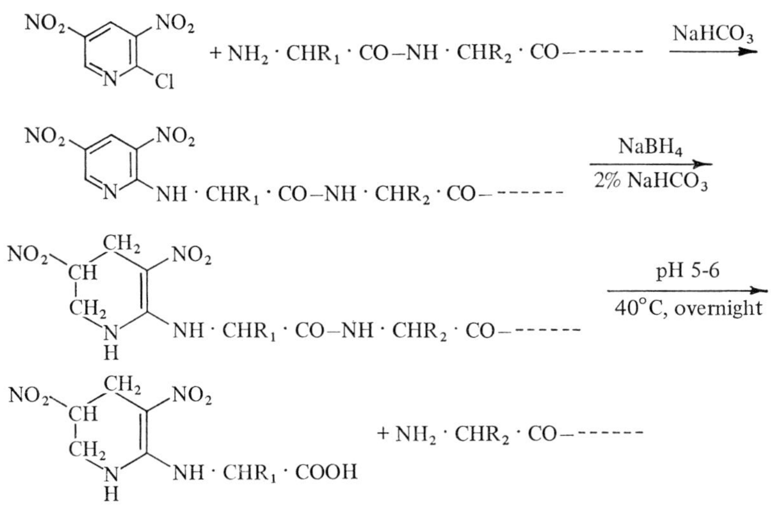

(9)

This interesting reaction can be carried out in the pH range 5 to 8 maintaining the native protein conformation. Rosmus and Deyl suggest this method to be of value in laboratories where no automatic Edman sequenator is available [1040]. However, no description has been presented on the cleavage yield and the chromatographic system used for identification of the resulting dinitrotetrahydropyridyl amino acids.

Alkylation reactions of the terminal amino group are also used for end group determination. N-Dimethylamino acid residues are formed by the action of formaldehyde on the free amino groups followed by catalytic reduction on palladium charcoal [606, 607]. On hydrolysis of the N-dimethylated peptides terminal residues are recovered as dimethylamino acids which can be identified on a paper chromatogram using the pH indicator, thymol blue.

KORMAN and CLARKE [723] carboxymethylated free amino groups in proteins with bromoacetic acid in a suspension of magnesium hydroxide. They observed that most α-amino groups were quantitatively dicarboxymethylated except glutamic acid, phenylalanine and aspartic acid. Proline was exclusively converted into the monosubstituted derivative. Dicarboxymethylated amino acids liberated, upon acid hydrolysis of the substituted peptides, were precipitated as the mercury salt with mercuric nitrate (or mercuric acetate in 0.1 M sodium chloride). The free amino acid derivatives resulting from hydrogen sulfide treatment are characterized by paper chromatography using the reduced ninhydrin reagent. KORMAN and CLARKE [723] determined the N-terminal aspartic acid of bovine serum albumin by this method.

CHIBNALL and SPAHR [231] found that methoxycarbonyl amino acids were readily decomposed in hot strong acid, but upon incubation in 8.6 N hydrochloric acid at 37° C the extent of hydrolysis of methoxycarbonyl amino acids was relatively low (14 to 20% in 10 to 21 days). On the basis of the above results, they reacted methoxycarbonyl chloride with proteins in an alkaline medium and incubated the methoxycarbonyl proteins in 8.6 N hydrochloric acid at 37° C for 21 days. After chromatographic separation, they estimated expected amounts of the terminal residues of insulin as the carbomethoxy amino acids. The results indicate that the introduction of methoxycarbonyl group into the terminal amino group labilizes the adjacent peptide bond. This effect was confirmed by using the dipeptide leucylglycine as a substrate for this reaction. The dipeptide was very stable but the corresponding methoxycarbonyl derivative was completely hydrolyzed in 10 days under the above conditions.

LEVY [766] demonstrated that carbon disulfide reacts reversibly with amino groups at pH 8.0 to yield substituted dithiocarbamates. Because of the extreme sensitivity of dithiocarbamates to oxidation, the reaction must be carried out under strict anaerobic conditions. CHERVENKA and WILCOX [224] showed that at neutral pH carbon disulfide reacts only with the α-amino group, but at an alkaline pH the ε-amino group of lysine also reacts. Upon acidification of the dithiocarbamates to pH 1 in aqueous solution at room temperature, they cyclize to 2-thio-5-thiazolidones simultaneous with cleavage of the terminal peptide bond. This is illustrated in scheme (10). The thiothiazolidone derivatives of amino acids absorb at 280 nm ($\varepsilon = 13500$), and the parent amino acids can be regenerated from the thiothiazolidone by heating with a mixture of 5 N hydrochloric acid-dioxane. With peptides, the overall yield of thiothiazolidone derivatives is 65% but only 3 to 5% with proteins [759]. Carbon disulfide is used for the reversible masking of amino groups of proteins and peptides [771] rather than as a reagent for the end group analysis.

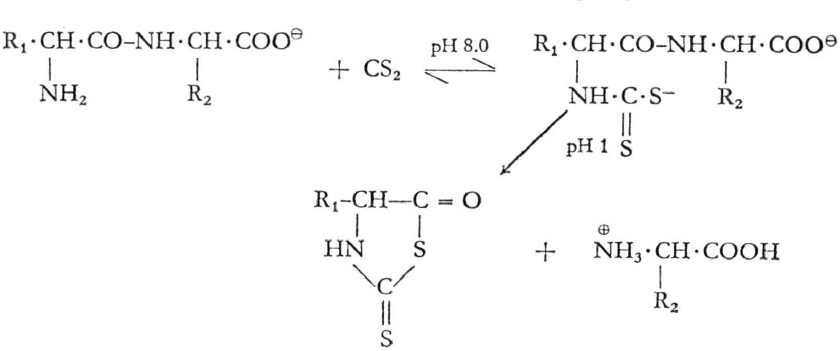

(10)

N-Thiocarboethoxy-peptides prepared from free peptides and methylethylxanthate at alkaline pH give 2-ethoxythiazol-5-one derivatives as intermediates with fission of the terminal linkage in anhydrous nitromethane saturated with dry hydrogen chloride [687]. This is shown in scheme (11). The intermediate is converted into the N-thiocarboxy anhydride of the amino acid, which yields the parent amino acid by successive treatment with alkali and acid.

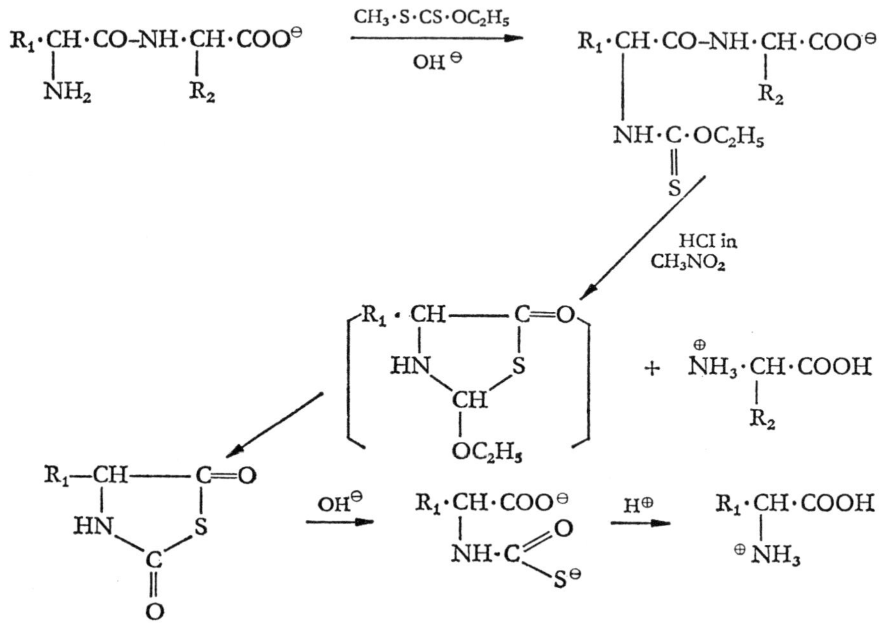

(11)

Elmore and Toseland [330] reported another route to cleavage of the terminal residue. N-Acyldithiocarbamates react at pH 8 with amino group to yield N-acylthiocarbamyl derivatives which are converted into the corresponding thiohydantoin with cleavage of the terminal peptide bond as shown in scheme (12).

Another sequential degradation method was reported by Barrett [64, 67]. N-thiobenzoyl peptides derived from peptides with methyldithiobenzoate or dimethyldithiobenzoate could be split off by the action of trifluoroacetic acid or phosphorous tribromide to give phenylthiazolones the reaction scheme of this reaction is shown in (13). Chromatographic identification of the split product was carried out on thiobenzoyl amino acid anilides. Previero and Pechere [998] used cyanoethyldithiobenzoate and methoxymethyldithiobenzoate as the thioacylating reagent and obtained similar results.

2. Enzymatic Methods

Leucine aminopeptidase (MW 300000) prepared from swine kidney, removes amino acids sequentially from the unblocked terminal amino end of peptides and of proteins [1175]. Rates of hydrolysis of the terminal amino acids differ, however, depending on the nature of side chain groups. Relative hydrolysis rates of amino acid amides by leucine aminopeptides are listed in Table 3-10, based on a hydrolysis rate

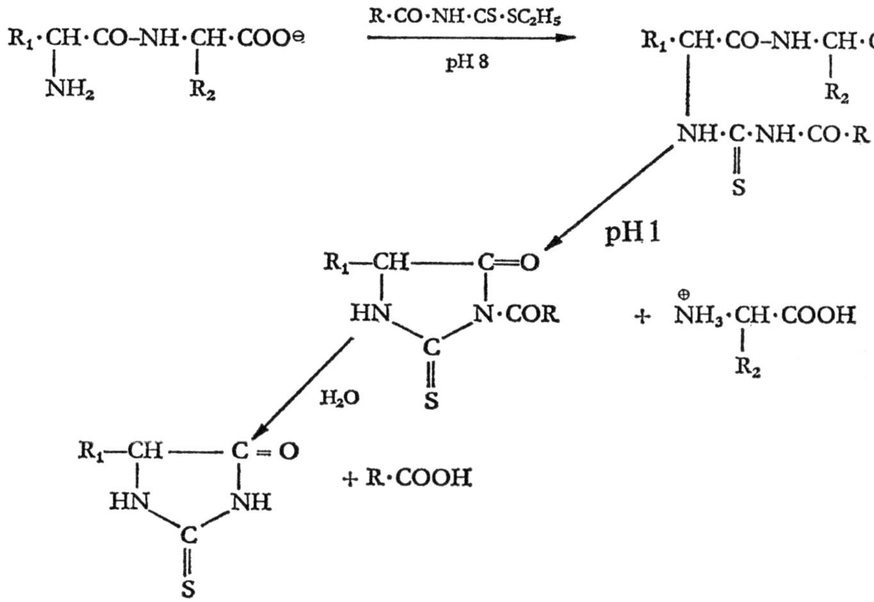

(12)

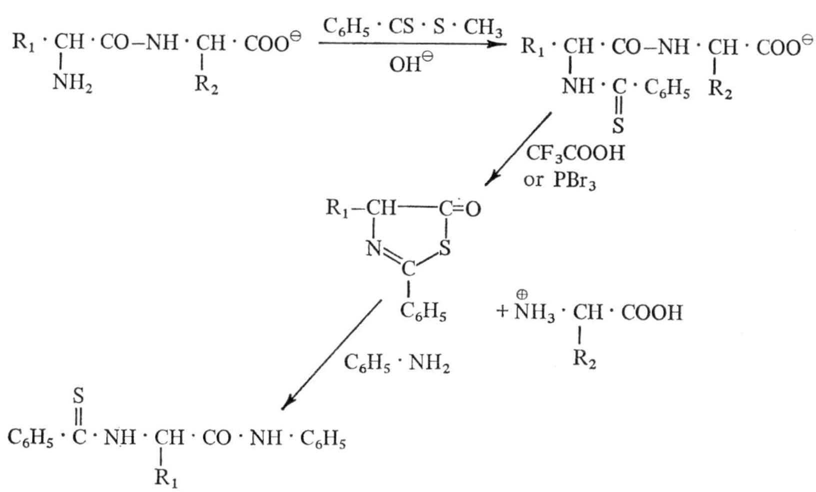

(13)

for leucineamide to be 100. Rates of hydrolysis for amino acid amides containing bulky aliphatic side chains are high, while that for glycine amide is extremely low. The enzyme catalyzes the hydrolysis of all terminal peptide bonds with the exception of the X-Pro linkage, where X may be any amino acid. When the released amino acids from peptides or proteins are estimated as a function of incubation time with leucine aminopeptidase, the amino acid sequence of small oligopeptides can be deduced. However, when proteins are hydrolyzed with the enzyme for the purpose of sequence analysis, the results are equivocal. Ambiguity derives from difficulties in interpreting the results and from impure enzyme preparations. Leucine aminopeptidase preparations often are contaminated with uncharacterized proteolytic enzymes. The enzyme preparation is usually treated before use with diisopropyl phosphorofluoridate to inactivate contaminating "active serine" proteolytic enzymes, and sometimes with iodoacetate to inactivate "SH" proteinases. Some preparations of leucine aminopeptidase exhibit carboxypeptidase action and it is recommended that a carboxypeptidase inhibitor, such as β-phenylpropionate, be added during hydrolysis to minimize unfavorable carboxypeptidase action. The procedures are described in detail by LIGHT [778].

Prolonged incubation of proline-free peptides with leucine aminopeptidase yields complete hydrolysis to a mixture of amino acids. The enzyme is specific for L-amino

Table 3-10. Relative hydrolysis rates of amino acid amides by leucine aminopeptidase[a]

Substrate	Mn^{2+} activation	Mg^{2+} activation
L-Leucinamide	100	100
L-Isoleucinamide	20	17
L-Valinamide	17	
L-Alaninamide	3.4	4.9
Glycinamide	0.13	0.10
L-Tryptophanamide	24	
L-Phenylalaninamide	26	17
L-Tyrosinamide	16	
L-Histidinamide	19.1	10
L-Argininamide	7.1	
L-Lysinamide	7.1	12
L-Isoglutamine	2.2	
L-Aspartic acid diamide	2.9	
L-Serinamide	0.76	
L-Prolinamide	0.71	

[a] Data from Ref. [1175].

acid residues and has been extensively used for the analysis of the optical purity of synthetic peptides. It is less frequently applied to the sequence analysis of peptides and proteins. Preparation of the enzyme from swine kidney is described by LIGHT [778] and HIMMELHOCH [542].

The enzyme is activated and stabilized by Mg^{2+} and Mn^{2+} and is irreversibly inactivated by chelating agents which remove these divalent cations. It is therefore necessary to dialyze the enzyme against 0.005 M $MgCl_2$ to 0.005 M Tris buffer, pH 8, at 5° C before use. Leucine aminopeptidase is labile at pH values above 9 or below 7 and generally is employed at pH 8.5. The hydrolysis of a peptide is performed with 0.1 μmole of peptide in 0.1 ml of 0.0025 M $MgCl_2$ 0.1 M Tris buffer of pH 8.5 and with 0.05 to 0.5 mg/ml of enzyme [778].

PFLEIDERER and his co-workers [974, 1321, 1322] isolated another aminopeptidase (MW 280000) from swine kidney in a homogeneous state by biochemical and physical criteria. This aminopeptidase is different from leucine aminopeptidase in various respects: the enzyme exists as a bound form with microsomal particle, it is solubilized by tryptic digestion, does not require any divalent metal ion, stable against temperature and extreme pH's, and has almost the same affinities for synthetic substrates of leucine-, alanine- and glycine-*p*-nitroanilide (see Table 3-11). Wachsmuth et al. [1321] reported that their aminopeptidase is a major aminopeptidase present in swine kidney. This enzyme hydrolyzes all the N-terminal amino acids with the exception of proline, at almost equal rates, and is used for test of optical purity of synthetic peptides [565]. The enzyme preparation is commercially available as aminopeptidase M. The optimum pH for amino acid *p*-nitroanilides is 9 but for peptides is 7. The hydrolysis of a peptide is performed with 0.1 μmole of peptide in 0.1 ml of 0.1 M sodium phosphate buffer of pH 7.0 and with 0.5 to 1 mg/ml of aminopeptidase M [778].

Table 3-11. Relative hydrolysis rates of amino acid *p*-nitroanilides and amino acid amides by aminopeptidase M

Substrate	Relative rate
L-Alanine *p*-nitroanilide	100
L-Leucine *p*-nitroanilide	71
L-Phenylalanine *p*-nitroanilide	83
Glycine *p*-nitroanilide	22
L-Alaninamide	1.15
L-Leucinamide	0.32
L-Phenylalaninamide	0.29
L-Serinamide	0.2
L-Valinamide	0.11
Glycinamide	0.085
L-Histidinamide	0.058
L-Prolinamide	0.002

Data from Ref. [1322].

The use of dipeptidyl aminopeptidase I (originally called cathepsin C) from either bovine spleen or rat liver is recently reported for sequence analysis of polypeptides [196, 836]. The enzyme removes dipeptides sequentially from the unblocked N-terminus with the exceptions of lysyl and arginyl termini. Peptide degradation by the enzyme is also blocked by a prolyl residue both in the N-terminus and adjacent position.

III. C-Terminal Group Determination

In contrast to N-terminal determination, only a few chemical methods are available for the determination of the C-terminal group of proteins. Because of this situation, the exopeptidases, carboxypeptidases A and B, are widely used. In carboxypeptidase experiments, it is difficult to draw definite conclusions concerning the C-terminal residue of a large protein unless the relative specificity of the enzyme for the terminal residues is carefully considered. The enzyme method can, however, effectively be used for oligopeptides. Only two chemical methods for the determination of the C-terminal end group are described here; one is the hydrazinolysis method and the other is the selective labeling of the C-terminal amino acid with tritium.

A. Hydrazinolysis Method

Akabori et al. [16] found in 1952 that heating proteins with anhydrous hydrazine at 100° C liberates amino acid hydrazides from the amino acid residues the carboxyl groups of which are involved in peptide linkages. The C-terminal amino acid residue is released as the free amino acid as shown in scheme (14).

$$\cdots\cdots - NH \cdot CHRm \cdot CO{-}NH \cdot CHRn \cdot COOH \xrightarrow[100\,°C]{\text{anhydrous } N_2H_4} \qquad (14)$$

$$\cdots\cdots + NH_2 \cdot CHRm \cdot CO \cdot NH \cdot NH_2 + NH_2 \cdot CHRn \cdot COOH$$

During hydrazinolysis at 100° C, arginine is decomposed and partly converted into ornithine and triaminoguanidine [893, 924], and cystine is decomposed into unknown product(s). Cysteic acid, S-carboxymethylcysteine and S-aminoethylcysteine are stable to the same extent as the other common amino acids, and any of the three residues as the C-terminus can be recovered as free amino acids after hydrazinolysis. It is therefore necessary to carry out the oxidation or reduction of disulfide bonds followed by alkylation prior to hydrazinolysis. The situation with aspartic acid, glutamic acid, asparagine and glutamine residues is complicated [934] as shown in Table 3-12.

Table 3-12. Hydrazinolysis product of acidic amino acids and their amides

	Original residue	Product after hydrazinolysis
C-Terminal aspartic acid and glutamic acid	-NH-CH-COOH | $(CH_2)_{1-2}$ | COOH	NH_2-CH-COOH | $(CH_2)_{1-2}$ | COOH
C-Terminal asparagine and glutamine	-NH-CH-COOH | $(CH_2)_{1-2}$ | $CONH_2$	NH_2-CH-COOH | $(CH_2)_{1-2}$ | $CO \cdot NH \cdot NH_2$
Non-terminal aspartic acid and glutamic acid	-NH-CH-CO- | $(CH_2)_{1-2}$ | COOH	NH_2-CH-$CO \cdot NH \cdot NH_2$ | $(CH_2)_{1-2}$ | COOH
Non-terminal asparagine and glutamine	-NH-CH-CO- | $(CH_2)_{1-2}$ | $CONH_2$	NH_2-CH-$CO \cdot NH \cdot NH_2$ | $(CH_2)_{1-2}$ | $CO \cdot NH \cdot NH_2$

Separation of the free amino acid derived from the C-terminal position from the bulk of amino acid hydrazides in the protein hydrazinolysate was originally carried out by benzaldehyde treatment [16]. Benzaldehyde reacts with the amino acid hydrazides in an aqueous solution to yield insoluble oil condensation products; free amino acids can be recovered from the supernatant. Aspartic β-hydrazide and glutamic γ-hydrazide derived, respectively, from the terminal asparagine and glutamine are also removed with benzaldehyde treatment and thus escape detection.

Amino acid hydrazides cannot be completely removed by a single treatment with benzaldehyde but the hydrazides remaining in the free amino acid fraction are absorbed tightly on the top of a resin column during amino acid analysis. These amino acid hydrazides are eluted together with the basic amino acids from the short column, hence, their separation and the analysis of C-terminal basic residues is rather difficult.

In place of benzaldehyde, isovaleraldehyde [14], enanthoaldehyde [144], and *p*-nitrobenzaldehyde [160] also may be used for the same purpose, but there exists no appreciable merit over benzaldehyde.

Akabori et al. [13, 14] modified the separation procedure by extending the DNP-method to the hydrazinolysate mixture. Hydrazides of amino acids yield neutral di- or tri-DNP-derivatives (substituted at amino group and hydrazide group). However, free amino acid and α- and ω-hydrazides of aspartic acid and glutamic acid give acidic DNP-derivatives, and thus neutral and acidic DNP-derivatives can be separated by fractional extraction with ether or ethyl acetate. Acidic DNP-derivatives are easily separated by two-dimensional paper chromatography with Levy's solvent system [768]. Kawanishi et al. [675] further modified the fractional extraction method of DNP-derivatives suitable for the characterization of di-DNP-aspartic and -glutamic monohydrazides.

Niu and Fraenkel-Conrat [924] reported an improved procedure for removing hydrazides by successive benzaldehyde treatment and dinitrophenylation, it being difficult to remove amino acid hydrazides quantitatively with a single aldehyde treatment of the hydrazinolysate. They determined C-terminal amino acids as their DNP-derivatives after characterization by paper chromatography with Levy's solvent system [768]. The recoveries of amino acids after heating at 100° C for 10 hrs with anhydrous hydrazine, benzaldehyde treatment, dinitrophenylation, paper chromatography and extraction of DNP-amino acids from the chromatogram with 1% $NaHCO_3$ were estimated by Niu and Fraenkel-Conrat and are listed in Table 3-13 (2nd column). This author (unpublished experiments) has confirmed their reported recoveries (3rd column in Table 3-13).

Alternatively, for the mutual separation of free amino acids from the bulk of the hydrazides, relatively short ion exchange columns, containing Amberlite CG-50 (H^+ form, 1 × 3 cm), Amberlite IR-120 (equilibrated with 0.2 M pyridine acetate buffer of pH 3.1, 1 × 18 cm) or phosphocellulose (equilibrated with 0.4 M pyridine formate buffer, pH 3.2, 1 × 30 cm) are used and the free amino acid fraction is subjected to amino acid analysis under the usual analysis conditions. Fraenkel-Conrat and Tsung [384] used an aqueous solution of a hydrazinolysate on a column of Amberlite CG-50 in the H^+ form. They recovered acidic and neutral amino acids in 15 ml of the water eluate together with the monohydrazides of aspartic and glutamic acids. The basic amino acids were recovered in 15 ml of the subsequent eluate with 0.1 M ammonium acetate (pH 7). BRAUN and SCHROEDER [159A] used an Amberlite IR-120 column which had been equilibrated with 0.2 M pyridine acetate, pH 3.1, and a phosphocellulose column which had been equilibrated with 0.4 M pyridine formate, pH 3.2. With the former column, acidic and neutral amino acids are eluted in the pH 3.1 buffer and lysine in 0.2 M pyridine acetate, pH 5.2. From the latter column, all free amino acids and aspartic and glutamic monohydrazides are

Table 3-13. Recoveries of amino acids after heating with hydrazine[a]

	100° C, 10 hrs without catalyst		60° C, 16 hrs 1M hydrazine sulfate		80° C, Amberlite CG-50[f]					
	c	d	e	d	11 hrs	23 hrs	40 hrs	60 hrs	82 hrs	112 hrs
Lysine	47	57	90	97	41	36	37	29	31	29
Histidine	43	—	53	—	88	83	85	54	73	60
Arginine[b]	25	32	60	46	—	—	—	—	—	—
Aspartic acid	19	8	99	78	77	63	40	26	24	20
Threonine	52	45	105	92	92	95	80	72	79	68
Serine	39	35	96	90	84	89	67	62	67	51
Glutamic acid	16	8	78	59	81	74	44	41	39	35
Proline	52	68	97	94	100	102	96	80	88	75
Glycine	37	30	—	77	92	88	69	58	60	57
Alanine	45	48	—	79	96	95	85	76	79	74
Half cystine	—	—	78	—	77	52	52	41	25	0
Valine	76	70	—	94	101	108	100	104	108	96
Methionine	56	54	66	72	90	90	86	59	87	79
Isoleucine	54	—	—	86	101	103	104	108	107	103
Leucine	54	60	—	91	99	99	94	94	90	93
Tyrosine	44	—	—	—	95	97	91	89	92	83
Phenylalanine	59	65	—	83	96	97	91	74	93	85
Tryptophan	39	47	46	74	—	—	—	—	—	—

[a] Recoveries of amino acids after hydrazine treatment without catalyst and with hydrazine sulfate are estimated by DNP-technique, and with CG-50 resin by the automatic amino acid analyzer. No value in the table means that estimation has not been carried out.
[b] Arginine was characterized and estimated as ornithine.
[c] Data from Ref. [924].
[d] Unpublished experiments.
[e] Data from Ref. [219].
[f] Data from Ref. [159A].

eluted with the equilibrated buffer. The monohydrazides of the acidic amino acids give no clear peak on the chart using an automatic amino acid analyzer.

Hydrazinolysis of proteins is complicated by the fact that the amino acid hydrazides are not very stable, in particular the dihydrazides of aspartic and glutamic acids, and may undergo hydrolytic and oxidative decomposition. Further, free amino acids with small side chains slowly undergo partial conversion to their hydrazides. It is necessary therefore to find conditions which minimize the conversion of free amino acids derived from the C-terminus into their hydrazides and the regeneration of amino acid from the hydrazides on the one hand, and to maximize the extent of hydrazinolytic cleavage of the C-terminal residues on the other. The longer the time of hydrazinolysis and the higher the hydrazinolysis temperature, the greater the extent of the release of the C-terminal amino acids. However, under these conditions, the unfavorable side reactions proceed more extensively. BRADBURY [155] and KAWANISHI et al. [675] introduced the catalytic hydrazinolytic technique to include

Table 3-14. Yields of C-terminal amino acids obtained from proteins with amberlite CG-50 catalyzed hydrazinolysis at 80° C [a]

Protein	C-Terminal amino acid and bond to be cleaved	Molecular weight	Hydrazinolysis time (hours)	Yield
Lysozyme	-Arg-Leu-OH	14600	20	95%
Ribonuclease	-Ser-Val-OH	13600	95	100%
Myoglobin	-Gln-Gly-OH	17800	88	100%
Cytochrome *c*	-Asn-Glu-OH	12400	16	90%
Tobacco mosaic virus protein	-Ala-Thr-OH	17400	46	72%
Hemoglobin β-chain	-Tyr-His-OH	16000	95	68%
Insulin	-Cys-Asn-OH	5700	45	52% [b]
	-Lys-Ala-OH		22	64%
Chironomus hemoglobin	-Met-OH	16500	72	78%
Bence-Jones Protein	-Glu-Cys-OH	23000	45	67% [c]
Glucagon	-Asn-Thr-OH	3483	75	92%

[a] Data from Ref. [159A].
[b] Aspartic monohydrazide fraction was hydrolyzed and estimated.
[c] Determined as cysteic acid after performic acid oxidation of the protein.

hydrazine sulfate. They reported increased yields of the C-terminal amino acids when proteins are heated with anhydrous hydrazine at 60 to 80° C for 16 hrs in the presence of 1 to 2 M hydrazine sulfate. In this author's experience, the recoveries of amino acids after heating the amino acid mixture under the above conditions increase appreciably, but yields of the C-terminal amino acids from several oligopeptides have not been significantly improved suggesting that hydrazinolytic cleavage of the C-terminal residues still is complete.

BRAUN and SCHROEDER [159A] reported the effectiveness of the H^+ form of Amberlite CG-50 resin as a catalyst for hydrazinolysis. They heated several proteins with anhydrous hydrazine at 80° C for 60 to 100 hrs in the presence of the resin and estimated the C-terminal residues in yields of 60 to 100% without any corrections, as shown in Table 3-14. Recoveries of amino acids after heating at 80° C under the same conditions as in the resin-catalyzed hydrazinolysis of proteins are listed also in Table 3-13. The resin-catalyzed technique seems to be superior to the sulfate-catalyzed method since the reported yields of C-terminal residues of several proteins increased significantly in comparison with those of uncatalyzed hydrazinolysis. Furthermore, the catalyst can easily be removed by centrifugation. BRAUN and SCHROEDER [159A] described difficulties in determining C-terminal asparagine and glutamine residues with confidence in contrast to the results of KAWANISHI et al. [675]. The situation of the C-terminal asparagine and glutamine is complicated because acidic amino acids and their amides seem to follow the reactions (15) as suggested by NARITA and OHTA [900].

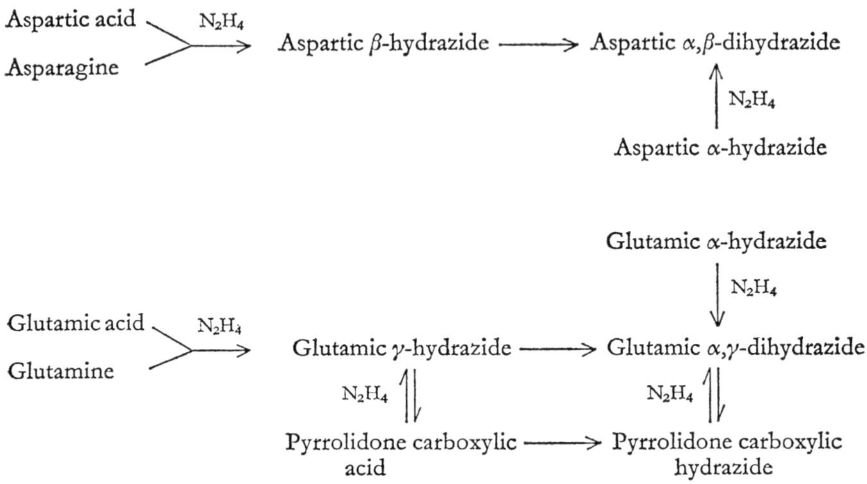

(15)

Furthermore, both aspartic monohydrazides seem to be interconvertible through an assumed intermediate of succinohydrazide as shown in scheme (16).

$$\underset{\displaystyle NH_2}{HOOC\cdot CH_2\cdot CH\cdot CO\cdot NH\cdot NH_2} \rightleftharpoons \begin{array}{l} CH_2\cdot CO \\ CH_2\cdot CO \end{array}\!\!>N\cdot NH_2 \rightleftharpoons \underset{\displaystyle NH_2}{H_2N\cdot NH\cdot CO\cdot CH_2\cdot CH\cdot COOH} \quad (16)$$

Another important point toward minimizing the unfavorable hydrolytic cleavage of amino acid hydrazides which results from hydrazinolysis of proteins is to avoid prolonged contact with water. Amino acid hydrazides, especially hydrazides of glycine, serine and alanine, are labile in aqueous solution and yield the corresponding free amino acids by hydrolytic cleavage at room temperature. Typical examples are shown in Figs. 3-12 and 3-13 [899]. Fig. 3-12 shows the released amino acids from Taka-amylase A by the uncatalyzed hydrazinolysis at 100° C over various periods. Their amounts are estimated as the DNP-derivatives as mentioned above. The results suggest that serine and glycine apparently are the C-terminal residues. In Fig. 3-13, the released amino acids are estimated by an automatic amino acid analyzer and only serine seems to be in the C-terminal group. C-Terminal serine was confirmed by the used of carboxypeptidase A. Hydrazinolysis conditions are the same for the above two experiments, but procedures after benzaldehyde treatment are different from another. In the experiment shown in Fig. 3-12, the supernatant, after the aldehyde treatment, was reacted with FDNB at room temperature for 1.5 h in $NaHCO_3$ solution, but for the experiment shown in Fig. 3-13, the supernatant was quickly lyophilized to minimize the hydrolysis of amino acid hydrazides and was subjected to amino acid analysis immediately after dissolving the lyophilisate in citrate buffer of pH 2.2. These results clearly suggest that prolonged contact of the hydrazinolysate with an aqueous solution, especially in an alkaline medium, is not favorable.

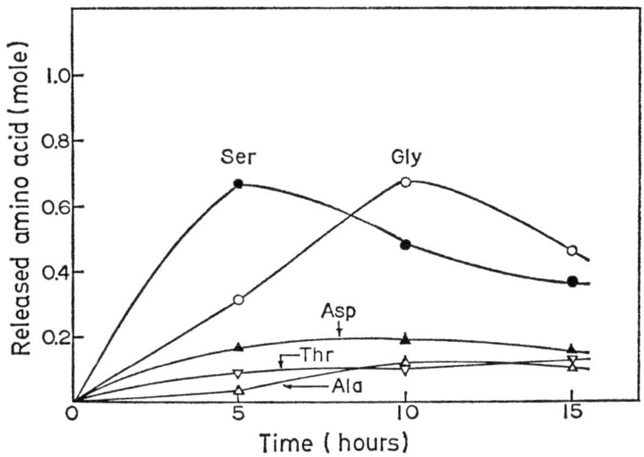

Fig. 3-12. Liberated amino acids from Taka-amylase by uncatalyzed hydrazinolysis at 100° C as a function of time. Released amino acids were estimated as their DNP-derivatives. Compare difference from Fig. 3-13. (Reprinted from Ref. [899], by courtesy of the Japanese Biochemical Society)

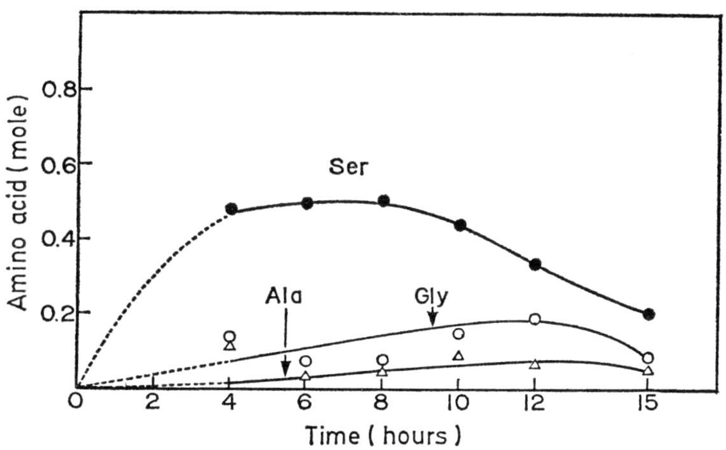

Fig. 3-13. Liberated C-terminal amino acid from Taka-amylase by uncatalyzed hydrazinolysis at 100° C as a function of time. Amino acids were estimated by an automatic amino acid analyzer. Compare difference from Fig. 3-12. (Reprinted from Ref. [899], by courtesy of the Japanese Biochemical Society)

When a protein contains small amounts of glycine, serine or alanine, the dinitrophenylation technique can be used without difficulties, especially for small oligopeptides. Hydrazinolysis can therefore be performed even if an amino acid analyzer is not available. Introduction of dansyl chloride in place of DNFB for the identification of the C-terminal amino acid released by hydrazinolysis makes possible to handle 5 nmole of sample [904].

1. Preparation of Anhydrous Hydrazine

A conventional method is to utilize the characteristic of hydrazine to form an azeotropic mixture with toluene. After 50 g of hydrazine (95% +), 300 g of toluene, and 50 g of CaO (or 50 g of 80% hydrazine hydrate, 500 g of toluene and 500 g of CaO) are mixed in a 500 ml round-bottom flask and kept overnight, the mixture is refluxed for 3 h while excluding the uptake of moisture with a $CaCl_2$ tube. The azeotropic mixture is distilled at 93 to 94° C [133]. The lower phase of the mixture is anhydrous hydrazine, of about 98.5% purity, which can be used for the hydrazinolysis of proteins and peptides. It is recommended that distillation be discontinued while the CaO is still wet, otherwise an explosion is likely. Hydrazine absorbs moisture quickly and the condensed hydrazine must be protected from moisture during distillation. The hydrazine thus prepared is best sealed immediately in a number of ampules of 1 to 2 ml and stored in a refrigerator.

Braun and Schroeder [159 A] dehydrated hydrazine with NaOH. A mixture of 100 ml of hydrazine (95% +) and 40 g of NaOH pellets in a 250 ml round-bottom flask is heated to the boiling point under nitrogen atmosphere and then cooled to room temperature. During vacuum distillation with an oil pump, nitrogen that has passed through concentrated H_2SO_4 is introduced through a nonflexible capillary. The hydrazine is distilled between 16 to 18° C and the distillate is condensed in a 100 ml round-bottom flask at the temperature of dry ice.

2. Hydrazinolysis

a) Uncatalyzed Method

To 0.1 to 0.5 μmole of a dried protein or peptide in a small (0.5 cm × 6 to 10 cm) constricted test tube is placed 0.2 to 0.5 ml of hydrazine by drawing the reagent through the constricted neck under cooling in an alcohol-dry ice bath. Hydrazinolysis is then carried out at 100° C for 6 to 10 h in a sealed tube. During heating, the sample tube should be shaken gently three or four times. The reaction tube is then opened and the contents are lyophilized. For a small peptide, the dried hydrazinolysate can be applied directly to an amino acid analyser after dissolving it in citrate buffer of pH 2 followed by adjustment of the pH to about 2. When the reaction mixture cannot be examined immediately, the reaction tube is best stored unopened in a freezer until analysis is begun [159 A].

b) Resin-Catalyzed Method [159 A]

About 50 mg of dry Amberlite CG-50 (H^+ form, less than 200 mesh, stored under vacuum in a desiccator over P_2O_5 after drying in an oven at 80° C) is placed in a heavy-walled Pyrex tube (1.5 × 12.5 cm). A weighed amount of the dried protein (usually around 0.25 μmole) and 2 ml of anhydrous hydrazine are added into the tube, as described in the uncatalyzed method. After hydrazinolysis at 80° C for 60 to 100 h, the reaction mixture is transferred into a 25 ml round-bottom flask, the reaction tube is rinsed twice with 1 ml of undistilled hydrazine (95% +), and the entire material is then lyophilized. Between rinses, the flask is stoppered to minimize the absorption of moisture. The totally dry mixture, which contains CG-50 resin, the hydrazides, and the released amino acids, is suspended in 3 ml of water; the resin is centrifuged and washed twice with 1 ml of water. A solution of the hydrazinolytic products in water has a pH between 8.5 and 9, which is lowered to a pH below 2 with a few drops of 2 N HCl before the reaction mixture is chromatographed for the removal of the hydrazides.

3. Removal of Amino Acid Hydrazides from the Released Free Amino Acid

a) Column Chromatography

According to the procedures of BRAUN and SCHROEDER [159 A], Amberlite IR-120 is equilibrated with 0.2 M pyridine acetate buffer of pH 3.1 and a 1 × 18 cm column is prepared. The reaction mixture derived from the above hydrazinolysis is placed on the column at pH 1.5 to 2.0 and the free amino acids are developed with 0.2 M pyridine acetate buffer at 30° C with a flow rate of 30 ml/hr. 2 ml fractions are collected. With the pH 3.1 buffer, all amino acid hydrazides are completely retained by the resin, whereas the acidic and neutral amino acids appear between fractions 8 and 30, except tyrosine, which emerges later between fractions 54 and 60. The buffer is then changed to 0.2 M pyridine acetate, pH 5.2. Lysine emerges between fractions 110 and 120, but histidine cannot be separated from the amino acid hydrazides by the IR-120 column. Since the hydrazides are strongly fixed on the resin, after chromatography, the resin must be regenerated by boiling with 0.5 N NaOH for 1 hr, washed successively with water and acid, and, finally, equilibrated with the pH 3.1 buffer.

For the separation of histidine from the hydrazides, a phosphocellulose column is employed. A hydrogen form of phosphocellulose (Whatman cellulose phosphate powder P-70, unsifted) is equilibrated with 0.4 M pyridine formate buffer of pH 3.2. The sample, at pH 1.5 to 2.0, is placed on a 1 × 30 cm column of the phosphocellulose and free amino acids are eluted with the 0.4 M buffer at room temperature with a flow rate of 30 ml/h. Acidic and neutral amino acids appear between fractions 12 and 22 (1.7 ml fraction), lysine and ammonia between fractions 33 and 43, and histidine between fractions 47 and 53. The monohydrazides of the acidic amino acids appear shortly before the lysine peak and overlap with the lysine position. The effluents are lyophilized and subjected to amino acid analysis by the standardize procedures.

Alternatively, FRAENKEL-CONRAT and TSUNG [384] used the hydrogen form of an Amberlite CG-50 column (1 × 3 cm) for separation. Neutral and acidic amino acids appear in 15 ml of the water eluate together with the monohydrazides of the acidic amino acids. Subsequent elution of the CG-50 column with 15 ml of 0.1 M ammonium acetate (pH 7) yields the basic amino acids. They recommend the use of a column of 0.9 × 29 cm (pH 4.26, 0.38 N sodium citrate, 0 to 185 min at 33° C, followed by pH 5.28 citrate, 185 to 395 min at 55° C) for the analysis of the basic amino acid fraction including ornithine by the accelerated system of an automatic amino acid analyzer.

b) Aldehyde Treatment

The lyophilized hydrazinolysate is dissolved in 2 ml of water, and 0.5 ml of freshly distilled benzaldehyde is added. The mixture is immediately shaken vigorously for 1 h at room temperature and then centrifuged. The supernatant is transferred to a lyophilization flask and the oily residue in the centrifuge tube is washed twice with 1 ml of water. The supernatant and washings are immediately frozen and dried. Just before analysis, the lyophilized material is dissolved in 1 ml of citrate buffer, pH 2.2, and the solution is subjected to amino acid analysis. In this case, the analysis of basic amino acids is rather difficult, because some of the hydrazides are eluted with basic amino acids by the usual procedures.

c) Dinitrophenylation Technique

The lyophilized hydrazinolysate (or effluent of the above chromatography) is dissolved in 3 ml of water, shaken with 0.3 g of $NaHCO_3$ and 6 ml of 2.5% DNFB

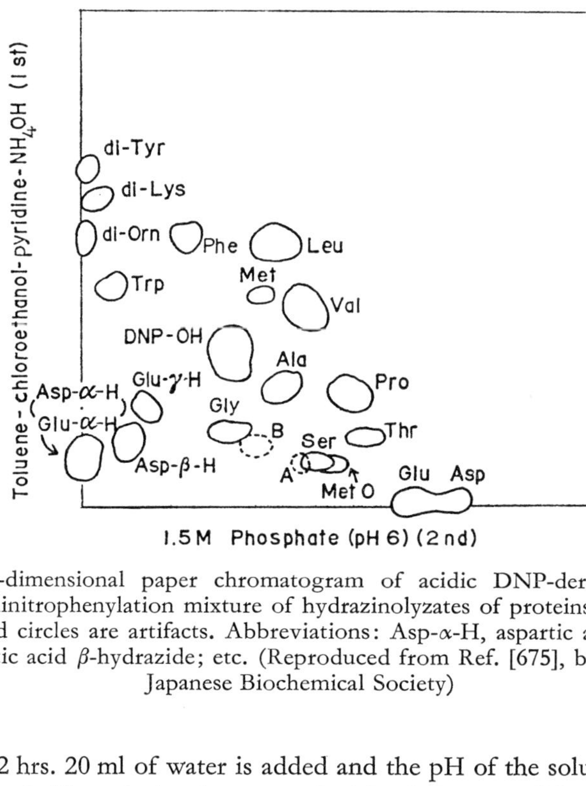

Fig. 3-14. Two-dimensional paper chromatogram of acidic DNP-derivatives possibly present in the dinitrophenylation mixture of hydrazinolyzates of proteins. Spots *A* and *B* shown in dotted circles are artifacts. Abbreviations: Asp-α-H, aspartic acid α-hydrazide; Asp-β-H, aspartic acid β-hydrazide; etc. (Reproduced from Ref. [675], by courtesy of the Japanese Biochemical Society)

in ethanol for 2 hrs. 20 ml of water is added and the pH of the solution is adjusted with 2 N HCl to 1. The solution is extracted with ethyl acetate (10 ml and 3 × 5 ml) and the extract is diluted with 2 volumes of ether. The organic solution is back-extracted with 2% aqueous $NaHCO_3$ (3 × 15 ml) and the aqueous extract is washed with ether (2 × 10 ml). The aqueous phase is acidified with 4 N HCl to pH 1 and again extracted with ethyl acetate (10 ml and 3 × 5 ml). The yellow residue, freed of dinitrophenol by sublimation [866], is chromatographed two-dimensionally according to LEVY's procedures [768] or with *n*-butanol saturated with 2 N NH_4OH [706] and 1.5 M phosphate buffer of pH 6, as described in the DNP-method. Locations of the DNP-derivatives of the monohydrazides of aspartic and glutamic acids together with usual DNP-amino acids are shown in Fig. 3-14. Spots A and B are artifacts and can be differentiated from the typical DNP-amino acids (OD_{390}/OD_{360}

= 0.55 to 0.65 in 1% $NaCO_3$) by their different ultraviolet absorption characteristics (OD_{390}/OD_{360} in 1% $NaHCO_3$ are 1.2 and 0.9, respectively). Millimolar extinction coefficients of typical DNP-amino acids at 360 nm in 1% $NaHCO_3$ average 17 and those of di-DNP-derivatives of aspartic β- and glutamic γ-hydrazides are 26 and 24 at their absorption maximum, 345 nm, in 1 N HCl ($OD_{390}/OD_{360} = 0.55$ to 0.60 in 1 N HCl). The absorption maximum of the di-DNP-derivatives of the above two hydrazides shift to 415 nm in 1 to 2% $NaHCO_3$ solution, but their millimolar extinction coefficients at the maximum wave length remain unchanged, giving a broader band and OD_{390}/OD_{360} of 1.1.

B. Tritium-Labeling Method

It is well-known that acyl amino acids with the exception of acyl proline and acyl aspartic acid are easily racemized during the condensation reaction with amine components in the synthesis of polypeptides. NEUBERGER [911], in 1948, suggested as a possible mechanism that racemization of the acyl amino acid might take place through the intermediate compounds, oxazolone and oxazole as shown in scheme (17).

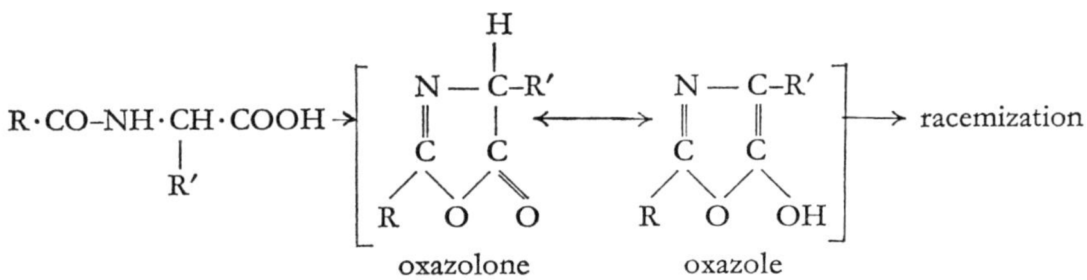

(17)

This racemization mechanism has been confirmed by YOUNG [1411] and GOODMAN and LEVINE [425]. MATSUO et al. [824] also tried to confirm the above mechanism by using heavy water containing isotopic hydrogen (deuterium or tritium) for the ring-opening reaction of the oxazolone derivative.

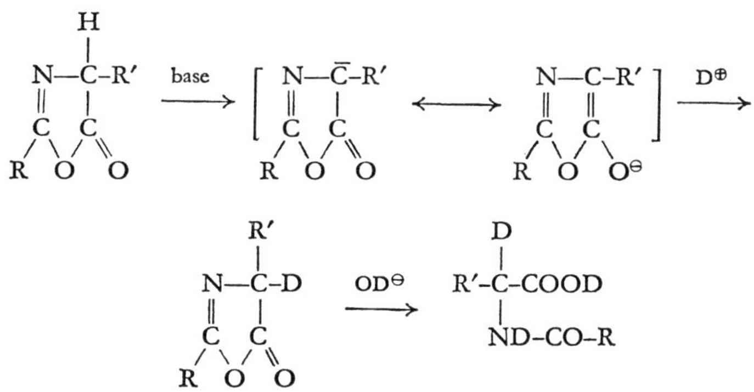

(18)

They originally studied the ring-opening reaction at room temperature with D_2O and the oxazolone derivatives derived from acetyl amino acids and acetic anhydride and found that acetyl derivatives of all of the protein-constituting amino acids, except proline, hydroxyproline and aspartic acid, had undergone racemization via oxazolone derivatives to yield the deuterated acetyl amino acids as shown in scheme (18). Further, they demonstrated by means of NMR that the release of the hydrogen at the four position of the oxazolone was a base-catalyzed reaction and that the addition of one drop of pyridine or triethylamine made racemization quantitative regardless of the acyl amino acids employed except for the three acyl amino acids mentioned above.

Under the same conditions of base-catalyzed deuteration of other acetyl amino acids, acetyl aspartic acid gave the intramolecular anhydride, but acetyl asparagine gave the corresponding oxazolone. Unlike acetyl aspartic acid, acetyl glutamic acid was converted into the corresponding oxazolone in a manner similar to acetyl glutamine. Acetyl proline gave the mixed anhydride with acetic acid and escaped the base-catalyzed deuteration reaction under the above conditions. When acetyl aspartic acid and acetyl proline were heated in a mixture of deuterated acetic acid and acetic anhydride, however, they underwent deuteration. Although the mechanism of the isotope labeling reaction of the latter two acetyl amino acids is not clear, the above results seem to provide a novel method of C-terminal characterization of peptides and proteins. The reaction mechanism is illustrated in scheme (19).

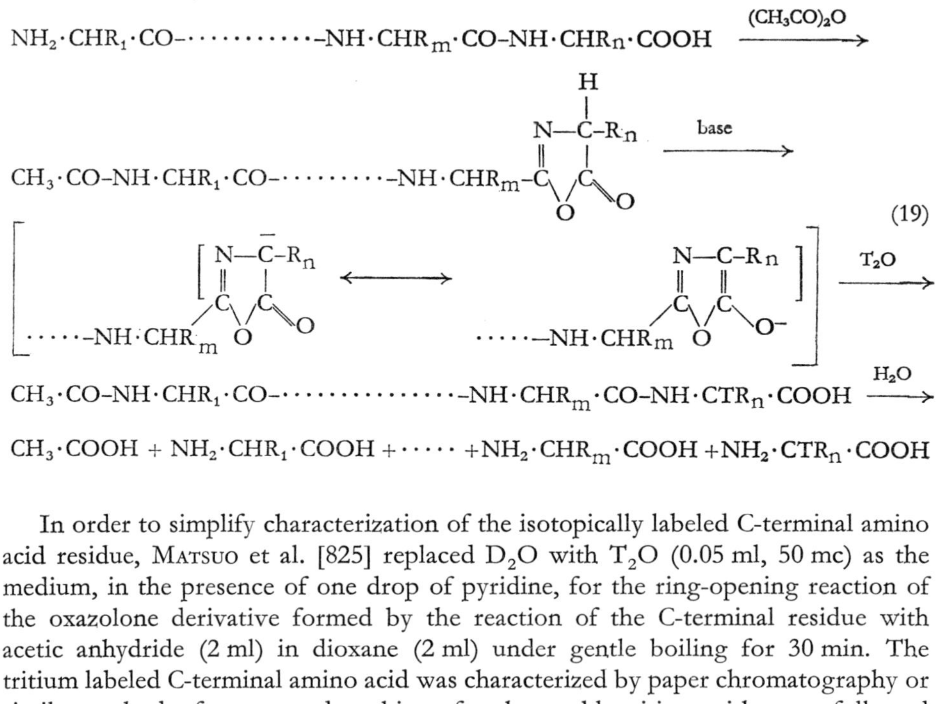

(19)

In order to simplify characterization of the isotopically labeled C-terminal amino acid residue, Matsuo et al. [825] replaced D_2O with T_2O (0.05 ml, 50 mc) as the medium, in the presence of one drop of pyridine, for the ring-opening reaction of the oxazolone derivative formed by the reaction of the C-terminal residue with acetic anhydride (2 ml) in dioxane (2 ml) under gentle boiling for 30 min. The tritium labeled C-terminal amino acid was characterized by paper chromatography or similar methods after repeated washing of exchangeable tritium with water followed by acid hydrolysis of the tritiated peptide. This method has proved to be useful as a C-terminal characterization with several synthetic oligopeptides and angiotensin II. Due to the insolubility of proteins in dioxane, Matsuo et al. modified the reaction conditions for proteins by dissolving them (ca. 2 mg) in T_2O-pyridine (0.1 ml : 0.2 ml), under warming, followed by the addition of acetic anhydride (0.05 ml) under ice-cooling and maintaining at room temperature for a few hours. Thus they succeeded in labeling selectively the C-terminal residues of bovine insulin and egg white lysozyme [825]. As one typical example, the radioautogram (by Radiochromatogram Scanner) of the paper chromatogram of the hydrolysate of the tritiated performic acid oxidized insulin is shown in Fig. 3-15. When the DNP-protein is used for selective tritium-labeling, both N- and C-terminal groups can be characterized with a single sample [826].

Isoaspartyl and isoglutamyl residues in peptide chains are also tritiated by the above base-catalyzed reaction. For example, glutathione gave radioactive glutamic acid and glycine with the base-catalyzed tritiation reaction and subsequent acid hydrolysis [826]. When DNP-glutathione (oxidized form) was used, however, only

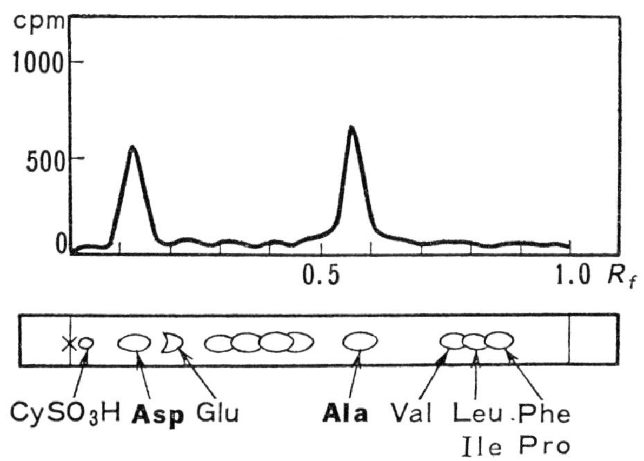

Fig. 3-15. Radio-chromatogram of the hydrolyzate of the tritiated insulin. Solvent used for paper chromatography is phenol saturated with water. Chromatogram was scanned by a Packard radio-chromatogram scanner (Model 7200). (Reproduced from a review by Matsuo [Kagaku, **36,** 244 (1966), in Japanese])

C-terminal glycine was tritiated, since the DNP-group is not an acyl function. The present selective tritium-labeling method, therefore, seems to be applicable to the identification of isoaspartyl and isoglutamyl residues in proteins.

The reaction conditions used for the tritiation of proteins also leads to tritium incorporation into a C-terminal aspartic acid [573], in contrast to the aforementioned conditions: heating a peptide with acetic anhydride in dioxane and subsequent addition of T_2O and pyridine. Co-existence of acetic anhydride, pyridine and T_2O may prevent possible formation of the intramolecular anhydride of acyl aspartic acid and may lead to the oxazolone formation followed by the tritiation of the regenerated aspartic acid residue.

Matsuo and Narita re-investigated the optimal conditions for the tritium-labeling suitable for quantitative C-terminal analysis of proteins [830]. Under the conditions mentioned above using a volume ratio of 4:2:1 for pyridine-T_2O-acetic acid, base catalyzed hydrolysis of acetic anhydride is shown to be fairly fast and thus acetic anhydride is consumed prior to the completion of the oxazolone formation, which results in a decreased reaction yield. Use of a great excess of acetic anhydride for raising yields of oxazolone formation did not give rise to better results because of serious disadvantages such as precipitation of proteins and unfavorable side reactions. With egg white lysozyme as a model protein, they examined various conditions, varying the ratio of the reagents. When a ratio of 3:1:3 (by volume) for pyridine-T_2O-acetic anhydride is used, the extent of tritium incorporation into the C-terminal leucine of lysozyme increased by about 20-fold in comparison with that under the original conditions mentioned above, and no side reaction was observed. Undeı these improved conditions, the pH of the reaction mixture can be maintained at 4.6. Cautions addition of acetic anhydride, in two portions, to the protein solution at 0° C in a manner as described later gave the best results with satisfactory reproducibility.

Comparison was made as to the extent of tritium incorporation into various acetyl amino acids under the improved conditions. The comparison was carried out by the following internal standard method [830]. For instance, a mixture of exactly

Table 3-15. Relative ratios of tritium incorporation into various acetyl amino acids to that into acetylalanine under the improved conditions

Acetyl amino acid	Relative ratio	Acetyl amino acid	Relative ratio
Alanine	1.00	Isoleucine	1.18
Arginine	1.96	Leucine	2.27
Asparagine	2.30	Lysine	1.52
Aspartic acid	0.32	Methionine	2.32
Glutamic acid	0.82	Phenylalanine	2.01
Glutamine	1.16	Threonine	0.09
Glycine	0.68	Tyrosine	1.95
Histidine	0.61	Valine	0.93

weighed acetylleucine (a µmole) and acetylalanine (b µmole), which served as an internal standard, was subjected to the tritium-labeling reaction in one vessel. After acid hydrolysis followed by separation of the resulting free amino acids by paper electrophoresis at pH 1.9, the radioactivities of leucine (x cpm) and alanine (y cpm) were measured, respectively. Relative tritiation ratio, x/y, was found to be proportional linearly to the molar ratio, a/b, in mixtures. Thus the relative value of specific activity of leucine to that of alanine, internal standard, can be determined as $\frac{x/a}{y/b}$. Such values for various acetyl amino acids were obtained as shown in Table 3-15. It can be seen from Table 3-15 that the amounts of tritium incorporation are significantly affected by the natures of side chain groups of acetyl amino acids. Relative values of tritium incorporation into the C-terminal residues were nearly the same as those listed in Table 3-15 when examined in any sets of acetyl amino acids and of several dipeptides, with the exception of glycyl amino acids. For example, tritium incorporation into the C-terminal phenylalanine of Gly-Phe was 83% of that of Ala-Phe. This yield-lowering effect by glycine at the penultimate position of the C-terminus may be explained in the following way. Abstraction of the amide hydrogen of the C-terminal residues to yield oxazolone derivatives is accelerated by steric hindrance due to side chain groups of penultimate residues, but penultimate glycine is devoid of such a steric effect.

Values listed in Table 3-15 can be used as C-terminal factors for quantitative C-terminal analysis of proteins. In Table 3-16 some quantitative data are shown, when bovine insulin or hen egg white lysozyme are used as internal standards with sample proteins. The listed data in Table 3-16 fall in the range of 0.97 ~ 1.13 mole of C-terminal residue per mole of protein in agreement with the theoretical value. Only in the case of human milk lysozyme [644], ending in -Gly-Val, is the analytical value appreciably lower than the theoretical one. This low value can be explained by the effect of the penultimate glycine as mentioned above.

Table 3-16. Quantitative C-terminal analysis of several proteins by internal standard method

Sample protein (C-Terminal sequence)	Internal standard (C-Terminal sequence)	Comparison	Observed ratio	C-Terminal factor (Table 15)	Molarity of C-Terminus
Lysozyme (egg white) (-Arg-Leu)	Insulin (bovine) (-Lys-Ala)	Leu/Ala	2.25	2.27	0.97 Leu
Ribonuclease A (bovine) (-Ser-Val)	Insulin (bovine) (-Lys-Ala)	Val/Ala	0.94	0.93	1.01 Val
Chymotrypsinogen A (bovine) (-Ala-Asn)	Lysozyme (egg white) (-Arg-Leu)	Asn/Leu	0.89	0.78	1.13 Asn
Cardiotoxin (formosan cobra venom) (-Cys-Asn)	Lysozyme (egg white) (-Arg-Leu)	Asn/Leu	0.82	0.78	1.04 Asn
Lysozyme (human milk) (-Gly-Val)	Insulin (bovine) (-Lys-Ala)	Val/Ala	0.39	0.93	0.42 Val

In quantitative analysis of peptides ending with C-terminal serine, it is difficult to obtain reproducible results, though qualitative analysis is possible. C-terminal threonine residues in peptides or in proteins incorporate radioactivity poorly as seen in Table 3-15. These difficulties involving C-terminal serine and threonine may arise from interfering dehydration which takes place in the side chain groups by the action of acetic anhydride.

C-terminal proline which does not yield the corresponding oxazolone cannot be determined even under the present improved conditions. Although quantitative experiments with peptides terminating in cystine or its derivatives and tryptophan have not yet been carried out, labeling of such C-terminal residues with tritium probably is possible. In the case of C-terminal tritiated tryptophan, acid hydrolysis must be carried out in the presence of thioglycolic acid [820] or *p*-toluenesulfonic acid containing 0.2% 3-(2-aminoethyl)-indole [786].

There still remains the serious problem of the sluggish rate of the tritiation reaction when proteins are insoluble in the reaction mixture. Addition of sodium dodecyl sulfate or urea to the reaction mixture sometimes is effective in solubilizing proteins, but conditions generally applicable to sparing soluble proteins have not yet been established.

Another problem which has been reported is the tritiation of non-C-terminal glutamic and aspartic acids. Such unfavorable "interior labeling" may lead to inconclusive results as in the case of rat liver lactate dehydrogenase [589]. In this case HSIEH et al. incubated the tritiation mixture at 37° C and 17% and 25% of the total radioactivity presumed to be due to the C-terminal phenylalanine could be attributed to glutamic and aspartic acids instead after acid hydrolysis. Generally the extent of "interior labeling" is not so high. According to CAPPUGI et al. [205], radioactivities of non-C-terminal glutamic and aspartic acids after tritiation of bovine ribonuclease A did not exceed 0.7% and 2.6% of that for its C-terminal, valine. It is not yet clear why non-C-terminal aspartic and glutamic acids undergo such "interior labeling" under the original reaction conditions. BABA et al. [52] examined "interior labeling" with synthetic peptides possessing α-aspartyl linkages and reported that only the aspartyl residues involving either in -Asp-Gly- or -Asp-His- sequence undergo "interior labeling" under the conditions similar to those of the original procedure.

The present authors investigated "interior labeling" under the improved tritiation conditions, but in all cases using egg-white lysozyme, bovine pancreatic ribonuclease A, and eight kinds of peptides having either -Asp-Gly-, -Asp-His-, -Asp-Phe-, -Asp-Val-, -Asp-Glu-, -Asp-Arg- or -Asp-Tyr- groups, no tritium incorporation was observed in aspartyl and glutamyl residues at all [831]. Nevertheless, careful consideration should be given to those cases in which aspartic or glutamic acid is found to be radioactive in the tritiation method.

Examination of released radioactive amino acid from the C-terminal tritiated protein by digesting with carboxypeptidases may provide confirmatory evidence [589]. It must be kept in mind, however, that the C-terminal residue is racemized to some extent as discussed in the reaction mechanism of the C-terminal tritiation above.

1. Tritiation of Protein

Tritiated water of high specific activity, e.g., 1 Ci/ml, is diluted with distilled water to a total of 10 ml and it is stored in a freezer (10 mCi/0.1 ml). Pyridine and acetic anhydride are stored in a cold room after redistillation of analytical grade reagents. In order to handle the tritiation reaction and subsequent hydrolysis in one vessel, use of an apparatus shown in Fig. 3-16 is convenient [823]. The apparatus consists of a small test tube (A, 0.8 × 1 cm) for the reaction and hydrolysis, a connecting glass adaptor with an arm tube (B) leading to a vacuum pump and a glass trap (C).

Protein or peptide (0.05 to 0.1 μmole) sample plus an internal standard (0.01 to 0.1 μmole) are placed in a test tube (A) and 5 μl of T_2O (500 μCi) and 10 μl of pyridine are added. It is important to insure a clear solution by gentle shaking; otherwise tritium incorporation sometimes becomes unsatisfactory. If additional volume is necessary to clear the solution, amounts of reagents to be added must be adjusted so as to keep the same relative ratio as described here. To the above protein solution, 10 μl of acetic anhydride is added with cooling in an ice bath. The mixture, in a parafilm-sealed tube, is kept at 0° C for 5 min and then at 20° C for an additional 15 min. After another addition of 20 μl of pyridine and 20 μl of acetic anhydride at 20° C, the reaction mixture is kept at 0° C for 5 min and then at 20° C for 1 hr more. After a further addition of 5 μl of T_2O, it is allowed to stand at 20° C for 1 hr to decompose excess acetic anhydride.

After completion of the tritiation reaction, the test tube is connected by means of a thick rubber tube with B which has been connected with C with silicone grease shown in Fig. 3-16. The solution in A is frozen in a dry ice-acetone bath and whole apparatus is evacuated by a vacuum pump through the arm of B. Then C is rotated to close the vacuum line and the dry ice-acetone bath is removed. Trap C is inserted in the dry ice bath and, T_2O, pyridine and acetic acid in A are distilled into C. During this distillation process, care should be taken to avoid foam formation. The highly radioactive distillate in C is discarded carefully. To the residue in A, 0.1 ml of 10% acetic acid is added to remove exchangeable tritium introduced into carboxyl, amide, etc. and above distillation procedure is repeated. This washing process is repeated five times to insure complete removal of labile tritium.

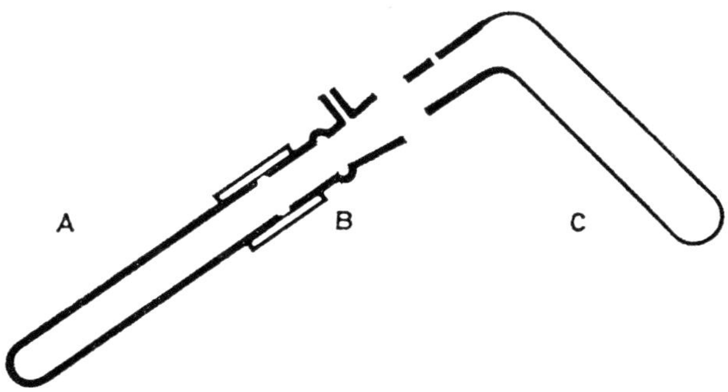

Fig. 3-16. Apparatus for C-terminal tritiation of proteins (see the text). (Reproduced from Ref. [823])

2. Characterization and Estimation of the Tritiated Amino Acid

Protein tritiated, at the C-terminus, which remains in test tube A (Fig. 3-16), is directly hydrolyzed with 6 N HCl as usual. Several methods, such as chromatography, thin layer chromatography on cellulose powder plates [513], gas-liquid chromatography [823] or ion-exchange chromatography [205] can be used for characterization of the tritiated amino acid. It is convenient to apply about 1/5-portion of the hydrolysate to a paper-strip (2 × 60 cm, Whatman No. 1 or Toyo Roshi No. 51) which is then subjected to electrophoresis (50 V/cm, 1 hr) at pH 6.4. The hydrolysate is separated into four portions, aspartic acid, glutamic acid, neutral amino acids and basic amino acids. The neutral amino acids are further separated on a paper-strip by electrophoresis at pH 1.8 at 50 V/cm, while the basic amino acids are separated by paper chromatography with pyridine-acetone-28% NH_4OH-water (50:20:5:50, v/v).

The chromatogram or electrophoretogram, without staining with ninhydrin, is cut into strips of 1 cm width which are submitted directly to radioactivity measurement in a liquid scintillation counter in 10 ml toluene scintillation mixture containing 0.5% 2,5-diphenyloxazole (PPO) and 0.01% 2,2'-*p*-phenylene-bis(5-phenyloxazole) (POPOP). Of course, as a blank, radioactivity also is measured in strips of similar size where no tritiated amino acid is located, and this value is subtracted from the counts for the sample. Approximate identification of the tritiated amino acids can be made by comparing stained paper chromatograms or electrophoretograms on which standard amino acids are developed simultaneously with the tritiated protein hydrolysate. Exact characterization of the tritiated amino acid can be made by re-chromatography of the amino acid extracted from the filter paper cut from the area where radioactivity is located.

Under the above reaction conditions, the radioactivity counted in the C-terminal residue in proteins ranges from 300 to 1000 cpm/0.01 μmole, and the specific radioactivity of the C-terminus depends linearly upon the specific activity of T_2O used. The highest specific radioactivity expected for the C-terminus is 300 ~ 1000 cpm per 0.2 nmole, with T_2O of 5 Ci/ml, which is the highest specific activity commercially available. Accordingly, microanalysis in 0.2 nmole scale is possible under the above conditions. Matsuo et al. [829], in their structural characterization of porcine luteinizing hormone-releasing hormone, were able to analyze successfully the C-terminal residues of the peptides of thermolysin and chymotrypsin digests in less than 1 nmole quantities.

It cannot be concluded that aspartic or glutamic acids found by the base-catalyzed tritiation method had arisen from the C-terminal acidic amino acid. Discrimination of C-terminal asparagine or glutamine from aspartic or glutamic acid can be made by digesting tritiated protein with carboxypeptidase A as usual. The C-terminal radioactive asparagine or glutamine is rather quickly released from the tritiated protein, but aspartic or glutamic acid only slowly under the usual conditions.

C. Carboxypeptidases

Carboxypeptidases remove amino acids sequentially from the C-terminus of a polypeptide chain. At present carboxypeptidases are classified into three groups, A, B and C, depending on their substrate specificities. Carboxypeptidase A (Cpase A), which was originally isolated by Anson from bovine pancreas, releases C-terminal residues with an aromatic or large aliphatic side chain (tyrosine, phenylalanine, tryptophan, leucine, isoleucine, methionine, threonine, glutamine, histidine, alanine, valine and homoserine) most rapidly asparagine, serine, lysine and methionine sulfone at moderate rates, and, very slowly, glycine, aspartic acid, glutamic acid, S-carboxymethylcysteine and cysteic acid [28]. The enzyme does not release C-terminal proline and arginine. The end products after digestion of oligopeptides composed of amino acid residues having strong affinity to Cpase A are the N-terminal dipeptide and the constituent amino acids of the parent peptide, unless extremely high concentrations of the enzyme are used. If high concentrations of the enzyme are used, Cpase A completely hydrolyzes oligopeptides into the constituent amino acids. Relative rates of hydrolysis of some synthetic substrates by Cpases A and B are listed in Tables 3-17 and 3-18. Cpase B, which was isolated by Folk and Gladner [365] also from bovine pancreas, releases C-terminal arginine and lysine rapidly, but it also releases neutral amino acids. All of the protein-constituent amino acids at the C-terminus of polypeptide chain can be removed by the action of either Cpase A or Cpase B with the exception of the C-terminal proline. Zuber isolated Cpase C from the peel of citrus fruits *(Citrus sinensis, C. medica* and *C. maxima)* and demonstrated that the citrus enzyme released C-terminal proline at pH 3.5 as well as other amino acids at roughly similar rates [1428, 1429]. Cpases C also have been isolated from plant sap (*Phaseolus vulgaris* [1346, 1347] and barley malt [1319]), molds (*Aspergillus saitoi* [595, 596], *A. oryzae* [594], *Penicillium janthinellum* [1150]) and baker's yeast [520, 521]. These Cpases C release all of the C-terminal amino acids from synthetic substrates optimally at pH 3 to 5, are inhibited by diisopropyl phosphorofluoridate (DFP) and inactivated by lyophilization. However, when either glycine or proline occupies the penultimate position of the C-terminus, hydrolysis rate of the terminal residue is greatly diminished; peptides with C-terminal sequence of -Pro-Pro-Gly-OH, -Gly-Pro-OH, -Gly-Gly-OH, -Gly-Gly-Leu-OH, and -Gly-Pro-Leu-OH are practically not hydrolyzed. These Cpases C have not yet become commercially available.

Table 3-17. Relative hydrolysis rates of synthetic substrates by carboxypeptidase A[a]

Substrate	Relative rate	Reference	Substrate	Relative rate	Reference
Cbz-Gly-L-Phe-OH	100	[1174]	Cbz-L-Ala-L-Phe-OH	85.9	[90]
Cbz-Gly-L-Tyr-OH	47.9	[1174]	Cbz-L-Ala-L-Tyr-OH	49.5	[90]
Cbz-Gly-L-Trp-OH	36.2	[1174]	Cbz-L-Glu-L-Phe-OH	6.11	[90]
Cbz-Gly-L-Leu-OH	20.0	[1174]	Cbz-L-Glu-L-Tyr-OH	3.76	[90]
Cbz-Gly-L-Met-OH	9.22	[274]	Cbz-L-Trp-L-Ala-OH	9.25	[1174]
Cbz-Gly-L-Ile-OH	4.15	[1174]	Cbz-L-Trp-L-Glu-OH	0.052	[1174]
Cbz-Gly-L-Ala-OH	0.29	[1174]	Cbz-L-Trp-L-Pro-OH	0.019	[1174]
Cbz-Gly-Gly-OH	0.018	[1174]	Cbz-L-Arg-L-Leu-OH	46.2	[44]
Cbz-Gly-L-Arg-OH	0	[44]	Cbz-L-Phe-L-Arg-OH	0	[44]
			Bz-Gly-L-Lys-OH	0	[365]

[a] Relative values of the first order proteolytic coefficients of the peptides are listed with that of Cbz-Gly-L-Phe-OH selected to be 100. Abbreviations of Cbz and Bz denote carbobenzoxy and benzoyl, respectively.

Table 3-18. Relative hydrolysis rates of synthetic substrates by carboxypeptidase B[a]

Substrate	Relative rate
Bz-Gly-L-Lys-OH	100
Bz-Gly-L-Arg-OH	71.5
Bz-Gly-L-Orn-OH	29.5
Bz-Gly-L-Homoarginine	1.19
Bz-Gly-L-Pro-L-Lys-OH	0.040
Bz-Gly-D-Lys-OH	0
Bz-Gly-L-Lys-NH_2	0

[a] Relative values of the zero order proteolytic coefficients of the peptides are shown with that of Bz-Gly-L-Lys-OH selected to be 100. Bz and Cbz are abbreviations of benzoyl and carbobenzoxy, respectively. (Data from Ref. [365]).

Preparations of Cpases A and B are commercially available. Cpase A is practically insoluble in water and is usually available as a crystalline suspension in water. Crystalline Cpase A is contaminated with other pancreatic proteolytic enzymes and treatment with DFP is required to inactivate contaminating trypsin and chymotrypsin prior to use. The DFP-treated enzyme also is commercially available and this preparation can be used after washing the crystals with water to remove contaminating amino acids [381].

A crystalline Cpase B has been reported [1384] but the commercial preparation is an aqueous solution. The B enzyme is rather unstable and must be kept frozen

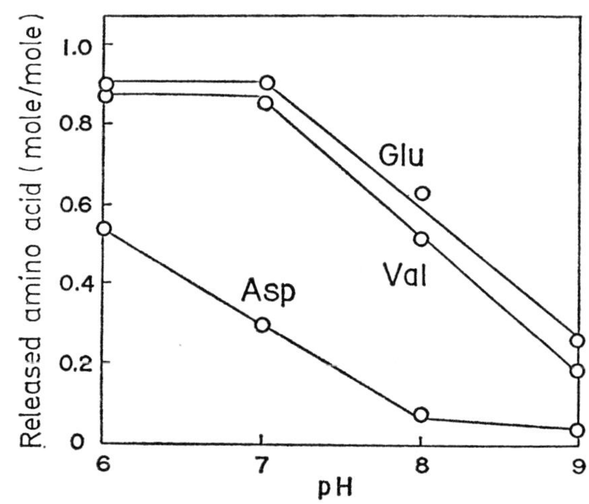

Fig. 3-17. The effect of pH in digestion of N-acetyl-Gly-Asp-Val-Glu-OH (N-terminal peptide of bovine heart cytochrome *c*) with carboxypeptidase A. The digestion was carried out at room temperature and the three pH's indicated in the figure with the enzyme-substrate ratio of 1:1 by weight. The released amino acids after 15 hrs incubation were plotted. (Reproduced from Ref. [1283], by courtesy of the Japanese Biochemical Society)

since lyophilization leads to inactivation. The enzyme preparation is generally contaminated with proteolytic enzymes and free amino acids; thus DFP-treatment is essential. The DFP-treated Cpase B is also available. The enzyme is best used after purification by gel filtration on a Sephadex G-25 column in 0.01 M tris-HCl buffer of pH 8.0 to remove free amino acids [28].

The two Cpases are stable between pH 7 and 10 and the enzyme digestion is normally carried out at pH 7.5 to 8.5. However, when a peptide terminates with glutamic or aspartic acid, digestion with Cpase A at lower pH's (pH 5 to 6) increases the rate of release of the acidic amino acid from the C-terminus [442, 1283] as shown in Fig. 3-17.

When the released amino acids in the incubate of a protein or peptide with Cpase A or B are characterized and estimated as a function of time by a suitable method, for example by an automatic amino acid analyzer, the amino acid sequence of the C-terminus can be deduced. When only one amino acid is released from the peptide chain in a stoichiometric amount, the interpretation of the results is quite simple. Tobacco mosaic virus and egg white lysozyme are such typical examples. Stoichiometric amounts of threonine and leucine are released from tobacco mosaic virus [509] and lysozyme [1274], respectively, by the action of Cpase A. The C-terminal sequences of the two proteins are, respectively, -Pro-Ala-Thr-OH and Gly-Cys-Arg-Leu-OH. In the former case, the alanine, which is the second residue from the terminus, should be liberated after the terminal threonine is removed, but no alanine is released. This fact suggests that the release rate is affected predominantly by the C-terminal amino acid and influenced also by the penultimate residue. Such a tendency is evident with synthetic substrates as shown in Table 3-17. In contrast to tobacco mosaic virus, even when the penultimate residue is proline, the terminal amino acid can sometimes be released. Adrenocorticotropin, possessing the sequence, -Pro-Leu-Glu-Phe-OH, is such a case [511]; C-terminal phenylalanine is released very quickly, glutamic acid at a moderate rate and leucine at a slightly slower rate. Isoleucine on the carboxyl side of a proline residue, was reported to be partially removed from Ac-Ala-Tyr-Asn-Pro-Ile-Thr-Ser-OH after liberation of serine and threonine [896]. This peptide was isolated from the protein of cucumber virus 4. Generally the C-terminal amino acid which is linked to a penultimate proline residue is difficult to remove quantitatively with these enzymes.

When several amino acids are released at similar rates from relatively large peptides or proteins, the interpretation of the results is quite difficult even when the relative specificity of Cpase is carefully considered. In such a case, it is dangerous to draw definite conclusions regarding the C-terminal amino acid of a protein based upon only Cpase experiments. Conclusion should be carefully made considering the results of other chemical methods for the C-terminal analysis.

An excellent review of Cpases A and B has been written by NEURATH [919], and procedures for the end group and sequence analyses with Cpases have been presented by FRAENKEL-CONRAT et al. [381], and by AMBLER [26, 28].

1. Preparation of the Enzyme Solution

Commercial preparations of Cpase A are normally aqueous suspensions of crystals at a concentration of 50 mg/ml. The suspension is stable at 4° C for years, but lyophilization will cause inactivation. The enzyme is active in crystalline suspension but usually the crystals are dissolved immediately before use. The enzyme solution cannot be stored for a long time prior to use. A suspension containing 1.25 mg protein (normally 25 μl) is diluted with 1 ml water in a centrifuge tube. The crystals sediment very quickly and the supernatant containing free amino acids is discarded. Into the sedimented crystals is added 0.1 ml of 1% $NaHCO_3$ with ice-cooling and 0.1 N NaOH (2 to 3 drops) is added with shaking until the crystals dissolve. The pH of the solution is then brought back to 8.0 with 0.1 N HCl (2 to 3 drops), and the volume is made up to 1.25 ml such that the enzyme concentration becomes 1 mg/ml with 0.2 M buffer of pH 8.0. When the digestion is followed by the ninhydrin reaction, buffers containing ammonia, amines or phosphate must be avoided. Phosphate buffers result in precipitation during the usual ninhydrin analysis. AMBLER [26] recommended the use of an N-ethyl morpholine acetate buffer. Care must be taken to prevent over-addition of hydrochloric acid during neutralization of the enzyme alkaline solution since the enzyme inactivates rapidly below pH 6 and precipitates.

The enzyme (molecular weight, 34300) solution, if necessary, is treated with a 50-fold molar excess of a solution of DFP (15 μl of 0.1 M DFP in dry isopropanol per mg of the enzyme) for 1 h at 25° C. The concentration of the enzyme solution is measured spectrophotometrically, assuming that the enzyme has a molar absorbancy of 8.6×10^4 at 278 nm [918].

Commercial Cpase B solution normally contains free amino acids and other proteolytic enzymes of pancreatic origin as discussed. If the preparation has not been treated with DFP, DFP-treatment should be carried out as described for Cpase A. In order to remove free amino acids, gel filtration on a Sephadex G-25 column or dialysis must be performed before use. The enzyme concentration may be measured spectrophotometrically by using a molar absorbancy of 7.4×10^4 at 278 nm [367].

2. Reaction of Carboxypeptidases with Protein

Native proteins generally are resistant to the action of Cpases and prior denaturation of proteins is necessary. Proteins containing disulfide bonds or cysteine residues usually require prior reduction and subsequent alkylation or performic acid oxidation. Proteins thus treated are of course denatured. Cpase A is not denatured in a solution of 6 M urea [479] or 0.056 M sodium dodecyl sulfate [455] and such a denaturant solution can be used during the reaction. Denaturant solutions can be used effectively to solubilize insoluble protein substrates at pH 8.0.

The amount of denatured protein to be digested depends upon the sensitivity of the analytical method for the amino acids released and upon the number of aliquots to be taken as a function of suitable time intervals. Usually 0.1 to 2 μmoles of denatured protein are dissolved in a volatile 0.2 M N-ethyl morpholine acetate buffer of pH 8.0 to a concentration of 0.2 to 0.5%. The enzyme solution is then added in a substrate: enzyme ratio of 1/20 to 1/200 (molar basis) and the mixture, in a screw-capped or stoppered tube, is incubated at 25 to 37° C. At scheduled time intervals (for example, 0, 0.2, 0.5, 1, 2, 4 and 8 hrs) 0.05 to 0.25 μmole portions of the mixture are removed, and acidified (trichloroacetic, acetic or hydrochloric acid) to pH 2 to stop the reaction. Any resulting precipitate is removed by centrifugation. The supernatant is subjected to amino acid analysis.

Under standard analysis conditions, the peaks of asparagine and glutamine overlap that of serine and the three amino acids cannot be estimated separately. For the quantitation of the three amino acids, asparagine and glutamine must be hydrolyzed to aspartic and glutamic acids and then estimated. For this purpose the released amino acids must be separated from the protein and the enzyme prior to the hydrolysis. Utilization of the molecular sieve effect of ion exchange resins is applicable to such separations [1274]. The hydrogen-form of sulfonate polystyrene beads (Dowex 50-X8-X10, 20 to 60 mesh) should be added until the pH becomes 2 to 3 [26]. Fine beads should not be used to minimize adsorption of proteins. The mixture is then shaken for 20 min, the supernatant is removed and the resin beads are washed twice with two resin volumes of water. The adsorbed amino acids on the resin are extracted three times with two resin volumes of 5 M NH_4OH and the combined extracts are evaporated to dryness. The residue is dissolved in an appropriate volume of pH 2.2 citrate buffer and the solution is divided into three portions. One portion is dried, hydrolyzed with acid and analyzed for neutral and acidic amino acids. The other two portions are analyzed without hydrolysis for acidic and neutral amino acids, and for basic amino acids, respectively. The increase in aspartic and glutamic acid values after acid hydrolysis gives the amide content. For mutual separation of asparagine, glutamine and serine by an automatic amino acid analyzer, use of the lithium form of the resin and lithium citrate buffer is proposed.

D. Other Methods

The first attempt to characterize the C-terminal group of a protein based on an organic chemical reaction was made by FROMAGEOT et al. [395, 641] in 1950. They reduced the carboxyl groups of insulin to the corresponding alcohol groups with $LiAlH_4$ in N-ethyl morpholine at 55° C for 8 hrs. The amino alcohols derived from the C-termini of reduced insulin in ether extracts of the acid hydrolysate were characterized by paper chromatography. CHIBNALL and REES [229] reduced insulin methyl ester with $LiBH_4$, a much milder reducing agent, and identified the amino alcohols derived from the C-terminal alanine and asparagine in the expected amounts. In addition to the two amino alcohols, they obtained ethanolamine derived from one of the glycyl peptide bonds. CHIBNALL and REES extensively studied the reduction method, but this method has not been widely used for end group determination because of two side reactions: reductive cleavage of peptide bonds and reduction of peptide carbonyls to methylene groups. Despite the problems of the reduction method, BAILEY [55] proposed an interesting procedure for the sequential removal of the C-terminal residue as the amino alcohol based on the coupling of reduction with an N, O-acyl shift reaction as illustrated in scheme (20).

$$
\begin{array}{ll}
R_1\cdot CO\text{–}NH\cdot CHR_2\cdot CO\text{–}NH\cdot CHR_3\cdot COOH & \xrightarrow{CH_3OH+HCl} \\
R_1\cdot CO\text{–}NH\cdot CHR_2\cdot CO\text{–}NH\cdot CHR_3\cdot COOCH_3 & \xrightarrow{AlH_3} \\
R_1\cdot CO\text{–}NH\cdot CHR_2\cdot CO\text{–}NH\cdot CHR_3\cdot CH_2\cdot OH & \xrightarrow{H^{\oplus}} \\
R_1\cdot CO\text{–}NH\cdot CHR_2\cdot CO\text{–}OCH_2\cdot CHR_3\cdot NH_3^+ & \xrightarrow{LiBH_4} \\
R_1\cdot CO\text{–}NH\cdot CHR_2\cdot CH_2OH + HO\cdot CH_2\cdot CHR_3\cdot NH_2 &
\end{array}
\qquad (20)
$$

He reduced the *p*-toluenesulfonyl peptide ester in tetrahydrofuran with AlH_3 at −40° C and carried out the N,O-acyl shift with phosphorous oxychloride at room temperature. Reductive cleavage of the *β*-amino esters with $LiBH_4$ gave the terminal amino alcohol in 85 to 90%. Although the principle of Bailey's method is of interest, side reactions have obviated its use with proteins.

Another sequential determination method for the C-terminus is the thiohydantoin method originally proposed by SCHLUCK and KUMPF [1087] in 1926. Peptides substituted at the terminal amino group are converted to the peptidyl thiohydantoin by heating for 30 min at 100° C with ammonium thiocyanate in acetic anhydride. The thiohydantoin corresponding to the C-terminal residue is released by treating with 0.1 N NaOH at room temperature for a few minutes. The reaction is shown in scheme (21).

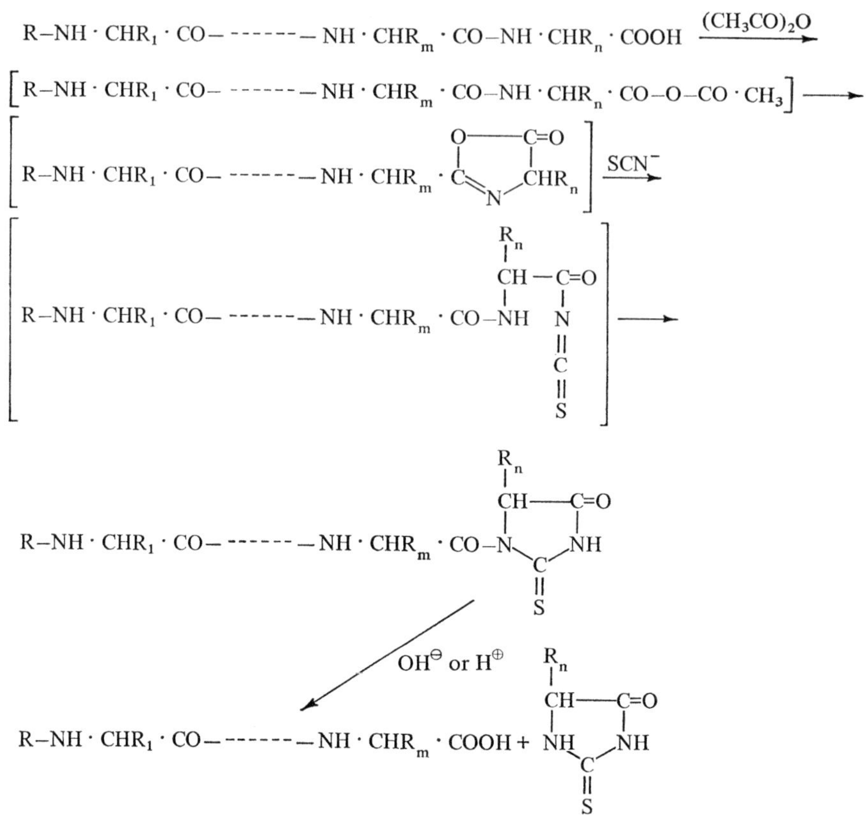

(21)

The released thiohydantoin is extracted with organic solvents and the parental amino acid is regenerated by hydrolyzing with hydrogen bromide. Thiohydantoins of amino acids can be characterized directly by paper chromatography [264, 323]. This method has been successfully applied to smaller oligopeptides but has not always been successful with proteins [60, 365, 436, 1255, 1280, 1324] without modification of the procedures. Stark [1211, 1213] reinvestigated this sequential degradation method and applied his modified procedures to a number of peptides. 100 mg of recrystallized ammonium thiocyanate were dissolved in a mixture of 4.0 ml of redistilled acetic anhydride and 1.0 ml of glacial acetic acid. To this solution was added 0.5 ml or less of 50% acetic acid solution of a peptide and the mixture was kept for 6 hrs at 50° C. After incubation for 18 hrs with an additional 100 mg of ammonium thiocyanate at 50° C, the peptidylthiohydantoin was isolated by the aid of a Sephadex G-25 column. For the release of the thiohydantoin derivative of the C-terminal amino acid from the peptidylthiohydantoin, he recommended the use of 0.1 M acetohydroxyamate in 50% pyridine solution at 50° C for 2 hrs. A portion of 50% acetic acid extract of the dried mixture was subjected to amino acid analysis and the amino acid removed was characterized and estimated in comparison with the amino acid composition of the parent peptide, without attempting direct characterization of the released thiohydantoin of the C-terminal amino acid. Even with these modified procedures, the reaction was not quantitative. Only two to three residues from the C-terminus could be determined with confidence. Aspartic acid and proline at the C-terminal end could not be determined by this thiohydantoin method in similar procedures to the tritum labeling method described above.

YAMASHITA [1407] reported further modification of this stepwise C-terminal determination method and demonstrated that the C-terminal tetrapeptide sequence of bovine pancreatic ribonuclease A could be determined in about 80% yield. He dissolved about 1 μmole of a peptide in a mixture of 0.5 ml of acetic anhydride, 0.5 ml of trifluoroacetic anhydride and 0.5 ml of acetic acid. After solution was complete, 100 mg of crystalline sodium thiocyanate were added with vigorous stirring and the mixture was kept for 6 hrs at 60° C. After the reaction mixture was lyophilized and redissolved in water, the peptidylthiohydantoin was isolated by lyophilization of the dialyzed aqueous solution. For the cleavage of the peptidylthiohydantoin, its aqueous or 0.1 N HCl solution was shaken vigorously with Amberlite IR 120 (H^+ form, 1 g) at room temperature for 2 to 6 h. The C-terminal thiohydantoin and the peptide with one residue less than the parent peptide were extracted from the resin three times with 5 ml each of ethyl acetate and water after the resin was washed with water. The combined extract was evaporated and the dissolved residue in 2 ml of 50% acetic acid was subjected to gel filtration on a column of Sephadex G-25. The peptide fraction eluted just after the void volume was lyophilized and used for the next degradation. The thiohydantoin fraction (strong absorption at 260 nm) was concentrated and subjected to thin layer chromatography on a silica gel plate containing fluorescent indicator with heptane-*n*-butanol-formic acid (10:7:3, v/v).

Another modification of the thiocyanate method was reported by KUBO et al. [734]. About 1 μmole of a peptide was dissolved in 10 μl of trifluoroacetic acid and 200 μl of acetyl chloride was added to the solution. The mixture was kept for 15 min at 30° C in order to activate the C-terminal residue. Then 200 μl of 3% hydrogen thiocyanate in dioxane was added and the mixture was kept for 90 min at 30° C. After evaporation of the mixture, the residue was dissolved in 100 μl of water and again evaporated. The peptidylthiohydantoin thus obtained was hydrolyzed with 0.5 M triethylamine aqueous solution at 30° C for 90 min. The released thiohydantoin was extracted with ethyl acetate from the dried hydrolysate and was subjected to thin layer chromatography. In this modification the reagents used can be removed by evaporation. They reported successful analysis of several synthetic substrates including C-terminal proline, but they observed that the C-terminal proline and penultimate residue were released as their thiohydantoins in a single step. For preparation of hydrogen thiocyanate, triethylammonium thiocyanate was treated with Amberlyst-15 in dioxane. Hydrogen thiocyanate in dioxane thus prepared can be stored in a refrigerator for a month.

A similar method has been proposed by KENNER et al. [681] which was diphenyl phosphoroisothiocyanatidate. An acylated peptide is reacted with the reagent in anhydrous acetonitrile in the presence of triethylamine to yield the peptidylthiohydantoin, from which thiohydantoin is released on alkaline treatment as shown in scheme (22).

In 1936 BERGMANN and ZERVAS [93] proposed a sequential degradation of peptides beginning at the C-terminal and using the Curtius rearrangement. The benzoyl peptide

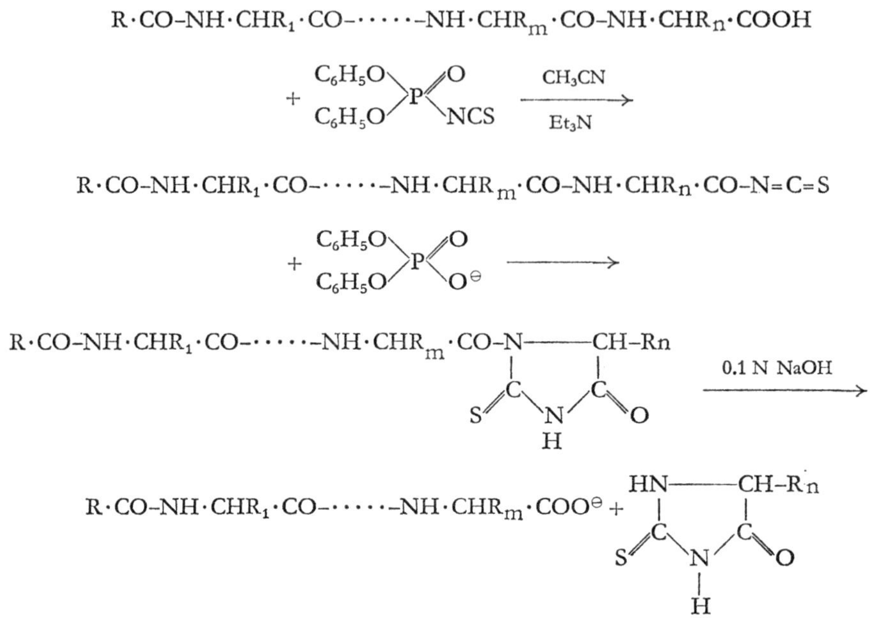

(22)

ester is converted into corresponding hydrazide and then into the peptide azide with nitrous acid. The peptide azide is then converted with benzyl alcohol, into the benzyl urethane derivative. Catalytic reduction yields the benzoyl peptide amide with one residue less than the parent peptide, an aldehyde derived from the C-terminal residue, ammonia, carbon dioxide and toluene as shown in scheme (23).

$$C_6H_5\cdot CO\text{–}NH\cdot CHR_1\cdot CO\text{–}\cdots\cdots\text{–}NH\cdot CHR_m\cdot CO\text{–}NH\cdot CHR_n\cdot COOC_2H_5 \xrightarrow{N_2H_4}$$

$$C_6H_5\cdot CO\text{–}NH\cdot CHR_1\cdot CO\text{–}\cdots\cdots\text{–}NH\cdot CHR_m\cdot CO\text{–}NH\cdot CHR_n\cdot CO\cdot NH\cdot NH_2 \xrightarrow{HNO_2}$$

$$C_6H_5\cdot CO\text{–}NH\cdot CHR_1\cdot CO\text{–}\cdots\cdots\text{–}NH\cdot CHR_m\cdot CO\text{–}NH\cdot CHR_n\cdot CO\cdot N_3 \xrightarrow{C_6H_5CH_2OH}$$

$$C_6H_5\cdot CO\text{–}NH\cdot CHR_1\cdot CO\text{–}\cdots\cdots\text{–}NH\cdot CHR_m\cdot CO\text{–}NH\cdot CHR_n\cdot NH\cdot COO\cdot CH_2\cdot C_6H_5 \xrightarrow[Pd]{H_2}$$

$$C_6H_5\cdot CO\text{–}NH\cdot CHR_1\cdot CO\text{–}\cdots\cdots\text{–}NH\cdot CHR_m\cdot CO\text{–}NH_2 + R_n\cdot CHO + NH_3 + CO_2 + C_6H_5\cdot CH_3 \quad (23)$$

Although this reaction can be repeated sequentially by converting the peptide amide into the peptide hydrazide, the procedures are very difficult when applied to proteins.

KHORANA [687] proposed a novel though difficult to apply route for selective removal of the C-terminal residues. He reacted an acyl peptide with di-(*p*-tolyl)- or di-(*p*-dimethylaminophenyl)-carbodiimide to yield the corresponding acyl peptidyl urea which undergoes degradative formation of N-*p*-tolyl-carbamyl amino acid *p*-toluidide on treatment with 0.01 N NaOH in 75% ethanol at room temperature as shown in the reaction scheme (24).

$$C_6H_5 \cdot CO\text{–}NH \cdot CHR_1 \cdot CO\text{–} \cdots\cdots \text{–}NH \cdot CHRm \cdot CO\text{–}NH \cdot CHRn \cdot COOH$$

$$\xrightarrow{CH_3 \cdot C_6H_4 \cdot N{=}C{=}N \cdot C_6H_4 \cdot CH_3} \qquad (24)$$

$$C_6H_5 \cdot CO\text{–}NH \cdot CHR_1 \cdot CO\text{–} \cdots\cdots \text{–}NH \cdot CHRm \cdot CO\text{–}NH \cdot CHRn{>}C{=}O \;\; H_3C \cdot C_6H_4 \cdot HN \cdot CO \cdot N \cdot C_6H_4 \cdot CH_3 \xrightarrow[\text{in } 75\%\ C_2H_5OH]{0.01\ N\ NaOH}$$

$$C_6H_5 \cdot CO\text{–}NH \cdot CHR_1 \cdot CO\text{–} \cdots\cdots \text{–}NH \cdot CHRm \cdot COO^{\ominus} + CH_3 \cdot C_6H_4 \cdot NH \cdot CO \cdot NH \cdot CHRn \cdot CO \cdot NH \cdot C_6H_4 \cdot CH_3$$

The product derived from the terminal residue can be converted into the parent amino acid by hydrolysis with acid or alkali.

IV. Masked Terminal Groups

Gramicidin S was the first oligopeptide for which the amino acid sequence was determined [244, 246]. Unlike the usual oligopeptides derived from proteins by partial hydrolyses, gramicidin S is a cyclic decapeptide devoid of both the terminal groups. Thereafter, the inability to find terminal residues by the usual terminal analysis methods led to the conclusion that the protein or peptide in question contained a cyclic chain or branched cyclic chain structure devoid of one or both of the free end groups. With the exceptions of some antibiotic and toxic oligopeptides cyclic and branched cyclic polypeptide chain structures have not been found with a frequency commensurate with such conclusions. The discovery of masked terminal residues has shed light on these ambigious problems.

A. N-Acylated Terminal Group

The many efforts to elucidate the N-terminal group of the protein of tobacco mosaic virus by the usual analytical methods had been unsuccessful leading to the postulation that the peptide structure was that of a branched cyclic chain having a C-terminal threonine [161, 383]. Narita [894, 895] eventually clearly demonstrated that the terminal α-amino group was blocked with an acetyl group. Since then, the proteins of cucumber virus 4 [896], turnip yellow mosaic virus [508], α-melanocyte stimulating hormone [506], ovalbumin [897, 898] and cytochromes *c* from vertebrates [732, 1283] have been shown to be blocked with an acetyl group. In Table 3-19 are listed the N-terminal acetyl proteins thus far reported. Proteins with terminal groups once reported to be acetylated (bovine fibrinopeptide B and α_1-acid glycoprotein) but later found to be pyrrolidone carboxyl (pyoglutamyl) residues are omitted from the table. Table 3-20 lists the terminal pyroglutamyl peptides and proteins.

Some oligopeptides, such as gramicidin A [1073] and bee venom, melittin [731] and the newly synthesized proteins in cells of *E. coli* [7, 1339] are found to be masked with a formyl group at their N-terminal α-amino groups. The terminal formyl group of *E. coli* proteins is believed to be split enzymatically during polypeptide chain synthesis in the bacterial cells. The only protein from eukaryote whose N-terminus is reported to be masked with a formyl group is lamprey *(Lampetra fluviatilis)* hemoglobin [401].

No method for end group determination can be directly applied to such terminal acylated proteins and pyroglutamyl proteins. In most cases peptides containing a

Table 3-19. N-terminal acetyl proteins

Protein	Sequence	Reference
Tobacco mosaic virus	Ac-Ser-Tyr-	[894, 895]
Cucumber virus 4	Ac-Ala-Tyr-Asn-Pro-Ile-	[896]
α-Melanocyte stimulating hormone	Ac-Ser-Tyr-Ser-Met-Glu-	[506]
Turnip yellow virus	Ac-Met-Gln-Ile-Asp-Lys-	[508]
Ovalbumin	Ac-Gly-Ser-Gly-Ile-Ala-	[897, 898]
Histone IV (calf thymus)	Ac-Ser-Gly-Arg-Gly-Lys-	[275]
Human fetal hemoglobin ($F_1\gamma$)	Ac-Gly-His-	[1109, 1219]
Chicken hemoglobin $A_1\beta$	Ac-Val-Thr-Leu-	[1075]
Enolase (rabbit muscle)	Ac-Ala-	[1383]
Myosin (rabbit muscle)	Ac-Ser-Ser-Asp-Ala-	[933]
Wool keratin	Ac-Ala-	[930]
Carbonic anhydrase (human erythrocyte)	Ac-Ala-Ser-Pro-Asp-	[599, 816]
Mitochondrial structural protein (beef heart)	Ac-Ser-	[256]
Carp hemoglobin α	Ac-Ser-Leu-Ser-Asp-Lys-	[540]
Actin (rabbit muscle)	Ac-Asp-Glu-[Thr, Ala]-	[23, 402]
Tropomyosin (rabbit muscle)	Ac-[Met, Asp]-Ala-	[24]
Apoferritin (horse spleen)	Ac-Ser-Ser-Gln-Ile-Arg-	[1247]
Hyaluronidase	Ac-[Tyr, Leu, Phe, Ala, Glu]-	[1194]
Glyceraldehyde 3-phosphate dehydrogenase (lobster)	Ac-Ser-Lys-Ile-Gly-Ile-	[265]
Frog hemoglobin α	Ac-Ala-Leu-	[219]
α-Crystallin (bovine)	Ac-Met-Asp-Ile-Ala-	[563]
Lactate dehydrogenase (M_4) (rat liver)	Ac-Ala-Ala-	[1064]
Lactate dehydrogenase (H_4) (beef, chicken heart)	Ac-Ala-Thr-	[1064]
Lactate dehydrogenase (M_4) (dog fish)	Ac-Thr-Ala-Leu-Lys-Asp-	[22]
Basic protein of myelin membrane (bovine)	Ac-Ala-Ser-Ala-Gln-Lys-	[344]
Cytochrome *c* (vertebrate)	Ac-Gly-Asp-Val-Glu-Lys-	[732, 1218]
Cytochrome *c* (wheat germ)	Ac-Ala-Ser-Phe-Ser-Glu-	[1228]
Cytochrome *c* (mung-bean, sun flower)	Ac-Ala-Ser-Phe-Ala-Glu-	[1012, 1278]
Ca-binding protein (carp muscle)	Ac-Ala-Phe-Ala-Gly-Val-	[925]

terminal acyl group or pyroglutamyl residue are isolated form their proteolytic digests and characterized separately as outlined in Fig. 3-18 and in Chapter 6. When acyl oligopeptides are isolated, mass spectrometry in combination with gas chromatography of the peptides provides information of the entity of acyl group attached as well as their amino acid sequences.

Table 3-20. N-terminal pyroglutamyl peptides and proteins[a]

Peptide	Structure
Eisenine	Pyrglu-Gln-Ala-OH
Fastigiatine	Pyrglu-Gln-Gln-OH
Eledoisin (octopus)	Pyrglu-Pro-Ser-Lys-Asp-Ala-Phe-Ile-Gly-Leu-Met-NH_2
Physalaemin (frog)	Pyrglu-Ala-Asp-Pro-Asn-Lys-Phe-Tyr-Glu-Met-NH_2
Caerulein (frog)	Pyrglu-Gln-Asp-Tyr-Thr-Gly-Trp-Met-Asp-Phe-NH_2
Phyllocaerulein (frog)	Pyrglu-Glu-Tyr-Thr-Gly-Trp-Met-Asp-Phe-NH_2
Phyllomedusin (frog)	Pyrglu-Asn-Pro-Asn-Arg-Phe-Ile-Gly-Leu-Met-NH_2
Ranatensin (frog)	Pyrglu-Val-Pro-Gln-Trp-Ala-Val-Gly-His-Phe-Met-NH_2
Bombesin (toad)	Pyrglu-Gln-Arg-Leu-Gly-Asn-Gln-Trp-Ala-Val-Gly-His-Leu-Met-NH_2
Gastrin II (human)	Pyrglu-Gly-Pro-Trp-Met-Glu-Glu-Glu-Glu-Glu-Ala-Tyr-Gly-Trp-Met-Asp-Phe-NH_2 (Tyr: | SO_3H)
Snake venom peptide	Pyrglu-Gln-Trp-OH
Snake venom peptide	Pyrglu-Asn-Trp-OH
Bradykinin-potentiator (snake venom)	Pyrglu-Gly-Leu-Pro-Pro-Arg-Pro-Lys-Ile-Pro-Pro-OH
α_1-Acid glycoprotein	Pyrglu-Pro-Ile-
Immunoglobulins (human)	
light chain (λ)	Pyrglu-Ser-Val-Leu-Thr-
heavy chain (γ)	Pyrglu-Val-Thr-Leu-Arg-
Fibrinopeptide B (human)	Pyrglu-Gly-Val-Asn-Asp-Asn-Glu-Glu-Gly-Phe-Phe-Ser-Ala-Arg-OH

[a] Abbreviation Pyrglu denotes pyrrolidone carboxylyl group. (Data from Ref. [268]).

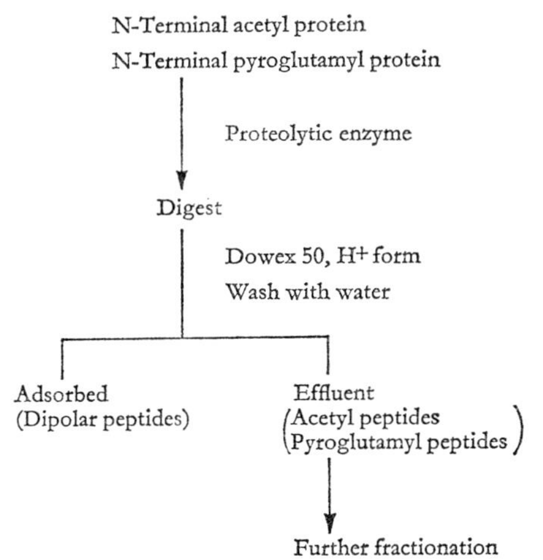

Fig. 3-18. Fractionation scheme for acyl peptides from proteolytic digests of the N-terminal acylated proteins

1. Acetylated Terminus

A protein for which an N-terminal residue cannot be found by the routine terminal analysis methods, is digested with proteolytic enzymes possessing a broad specificity, such as pronase, an extracellular proteinase produced by *Streptomyces griseus*. If the acetyl peptide is large, it probably will contain basic amino acids giving a neutral, or perhaps basic peptide in spite of the blocked N-terminal amino group which tends to make the peptide more acidic. Trypsin therefore cannot be used as a proteolytic enzyme, since tryptic peptides always contain lysine or arginine at the C-terminal end with the exeption of the C-terminal peptide of the parent protein. When the terminal acetyl peptide is thought to contain basic residues, the functional groups of the lysine and arginine residues in the protein must be blocked prior to proteolytic digestion. Suitable techniques for this purpose are succinylation of the amino groups with succinic anhydride at alkaline pH [469, 771] and benzylation of the guanido groups with dibenzoly reagents [622]. The latter reaction is carried out under strongly alkaline conditions and a possibility of partial hydrolysis of peptide bonds exists. This is however of no consequence at the present step. For the modification of arginine residues, reaction with 1,2-cyclohexanedione can also be used under much milder alkaline conditions [1287].

In general, acetyl peptides present in digests of the terminal acetylated protein are acidic. Other peptides in the digest are amphoteric, at least, based on the presence of free terminal amino and carboxyl groups. The hydrogen form of a strong cation exchanger such as Dowex 50 adsorbs these amphoteric peptides but does not retain the acetyl peptides or the pyroglutamyl peptides, if present. The digest is placed on a Dowex 50 column (X2-X16, H^+ form) and the column is washed thoroughly with water until the effluent is no longer acidic. The acidic peptides (acyl peptides, pyroglutamyl peptides, and peptides containing cysteic acid or phosphoserine residue) are contained in the water eluate together with large peptides not adsorbed on the resin column due to the molecular sieve effect of the resin [894, 898]. The water eluate is then applied to a column of a strongly basic anion exchange resin such as Dowex 1 (chloride or acetate form) having an identical or lower degree of cross-linking compared to that of the cation exchange resin previously used. The large peptides again are not adsorbed because of the molecular sieve effect. These are removed by washing with water. The adsorbed acidic peptides are then developed chromatographically and eluted with diluted hydrochloric, acetic or formic acid [894, 898, 1283]. A typical example is shown in Fig. 3-19. However, acetyl peptides in the effluent cannot be detected by the usual ninhydrin reaction but can be performed only after alkaline hydrolysis [549] of each fraction from the column [898, 1283]. The Folin reagent [794] containing cupric ion is also used for detection of acetyl peptides [894, 896]. The acetyl peptides in the water eluate can be fractionated directly by paper chromatography or electrophoresis if large peptides are not contained in the digest [896].

When glutaminyl peptides are present in the digest, they are quite labile and easily converted into pyroglutamyl peptides. This reaction is accelerated by contact with the hydrogen form of a sulfonic acid exchanger and also by heating at neutral or slightly acidic conditions [895, 912]. It is quite difficult to distinguish acetyl peptides from pyroglutamyl peptides without structural analysis since both groups of peptides are negative to the ninhydrin reaction and to other reactions based on amino groups. Acetyl and pyroglutamyl peptides can be revealed on paper chromatogram by chlorine-iodide-starch reaction of RYDON and SMITH [1051] or by a modification in which chlorine gas is replaced with *tert*-butyl hypochlorite in cyclohexane (1%) [306].

If the acetyl peptide is small enough for identification of the acetyl group, several simple methods are available. NARITA identified the acetyl hydrazide, amino acid hydrazide and C-terminal amino acid simultaneously by paper chromatography of the hydrazinolysate of acetyl dipeptides isolated from proteolytic digests of the proteins of tobacco mosaic [894, 895] and cucumber [896] viruses and ovalbumin [897,

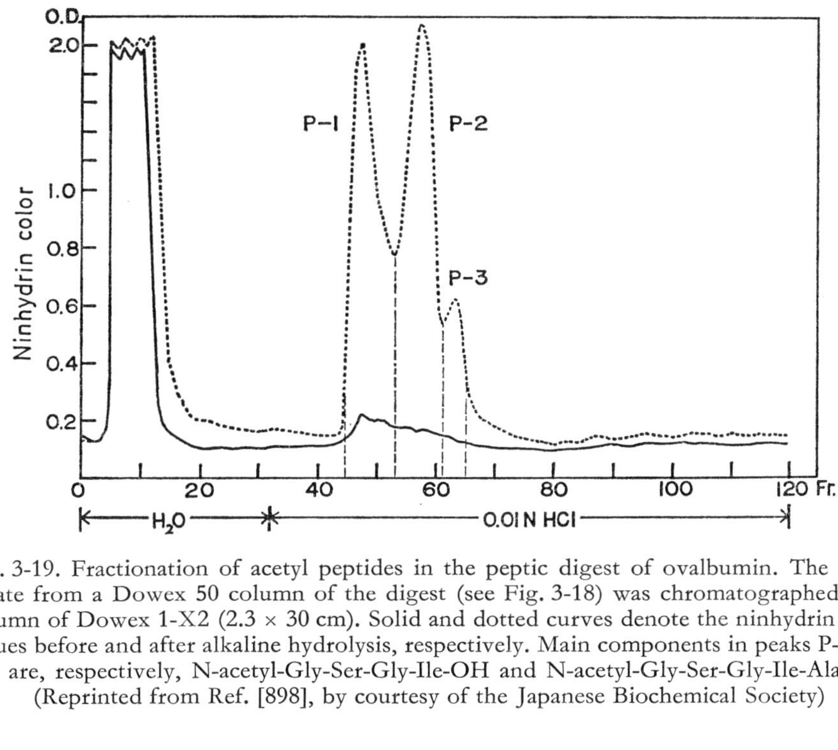

Fig. 3-19. Fractionation of acetyl peptides in the peptic digest of ovalbumin. The water eluate from a Dowex 50 column of the digest (see Fig. 3-18) was chromatographed on a column of Dowex 1-X2 (2.3 × 30 cm). Solid and dotted curves denote the ninhydrin color values before and after alkaline hydrolysis, respectively. Main components in peaks P-1 and P-2 are, respectively, N-acetyl-Gly-Ser-Gly-Ile-OH and N-acetyl-Gly-Ser-Gly-Ile-Ala-OH. (Reprinted from Ref. [898], by courtesy of the Japanese Biochemical Society)

898], respectively. Acetyl and amino acid hydrazides react positive by to the alkaline-silver nitrate reagent used for reducing sugars, and amino acids and their hydrazides are reactive to the ninhydrin reagent. *Rf* values of some of acyl and amino acid hydrazides are listed in Table 3-21. It should be noted that pyroglutamyl peptides yield a spot on paper chromatograms which cannot be distinguished from the spot of acetyl hydrazide even when the peptide does not contain any acetyl substituent. Therefore, when a ninhydrin negative peptide containing glutamic acid residues is isolated, the possibility of pyroglutamyl peptides must be carefully considered. The procedure for this will be described later. Acetyl hydrazide is also identified chromatographically as its DNP-derivative (dinitrophenylation at pH 3.0 and 38° C for 4 hrs) [975], extractable with ether from aqueous solution at pH 3. PHILLIPS [976] described the formation of propionyl hydrazide from cystine residues on heating with anhydrous hydrazine. An even more sensitive method is dansylation. SCHMER and KREIL [1088] dansylated the hydrazinolysate of a peptide at pH 3.0 for 24 hrs, conditions in which the reaction of amino groups with dansyl chloride was negligible. Dansyl derivatives of acyl hydrazides can be characterized by thin layer chromatography.

When acetyl peptides are subjected to methanolysis by heating (100° C, 5 hrs) with methanol (0.5 ml) and concentrated sulfuric acid (25 μl), methyl acetate is formed. Distilled methyl acetate (condensed in a dry ice-acetone bath) can be identified and estimated by gas chromatography or by chromatography of the acetohydroxamate formed with a neutralized concentrated hydroxylamine solution [796]. Fraction of volatile methyl esters from the methanolysis products of whole proteins, the N-terminal residues of which are known to be unblocked, were heated with hydrazine, and the products (fatty acid hydrazides) were examined by paper chromatography.

Table 3-21. *Rf* values of some acyl and amino acid hydrazides

	Pyridine-aniline-water[a] (9:1:4, v/v)	Collidine-water[b] (10:2, v/v)	Collidine-picoline-water[c] (5:3:2, v/v)
Formyl hydrazide	0.56	0.28	
Acetyl hydrazide	0.66	0.35	0.45
Propionyl hydrazide	0.75	0.50	
Glycine hydrazide	0.37	0.06	0.11
Serine hydrazide	0.51	0.07	0.16
Alanine hydrazide	0.46		
Valine hydrazide	0.65		0.34
Aspartic α-hydrazide	0.30		0.02
Glutamic α-hydrazide	0.34		0.02

[a] From Ref. [894—898, 1283].
[b] From Ref. [894].
[c] From Ref. [897, 898, 1283].

The paper chromatogram revealed several spots positive to the alkaline-silver nitrate reaction and *Rf* values of some of them corresponded to the formyl, acetyl and propionyl hydrazides. It is therefore recommended that the acidic peptide fraction rather than the whole protein, be used for the analysis of acyl groups.

Alternatively, acetate can be detected after partial hydrolysis of acetyl proteins or, preferably, acetyl peptides, in 6 N sulfuric acid at 100° C for 5 h. It is difficult to distill acetic acid quantitatively from the hydrolysate and PHILLIPS' method [975] is preferable. He extracted acetic acid with ether and subjected the ether extracts to paper chromatography by the method of BROWN and HALL [178]. WARD et al. [1333] extracted acetic acid with *tert*-butyl ethyl ether after addition of solid Na_2SO_4 (anhydrous) and estimated the extracted acid by gas chromatography.

The structures of acylated peptides may be determined from their C-terminal ends by the use of carboxypeptidase A. However if the acyl peptides contain residues (proline, aspartic acid, glycine, etc.) resistant to carboxypeptidase, the peptides must be further degraded by enzymes or by chemical methods, and structures of the resulting fragments elucidated separately. The acetyl-amino acid linkage, fortunately, is rather resistant toward concentrated hydrochloric acid at room temperature for 24 to 48 hrs [732, 1283]. Thus most acetyl amino acids can be released by this treatment from the parental acetyl peptides.

a) Fractionation of Acetyl Peptides

About 10 to 20 μmoles of a protein is digested with a proteolytic enzyme (pronase, chymotrypsin, pepsin, subtilisin, etc.) for 24 hrs and the digest is placed on the top of a Dowex 50-X2 (H^+ form) column, 1 × 15 cm (can handle about 1 g of protein digest). The column is then thoroughly washed with water until the effluent becomes neutral (about 200 ml). Acidic peptides, in general, are eluted at the break-through point. The water eluate is lyophilized and the lyophilizate is redissolved in 1 to 2 ml of water. The solution is chromatographed on a column of Dowex 1-X2 (chloride form, 200 to 400 mesh, 1 × 30 cm). The column is first washed with water and about 100 ml of the effluent is collected. This fraction is ninhydrin positive. Large peptides which have escaped adsorption on the Dowex 50 column are contained in this fraction. Acidic peptides on the column are developed with a gradient of HCl reaching the high concentration of 0.03 N and 2 to 3 ml fractions are collected with a fraction collector. Most acidic peptides are eluted within 100 fractions. Each fraction is then analyzed with the ninhydrin reagent before and after alkaline hydrolysis according to the method of Hirs et al. [549]. Fractions which are ninhydrin negative before alkaline hydrolysis but positive after hydrolysis are pooled and lyophilized for structural analysis.

b) Identification of the Acetyl Group of the Acetyl Peptide

About 1 μmole of the purified acetyl peptide is placed in a small test tube and 50 μl of anhydrous hydrazine is added. The tube is then sealed and heated at 100° C for 10 to 15 hrs. The hydrazinolysate is frozen and dried in an evacuated desiccator. The hydrazinolysate is examined by paper chromatography with the following solvent system: pyridine-aniline-water (9:1:4, v/v), collidine-water (10:2, v/v) or collidine-α-picoline-water (5:3:2, v/v). Acyl hydrazides and amino acid hydrazides are characterized by the alkaline silver nitrate reagent, and the C-terminal amino acid and amino acid hydrazides can be stained with ninhydrin reagent. The silver nitrate reagent is prepared by adding 0.1 ml of saturated aqueous $AgNO_3$ solution to 50 ml acetone. The paper is first dipped through the reagent and the acetone is blown off. The dried paper is then dipped through 0.5% KOH in ethanol and the solvent is again blown off. Hydrazides begin to appear at once. The background of the chromatogram prepared by the silver nitrate reagent gradually becomes brown with exposure to light. To prevent formation of Ag_2O from excess $AgNO_3$ by light, the paper chromatogram is quickly washed successively with 5% sodium thiosulfate and water and finally dried. The chromatogram can be stored over a year. The ninhydrin negative and alkaline silver nitrate positive spot is acyl hydrazide and the reverse one is the C-terminal amino acid of the acyl peptide. Positive spots for both reactions are derived from the amino acid residues between the termini.

Standard acyl hydrazides and amino acid hydrazides can be prepared from their esters by heating with hydrazine in ethanol at 60 to 100° C for 1 hr. The dried residue of the mixture can be used as a standard.

For the conversion of acyl hydrazides into their DNP- or dansyl derivatives [976, 1088], the dried hydrazinolysate is dissolved in 1 ml of 0.2 M citrate (pH is adjusted to 3 with 1 N HCl), and three volumes of ethanol and 0.5 ml of 5% FDNB in ethanol are added or 5 to 10-fold excess of dansyl chloride (w/w) in 0.3 ml ethanol is added. The reaction is allowed to proceed for 4 hrs or 24 hrs at 38° C. After evaporation of the mixture and addition of 2 ml of water, DNP-derivatives are extracted 5 to 6 times with 2 to 3 ml of ether. The ether extracts are dried and dinitrophenol is removed by vacuum sublimation as described in the DNP method. The DNP-acyl hydrazides (brown color by exposing to ammonia vapor) are characterized by paper chromatography or thin layer chromatography (silica gel G, hexane-ethyl acetate, 1:1, v/v) and estimated colorimetrically at 420 nm in comparison with known amounts of the authentic acyl hydrazides which are treated under the same conditions as the hydrazinolysate of the acetyl peptide.

For the dansylated products, the mixture is taken to dryness, redissolved in 0.3 ml of water and extracted three times with an equal volume of peroxide free ether or chloroform. Dansyl derivatives of acyl hydrazides are extractable into the organic layer, while the bulk of the dansylic acid and other by-products remain in the aqueous phase. A part of the organic layer is subjected to two-dimensional thin layer chromatography on silica gel G plates with chloroform-*n*-butanol-acetic acid (6:3:1, v/v) and then with isopropanol-methyl acetate-conc. NH_4OH (6:3:1, v/v).

2. Formylated Terminus

The N-formyl group is labile in comparison with N-acetyl and can be removed by incubating with 1 N hydrochloric acid in methanol for 1.5 hrs at room temperature without serious cleavage of peptide bonds in proteins [1157]. Isolation of formyl peptides and characterization of the formyl group are carried out by the same techniques as described for the terminal acetylated proteins.

Recently Gussin et al. [461] isolated deformylase from *E. coli* and they reported that this enzyme attacks N-formyl peptides and not formyl amino acids. When terminal formyl proteins or peptides are found, this enzyme is probably useful for the removal of the terminal formyl group in order to elucidate amino acid sequences of such proteins and peptides. Further studies demonstrated that this deformylase also hydrolyzes N-acetylated peptides at an appreciably slower rate [6].

3. Pyroglutamyl Terminus

Pyroglutamyl peptides are separated by the same techniques as those for the acyl peptides. However, it is rather difficult to conclude with confidence, that the isolated pyroglutamyl peptides were derived from the N-terminus of the parent protein. It is known that glutaminyl peptides derived from the internal part of the molecule by proteolytic enzymes can be transformed easily, sometimes quantitatively, into pyroglutamyl peptides during the digestion and fractionation. The complete amino acid sequence can resolve this dilemma. When negative results for terminal analysis and for the identification of acyl group are obtained, and when a pyroglutamyl peptide is isolated in quantitative yield, an N-terminal pyroglutamyl residue is then indirectly assumed for the parent protein in question. Ikenaka et al. [598] carefully investigated the terminal pyroglutamyl residue of α_1-acid glycoprotein and reported that isolation of acidic peptides with a Dowex 50 column, H^+ form, at lower temperatures such as 4° C minimized the transformation of N-terminal glutaminyl residue into pyroglutamyl groups. Furthermore, they showed that alkaline incubation of Pyrglu-Pro-Ile-OH, which is the N-terminus of the glycoprotein, at 24° C for 40 and 72 hrs achieved ring opening in 36 and 50%, respectively, without appreciable cleavage of the peptide bonds. However, when the same conditions were used for native α_1-acid glycoprotein, cleavage of several peptide bonds was observed.

Doolittle and Armentrout [289] reported the isolation of pyrrolidone carboxylyl peptidase from *Pseudomonas fluorescens* which hydrolyzes specifically the N-terminal pyroglutamyl linkage with the exception of pyroglutamyl-proline and -lysine [290, 291]. Incubation of N-terminal pyroglutamyl proteins with this enzyme, therefore, opens blocked proteins to usual terminal analytical methods.

B. C-Amidated Terminal Group

The only known substituent at the C-terminus of naturally-occurring polypeptides is the amide group. Glycinamide was first demonstrated as the C-terminus of oxytocin [305] and vasopressin [988] by Du Vigneaud and associates. Several amidated C-terminal groups in polypeptides are now known; glutamine amide for melittin and apamin, valine amide for α-melanocyte stimulating hormone and secretin, phenylalanine amide for gastrin, pancreozymin and caerulein, and methionine amide for eledoisin, physalaemin, phyllomedusin, bombesin and substance P [268]. The C-terminal tryptophan of Gramicidin A is known to be amidated with ethanolamine. This constitutes an exception to the general rule. The structures of these polypeptides have been completely or nearly completely elucidated and the nature of these amidated groups is established. It is difficult, however, to characterize the amidated C-terminal residues of large molecular weight proteins. Although this problem has not been extensively explored, one possible method for the characterization of such residues is the isolation of basic peptides (blocked C-terminus) from proteolytic digests of proteins in a manner similar to that described for N-terminal peptides.

The enzyme digest is passed through a basic anion exchange resin (free form) column, e.g., Dowex 1-X2, and the column is thoroughly washed with water. Peptides possessing free carboxyl groups are adsorbed on the column, while the basic peptide amide generally free of acidic residues is eluted from the column with water. The water eluate should contain the peptide amide together with arginine and arginine peptides, the basicities of which are similar to that of free arginine. The basic peptide fraction is then fractionated by suitable chromatographic techniques. A basic peptide, negative to the Sakaguchi reaction, locates the peptide amide. The amino acid sequence of the amidated peptide can be determined by the usual terminal analysis methods without appreciable difficulties.

Chapter 4

Improved Tritium-Labeling for Quantitative C-Terminal Analysis

Hisayuki Matsuo and Kozo Narita

I. Improved Reaction Conditions for Tritium-Labeling

The carboxyl-terminal amino acid residues of proteins are selectively tritiated through racemization *via* the oxazolone intermediate by the action of acetic anhydride in a medium containing ^3H_2O and pyridine. The characterization of the tritium-labeled C-terminal amino acid, after acid hydrolysis of the tritiated protein, can afford a method for C-terminal analysis [824, 825]. This method has successfully been applied to the qualitative C-terminal determination of proteins [828, 901]. Accumulating data reveals that there still remain several problems to be improved involving reaction conditions or characterization of the tritiated amino acids. The mechanistic features of this reaction are examined below under the improved reaction conditions together with the scope and limitations of this method. Other improvements in the characterization procedure of the radioactive C-terminal amino acid have recently been proposed.

As shown in Scheme 1, the tritium-labeling reaction is based on the following sequence of three different reactions [825]:

1. C-Terminal oxazolone formation by the action of acetic anhydride.
2. Tritium-labeling through base-catalyzed racemization at the C_4-position of the oxazolone in the presence of pyridine and ^3H_2O.
3. Ring-opening of oxazolone to regenerate the C-terminal amino acid which incorporates tritium in its α-position.

Non-polar solvents are suitable for the reaction of step 1 [824] but most proteins are insoluble in such media. An aqueous medium containing ^3H_2O, pyridine and acetic anhydride must be used for the tritiation of proteins in order to carry out the reaction in a homogeneous state. In this system, all three reactions mentioned above take place in one step [825]. The base-catalyzed hydrolysis of acetic anhydride also proceeds fairly fast under these conditions and thus acetic anhydride is consumed prior to completion of the oxazolone formation, which results in a decrease of the reaction yield. Use of a great excess of acetic anhydride for the purpose of raising the yield of the oxazolone does not provide better results because of serious disadvantages such as precipitation of proteins and unfavorable side-reactions.

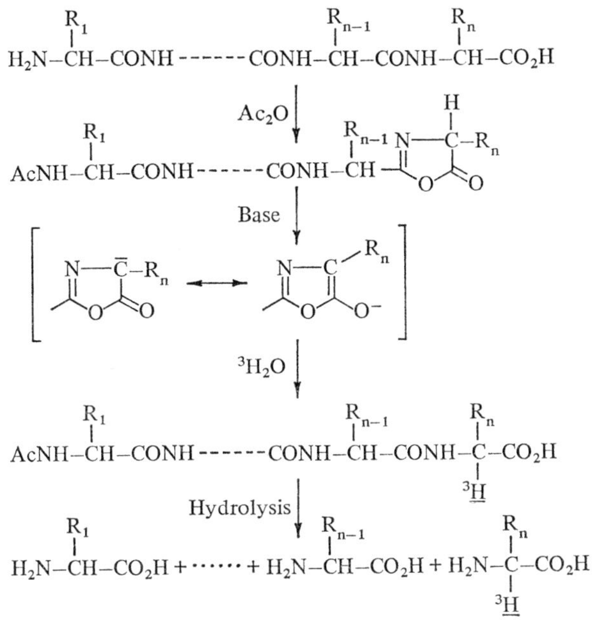

Scheme 1

Matsuo and Narita [830] have reinvestigated the conditions for C-terminal tritiation and have suggested improvements making the reaction suitable for quantitative analysis of proteins. With egg white lysozyme as a model protein, they examined various reaction conditions by varying the ratio of the reagents. When the ratio of pyridine:^3H_2O:acetic anhydride of 3:1:3 (by volume) was used, the extent of tritium-incorporation into the C-terminus of lysozyme increased by about 20-fold compared to that under the previous conditions [825] using a ratio of pyridine:^3H_2O: acetic anhydride of 4:2:1. In addition no side reaction was observed. Under these improved conditions, the pH of the reaction mixture can be maintained at 4.6. Cautious addition of acetic anhydride in two portions to the protein solution at 0° in the manner described below gave the best results with satisfactory reproducibility.

Improved Procedure [830, 831]: Tritiated water of high specific radioactivity (1 Ci/ml) is diluted 10-fold (100 mCi/ml) with distilled water and then stored in a freezer. Pyridine and acetic anhydride are stored in a cold room after redistillation of analytical grade reagents. A sample protein or peptide (0.05 to 0.1 μmole) is placed in a small test tube and ^3H_2O (5 μl; 500 μCi) and pyridine (10 μl) are added. It is important to make solution clear, otherwise the tritium-labeling does not proceed smoothly. If more volume is necessary for clear solution, amounts of reagents to be added must be adjusted so as to keep the same relative ratio as described here.

To the above protein solution, acetic anhydride (10 μl) is added at 0°. The mixture in a parafilm-sealed tube is kept at 0° for 5 min and then at 20° for further 15 min. After another addition of pyridine (20 μl) and acetic anhydride (20 μl) at 0°, the reaction mixture is kept at 0° for 5 min and then at 20° for an additional 1 hr. After addition of tritiated water (5 μl), it is allowed to stand at 20° for another hour to decompose the excess acetic anhydride. The resulting solution is then evaporated to dryness and removal of exchangeable tritium is performed by addition of 100 μl of 10% acetic acid, followed by evaporation repeatedly five times. Tritiated protein thus obtained in the test tube is directly hydrolyzed with 6 N HCl as usual.

Although several methods such as paper chromatography, thin layer chromatography [573], gas-liquid chromatography [828] or ion-exchange chromatography [205] can be used for characterization of the tritiated amino acid, it is convenient to apply about 1/5 of the hydrolysate to a paper strip (2 × 60 cm: Whatman No. 1 or Toyo Roshi No. 51) which is then subjected to electrophoresis (3000 V, 1 hr) at pH 6.4, by which the hydrolysate is separated into four portions such as aspartic acid, glutamic acid, neutral amino acids and basic amino acids. Neutral amino acids are further separated on a paper-strip by electrophoresis (3000 V, 1 hr) at pH 1.8, while basic amino acids by paper-chromatography (pyridine:acetone:28% NH_4OH: water = 50:20:5:50, by volume).

The chromatogram or electrophoretogram, without staining by ninhydrin, is cut into strips 1 cm wide, which are directly submitted to the radioactivity measurement in 10 ml toluene scintillation mixture containing 0.5% PPO (2,5-diphenyloxazole) and 0.01% POPOP (2,2'-*p*-phenylene-bis-(5-phenyloxazole)) by the aid of a liquid scintillation spectrometer. Of course, radioactivity in the same size of the papergram, where no tritiated amino acid located, is also measured as a blank and this value is subtracted from the counts for the sample. Approximate identification of the tritiated amino acids can be performed in comparison with the stained paper-chromatogram or electrophoretogram, on which standard amino acids are developed at the same time with the tritiated protein hydrolysate. Exact characterization of the tritiated amino acid can be made by rechromatography of the extracted tritiated amino acid from the cut filter paper where radioactivity occurs.

The relationship between side chain groups of the C-terminal residues and the extent of tritium-incorporation will be discussed in the following section. Under the above conditions, however, the radioactivity observed in the C-terminal amino acids in proteins ranges from 300 to 1000 cpm/0.01 μmole, and the specific radioactivity of the C-terminus derived by the tritiation reaction depends linearly on the specific radioactivity of tritiated water used. The highest specific radioactivity expected for the C-terminus may be 300 to 1000 cpm/0.2 nmole, when tritiated water of 5 Ci/ml, which is the commercially available highest specific radioactivity, is used. Accordingly, micro-analysis in 0.2-nmole scale is possible under the present improved conditions. In fact, Matsuo et al. [829] in their structural work of porcine luteinizing hormone-releasing hormone (LH-RH) successfully analyzed the C-termini of the peptides in the thermolysin and the chymotrypsin digests in less than 1 nmole scale by the improved tritiation method.

II. Structural Effect on Tritium Incorporation

After setting up the improved conditions as described, Matsuo and Narita [830] investigated the relationship between C-terminal amino acids having various side chain groups and the amounts of ^3H-incorporation into them, for the purpose of developing this qualitative technique to quantitative C-terminal analysis.

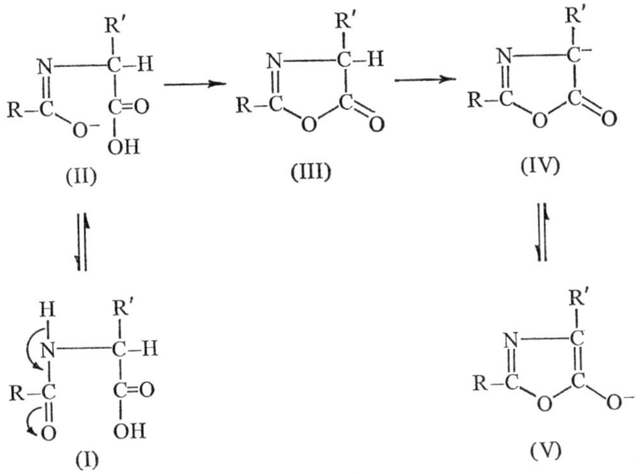

Scheme 2

As shown in Scheme 2, in the racemization reaction of N-acyl-amino acids (I) through the oxazolone intermediate (III), the first step is the base-catalyzed abstraction of the proton from amide nitrogen in (I) yielding an amide anion (II), where the equilibrium between (I) and (II) mainly depends on the acyl species (RCO-), as demonstrated by KEMPF [679]. In racemization through the oxazolone intermediate [(III) ⇌ (IV)], the effect of the substituent (R') at the α-carbon atom of the C-terminus may be predominant, since abstraction of the hydrogen atom attached to the α-carbon must have taken place [425]. Accordingly, the overall rate of racemization must be controlled by the substituents, R and R', in other words, by the structural effects of the penultimate and the C-terminal amino acids.

In order to analyze the structural effects on the tritium-labeling under the improved conditions, the amounts of tritium-incorporation into the C-terminal residues in various N-acetyl-amino acids and dipeptides were compared. The comparison was performed by the use of the following internal standard method [830]. For instance, a mixture of exactly weighed acetylleucine (a μmole) and acetylalanine as an internal standard (b μmole) was submitted to the tritium labeling reaction in a vessel. After acid hydrolysis, followed by separation of resulting tritiated amino acids by paper electrophoresis, the radioactivities of leucine (x cpm) and alanine (y cpm) were measured, respectively.

As shown in Fig. 4-1, observed relative tritiation ratio ($R = x/y$) was found to be linearly proportional to the molar ratio in a mixture (a/b), that is, relative ratio of specific radioactivity (f) is always constant for any molar ratio of a mixture. Thus, the relative ratio (f) of specific radioactivity of leucine compared with that of the internal standard alanine can be precisely determined by the following equation.

$$f = \frac{x/a}{y/b}\,.$$

As seen in the f-values listed in Table 4-1, the extent of C-terminal tritiation depends significantly on the side chain groups of the C-terminal amino acids. f-Values thus obtained from the tritiation of acetylamino acids are convenient indices for comparing the structural effect of the C-terminal amino acids on the tritiation rate, ignoring the effect of the penultimate acyl-group, all acetyl amino acids having a common penultimate group. Thus, we designate *f*-values as C-terminal factor.

Similar investigation was carried out by tritiation of various sets of two different dipeptides and the relative ratio of C-terminal tritiation (R^*) was determined in the same way as in the cases for *f*-values. R^*-values in typical runs are listed in Table 4-2.

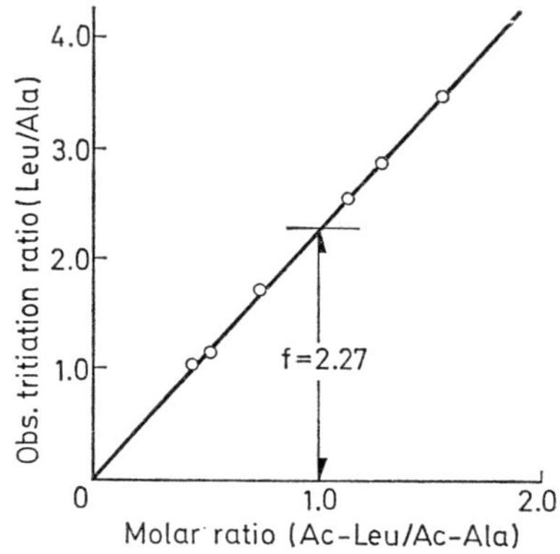

Fig. 4-1. The plot of observed relative ratios of tritium-incorporation into leucine and alanine *vs.* molar ratios of mixtures of acetylleucine and acetylalanine, which were tritiated under the improved reaction conditions

Table 4-1. *f*-Values for various C-terminal amino acids under the improved tritiation conditions[a]

Leucine	2.27	Asparagine	2.30
Isoleucine	1.18	Aspartic acid	0.32
Phenylalanine	2.01	Glutamine	1.16
Tyrosine	1.95	Glutamic acid	0.83
Methionine	2.32	Histidine	0.61
Valine	0.93	Lysine	1.52
Alanine	*1.00*	Arginine	1.96
Glycine	0.68	Threonine	0.09

[a] Relative ratios of tritium-incorporation into various acetylamino acids are listed with that of acetylalanine selected as 1.00.

As seen in sets (1) and (2) in Table 4-2, where both sample and standard dipeptides have a common penultimate glycine or leucine residue, R^*-values agree quite well with *f*-values for Ac-Phe/Ac-Ala, namely R^*/f agrees quite closely. Even in the case of set (3), where the sample peptide, Ala-Phe, has different amino acid residue at penultimate position from that of the standard dipeptide, Leu-Ala, the R^*/f value (0.95) is in close agreement, even though a slight deviation is observed as compared with sets (1) and (2).

These results revealed that the extent of tritium-incorporation is predominantly affected by the C-terminal amino acids. Accordingly, this fact suggests that quantitative analysis of the C-terminus is possible by the use of internal standard, when *f*-values for the C-terminal amino acids to be analyzed are available, because R^*/f value must express the molarity of the C-terminus.

It is striking, however, that R^*/f observed for set (4) shows more variation. As clearly shown in comparison of the result for set (2) with that for set (4), tritiation reaction of Gly-Phe proceeds more slowly than Leu-Phe, that is, penultimate glycine has a special slow-down effect on the C-terminal tritiation of the dipeptide. A similar effect of penultimate glycine is also observed in the tritiation of several glycyl-amino acids. This fact may be rationalized by considering a rate-acceleration effect on the amide proton abstraction caused by the steric hindrance due to the side chain of

Table 4-2. Quantitative C-terminal analyses of several dipeptides by internal standard method

Sample peptide	Internal standard	R^{*a}	Molarity of C-termini $(R^*/f)^b$
(1) Gly-Phe	Gly-Ala	2.06	1.03
(2) Leu-Phe	Leu-Ala	2.02	1.01
(3) Ala-Phe	Leu-Ala	1.90	0.95
(4) Gly-Phe	Leu-Ala	1.60	0.79

[a] R^* denotes relative ratios of specific radioactivities of C-terminal Phe of sample peptides with those of C-terminal Ala of internal standards.
[b] f (Phe/Ala): 2.01, which is obtained from Table 4-1.

penultimate residue. As discussed at the beginning of this section, the oxazolone formation reaction is initiated by the base-catalyzed proton abstraction from the amide nitrogen to form the amide anion. We number the atoms involved in penultimate amino acid amide grouping consecutively, starting with the amide hydrogen to be abstracted as 1, as shown in Scheme 3. All penultimate amino acids with the exception of glycine possess atoms at the six position. Since the atoms at the six position approach most closely to the amide hydrogen and are more effective in providing the steric hindrance [920], the abstraction of hydrogen from amide nitrogen may be significantly accelerated by the steric hindrance in all cases except glycyl-amino acids which lack such a hindrance.

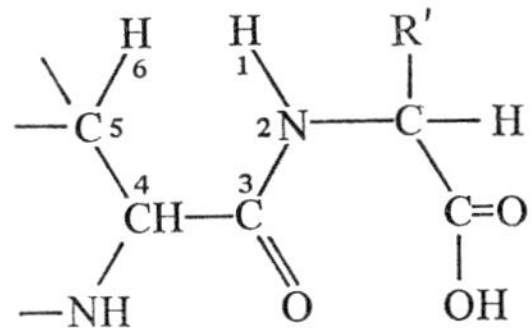

Scheme 3

III. Approach to Quantitative Analysis of the C-Terminal Residues (Internal Standard Method)

Preliminary results mentioned in the preceding section suggest that the internal standard method can be applied to the quantitative C-terminal analysis in proteins, with the exception of proteins having the penultimate glycine, when we choose an appropriate internal standard. The present authors [830] investigated the above possibility using a mixture of egg white lysozyme and bovine insulin as in the case of dipeptides. As summarized in Table 4-3, relative ratio of specific radioactivities of the C-terminal leucine of lysozyme *versus* the C-terminal alanine of insulin (R^*) falls in the range of 2.22 to 2.30 at any molar ratio of the two proteins. R^*-Value thus obtained is very close to the f-value of 2.27 for leucine *vs.* alanine, which is determined by the tritiation of a mixture of acetylleucine and acetylalanine under the identical conditions.

Table 4-3. Quantitative C-terminal analysis of egg white lysozyme with bovine insulin as an internal standard

	Lysozyme (μmole)	Insulin (μmole)	R^{*a}	Molarity of C-terminal leucine $(R^*/f)^b$
Run 1	0.0154	0.0447	2.23	0.98
Run 2	0.0338	0.0407	2.22	0.97
Run 3	0.0342	0.0683	2.30	1.01
Run 4	0.0451	0.0225	2.25	0.99
Run 5	0.0280	0.0514	2.23	0.95

[a] R^* denotes the relative ratio of specific radioactivities of C-terminal Leu of lysozyme with that of C-terminal Ala of internal standard, insulin.
[b] f (Leu/Ala): 2.27, which is obtained from Table 4-1.

Table 4-4. Quantitative C-terminal analyses of several proteins by internal standard method

Sample protein	Standard	R^*	f^a	Molarity of C-terminus (R^*/f)
Hen egg-white lysozyme (-Arg-Leu)	Bovine insulin	2.25	2.27 (Leu/Ala)	0.97 (Leu)
Bovine pancreatic ribonuclease (-Ser-Val)	Bovine insulin	0.94	0.93 (Val/Ala)	1.01 (Val)
Bovine chymotrypsinogen (-Ala-Asn)	Egg white lysozyme	0.89	0.78 (Asn/Leu)	1.13 (Asn)
Cardiotoxin (-Cys-Asn)	Egg white lysozyme	0.82	0.78 (Asn/Leu)	1.04 (Asn)
Human milk lysozyme (-Gly-Val)	Bovine insulin	0.39	0.93 (Val/Ala)	0.42 (Val)

[a] f-Values are obtained from Table 4-1.

Accordingly, the molarity of the C-terminal leucine for lysozyme is determined as $R^*/f = 0.92$ to 1.01. It is thus shown that C-terminal leucine of lysozyme can be determined quantitatively by the use of insulin as an internal standard.

Several model proteins are also submitted to the quantitative analysis with lysozyme or insulin as the internal standard and the results thus obtained are listed in Table 4-4.

In the cases of lysozyme, bovine pancreatic ribonuclease, bovine chymotrypsinogen and cardiotoxin from Formosan cobra venom [902], the analytical data fall in the range of 0.97 to 1.13, which shows satisfactory agreement in value. Only in the case of human milk lysozyme ending -Gly-Val as the C-terminal sequence, is the analytical value for the C-terminus (0.42) much lower than the theoretical value, as anticipated from the fact that glycine occupies the penultimate position.

These results show that the present procedure can be used for the quantitative determination of the C-terminal residues in proteins, even though further evaluation must await future study.

IV. Improvement in the Characterization Procedure of the Tritiated C-Terminus

For the characterization of the tritiated C-terminal amino acids, several methods can be used. As indicated in the preceding section, a combination of paper electrophoretic separation of the tritiated amino acids and counting of their radioactivities by liquid scintillation method gave satisfactory results. For unambiguous identification of the tritiated amino acids, however, it is desirable to carry out separation of the C-terminal amino acid from the hydrolysate and its identification in one step. Recently, Cappugi et al. [203] reported the use of amino acid analyzer for this purpose in the following manner: the hydrolysate of a tritiated protein is applied to a column of an amino acid analyzer, to which a fraction collector is connected by a Teflon capillary tube at the exit of the flow colorimeter, and 2-ml fractions of eluates are collected. The ninhydrin color of each fraction is measured at 570 nm or 440 nm. A 0.8 ml portion of each fraction is submitted to a liquid scintillation counter, after adding 0.5 ml of water and 10 ml of scintillation solution, whose composition is Triton-X-100: Liquifluor (1:2) diluted with toluene (1:10). Using this system, unambiguous identification of the C-terminal amino acid was accomplished with ribonuclease A in 0.035 μmole scale.

A combination of tritium labeling and carboxypeptidase digestion has been reported by Hsieh et al. [589] as another interesting improvement in characterization step of the C-terminal amino acid during their structural work on rat liver lactate dehydrogenase. Since attempts to determine the C-terminal amino acid of lactate dehydrogenase by carboxypeptidase method and by hydrazinolysis have led to inconclusive results, tritium labeling was carried out in a similar way to our original procedure with a modification that the reaction mixture was incubated at higher temperature such as 37°. Maximum radioactivity was found in phenylalanine, while glutamic and aspartic acids showed about 17% and 25% of the radioactivity of phenylalanine, respectively. This result demonstrates that phenylalanine is, possibly, the C-terminus, but it is not quite clear whether glutamic and aspartic acids are also the C-terminal residues. In order to eliminate such ambiguity, the tritiated protein was digested with carboxypeptidases A and B for a short period of time. The rates of release of the tritiated amino acids were measured by analyzing them on paper chromatography and it was concluded that phenylalanine was the C-terminal amino acid. Such a combination of two methods may be useful for the qualitative identification of the C-terminal residues. But it must be kept in mind that the C-terminal residue is racemized to some extent after the tritium-labeling reaction.

The "interior labeling" [589] i.e., abnormal tritium incorporation into non-C-terminal amino acids as observed above will be discussed in the following section.

V. Further Problems

It has been reported that radioactivity was observed in non-C-terminal glutamic and aspartic acids in some cases in the tritiation of proteins under the original conditions [825], and such unfavorable "interior labeling" may lead to inconclusive results, as in the case of rat liver lactate dehydrogenase [589] mentioned in the preceding section. Generally, the extent of "interior labeling" is not so high. According to the data by Cappugi et al. [205], radioactivities of non-C-terminal glutamic and aspartic acids after tritiation of ribonuclease A did not exceed 0.7% and 2.6% of that of C-terminal valine, respectively. The interior labeling observed in the case of lactate dehydrogenase seems to be the extreme case, which may be caused by the tritiation at higher reaction temperature. It is not quite clear yet why non-C-terminal aspartic and glutamic acids undergo such interior labeling under the original reaction conditions. But Holcomb et al. [573] have suggested a possible explanation that free carboxyl groups of non-C-terminal aspartic and glutamic acids form mixed anhydrides by the action of acetic anhydride and the resulting anhydrides undergo base-catalyzed tritiation at their α-position. Recently, another possible explanation has been proposed that non-C-terminal aspartic acid residues undergo the rearrangement to β-aspartyl linkage through a succinimide intermediate. Baba et al. [52] examined the interior labeling with synthetic peptides possessing α-aspartyl linkage and observed that only the aspartyl residue involved either in -Asp-Gly- or in -Asp-His- sequence underwent the interior labeling under the conditions similar to the original ones.

The present authors [831] investigated the interior labeling problem under the present improved tritiation conditions, but in all cases with egg white lysozyme, bovine pancreatic ribonuclease and eight kinds of peptides having -Asp-Gly-, -Asp-His-, -Asp-Phe-, -Asp-Val-, -Asp-Glu-, -Asp-Arg- and -Asp-Tyr-, no incorporation of tritium-radioactivity was observed in the aspartyl and glutamyl residues at all. Thus, as far as our experiment concerns, it revealed that the non-C-terminal acidic amino acid residues do not incorporate radioactivity under the present tritiation conditions, but the possibility of the interior labeling is not completely ruled out. Careful consideration must therefore be paid, when aspartic or glutamic acid was radioactive in the tritiation method.

Another problem accompanied with the tritiation procedure is observed in the case of C-terminal serine or threonine [831]. In the quantitative analysis of the peptides whose C-termini are serine, it is difficult to obtain reproducible results, though the qualitative analysis is possible. On the other hand, the C-terminal threonine residues either in peptides or in proteins incorporate radioactivity poorly, as predicted from low *f*-value for threonine, shown in Table 4-1. These problems may arise from the unfavorable dehydration in the side chain group of the two β-hydroxy-amino acids, which may occur by the action of acetic anhydride.

The C-terminal proline, which does not afford the corresponding oxazolone intermediate, cannot be determined even under the conditions recommended here. Although quantitative experiments with peptides or proteins ending with tryptophan and cystine or its derivatives have not yet been carried out, their C-terminal labelings with tritium are probably possible. It is also noteworthy that acid hydrolysis of tritiated protein or peptide must be carried out in the presence of thioglycolic acid [821] in the tritiation of C-terminal tryptophan [829].

There still remains a serious problem that the tritiation reaction is very sluggish when the proteins are insoluble in the reaction mixture. Addition of sodium dodecyl-sulfate or urea in the reaction medium to solubilize proteins is sometimes effective but the conditions generally applicable to insoluble proteins have not yet been established.

Chapter 5

Fragmentation of Proteins for Sequence Studies and Separation of Peptide Mixtures

Charles B. Kasper

I. Introduction

When a protein has been judged homogeneous and the end groups and amino acid composition are known, the next step in the elucidation of the covalent structure is the degradation of the macromolecule into low molecular weight peptides which can be completely characterized with respect to composition and sequence. The entire composition of the protein should be accounted for by the summation of unique peptides. Once this is achieved, the sequenced peptides must be arranged in the same order as they occur in the intact protein. No single technique is ideally suited to accomplish this task; therefore, a combination of enzymic, or chemical and enzymic techniques, is required. The methods discussed in this chapter deal primarily with the fragmentation of proteins and the fractionation of the resulting peptide mixtures. With the exception of partial acid hydrolysis, the degradative procedures have been restricted to those which exhibit a reasonable degree of specificity and which are most often used to obtain sets of peptides necessary for the elucidation of the primary structure of a protein.

II. Non-Enzymic Cleavage of Peptide Bonds

As the structure of an increasing number of large proteins is examined, the need for methods capable of cleaving the polypeptide chain at selectively specific sites becomes a necessity. A variety of non-enzymic methods which involve peptide bonds formed by specific amino acids have been reviewed by Witkop [1386]. N-bromosuccinimide has been shown to cleave tryptophanyl bonds in proteins [100, 955, 1010]; however, oxidation will frequently occur without peptide bond cleavage. Tyrosyl and histidyl bonds are also sensitive to this reagent, and the peptide chain is often broken at positions where these residues occur. A procedure for the selective cleavage of tyrosyl bonds in the presence of tryptophan [994] is based on the prior conversion of tryptophanyl residues to N-formylkynurenine by ozonolysis before treatment with N-bromosuccinimide. Undesirable side reactions, such as oxidation of sulfhydryl groups and disulfide bonds, are also noted with this reagent. A second chemical modification which results in the cleavage of cysteinyl bonds has been described

[1191]. This method requires S-dinitrophenylation, intermediate formation of a dehydroalanyl derivative, and mild hydrolysis with dilute HCl to cleave the peptide chain. Still another procedure has taken advantage of the ease of reduction of tertiary amides. This has resulted in the preferential cleavage of acyl-proline bonds in proteins and peptides by a variety of reducing agents [80, 81, 957, 1050]. The above methods suffer from one or more of the following disadvantages: (a) low cleavage yields, (b) lack of specificity, (c) undesirable side reactions, and (d) wide variability in the reactivity of the sensitive bond. Because of these factors and since these methods have not been developed to the stage where they may be routinely applied in a predictable fashion to the chemical fragmentation of proteins, they will not be considered at this time. The most specific and generally applicable method available is the cyanogen bromide cleavage of the peptide chain at methionyl residues [451].

A. Cyanogen Bromide

Cyanogen bromide cleavage has been employed with excellent results in studies related to the elucidation of the primary structure of numerous proteins e.g., ribonuclease [451], myoglobin [321], azurin [27], cytochrome *c* [233, 526], subtilisin BPN′ [813], extracellular nuclease of *Staphylococcus aureus* [1264], and trypsinogen [572]. This method owes its success to its specificity for methionyl residues and the fact that methionine is usually rare in proteins. This increases the probability of obtaining large peptide fragments which provide valuable information important either for the correct positioning of peptides derived from enzymic digests or for the confirmation of previously established sequences. Other applications of this reaction include the determination of the identity or non-identity of polypeptide chains [1216], degradation of γ-globulin to obtain immunologically active fragments [193], and the location of oxidized methionyl residues [725].

1. General Properties of the Reaction

Cyanogen bromide reacts with the sulfur in the thioether side chain of methionine to yield a mixture of homoserine and homoserine lactone plus methyl thiocyanate. If this reaction occurs when methionine is in peptide linkage, the bond involving the carboxyl group of methionine is cleaved [Eq. (1)]. All fragments produced by the action of cyanogen bromide will contain C-terminal homoserine or its lactone except for the C-terminal peptide of the protein (unless methionine is C-terminal). Under acidic conditions, cysteinyl side chains are also attacked but peptide bond cleavage does not result. This reaction can be prevented by carboxymethylation or aminoethylation of the thiol prior to incubation with cyanogen bromide. Repetitive sequences of methionine, such as the methionyl methionine sequence in ribonuclease [451], and N-terminal methionine yield free homoserine lactone. If the methionyl side chains in a protein or peptide become oxidized to the sulfoxide or sulfone, reaction with cyanogen bromide will not occur. The presence of these oxidation products can be detected in acid hydrolysates by amino acid analysis; however, the sulfoxides are capable of reverting to methionine during the hydrolysis stage [1015], and the result may be incorrectly interpreted as incomplete reaction at certain methionyl residues.

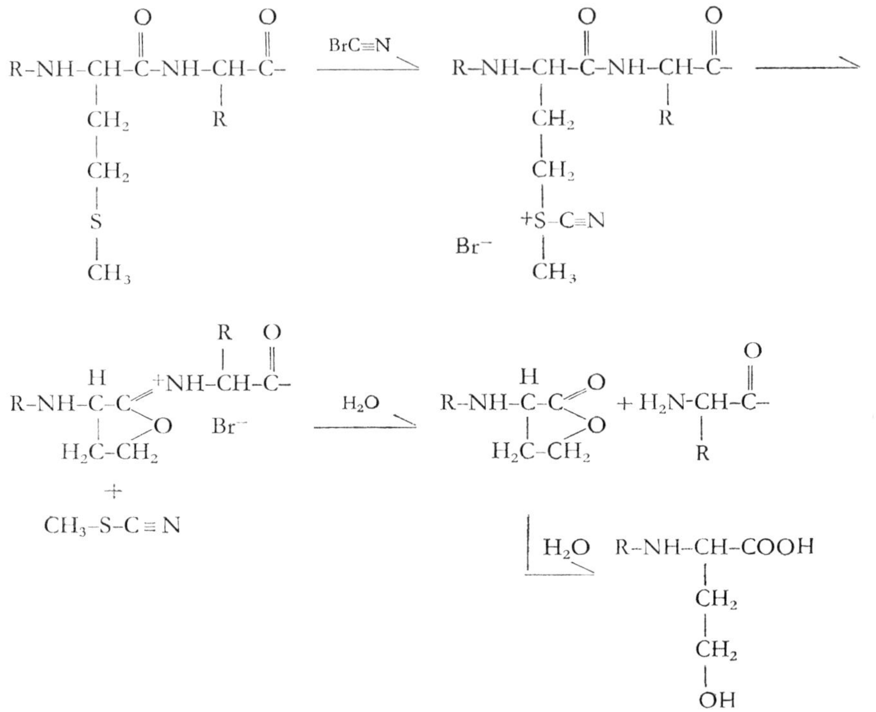

(1)

2. Method

The cleavage reaction has been successfully performed in either 0.1 N HCl, 70% formic acid, or anhydrous trifluoroacetic acid. A difference in the extent of cleavage has been noted when using the first two solvents; in the case of β-galactosidase, 80% of the susceptible bonds were cleaved in 0.1 N HCl while 96% cleavage was achieved in 70% formic acid [1216]. This concentration of formic acid has also been found to be optimal for the fragmentation of subtilisin BPN' (M.W. 27500; 5 methionyl residues) [813] and Staphylococcus nuclease (M.W. 17000; 4 methionyl residues) [1264]; greater than 90% cleavage was noted in each case. Formic acid has the decided advantage of being a better solvent than HCl for both proteins and peptides.

Cyanogen bromide[1] may be obtained from Eastman Organic Chemicals (Rochester, New York) as the analytical grade reagent and used directly. The general procedure involves dissolving or suspending the protein (approximately 10 mg/ml) in the appropriate acid solution and adding a weighed amount of the solid cyanogen bromide directly to the protein solution. Cyanogen bromide is usually present at a 30 to 100-fold molar excess with respect to methionine. A vial or tube which can be tightly stoppered makes a suitable reaction vessel. If the protein is not completely soluble, an even suspension should be maintained by using a magnetic stirrer or gentle shaking until solubility is achieved. After a reaction period of approximately

---

[1] Caution should be exercised when using this poisonous chemical. All operations involving handling of the solid should be carried out in a well ventilated hood.

24 hrs at room temperature, excess reagent and by-products are removed by lyophilization or gel filtration. To illustrate this technique, two examples are presented with an indication as to the methods used to fractionate the resulting fragments.

a) *β-Galactosidase* [1216]

The lyophilized, reduced carboxymethylated protein is dissolved in 70% formic acid, in which it is readily soluble at a concentration of 1 to 5 mg/ml. A 50-fold excess of crystalline cyanogen bromide is added and the reaction allowed to proceed for 16 to 20 hrs at room temperature in a tightly stoppered flask. After cyanogen bromide cleavage, the reaction mixture is diluted approximately 10-fold with distilled water, shell frozen, and lyophilized to dryness. The residue is dissolved in a minimum volume of 0.01 N NaOH and applied directly to a 4 × 225 cm jacketed column of Sephadex G-75 equilibrated with 0.2 M NH_4HCO_3, pH 8.6.

b) *f_2 Bacteriophage Coat Protein* [719]

Protein prepared by the hot KCl treatment is freeze-dried and dissolved in anhydrous trifluoroacetic acid. A 50-fold excess of cyanogen bromide is added and the reaction allowed to stand at 25° for 16 hrs. The trifluoroacetic acid is removed and 8 M urea containing 0.05 M Tris-HCl buffer, pH 8.0, is added to the residue. The suspension is centrifuged and the supernatant solution is chromatographed on DEAE-Sephadex A-25 equilibrated with the urea containing buffer. The recovery of peptides from 8 M urea is accomplished by dialysis against 4-hour acetylated 23/32 Visking membranes.

The fractionation of cyanogen bromide cleavage products by such procedures as paper electrophoresis and ion exchange chromatography is complicated by the fact that a peptide may exist in two distinct but interconvertible forms which differ only with respect to the presence or absence of a free terminal carboxyl group. This problem is circumvented by converting all C-terminal homoseryl residues to the lactone by treating the peptide residue obtained after lyophilization with anhydrous trifluoroacetic acid at 20° for 1 hr [25]. Homoserine lactone peptides can be purified by paper electrophoresis without detectable changes during the run. In peptides, the conversion of the lactone to homoserine occurs more slowly than in the case of free homoserine lactone. For example, incomplete conversion is noted even after 12 hrs at 37° in 2 M ammonium hydroxide [25].

In order to determine the extent of cleavage, an aliquot of the digest, after removal of excess reagent, is hydrolyzed with 6 N HCl at 110° for 24 hrs in an evacuated (< 50 μ Hg) sealed tube, and homoserine and homoserine lactone are quantitated. Homoserine lactone elutes from the 5-cm column (PA-35 Beckman resin) between the ammonia and arginine peaks while homoserine co-elutes with glutamic acid on the 60-cm column (AA-15 Beckman resin) with the pH 3.25 buffer. To separate homoserine from glutamic acid, the column is equilibrated with the usual pH 3.25 buffer, but development is started with pH 2.80 buffer (prepared by addition of acid to the pH 3.25 buffer) [1113]. Under these conditions homoserine elutes approximately 15 ml before glutamic acid. Homoserine lactone may be hydrolyzed to the free acid by dissolving the residue of amino acid hydrochlorides in a small amount of pyridine acetate buffer (100 ml pyridine + 4.0 ml glacial acetic acid/1000 ml, pH 6.5) and heating at 105° for 1 hr [25]. The volatile buffer is removed by rotary evaporation, and the sample is dissolved in the sample dilution buffer and immediately applied to the column of the amino acid analyzer. The recovery of homoserine is about 90% (uncorrected for hydrolytic losses).

B. Partial Acid Hydrolysis

The hydrolysis of proteins by acid under conditions which do not cleave all peptide bonds has proven to be a useful method in the study of protein structure. Basically two methods are employed: (a) partial hydrolysis with concentrated or constant boiling hydrochloric acid and (b) hydrolysis in dilute acid solution.

1. Specificity

The initial stages of partial hydrolysis with concentrated acid are characterized by a rapid cleavage of peptide bonds while the remainder of the breakdown becomes progressively slower as free amino acids and small peptides (di- and tripeptides) accumulate in the reaction mixture [187, 1221]. Thus, the complexity of the hydrolysate in the early phase is considerably greater than that at the end of the reaction; consequently, examination of the peptides near the terminal stage of hydrolysis provides the most useful information.

Various factors, such as electrostatic and steric effects, influence the rate at which a particular peptide bond is hydrolyzed. The presence of charged groups in the vicinity of the amide bond affects the approach of hydrogen ions and therefore the rate of hydrolysis. In strong acid, the carboxyl groups are unionized and the only charged species are the protonated forms of the α- and ε-amino, guanido, and imidazole groups. Positively charged groups would be expected to repel hydrogen ions and therefore decrease the rate of cleavage of peptide bonds in the immediate vicinity. Of the basic groups mentioned, only the α-amino group appears to retard hydrolysis, presumably owing to its proximity to the susceptible bond. Failure of the lysyl, arginyl, and histidyl side chains to behave in a similar fashion [430] is probably related to their increased distance from the peptide bond which would greatly diminish any electrostatic influence. The rates of hydrolysis of various dipeptides relative to glycylglycine are presented in Table 5-1. It is clear that as the bulk of the side chain increases the rate of hydrolysis decreases. The fact that glycylvaline is less stable than valylglycine suggests that the bond formed by the carboxyl group of the branched chain amino acids is most resistant to hydrolysis. It is evident from a comparison of the relative rates of hydrolysis of leucylglycine and valylglycine that, when branching of the side chain occurs close to the polypeptide backbone, the reduction in rate is pronounced. Peptide bonds formed between two β-branched amino acids (e.g., Val-Ile, Ile-Ile) are extremely difficult to hydrolyze and probably form the most acid stable type of peptide bond. It is not uncommon for such bonds to require up to 60 to 70 hrs in 6 N HCl at 110° C in order to obtain a quantitative cleavage. The generally accepted explanation for this phenomenon is that the approach of the hydronium ion to the peptide bond is hindered by steric factors.

Peptide bonds involving the amino groups of the aliphatic β-hydroxy-amino acids, serine and threonine, are quite susceptible to hydrolysis by concentrated acid at

Table 5-1. Hydrolysis of dipeptides in acid solution [1254]

Peptide	Relative rate of hydrolysis (Gly-Gly = 1)
Gly-Gly	1
Gly-Ala	0.62
Ala-Gly	0.62
Gly-Leu	0.40
Gly-Tyr	0.35
Gly-Val	0.31
Leu-Gly	0.23
Leu-Leu	0.048
Leu-Tyr	0.041
Val-Gly	0.015

The kinetic study was performed in a solution containing equal volumes of 10 N HCl and glacial acetic acid at 37°.

room temperature. This is presumably due to the following sequence of reactions [Eq. (2)]: (a) intermediate formation of an oxazoline derivative, (b) hydrolysis of the $-\underset{|}{C}=N-$ bond to yield the ester, and (c) cleavage of the ester bond to sever the peptide chain [279].

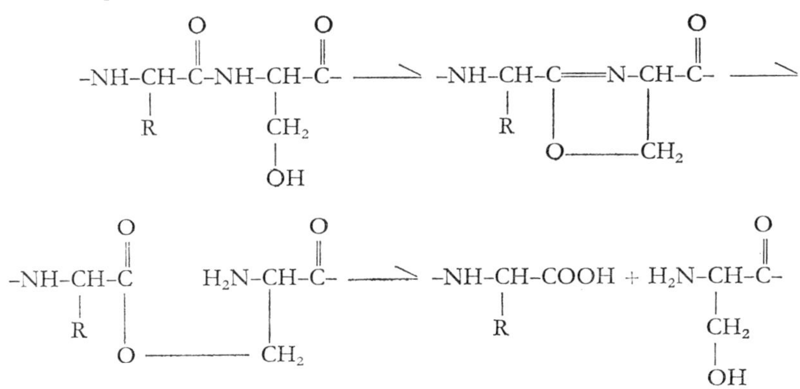

(2)

Although hydrolysis with concentrated acid has been more widely used, dilute acid hydrolysis exhibits a different type of specificity and is occasionally useful in obtaining additional information about the structure of a protein or peptide. PARTRIDGE and DAVIS [952] first demonstrated that when a protein was refluxed with dilute acetic or oxalic acid, aspartic acid was preferentially released. Furthermore, the specificity is a function of the pH rather than the nature of the acid employed [113]. This opened the possibility that a polypeptide chain could be cleaved specifically by the excition of aspartyl residues leaving other peptide bonds intact. Application of this method to proteins has resulted in the expected preferential release of aspartic acid [434, 837, 1115]; however, hydrolysis also occurred at other sensitive bonds involving predominantly glycine, serine, and threonine. In fact, certain bonds formed by aspartic acid are more resistant to hydrolysis than others; this suggests that the rate of release can be markedly affected by neighboring side chains [1115]. Dilute acid hydrolysis of small peptides containing aspartic acid appears to proceed in a more specific manner [592, 1114, 1296], since fewer secondary points of cleavage are noted. In both peptides and proteins, asparaginyl residues are converted to aspartyl residues with the subsequent release of aspartic acid.

2. Side Reactions

Certain acid-catalyzed reactions may adversely affect the results obtained from a partial acid hydrolysate of a protein or peptide. These include (a) disulfide interchange, (b) diketopiperazine formation, (c) cyclization of amino terminal glutamine to pyrrolidone carboxylic acid, (d) rearrangement at an aspartic acid residue, (e) destruction of tryptophan, and (f) hydrolysis of asparagine and glutamine to the corresponding carboxylic acids.

The occurrence of disulfide interchange [Eq. (3)] during partial acid hydrolysis has been examined in detail by Ryle and Sanger [1052].

$$R^1\text{–S–S–}R^1 + R^2\text{–S–S–}R^2 \rightleftharpoons 2\,R^1\text{–S–S–}R^2 \qquad (3)$$

Insulin was allowed to hydrolyze in the presence of *bis*-DNP-cystine under a variety of experimental conditions, and disulfide interchange was detected by the incorporation of DNP-cysteine into peptides with the general structure:

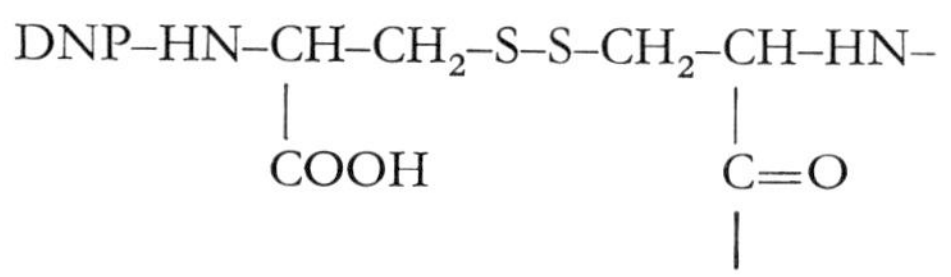

Table 5-2. Disulfide interchange during partial acid hydrolysis of insulin [1052]

Acid	Temperature	Additions	Time	Number of disulfide bonds reacting per molecule of insulin (MW = 6000)
12 N HCl[a]	37°	None	1 hr	0.3
			2 hrs	0.65
			5 hrs	1.2
5.7 N HCl in 50% acetic acid	37°	10^{-3}M cysteine	1 day	0.35
			4 days	1.0
			10 days	1.35
10 N H_2SO_4 in 50% acetic acid[b]	37°	2 × 10^{-3}M cysteine	5 days	0
			10 days	0.01
10 N H_2SO_4 in 30% acetic acid[b]	100°	—	30 min	0.35
			90 min	0.7
10 N H_2SO_4 in 30% acetic acid[b]	100°	5 × 10^{-3}M thioglycollic acid	35 min	0.05
			2 hrs	0.15

[a] 2 × 10^{-3}M *bis*-DNP-cystine; 1 mg insulin/ml.
[b] 2 × 10^{-2}M *bis*-DNP-cystine; 10 mg insulin/ml.

The amount of DNP derivative remaining in the aqueous phase after ether extraction was taken as a measure of the interchange reaction. A summary of the results appears in Table 5-2. No appreciable interchange occurred when insulin was hydrolyzed in the presence of 2×10^{-3} M cysteine for 10 days at 37° with 10 N H_2SO_4; the same solution maintained at 100° for 2 hrs also showed little evidence of disulfide interchange. The notable features of this study are (a) the inhibition of the rearrangement by thiols and (b) the finding that the rate of the reaction is markedly reduced in 10 N H_2SO_4. The latter observation may be associated with the rapid increase in the activity of HCl solutions with increasing concentration; the activity coefficient of 10 N HCl is substantially greater than the corresponding value for 10 N H_2SO_4. It should be emphasized that when partial acid hydrolysis is used to obtain cystine-containing peptides, conditions must be employed which will preserve the integrity of the disulfide bridges in the native protein. The recovery of a cystine peptide in high yield is a good indication that it is not a product of an interchange reaction.

Rearrangement at aspartic acid involves the interconversion depicted by Eq. (4). SWALLOW and ABRAHAM [1249] first described this reaction and demonstrated in

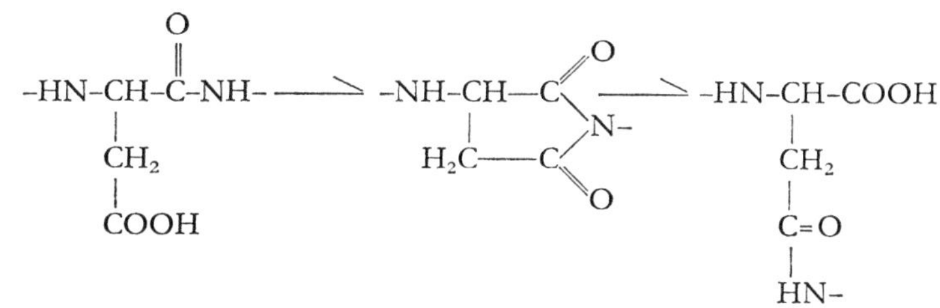
(4)

model experiments the interconversion of *E*-(α-aspartyl) lysine and *E*-(β-aspartyl) lysine *via* the α,β-aspartyl derivative. Difficulties due to this reaction were encountered by NAUGHTON et al. [903] when investigating the amino acid sequence around the reactive seryl residues of several proteases. Because of the aspartyl interconversion reaction, each aspartic acid peptide can occur in three different forms, which greatly increases the complexity of the hydrolysate. Conditions could not be found that completely avoided the α-β-shift; however, the reaction was less marked in 5.7 N HCl than in 12 N HCl. The presence of the α,β-aspartyl or β-aspartyl structure in a peptide renders the peptide bond resistant to hydrolysis by exopeptidases such as leucine aminopeptidase and carboxypeptidase. Furthermore, the Edman degradation is blocked at the cyclization stage by the imide or β-aspartyl structure. Hydrolysis of the imide occurs readily at alkaline pH values to yield predominantly the β-peptide bond [639, 1249].

Diketopiperazine formation is another problem encountered during the course of partial acid hydrolysis. The cyclization of a dipeptide through the elimination of water between the terminal amino and carboxyl groups is favored in dilute acid solution at elevated temperatures [1] [Eq. (5)]. Hydrolysis of the anhydride may yield either the parent dipeptide (I) or a new dipeptide which is a structural isomer (II). In general, diketopiperazines are less stable to acid hydrolysis than the corresponding dipeptides. The possibility of obtaining an inverted dipeptide sequence under conditions of dilute acid hydrolysis markedly diminishes the reliability of a sequence deduced from a dipeptide isolated from such a hydrolysate. SANGER and

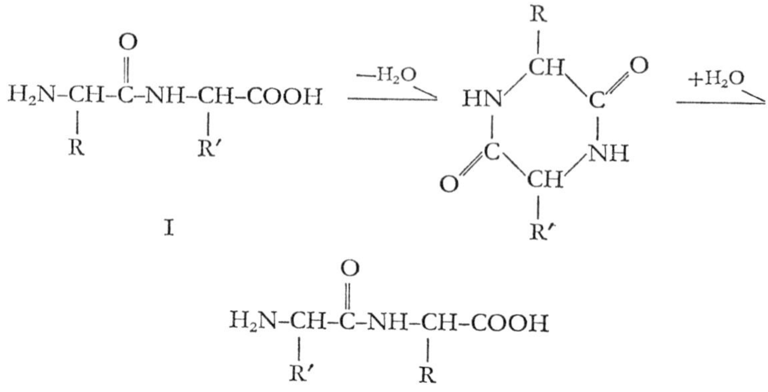

(5)

II

THOMPSON [1068] noted that when glycylvaline was treated with 0.1 N HCl under reflux conditions for 24 hrs the reaction mixture contained both glycylvaline and valylglycine as well as some of the free amino acids. No inversion was detected in 12 N HCl at 37°.

A second type of cyclization reaction involves the conversion of amino terminal glutamine to a pyrrolidone carboxyl residue [Eq. (6)].

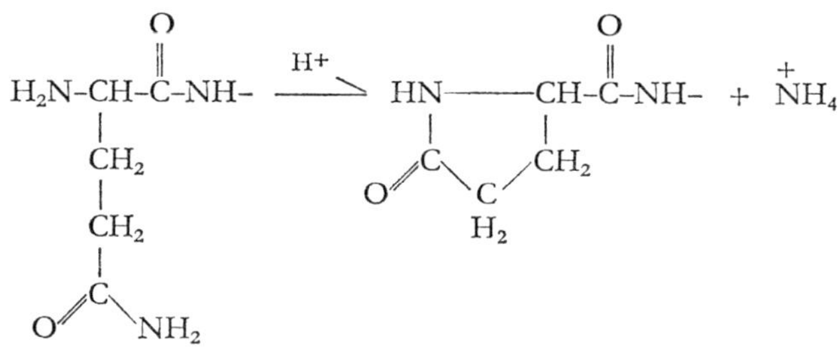

(6)

This reaction occurs in acid solutions and has been found in certain tryptic and chymotryptic peptides of ribonuclease [548, 1184]. Since peptides are frequently exposed to acid conditions, the occurrence of this cyclization reaction must be recognized in order to avoid errors. During the fractionation of the partial acid hydrolysate of papain [837] on Dowex 50-X2 the first component to be eluted was ninhydrin-negative and yielded only glutamic acid on acid hydrolysis. Such properties are typical of pyrrolidone carboxylic acid. Glutamic acid may also form the cyclic amide, but this conversion requires vigorous conditions generally not employed in protein structure studies.

Side reactions involving the destruction of tryptophan and the conversion of asparaginyl and glutaminyl residues to the corresponding carboxylic acids require that milder degradative techniques be used where these residues are concerned.

3. Method

Various experimental conditions that have been employed for partial acid hydrolysis of proteins are presented in Table 5-3. Hydrolyses conducted at 100° are carried out conveniently in a boiling water bath while a thermostated bath suffices for lower reaction temperatures. In either case, a test tube equipped with a ground glass stopper makes a suitable reaction vessel. The use of concentrated HCl requires long term incubation; however, treatment with 5.7 N HCl for 10 min at 100° has been reported to be equivalent to a 24-hrs hydrolysis period with 12 N HCl at 37° [903]. The choice of optimal hydrolysis conditions depends to some degree on the sample, and trial experiments with small amounts of protein or peptide are recommended.

The conditions used for dilute acid hydrolysis of proteins and peptides differ from those discussed for concentrated and constant boiling HCl in that the sample is usually hydrolyzed in an evacuated, sealed, heavy-walled Pyrex tube at 105 to 110°.

Table 5-3. Conditions for the partial acid hydrolysis of some proteins

Acid	Temperature	Protein	Time of hydrolysis	Reference
12 N HCl	37°	Insulin	3 days	[1069]
		DIP[a]-chymotrypsin	3 days	[903, 1082]
		Phosphoglucomutase	30 to 40 hrs	[871]
		IMP[b]-trypsin	3 days	[1081]
		DIP-trypsin, DIP-elastase	2 to 3 days	[903]
		Dinitrophenylated subtilisin BPN′	6 days	[660]
		Lysozyme	4 days	[1275]
		Performic acid oxidized papain	7 days	[690]
5.7 N HCl	37°	DIP-trypsin, DIP-chymotrypsin, DIP-elastase	3 days	[903]
5.7 N HCl	100°	DIP-trypsin, DIP-chymotrypsin, DIP-elastase	20 to 30 min	[903]
		Phosphoglucomutase	30 min	[871]
2 N HCl	100°	Phosphoglucomutase	2 hrs	[514]
0.03 N HCl	105°	Ribonuclease, insulin, glucagon	8 to 48 hrs	[434, 1115]
0.03 N HCl	110°	Reduced-carboxymethylated papain	48 hrs	[837]

[a] Diisopropylphosphoryl-
[b] Isopropylmethylphosphonyl-

When very small amounts of material are being hydrolyzed, the sample dissolved in acid may be sealed in a glass capillary tube and heated in an oven for the required period of time. Protein concentrations ranging from 1 mg [1115] to 5 mg [837] per ml of acid have been used. Hydrolysates are taken to dryness either in a vacuum desiccator ($<100\ \mu$ Hg) over solid NaOH or by rotary evaporation at 40°. In order to remove the last traces of HCl, the residue is dissolved in distilled water and re-evaporated. The latter method is rapid and efficient, but if many samples are being processed at one time the use of a desiccator is most convenient. When sub-micromole amounts of sample are hydrolyzed it is advisable to carry out the hydrolytic and HCl removal steps in the same vessel to avoid loss of material on transferring.

Schroeder et al. [1114] have employed the following conditions for the release of aspartic acid from peptides derived from the γ-chain of fetal hemoglobin. A solution of several micromoles of peptide in 10 ml of 0.25 M acetic acid was refluxed for 8 hrs and the solvent then removed by rotary evaporation. Under these conditions, cleavage occurred on both sides of the aspartyl residue, and little or no effect was noted on asparaginyl, glutaminyl, or glutamyl residues. Dilute acid hydrolysis of several tryptic peptides from tobacco mosaic virus protein [1296] was accomplished by dissolving the peptide in 0.03 N HCl (approximately 1 mg/ml) and hydrolyzing in an evacuated sealed tube at 105° for the desired time. Aspartyl and asparaginyl bonds were cleaved on both sides with the liberation of free aspartic acid. The release was considerably slower from asparaginyl peptides than from aspartyl peptides indicating that deamidation of asparagine is required prior to the release of the free acid. It has been suggested that dilute acid hydrolysis of peptides may be advantageously conducted in two stages [1296]. The first step would involve a hydrolysis period of 6 hrs (0.03 N HCl, 105°), during which time cleavage would occur at the amino and carboxyl side of each aspartyl residue, and asparaginyl residues would be deamidated. In the second step, the aspartic acid peptides (formerly asparaginyl peptides) would be isolated and subjected to another 6-hour hydrolysis period. The net result of this procedure would be the stepwise degradation of a peptide with cleavage first at aspartyl residues followed by cleavage at asparaginyl residues.

Dilute acid hydrolysis (0.03 M HCl, 110°, 48 hrs) of reduced carboxymethylated papain was performed to obtain unique peptides which would permit the unequivocal assignment of overlapping sequences in the native protein [837]. No attempt was made to isolate and purify all of the peptides from the hydrolysate; instead, special attention was devoted to peptides containing arginine and tyrosine which could be readily detected by specific color tests on paper. Kinetic studies on the release of free amino acids indicated that aspartic acid was released more rapidly than any other, but extensive hydrolysis also occurred at peptide bonds involving glutamic acid, glycine, alanine, serine, and threonine. When hydrolysis was performed with 0.01 N and 0.003 N HCl, no change in selectivity was observed.

4. Assessments of the Method

Generally, the peptide mixtures obtained from a protein or large polypeptide by partial acid hydrolysis are exceedingly complex and present special problems in separation and purification. For example, the separated A and B chains of insulin (21 and 30 residues, respectively) yielded numerous di- and tripeptides which were resolved chiefly by preliminary fractionation on charcoal columns followed by paper electrophoretic and chromatographic techniques [1069, 1071]. Schroeder et al. [1110] used column chromatography to fractionate the products formed when gelatin was incubated with 3.6 N HCl at 37° for 7 days and found predominantly dipeptides and free amino acids. The partial acid hydrolysate of lysozyme (12 N HCl, 37°, 4 days) also yielded complex mixtures of peptides which did not permit the establishment of sequences beyond the hexapeptide stage [1275]. Although determining the sequence of small peptides is not difficult, limited information is gained regarding the unambiguous positioning of these units in the polypeptide chain.

Dilute acid hydrolysis is not a suitable method for the selective degradation of proteins. The specificity of the method is not as great as the release of free amino acids would suggest, since extensive peptide bond cleavage occurs at numerous sites along the polypeptide chain. In the case of small peptides, hydrolysis at aspartyl residues is reasonably specific, and its use is recommended.

III. Enzymic Degradation

Enzyme catalyzed cleavage of a polypeptide chain offers many distinct advantages over partial acid hydrolysis. In addition to being required in only catalytic amounts, certain enzymes exhibit a high degree of specificity hydrolyzing only peptide bonds formed by specific amino acids; consequently, the resulting peptide mixtures are comparatively less complex and more amenable to fractionation. Peptides are usually obtained in high yields, and side reactions are negligible. Glutamine, asparagine, and tryptophan are preserved, as are other acid-labile structures such as esters of serine, threonine, and tyrosine. Carbohydrate chains and lipid moieties which are integral parts of some proteins but which may be sensitive to acid are recovered intact. In this section, the three enzymes most commonly used to obtain sets of peptides for the determination of the amino acid sequence of proteins are discussed. For the purpose of illustration, the primary structure of subtilisin BPN′ [813] is presented and the sites of attack by trypsin, chymotrypsin, and pepsin are indicated (Fig. 5-1).

A. General Considerations

The exact conditions for digestion depend upon the enzyme employed, but the following factors are important considerations irrespective of the type of digest: (a) pH, (b) temperature, (c) enzyme to substrate ratio, and (d) time of incubation. The optimal conditions for complete digestion should be determined by performing trial digests using sub-micromole amounts of protein. Examination of the digest by a two-dimensional separation technique ("fingerprint") provides useful information regarding the number of basic, neutral, and acidic peptides as well as the relative complexity of each of these electrophoretic groups. Additional information may be obtained by staining the peptide map with reagents specific for certain amino acids, e.g., the Ehrlich (tryptophan), Pauly (tyrosine and histidine), Sakaguchi (arginine), platinic acid (methionine and cysteine), and nitroprusside (cystine) stains.

Frequently the denatured protein is insoluble under the initial hydrolysis conditions, but the suspension will generally clarify during the early stages of digestion. It is desirable to obtain a suspension of finely divided particles prior to the addition of enzyme; this may be easily accomplished by either trituration or sonication. Whenever possible, it is advantageous to perform the digestion in either a volatile buffer (e.g., ammonium bicarbonate or trimethylammonium acetate), so that salts can be removed by lyophilization, or water adjusted to and maintained at the correct pH by the addition of titrant. With the latter system lyophilization of the digest is not required prior to fractionation, since salts are not present at a level which would interfere with either column chromatography or preparative paper separation methods. When digestion is carried out for an extended period of time, it is advisable to add thimerosal (1:5000) or a crystal of thymol to discourage the growth of microorganisms. Hydrolysis is terminated by the addition of a specific enzyme inhibitor, freezing, or a pH adjustment.

B. Preparation of a Protein for Enzymic Digestion

In general, native proteins are not good substrates for extensive proteolytic degradation. Peptide bonds on the surface of the three-dimensional structure of the

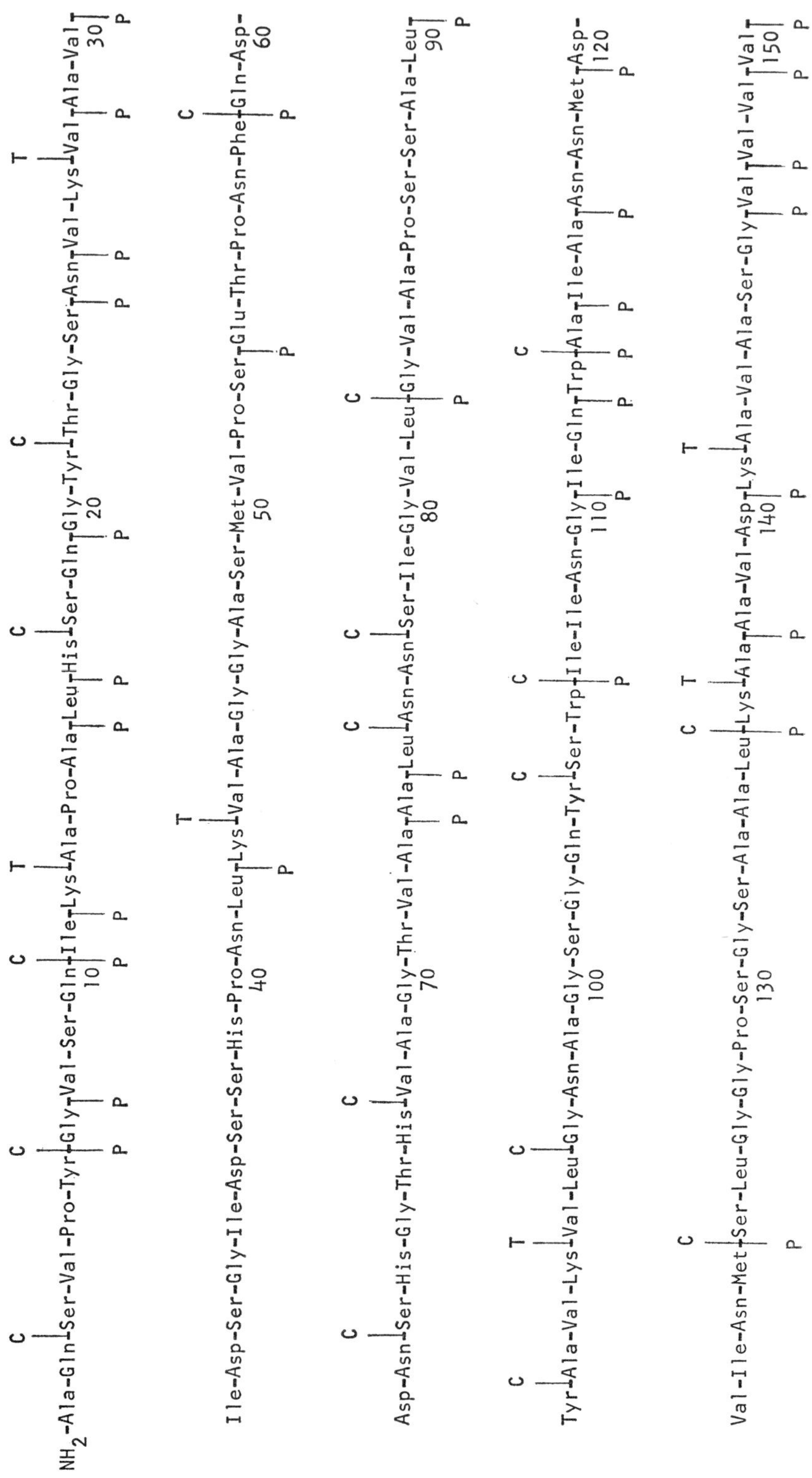
NH2-Ala-Gln-Ser-Val-Pro-Tyr-Gly-Val-Ser-Gln-Ile-Lys-Ala-Pro-Ala-Leu-His-Ser-Gln-Gly-Tyr-Thr-Gly-Ser-Asn-Val-Lys-Val-Ala-Val
Ile-Asp-Ser-Gly-Ile-Asp-Ser-Ser-His-Pro-Asn-Leu-Lys-Val-Ala-Gly-Gly-Ala-Ser-Met-Val-Pro-Ser-Glu-Thr-Pro-Asn-Phe-Gln-Asp-
Asp-Asn-Ser-His-Gly-Thr-His-Val-Ala-Gly-Thr-Val-Ala-Ala-Leu-Asn-Asn-Ser-Ile-Gly-Val-Leu-Gly-Val-Ala-Pro-Ser-Ser-Ala-Leu
Tyr-Ala-Val-Lys-Val-Leu-Gly-Asn-Ala-Gly-Ser-Gly-Gln-Tyr-Ser-Trp-Ile-Ile-Asn-Gly-Ile-Gln-Trp-Ala-Ile-Ala-Asn-Asn-Met-Asp-
Val-Ile-Asn-Met-Ser-Leu-Gly-Gly-Pro-Ser-Gly-Ser-Ala-Ala-Leu-Lys-Ala-Ala-Val-Asp-Lys-Ala-Val-Ala-Ser-Gly-Val-Val-Val-Val

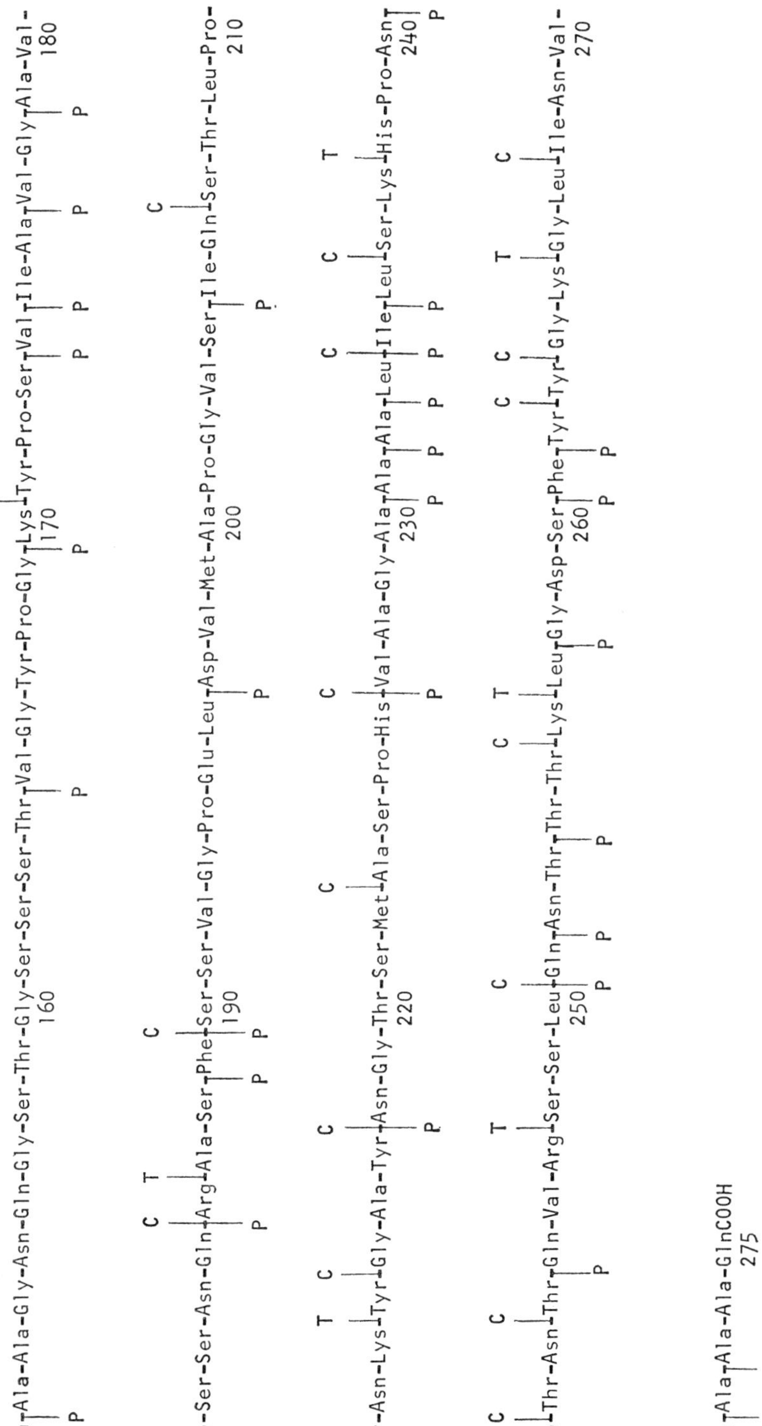

Fig. 5-1. The amino acid sequence of subtilisin BPN′ showing the bonds hydrolyzed by trypsin (*T*), chymotrypsin (*C*), and pepsin (*P*)

protein in contact with the external environment are accessible and may be susceptible to enzymic attack, but peptide bonds buried in the protein matrix are not readily hydrolyzed. Since the main objective in structural investigations is to cleave as quantitatively as possible all sensitive bonds, denaturation of the substrate is a prerequisite for optimal proteolytic degradation. Denaturation disrupts the native secondary and tertiary structure of the polypeptide chain so that its conformation approaches that of a random coil.

Of the methods available, heating and trichloroacetic acid precipitation are simple and effective treatments for preparing protein substrates. When the native structure of the protein is stabilized by disulfide bridges, these methods are inadequate, and it is first necessary to cleave the -S-S- bonds by either a reductive or oxidative procedure. When reduction is used the newly formed cysteinyl side chains must be converted to a stable derivative for further manipulation. Reduction followed by S-carboxymethylation has been used in structural studies involving papain [781], lysozyme [201], f_2 bacteriophage coat protein [719], and Bence-Jones proteins [539]. The oxidative procedure, on the other hand, yields stable derivatives in a single step. Treatment of the intact protein with performic acid converts cysteine and cystine to cysteic acid, and methionine is oxidized to the stable sulfone. This method has been applied to numerous proteins among which are insulin [1067], bovine ribonuclease [549], papain [690, 781], ribonuclease T-1 [1257] and azurin [27]. It should be emphasized that oxidation destroys tryptophan and also renders the protein resistant to degradation by cyanogen bromide. Procedures such as S-carboxymethylation and performic acid oxidation introduce charged groups into the protein molecule, thereby increasing the solubility properties of the peptides derived from those modified regions of the polypeptide chain. The derivatized protein is frequently insoluble in non-urea containing buffers because of the denaturing conditions employed in its preparation.

1. Heating

A Pyrex test tube containing the solution of native protein (approximately 1% w/v) is placed in a boiling water bath for usually 2 to 10 min. The tube is then transferred to an ice bath and the contents cooled to 0°. Frequently, the heat-denatured protein will be insoluble and may be isolated by centrifugation, but it is most convenient at this stage to suspend the protein and proceed with the enzymic digestion. Care should be taken to avoid pH extremes during the heating process so as to minimize the possibility of hydrolytic cleavage of peptide bonds as well as oxidation of sulfur containing amino acids. The temperature and time of heating may be varied, depending upon the characteristics of the protein under investigation. Conditions which have proven satisfactory for the thermal denaturation of certain proteins are the following: hemerythrin, 65° for 5 min at pH 8 [448]; calf rennin, 100° for 15 min in distilled water [368]; phosphoglucomutase, 100° for 30 min [870]; hemoglobin F, 90° for 4 min at pH 8 [1114].

2. Trichloroacetic Acid Precipitation

To a 1% solution of protein, an equal volume of 10% (w/v) trichloroacetic acid is added at room temperature. After 10 to 15 min, the protein suspension is cooled to ice-bath temperature, and the insoluble protein is collected by centrifugation. TCA is removed from the pellet by washing five times with 20 volumes of reagent grade acetone followed by an ether extraction (peroxide free)[2] with similar volumes. Residual amounts of organic solvent may be removed by drying the pellet in a gentle stream of nitrogen. Trichloroacetic acid precipitation is an efficient method for removing bound amino acids and peptides; this is particularly useful when a proteolytic enzyme is being examined and autolysis products are possible contaminants.

3. Reduction and S-Carboxymethylation

The preparation of reduced and alkylated proteins by the method of Crestfield et al. [254] is described. To 5 to 100 mg of protein in a 12-ml screw cap vial maintained under a nitrogen barrier, add 3.61 g of deionized, crystalline urea, 0.30 ml EDTA solution (50 mg of disodium EDTA per ml), 3.0 ml of Tris buffer, pH 8.6, (5.23 g of Tris and 9 ml of 1.0 N HCl diluted to 30 ml with water), and finally 0.10 ml of mercaptoethanol (Eastman Organic Chemicals). The solution is made up to a 7.5 ml mark with water, and a solution 8 M in urea and 0.2% in EDTA is used to fill the vial completely. A disk of polyethylene is placed over the top to exclude air and is held in place with the screw cap. After 4 hrs at room temperature (22 to 25°), the contents of the vial are transferred to a 25-ml beaker under a nitrogen barrier. A freshly prepared solution of 0.268 g of iodoacetic acid in 1.0 ml of 1.0 N NaOH is added to the reaction mixture. The iodoacetate added is slightly less on a molar basis than the amount of mercaptoethanol. The -SH groups of cysteinyl residues react most rapidly; since the excess iodoacetate reacts faster with mercaptoethanol above pH 8 than it does with thioether sulfur, alkylation of methionine is minimized. The buffering capacity of the system is adequate to maintain the pH above 8.3.

After the addition of iodoacetic acid, the solution should be protected from light to prevent the formation of iodine which may react with tyrosine, tryptophan, and histidine residues. Fifteen minutes after the addition of iodoacetate, the sample is applied to a 4 × 40-cm column of Sephadex G-75, medium particle size, equilibrated with 50% acetic acid and wrapped in aluminum foil. Fractions of 8 ml are collected at a flow rate of 20 ml per hour. Aliquots equivalent to approximately 2 mg of protein are transferred to plastic tubes (115 × 19 mm) in a rack fitting in a heated desiccator [546]. The acetic acid is removed in 2 hrs with a water pump and a voltage setting such that the temperature of the desiccator wall is 70°. Alkaline hydrolysis is performed (Section 5. IV. A. 1. d), and the hydrolysates are analyzed by the ninhydrin method (Section 5. III. C. 1). The fractions containing the reduced and alkylated protein are pooled and lyophilized. The protein is stored in the cold.

4. Performic Acid Oxidation

The method as applied to ribonuclease is described [543]. It is necessary to perform the oxidation on halide-free protein, since free halogen formed in the presence of performic acid results in the production of halotyrosines. The fact that tryptophanyl residues are converted by this oxidizing agent to kynurenine and other degradation products particularly limits the applicability of this method. Removal of halide ion is accomplished by passing a 10% solution of protein in 0.01 M acetic acid

---

[2] Ether is shaken with moist $FeSO_4$, distilled, and stored in a ground glass stoppered bottle over dry $FeSO_4$.

over a bed of Dowex 2-X10 in the acetate form equilibrated with the same solvent. One ml of resin bed is used per ml of protein solution. The halide-free preparation is recovered by lyophilization of the filtrate. Performic acid is prepared by adding 0.50 ml of 30% H_2O_2 (reagent Superoxol, Merck) to 9.5 ml of 99% formic acid. The resulting solution is allowed to stand at room temperature for 2 hrs in a closed vessel prior to use. Solutions of performic acid are not stable and must be prepared each time. In a second flask 200 mg of ribonuclease (approximately 8% moisture) are dissolved in 5.0 ml of 99% formic acid, after which 1.0 ml of anhydrous methanol is added with stirring; the methanol prevents freezing of the solution during the oxidation. The quantity of performic acid used is 12 times that required to transform all of the cystine and methionine to cysteic acid and methionine sulfone, respectively. The protein and performic acid solutions are cooled for 30 min in a bath maintained between −7° and −10° before mixing the contents of the two flasks. The reaction is allowed to proceed for 2.5 hrs at the indicated temperature. The contents of the tube are rinsed with 50 ml of ice water into a flask containing 350 ml of water at 0°. The diluted solution is shell frozen and the solvents removed by lyophilization. The residue is dissolved in 40 ml of water, transferred to a smaller flask with the aid of an additional 40 ml of water and again lyophilized.

In studies performed on papain [690] the lyophilization step was omitted and the oxidized protein was isolated by the addition of trichloroacetic acid to a final concentration of 10%. The precipitated protein was then washed with 50 ml portions of 10% trichloroacetic acid until the washings gave a negative test for peroxide with KI-starch paper. The precipitate was further washed with three 50 ml volumes of absolute ethanol followed by three 50 ml volumes of ether.

C. Methods of Measuring Hydrolysis

1. Ninhydrin Method [880]

This method is based on the colorimetric measurement of the α-amino groups released during proteolysis. The ninhydrin solution is prepared in a 1-l bottle appropriately equipped so that nitrogen may be continuously bubbled through the solution during the preparation procedure and which permits the transfer of the solution to a dark colored reagent bottle under nitrogen pressure. Dissolve 20 g of ninhydrin and 3 g of hydrindantin in 750 ml of methyl Cellosolve[3] with stirring. Add 250 ml of 4 N sodium acetate buffer (pH 5.5)[4], mix thoroughly, and transfer to a dark bottle for storage in the cold room under a nitrogen atmosphere; not more than a 2-week supply should be prepared at one time.

---

[3] These reagents are obtainable from Pierce Chemical Co. in a sufficiently pure state so that they may be used without further purification. The methyl Cellosolve (monomethyl ether of ethylene glycol), however, should always be tested for peroxides by adding 2 ml of the solvent to 1 ml of freshly prepared 4% aqueous KI. A colorless solution indicates the absence of peroxides. Methyl Cellosolve is quite toxic and should be handled only in a well ventilated hood. The fact that it is odorless emphasizes this precaution.

[4] Dissolve 2720 g of $NaOAc \cdot 3\,H_2O$ (reagent grade) in 2 l of water; gentle heat may be used to hasten solubility. Cool to room temperature, add 500 ml of glacial acetic acid, and make up to a final volume of 5 l. The pH of the undiluted buffer should measure 5.51 ± 0.03. If a final adjustment of pH is necessary, 5 g of NaOH correspond to about 0.04 pH unit.

In order to follow the course of proteolysis, aliquots are removed from the digestion mixture at various periods of time and added to 1.0 ml of ninhydrin solution; this immediately stops the enzymic reaction. The capped tubes are shaken for a few seconds to insure adequate mixing and then heated 15 min (accurately timed) in a covered boiling water bath [877]. Transfer the rack containing the tubes to a pan of cold water and allow the reaction solution to cool to room temperature. Add 5 ml of 50% ethanol to each tube and shake vigorously for 30 to 60 sec to air oxidize the residual hydrindantin. The optical density of each solution is measured at 570 mμ.[5] The rate of hydrolysis may be expressed as the relative increase in ninhydrin color value as a function of time. Throughout the procedure, care must be taken to prevent contamination with ammonia.

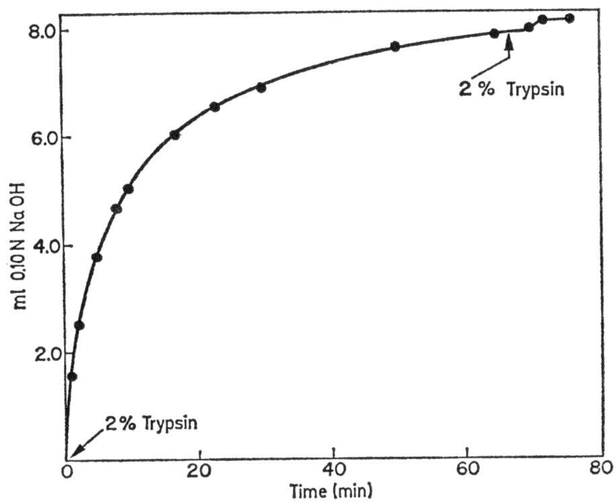

Fig. 5-2. Time course of the tryptic digestion of subtilisin BPN′ (88 μmoles) at pH 8.1. (Reprinted from Ref. [661] by courtesy of The Journal of Biological Chemistry)

2. Titrimetric Method

This procedure takes advantage of the fact that, in the pH range of 7.4 to 9.0 where the titrimetric method has the greatest application, each newly liberated carboxyl group will be fully ionized while the α-amino group will be only partially protonated; thus, a decrease in the pH of the solution will result. Titration of the unbuffered enzyme-substrate solution with standardized alkali permits a direct means of measuring the rate and extent of proteolysis. The alkali uptake is related to the average pK of the α-amino groups and the pH at which the hydrolysis is performed ($pH = pK + Log [-NH_2]/[-NH_3^+]$). This is most easily measured with an automatic titrator such as the pH-stat [631].

[5] Although the color yield will vary with the amino terminal residue of the peptide, the molar absorption coefficient of the colored product of the ninhydrin reaction is close to 1.8×10^4 for a 1-cm light path. This value is useful in determining the size aliquot required for analysis.

The precise titrimetric monitoring of the hydrolytic reaction requires a titration vessel which allows temperature regulation and efficient stirring. When digestions are carried out in the alkaline range of the pH scale, CO_2 from the air should be excluded by continuously passing a gentle stream of nitrogen through the solution; a flat base line must be established before the enzyme is added. When a plateau region is reached, the digestion should be terminated since beyond this point a slow hydrolysis of peptide bonds occurs that usually increases the complexity of the digest and does not improve the yield of major peptides. A typical result is presented in Fig. 5-2. If it is assumed that the average pK of the liberated α-amino groups is near 7.8, then the consumption of alkali is that expected for the scission of the 13 trypsin-susceptible peptide bonds in the subtilisin molecule.

D. Trypsin

1. Specificity

Trypsin exhibits the highest degree of substrate specificity known for an endopeptidase; only those bonds involving the carboxyl groups of lysyl and arginyl residues are hydrolyzed. This permits the estimation of the number of tryptic peptides obtainable from a protein when the amino acid composition is known. In theory, this number is equal to the sum of lysyl and arginyl residues present in the molecule plus a C-terminal peptide. The relative rates of cleavage of susceptible peptide bonds are influenced by the chemical nature of the side chains in the immediate vicinity. Generally, the presence of polar groups close to the susceptible bond decreases the rate of hydrolysis. Side chains with a net negative charge (e.g., aspartic acid, glutamic acid, cysteic acid, or S-carboxymethyl cysteine) which are adjacent to a lysyl or arginyl residue tend to reduce significantly the rate of hydrolysis. Thus, in oxidized ribonuclease, the sequence -Arg-$CySO_3H$- is only partially cleaved, as are the -Asp-Lys-Leu and -Asp-Lys-Ala- sequences found in the γ-chain of human hemoglobin. In the copper-containing protein, azurin, incomplete hydrolysis at lysine 92, which is positioned between an aspartic acid and glutamic acid residue, has been noted [27]. Similarly, if the α-amino group of lysine or arginine is free, as is the case for amino terminal lysine in ribonuclease [1184] and lysozyme [200], free lysine is released very slowly. Repetitive sequences of lysine and/or arginine are also attacked by trypsin at a considerably slower rate. Peptide bonds involving the imino nitrogen of proline appear to be totally resistant to tryptic hydrolysis, i.e., Lys-Pro or Arg-Pro. Examples of this type of linkage may be found in ribonuclease [1184], corticotropin [773], whale myoglobin [321], lobster glyceraldehyde 3-phosphate dehydrogenase [268], Staphylococcal nuclease [1266], azurin [27], and α-MSH [510].

2. Removal of Chymotryptic Activity

Commercial preparations of crystalline trypsin may contain trace quantities (less than 0.5%) of chymotrypsin, which, if present during a prolonged incubation period, could significantly decrease the yield of tryptic peptides. Early attempts to eliminate chymotryptic-like activity from trypsin preparations took advantage of the fact that trypsin is more stable than chymotrypsin when exposed to 8 M urea [505] or dilute acid [1018]. The most promising technique, however, involves treating the enzyme with an inhibitor specific for chymotrypsin such as L-(1-tosylamido-2-phenyl)ethyl chloromethyl ketone (TPCK) [1095]. Diphenylcarbamyl chloride [337] and phenylmethanesulfonyl fluoride [346] also inhibit chymotrypsin but do not appear to be as effective as TPCK in eliminating chymotryptic activity from trypsin preparations.

Treatment of trypsin with TPCK [726, 1332]: one gram of crystalline trypsin is dissolved in 330 ml of 0.001 M $CaCl_2$, and the pH is adjusted to 7.0 with 1 N NaOH. TPCK solution (30 mg in 6 ml of anhydrous methanol) is added dropwise to the enzyme solution with gentle stirring. The reactants are stirred for 5.5 hrs at room temperature during which time titrant is added to maintain the pH of the solution at 7.0. The reaction is terminated by the addition of 1 N HCl to pH 3.0. The acidified enzyme solution is dialyzed against 25 l of water (pH 3.0) for 24 hrs at 4°, and the solvent is finally removed by lyophilization. The yield of TPCK-trypsin is approximately 70 to 72% on a weight basis. The treated enzyme retains full tryptic activity as judged by its ability to hydrolyze benzoyl-L-arginine ethyl ester. TPCK-trypsin is still capable of slowly hydrolyzing N-acetyl-L-tyrosine ethyl ester but does not exhibit chymotryptic activity toward peptide substrates.

3. Chemical Modifications Altering Trypsin Specificity

Masking of the ε-amino groups of lysine restricts tryptic hydrolysis specifically to arginyl residues in the polypeptide chain. This may be accomplished by carbamylation [1214], carbobenzoxylation [46], dinitrophenylation [1018], succinylation [771], guanidination [1343], dithiocarbamylation [849], trifluoroacetylation [420], and maleylation [192]. Of the methods mentioned, trifluoroacetylation and maleylation have two main advantages: (a) they exhibit a high degree of specificity for the ε- and α-amino groups of the protein, and (b) the blocking groups may be removed under mild conditions to regenerate the unmodified amine functions. The advantage of such reactions are obvious. Proteins with blocked ε-amino groups, when digested with trypsin, will yield fragments with C-terminal arginine. These large fragments can be purified, the blocking groups removed, and then further digested with trypsin to cleave at the lysyl bonds. The stepwise use of trypsin in this manner provides the most specific means for the controlled proteolytic degradation of proteins.

In addition to modifications which restrict tryptic hydrolysis, it is also possible to expand the specificity of trypsin by creating new points of cleavage. The S-aminoethylation of cysteinyl residues is an example of this type of alteration.

a) Trifluoroacetylation

The preparation of trifluoroacetylated ribonuclease [420] is presented below. The basic reaction is represented by Eq. (7):

$$F_3C{-}\overset{\displaystyle O}{\overset{\|}{C}}{-}S{-}C_2H_5 + \text{Protein-}NH_2 \underset{\text{1 M piperidine, } 0^\circ}{\overset{\text{pH 10, 25 °C}}{\rightleftarrows}} \text{Protein-}\underset{\displaystyle H}{N}{-}\overset{\displaystyle O}{\overset{\|}{C}}{-}C{-}F_3 + C_2H_5{-}SH \tag{7}$$

In a 125-ml Erlenmeyer flask 0.5 g of ribonuclease is dissolved in 50 ml of water and the pH of the solution is adjusted to 10.0 with 1.0 N KOH. Base is added from a 25-ml syringe fitted with a No. 19 needle attached to polyethylene tubing. The flask is placed on a magnetic stirrer in a well ventilated hood, and the electrodes of a pH meter are immersed in the protein solution so that the pH can be continuously monitored during the course of the reaction. Ethyl thioltrifluoroacetate (2.5 ml), prepared by the method of HAUPTSCHEIN et al. [519] or obtained from the Pierce Chemical Company (Rockford, Illinois), is added to the solution with rapid stirring. The reagent, which is immiscible with water, is divided into small droplets by the revolving magnet, exposing a large surface area for reaction with protein in the aqueous phase. The pH is maintained between 9.95 and 10.00 by the addition of KOH from the syringe. The reaction is conducted at room temperature and allowed to proceed for about 1 hr, after which the base uptake (approximately 18 ml) is essentially complete. A large amount of alkali is consumed because of the spontaneous hydrolysis of the thioester. To the reaction mixture is added 5.0 ml of 0.1 M sodium acetate buffer, pH 5.0. The pH is lowered to 6.0 by the addition of a few drops of concentrated HCl; this treatment causes the derivatized protein to precipitate. The reaction mixture is added to 4 volumes of cold (−20°) absolute ethanol. The precipitate is collected by centrifugation and washed twice with a mixture of ethanol and sodium acetate buffer (5:1 by volume) at 3°. The final residue is suspended in 50 ml of the ethanol-sodium acetate mixture and dialyzed against 100 volumes of distilled water at 3° for 24 hrs with repeated changes of the outer solution. The contents of the dialysis sac are centrifuged, and the supernatant, containing a few milligrams of protein, is discarded. The residue is lyophilized and stored as the dry powder. The yield of trifluoroacetylated ribonuclease is approximately 75%. An alternate procedure has been described in which excess reagents and hydrolysis products are removed from the derivatized protein by dialyzing the reaction mixture against repeated changes of 0.1 M acetic acid [1266].

The degree of trifluoroacetylation is easily determined by treating the derivatized protein with nitrous acid [46, 1266]. Only free amino groups undergo deamination; acylated amino groups are unaffected. After nitrous acid treatment, the protein is subjected to acid hydrolysis in the usual manner and the hydrolysate analyzed for lysine. The lysine level is a direct measure of the number of ε-(trifluoroacetyl)lysine residues in the modified protein.

Detrifluoroacetylation is readily accomplished by the following method [420]. Ten mg of TFA-ribonuclease are dissolved in 0.5 ml of 1 M piperidine at room temperature and the solution immediately chilled in an ice bath. After incubation at 0° for 2 hrs, the reaction mixture is added slowly, with constant stirring, to 1.5 ml of precooled 0.5 N acetic acid (0 to 3°). The deblocked protein is recovered by lyophilization after removal of reagents by either gel filtration or dialysis.

Caution should be exercised if trifluoroacetylation is carried out on a protein containing disulfide bonds; the formation of ethylmercaptan in alkaline solution favors both disulfide bond reduction and interchange. This problem may be circumvented by acylating the performic acid oxidized or reduced-carboxymethylated protein.

b) Maleylation

The reversible blocking of amino groups with maleic anhydride has been recently described [192, 192b]. Maleic anhydride rapidly reacts at pH 8.5 to 9.5 with the α- and ε-amino groups of proteins according to the following reaction [Eq. (8)]:

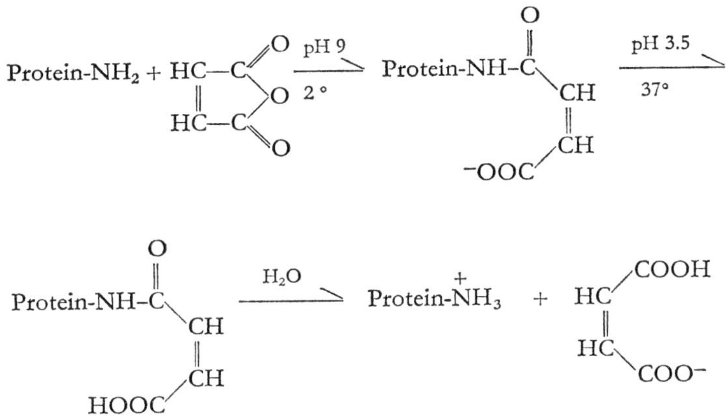

(8)

Maleyl proteins are frequently soluble at neutral or alkaline pH in the absence of urea or guanidine since the introduction of a negative charge minimizes aggregation through electrostatic repulsion. The maleyl amino bond is very stable above pH 6 but is readily hydrolyzed below pH 5 by intramolecular catalysis involving the unionized carboxyl group. The half-life of ε-maleyl lysine is 11 hrs at 37° and pH 3.5.

The application of this derivatization procedure to chymotrypsinogen is presented [192 a]. Bovine chymotrypsinogen A (20 mg) is dissolved in 2.0 ml of 0.1 M sodium pyrophosphate buffer, pH 9.0, and treated at 2° with 300 μl of 1.0 M maleic anhydride in redistilled dioxane. The maleic anhydride solution is added in six additions and the pH of the mixture is maintained at 9.0 by addition of 0.1 M NaOH. When alkali uptake ceases, the maleylated protein is desalted by passing it through a column (40 × 3 cm) of Sephadex G-25 in 0.01 M NH_3. After enzymic digestion, peptides may be demaleylated by incubation in an aqueous solution of 5% (v/v) acetic acid containing 1% (v/v) pyridine at 60° for 6 hrs.

The above procedure can be modified as in the case of the maleylation of S-carboxymethylated glutamic dehydrogenase [159 b]. A finely dispersed suspension (100 ml) of protein in H_2O (10 mg/ml) is adjusted to pH 9.0 and with efficient stirring, 2.0 g of solid maleic anhydride are added slowly over a period of 40 min at room temperature, while the pH is maintained between 8.5 and 9.5 by the addition of 2 M NaOH. The protein is completely solubilized during maleylation. Low molecular weight materials are removed by dialysis against distilled water adjusted to a slightly alkaline pH with NH_4OH. After tryptic digestion, the peptides are deblocked by incubation in 30% acetic acid at 40° for 48 hrs.

c) S-Aminoethylation

It is possible through the reaction of a reduced protein with ethylenimine to quantitatively convert cysteinyl side chains to the S-(β-aminoethyl) cysteinyl derivative [1005] [Eq. (9)]:

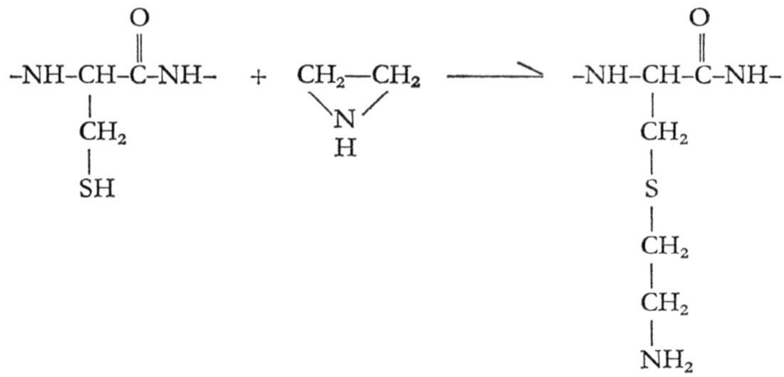

(9)

Reactions employing β-bromoethylamine [783] yield the same product; however, the intermediate formation of the imine is required before the reaction with the mercaptide anion occurs. Since the cyclization step has been shown to be rate limiting [1005], the use of ethylenimine is most efficient and direct. The introduction of this basic side chain into the protein creates a new point for tryptic hydrolysis [783, 1005]; the rate at which this bond is cleaved is markedly reduced and in order to insure maximal hydrolysis the level of trypsin and time of incubation must be increased. It should be remembered that prolonging the digestion time increases the possibility that non-tryptic cleavages will occur. A quantitative study on the tryptic digestion of S-aminoathylated ribonuclease revealed that the cleavage of S-aminoethylcysteinyl bonds averaged approximately 56% while the corresponding hydrolysis of lysyl and arginyl bonds was close to 83% [986]. The derivatized cysteinyl residue at position 40 in the ribonuclease molecule apparently was not cleaved; this is not too surprising since the sequence is -Lys-Asp-Arg-Cys-Lys-Pro-. Aminoethylation has been success-

|
AE

fully employed in structural investigations on trypsinogen [572], tryptophan synthetase [454], ferridoxin [820, 1260], and tobacco mosaic virus protein [1297]. SCHROEDER et al. [1113] have recently reported that aminoethylation of apocatalase resulted in the modification of methionyl residues when excess reagent was removed under acidic conditions. This alteration interfered with the subsequent cyanogen bromide degradation of the protein.

Procedure: The preparation of aminoethyl trypsinogen is described [1004]. Disulfide bonds are reduced by the addition of 200 µl (2.88 mmoles) of β-mercaptoethanol to 200 mg of the protein dissolved in 20 ml of 8 M urea containing 1.0 ml of EDTA solution (2 mg/ml) and 4 ml of 2-amino-2-methyl-1,3-propanediol buffer (4 M, pH 8.6). Alternatively, Tris buffer may be employed at a final concentration of 1.0 M and at pH 8.6. The mixture is allowed to stand at room temperature for 4 hrs. Alkylation is accomplished by the addition of ethylenimine in three portions of 0.250 ml each (4.8 mmoles) at 10-min intervals; the progress of the reaction may be measured by examining aliquots of the reaction mixture for the presence of free thiol with the nitroprusside reagent. Thirty minutes is usually ample time for the completion of the reaction. After a negative nitroprusside test is obtained, the pH is adjusted to 5.0, and the entire mixture is dialyzed for 20 hrs against changes of distilled warer and finally lyophilized.

The degree of substitution is determined from the recovery of S-(β-aminoethyl) cysteine from an acid hydrolysate of the modified protein. Hydrolysis is performed in 6 N HCl at 110° C for 20 hrs in a sealed evacuated glass tube ($<50\ \mu$ of Hg). S-(β-Aminoethyl) cysteine is eluted between lysine and histidine under the standard conditions of SPACKMAN et al. [1198]. The length of the column must be increased to 15 cm, however, in order to sufficiently resolve the alkylated cysteine derivative from lysine and histidine. The color value of S-(β-aminoethyl) cysteine is close to 91% that of lysine [986].

4. Method of Hydrolysis

A variety of conditions has been employed for the tryptic hydrolysis of proteins with respect to enzyme-to-substrate ratios, temperature, pH, and time of digestion. Trypsin preparations of a high degree of purity, which have been treated with TPCK, should be used in structural investigations in order to assure the highest possible yields of tryptic peptides. Trypsin is usually employed at the level of 1 to 4% by weight of substrate with the temperature and pH ranging from 23° to 40° and 7 to 9 respectively. A fingerprint of the digest quickly indicates whether the predicted number of peptides is being produced and also reveals any undigested material. Staining the peptide map with the Sakaguchi reagent will identify the position and number of arginine-containing peptides, which should correspond to the value expected from the arginine content of the intact protein.

The following examples are presented to illustrate conditions employed for tryptic degradation.

a) Hydrolysis of Egg White Lysozyme [199]

Reduced-carboxymethylated lysozyme is suspended in distilled water at a 1% concentration. A few drops of phenol red are added together with sufficient NH_4OH to adjust the suspension to a color corresponding to approximately pH 8. Trypsin, in an amount equal to 2% of lysozyme by weight, is added and the solution stirred for varying periods of time at either 23° or 37°. Aliquots of dilute NH_4OH are added to keep the phenol red color constant. The suspension usually clears in 10 to 15 min; however, the enzymic digestion is generally allowed to proceed for 120 min. A small amount of insoluble material is occasionally present at the end of the digestion which has an amino acid composition similar to that of reduced-carboxymethylated lysozyme. More complete digestion is obtained if the 1% suspension of the derivatized protein is first adjusted to pH 10 with NH_4OH, heated to 60°, and then digested with trypsin at 37°. Proteolysis is terminated by adjusting to pH 2.8 with formic acid.

b) Hydrolysis of Diisopropylphosphoryl-Subtilisin BPN′ [661]

The TCA denatured protein is triturated with a small volume of water until it is evenly wetted and all lumps are dispersed. Additional water is added so that the suspension is equivalent to a 0.8% protein solution. The denatured protein is uniformly suspended by use of a magnetic stirrer, and 0.1 N NaOH is added to pH 10 to 10.5 in order to solubilize the substrate. The pH is immediately adjusted to 8.1 with 30% acetic acid, and the solution of denatured protein (323 ml), in an appropriate flask, is placed in a constant temperature bath at 30°. A radiometer automatic titrator equipped with a magnetic valve to regulate delivery of titrant (0.10 N NaOH) is used to maintain the pH at 8.1. Stirring is accomplished with a teflon-coated magnetic bar and an external magnetic stirrer. The protein solution is thoroughly flushed with nitrogen gas prior to the addition of trypsin and during the digestion period. Digestion is initiated by adding 4.3 ml of trypsin solution (12 mg per ml Worthington twice crystallized trypsin in 0.001 N HCl). Hydrolysis is followed by measuring alkali consumption (Fig. 5-2). After 67 min, an additional 51 mg of trypsin is added to give a total enzyme concentration of 4%. The digestion is terminated after 76 min by adding 17 ml of glacial acetic acid to pH 2.5. The digest is not lyophilized but applied directly to a column of AG 50W-X2.

c) Hydrolysis of f_2 Bacteriophage Coat Protein [719]

The alkylated protein is suspended in an aqueous solution of ammonium hydroxide, pH 10.5, and sonically treated for 30 sec to give a finely divided colloidal suspension. A 1% solution of trypsin, pretreated with TPCK, is added to give a final trypsin concentration of 0.01% and a total protein concentration of 1%. Solid CO_2 is used to adjust the pH to 8.5. The digestion is allowed to proceed for 8 hrs at 37°, with the addition of another equal portion of trypsin after 4 hrs. The digest is lyophilized twice prior to column fractionation of the tryptic peptides.

E. Chymotrypsin

1. Specificity

Although chymotrypsin exhibits a rather broad specificity, peptide bonds involving the carboxyl groups of tyrosine, phenylalanine, tryptophan, and leucine are hydrolyzed preferentially. Protein structure studies involving proteolytic degradation with chymotrypsin have revealed that cleavage may also occur at the carboxyl group of methionine, glutamine, asparagine, histidine, and threonine; however, many of these bonds are split slowly and do not represent major points of hydrolysis. Cleavage has also been reported to occur at bonds formed by serine [781], glycine [761], valine [32, 761], and isoleucine [761], but the extent of hydrolysis was low, and cleavage at these bonds is unexpected. The use of high ratios of enzyme to substrate and long digestion periods may increase the likelihood of cleavage at sites normally not susceptible to chymotrypsin. Frequently, within the same substrate certain peptide bonds formed by the same amino acid are susceptible while others are not attacked. In the case of subtilisin BPN′, hydrolysis occurred at a His-Ser (17–18) and two His-Val (67–68 and 226–227) sequences while a His-Gly (64–65) bond was not noticeably attacked (Fig. 5-1). In the same protein chymotryptic cleavage was detected at Met-Ser (124–125) and Met-Ala (222–223) sequences but three other methionyl bonds were apparently unhydrolyzed (Met-Val, 50–51; Met-Asp, 119–120; Met-Ala, 199–200). Thus, it appears that factors other than the immediate environment of the susceptible bond may influence chymotryptic action. Neighboring groups are known to have a marked effect; the presence of a basic residue (Lys, Arg, or His) adjacent to a weakly susceptible bond seems to enhance cleavage at that site. The chymotryptic hydrolysis of the Thr-Lys bond (255–256) in subtilisin BPN′ [811] as well as hydrolysis of the threonyl bond in the sequence Arg-Lys-Thr-Gly which occurs in human cytochrome *c* [822] illustrates this point. Furthermore, cleavage at the carboxyl group of lysine has been noted when it occurs in a sequence of basic amino acids [53, 761, 822]. Peptide bonds involving the imino nitrogen of proline appear to be resistant to chymotryptic hydrolysis; only the Phe-Pro bond in ovine corticotropin [761] has been reported to be hydrolyzed to a small extent by chymotrypsin. Even though the sites of attack of chymotrypsin on proteins of known primary structure have been examined, it is not possible to predict clearly the course of chymotryptic hydrolysis except in the case of the aromatic amino acids.

2. Method of Hydrolysis

The experimental conditions used for the chymotryptic digestion of a protein are very similar to those employed with trypsin. Chymotrypsin has been used in structural investigations with a number of proteins some of which are the following: insulin [1069], ribonuclease [549], lysozyme [200], papain [781], hemoglobin [534, 1114], cytochrome *c* [53, 810, 822], subtilisin BPN′ [811], Staphylococcal nuclease [1265], azurin [27], and myoglobin [321]. Three examples are presented.

a) Hydrolysis of the α-Chain of Human Hemoglobin [534]

The α-chain (1 g) is dissolved in 90 ml water and 10 ml of 10% (w/v) NH_4HCO_3, and 10 mg of chymotrypsin, dissolved in 1 ml of 0.001 N HCl, are added. After the pH is adjusted to 8.0, digestion of the milky suspension is allowed to proceed at 25° for 6 hrs with constant stirring. The precipitate is removed by centrifugation, and 10 ml of glacial acetic acid are added to stop the reaction. The solution is lyophilized twice to remove ammonium acetate.

b) Hydrolysis of Diisopropylphosphoryl-Subtilisin BNP′ [811]

About 60 μmoles of denatured DIP-subtilisin are suspended in 250 ml of water and the pH is raised to 10 by dropwise addition of 40% NaOH. The solution is stirred overnight to completely wet and evenly disperse the protein that remained insoluble. Acetic acid (1 N) is slowly added to lower the pH to 8.15. Digestion is performed in a jacketed vessel at 40° under a steady stream of nitrogen. The pH is maintained at 8.15 by the addition of 0.10 N NaOH with a Radiometer automatic titrator. Hydrolysis is initiated by the addition of 35 mg of α-chymotrypsin (Worthington, three times crystallized), and the reaction is followed by measuring alkali consumption. After 70 min, an additional 35 mg of α-chymotrypsin are added. The reaction is terminated after 130 min by placing the reaction vessel in a boiling water bath for 15 min. The pH of the digestion mixture is then raised to 9.8 with 1 N NaOH in preparation for chromatographic fractionation on AG 1-X2.

c) Hydrolysis of Calf Rennin [368]

Crystalline rennin (140 mg) is dissolved by dialysis against distilled water and diluted to 8 ml in the reaction vessel of a pH-stat. The solution is adjusted rapidly to pH 10.6 with 0.1 N NaOH and incubated for 30 min at room temperature to denature the rennin. The pH is then adjusted to 8.5 with HCl and 2 ml of a solution containing α-chymotrypsin (1 mg/ml) and soya-bean trypsin inhibitor (1 mg/ml) are added. The solution becomes turbid after 2 hrs digestion and after 5 hrs the alkali uptake indicates that about 20 bonds per molecule of rennin have been split. The pH is then adjusted to 6.5 with HCl and the insoluble precipitate is removed by centrifugation. The soluble chymotryptic peptides containing cystine were separated by the appropriate diagonal paper electrophoretic procedure [179].

F. Pepsin

Commercial preparations of crystalline pepsin are inhomogeneous on the basis of a number of physicochemical criteria [1006]; this is primarily the result of autolysis products produced throughout various stages of preparation. Several of the main components possess proteolytic activity, and in practice this mixture of native and modified pepsin may be used without the need of further purification being apparent. However, to achieve the highest degree of specificity and the most reproducible hydrolysis possible, the use of the purified enzyme seems desirable.

1. Preparation of Pepsin from Pepsinogen [1006]

A 75 mg sample of pepsinogen is dissolved in 8 ml of glass-distilled water to yield a slightly turbid solution, pH 5 to 6. The solution is transferred to a jacketed vessel equipped with a magnetic stirrer and fitted with electrodes from a Radiometer TTT-1 autotitrator. Chilled water from a thermostated bath is circulated through the jacket of the vessel to maintain the solution at 14 $\pm$ 1°. Temperature equilibrium is attained within 2 to 3 min. The solution is brought to pH 2.0 by the addition of 0.75 ml of a chloroacetic acid — HCl mixture (9 ml of 2 N chloroacetic acid + 1 ml of N HCl). The acid must be added rapidly, in 20 to 25 sec. The amount of acid needed is determined in a trial experiment with a small sample. With preparations which contain a large amount of inorganic salt, more acid is needed; when this is the case, the concentration of HCl is increased, rather than that of monochloroacetic acid. The solution is maintained at pH 2.0 and 14° for exactly 20 min, at which time 0.75 ml of 4 N acetate buffer, pH 5.0, is added to bring the mixture to pH 4.40. The activation mixture is applied promptly (within 5 min) to a column, 2.5 $\times$ 30 cm, of sulfoethyl Sephadex C-25 equilibrated with a 0.4 N sodium acetate buffer, pH 4.38 $\pm$ 0.03 (40.29 g of sodium acetate $\cdot$ $3H_2O$ plus 29 ml of glacial acetic acid, made to a volume of 2 l). The column is operated at 4° under gravity at a flow rate of about 60 ml per hour. Fractions of 4 ml are collected. The effluent fractions which contain the material emerging at the void volume, as shown by the recorder, are pooled within a few minutes after elution. The solution is brought to pH 5 by the addition of a few drops of 2 N sodium acetate. The pepsin solution is promptly frozen in 10-ml lots and stored at −20°. The frozen samples are thawed immediately before use and converted to a solution of the desired pH and ionic composition by passage at 4° through a column, 2 $\times$ 30 cm, of Sephadex G-25.

2. Specificity

Pepsin has proved to be an extremely useful enzyme for the degradation of proteins. Peptic digests are usually used to supplement information derived from a characterization of the tryptic and chymotryptic peptides. Studies with synthetic substrates [57, 398, 504] have demonstrated that pepsin can hydrolyze peptide bonds formed by either the amino or carboxyl groups of phenylalanine, tyrosine, glutamic acid, cystine, and cysteine. In addition to protease activity, pepsin is also capable of hydrolyzing ester bonds [611]; in both proteolysis and esterolysis, activity is favored by hydrophobic side chains on both sides of the sensitive bond. An assessment of the susceptible bonds in proteins reveals that hydrolysis has occurred at bonds formed by the carboxyl groups of all other L-amino acids except proline. It should be emphasized, however, that pepsin acts mainly at the amino and carboxyl side of aromatic residues and leucine. The presence of an α-carboxylate group or a protonated α-amino group adjacent to the sensitive bond is inhibitory to peptic cleavage [613]. Thus, residues in the amino-terminal or carboxyl-terminal position of the polypeptide chain would be released slowly by pepsin. Recent studies employing *p*-bromophenacyl bromide [450], diazoacetyl norleucine methyl ester [1007], and L-1-diazo-4-phenyl-3-tosylamidobutanone [276] have indicated that the esterification of a single carboxyl group in the pepsin molecule results in loss of proteolytic activity.

With low concentrations of enzyme and short digestion periods (0.002% pepsin, 4 min, 25°), it was found that certain tryptic peptides from the β-chain of hemoglobin were hydrolyzed in a very specific manner by pepsin to give high yields of the resulting peptic peptides [421]; when higher enzyme concentrations and longer hydrolysis times were employed, more extensive cleavage was obtained. Our knowledge of pepsin action on large protein substrates has been extended appreciably by the detailed characterization of the peptic digest of subtilisin BPN′ [811]. Of particular interest is the hydrolysis of an Ile-Leu bond (234–235) and of several Ala-Ala bonds. Cleavage at peptide bonds involving basic amino acids also occurred in several instances, e.g., between Leu-His (16–17), Leu-Lys (42–43; 135–136), Asp-Lys (140–141), and Gly-Lys (169–170).

3. Method of Hydrolysis

For synthetic substrates pepsin has a pH optimum of 4.0, whereas for protein substrates the maximum hydrolysis rate occurs near pH 2. The general procedure is to dissolve or suspend the protein in an acidic solvent which will provide a pH of approximately 2 (5% v/v aqueous formic acid, 30% acetic acid, dilute hydrochloric acid, or formic acid-acetic acid-water, 1:4:45 by volume) and then to start the reaction by the addition of an aliquot of a freshly prepared enzyme solution. Proteolysis is terminated by raising the pH above 5. Conditions for the peptic digestion of several proteins appear in Table 5-4, and the use of pepsin is illustrated below.

Hydrolysis of diisopropylphosphoryl-subtilisin BPN′ [811]: Subtilisin (75 μmoles in a volume 280 ml) is digested in a thermostated vessel at 25° and pH 2.25. The pH is maintained at 2.25 by the automatic addition of 0.1 N HCl by a Radiometer titrator (type TTT 1C) equipped with a magnetic valve. Hydrolysis is initiated by the addition of 20 mg of pepsin (Worthington, twice crystallized) in 1.4 ml of 10^{-3} M HCl, and the reaction is followed by measuring acid consumption. After 58 min the digestion is stopped by raising the pH to 6.0 with pyridine and the solution is allowed to stand at room temperature for 2 hrs. Acetic acid is added to pH 4.0 followed by hydrochloric acid to a final pH of 2.2. A slight precipitate is removed by centrifugation, and the digest (350 ml) is applied to a Dowex 50-X2 column.

Table 5-4. Conditions for peptic hydrolysis of some proteins

Substrate	Enzyme: substrate molar ratio[a]	Conditions of hydrolysis				
		pH	Solvent	Temp.	Time	Reference
DIP-subtilisin BPN′	1:123	2.25	Dilute HCl	25°	58 min	[811]
Reduced-carboxy-methylated lysozyme	1:195	—	5% aqueous formic acid	23°	1 hr	[201]
Ribonuclease	1:230 1:50	2.2 1.9	0.2 M Na citrate buffer	25°	4 hrs 24 hrs	[56]
Human hemoglobin, α-chain	1:188	2.8	0.01 N HCl	Room temp.	16 hrs	[718]
Human hemoglobin, β-chain	1:188	2	0.01 N HCl	25°	1 hr	[716]
Human hemoglobin, γ-chain	1:100	2	—	Room temp.	2 hrs	[1114]
Papain	1:71	2	—	40°	10 hrs	[779]
Insulin, A chain	1:330	—	0.1 N acetic acid	37°	48 hrs	[1070]
Reduced-carboxy-methylated γ-globulin	1:100	2	30% acetic acid	38°	18 hrs	[539]

[a] The molecular weight of pepsin is taken as 32700 [1380]

G. Thermolysin

1. Properties and Specificity

The enzymic and physiochemical properties of thermolysis have been recently reviewed [822b] and the method of purification outlined [822a]. Both the amino acid sequence [1283a] and the three dimensional structure [831a] of this bacterial metallo-endopeptidase have been established. The crystalline enzyme contains 1 g atom of zinc per 35000 as well as between 3 or 4 g atoms of calcium per mole. The zinc ion is an integral part of the active site while calcium confers thermostability but is not essential for activity. Recent chemical studies [189a] employing ethoxy-formic anhydride have identified a reactive histidyl residue as being necessary for proteolytic activity. Studies on proteins and synthetic peptide substrates (see review by Matsubara) indicate that thermolysin exhibits a preference for cleavage at the amino side of hydrophobic amino acids, e.g. Ile, Leu, Val, Phe, Ala, Met, and Tyr. If the carboxyl group of the sensitive residue is in peptide linkage with proline, cleavage will not occur at the expected site. For example, in the sequence ··· Thr-Phe-Pro ···, the Thr-Phe bond is resistant to hydrolysis. Thermolysin is inhibited by ethylenediaminetetraacetic acid but is not affected by the conventional inhibitors of "serine" proteases.

2. Method of Hydrolysis

The conditions for hydrolysis are quite similar to those employed with chymotrypsin and trypsin. The peptide residue is either dissolved or suspended at a concentration of 0.1 to 1 μmole/ml in 0.1 M NH_4HCO_3 at pH 8.0. Thermolysin is added to the level of 3 to 5% by weight of substrate and digestion is allowed to proceed at 37° for 4 to 8 hrs. Peptides that are only sparingly soluble in the bicarbonate buffer usually become completely soluble with increasing time of proteolysis.

IV. Fractionation of Peptides

The successful separation of peptide mixtures resulting from either the chemical or enzymic degradation of a protein frequently requires the application of more than one technique. Several different approaches are possible. Initial fractionation may be performed on a cation exchange resin such as Dowex 50, followed by rechromatography of impure fractions on the anion exchanger Dowex 1 as recommended by SCHROEDER et al. [1111]. Further subfractionation of unresolved peptide mixtures may be accomplished by gel filtration, chromatography on derivatized cellulose or dextran gels, or paper electrophoresis and chromatography. Alternatively, the digest may be "sized" by a molecular sieving process to yield peptide fractions of varying chain lengths which in turn can be further resolved by ion-exchange chromatography.

In fractionating peptide mixtures, it is useful to know the number of peptides present, their relative electrophoretic and chromatographic behavior, and the distribution among the various peptides of those amino acids detectable by spot tests (e.g., tyrosine, histidine, tryptophan, methionine, cysteine, arginine). This information is obtained by fingerprinting the mixture and using the appropriate method of staining. The results are often instructive in selecting that system capable of best resolving the constituent peptides, since properties other than net charge are important considerations in the separation process. This also provides a means of bookkeeping, for all peptides noted on the peptide map should be recovered from the column.

There are certain problems encountered in handling peptides that can produce difficulties in sequence studies. Several modifications which may result during the purification of a peptide are cyclization of N-terminal glutamine to pyrrolidonecarboxylic acid, the α,β-transformation at an aspartyl residue, and deamidation; each of these reactions has been discussed in Section II.B.2. Recently the conversion of N-terminal S-carboxymethylcysteine to a thiazane carboxyl residue has been described [Eq. (10)] [1186].

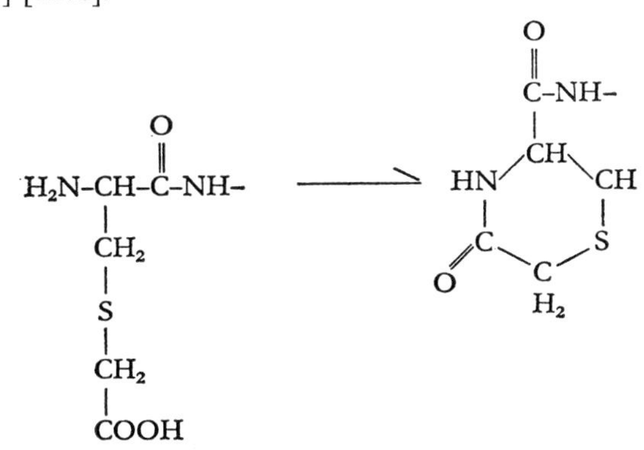

(10)

Ring closure occurred under rather mild conditions, e.g., chromatography at pH 3.2.

Difficulties are frequently encountered during the fractionation of peptides containing sulfur in thioether linkage. Exposure to mild oxidizing conditions can result in the formation of the corresponding sulfoxide, which markedly influences the chromatographic properties of the parent peptide. Thus, when multiple forms of the same peptide are encountered, low yields of the pure methionyl, S-carboxymethyl cysteinyl, or S-aminoethyl cysteinyl peptides result. Two corrective measures are available. Either the native protein can be oxidized with performic acid to convert methionine to the stable sulfone and cysteine and cystine to cysteic acid, or an antioxidant such as thiodiglycol may be incorporated into the buffer systems used for fractionation of the peptides. As mentioned previously, performic acid destroys tryptophan as well as rendering methionyl linkages insensitive to cyanogen bromide cleavage. Thiodiglycol, at a level of 0.1 to 0.3%, is quite effective in maintaining the thioether sulfur in the reduced state [411, 457, 785] and its use is recommended when fractionating peptides with oxidizable sulfur.

A. Ion-Exchange Chromatography

1. Dowex 50

This resin is a strongly acidic cation exchanger which is prepared by the sulfonation of a styrene-divinylbenzene copolymer. The per cent crosslinkage is indicated by an "X" number following the name of the particular Dowex resin. For example, Dowex 50W-X2 is made from a copolymer containing 2% divinylbenzene. Because of the highly ionized character of the sulfonic acid resin, ion-exchange behavior is exhibited over a wide pH range.

a) Buffers

Only volatile developers will be considered. The composition of aqueous buffers used with Dowex 50 resins appears in Table 5-5. Pyridine (reagent grade) must be distilled prior to use owing to the presence of ninhydrin-positive contaminants. Distillation from phthalic anhydride reduces the level of these contaminants, but the pyridine so obtained is frequently not of adequate purity for use in chromatographic solvents. The usual procedure involves distillation of reagent grade pyridine after refluxing with ninhydrin [85]. Not more than a 2-month supply of pyridine should be purified at one time, since some deterioration occurs with time. Amber bottles that can be tightly sealed should be used for storage and exposure to direct sunlight avoided. Glacial acetic acid is reagent grade and does not require further purification.

Table 5-5. Volatile buffers for Dowex 50 chromatography

Pyridine concentration M	pH	Pyridine (ml/1000 ml solution)	Glacial acetic acid (ml/1000 ml solution)	Reference
0.20	3.1	16.1	260	[690]
0.5	4.1	40.2	138	[690]
2.0	5.1	161	145	[690]
4.0	5.6	318	90	[822]
8.5	5.6	684	180	[717]

b) Preparation of Resin and Chromatographic Column

Dowex 50-X2 may be obtained as the analytical grade resin (AG 50W-X2) from the Bio-Rad Laboratories (Richmond, Calif.). The most commonly used particle size range has the dry mesh designation of 200 to 400 mesh. This highly purified resin is ready for use after it has been cycled through the following regeneration scheme. The resin contained in a Büchner funnel is washed successively with 5 to 6 volumes of 2 N NaOH, distilled H_2O, 4 N HCl, and distilled H_2O. Before use, the resin is converted to the pyridinium form by washing with several volumes of 2 M pyridine followed by a final equilibration with 0.2 M pyridine-acetic acid, pH 3.1.

Heavy-walled, jacketed, chromatographic columns of varying diameters and lengths as well as the appropriate connectors are available from Beckman Instruments (Palo Alto, Calif.). Manufacturers of other suitable columns are Technicon Corp. (Chauncey, New York) and Phoenix Precision Instrument Co. (Philadelphia, Pa.). The column is mounted vertically and the bottom outlet closed. Water from a constant temperature bath maintained at 40° is circulated through the jacket, and a small amount of the pH 3.1 buffer is poured into the column to establish a column height of approximately 10 cm. A slurry of resin (three volumes of buffer to one volume of settled resin) is introduced down the inside of the column and about a 1- to 2-cm layer allowed to settle by gravity before the column outlet is opened. The remainder of the packing operation is carried out under a pressure of 10 to 15 cm Hg as directed by Moore and Stein [879]. Additional portions of slurry are added to obtain the desired bed height. The resin bed is washed with 1 to 2 column volumes of starting buffer before the sample is applied. It is advisable not to prepare a column more than a few days in advance, since distortion of the resin may occur and a non-volatile colored material is eluted with the gradient.

c) Chromatographic Procedure

If the sample has been lyophilized, the residue is dissolved in the minimum volume of either 0.2 M, pH 3.1 equilibrating buffer or 2% acetic acid and the pH adjusted to 2.2 to 2.5 with HCl before application to the column. When lyophilization is not necessary, the digest may be acidified directly by adding glacial acetic acid. The buffer above the surface of the resin is carefully removed and the sample layered on top of the column. Air pressure may be used to facilitate sample application; however, care should be exercised to prevent air from entering the resin. Whenever pressure is applied to the column, it should be released slowly to prevent cracking of the resin bed or the introduction of air at the bottom of the column on decompression. The sample is rinsed in with several small volumes of equilibrating buffer before the column is filled.

Generally, one of two types of gradients may be used to develop the chromatogram, convex or linear (Fig. 5-3). The system generating a linear gradient of increasing pH and pyridine concentration will be described. Cylinders of equal cross sectional area, similar to those illustrated in Fig. 5-3, are joined by two short pieces of Tygon tubing with a 5-cm section of heavy-walled glass capillary tubing (2 mm internal diameter) in the center. The connecting tubing is filled with the pH 3.1 buffer and clamped shut on the side distal to the mixing cylinder, which is then partially filled with the starting buffer. Next, the tubing leading from the mixing cylinder to the pump and to the column is filled with the same buffer. Connection is made to the top of the column, and elution is begun by pumping approximately 1 to 2 column volumes of the pH 3.1 buffer through the column at the appropriate flow rate. To establish the gradient, equal volumes of 0.2 M pyridine-acetic acid, pH 3.1 and 2 M pyridine-acetic acid, pH 5.1 are placed in the mixing and adjoining cylinders, respectively. The tops of the two buffer columns are adjusted to the same height by raising or lowering one of the cylinders. With the pump shut off, the clamp between the two cylinders is carefully opened and a final adjustment is made at this time to attain hydrostatic equilibrium. Magnetic stirring is started and pumping is resumed. At the end of the gradient, limiting buffer (2 M pyridine-acetic acid,

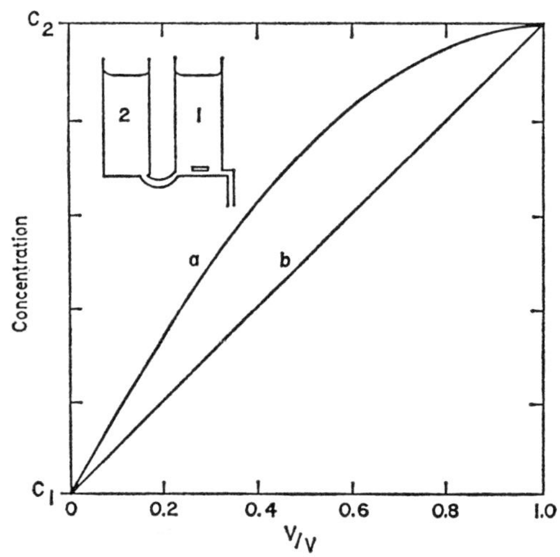

Fig. 5-3. Concentration gradient curves. Curve *a*: cross-sectional area of vessel *2* is twice that of mixing vessel *1*. Curve *b*: vessels *1* and *2* have the same cross sectional area. The abscissa indicates the fraction of the total volume of the system that has been delivered. (Reprinted in part from Ref. [124] by courtesy of Analytical Chemistry)

pH 5.1) is added to the mixing cylinder, and an additional three column volumes of effluent are collected. Finally, the column is eluted with either 4 M or 8.5 M pyridine acetic acid, pH 5.6 in order to remove tightly adsorbed peptides. During this phase, the resin will contract appreciably and the column back pressure will tend to increase. It may be necessary to compensate for this by decreasing the flow rate. The general operating conditions for columns of varying dimensions are presented in Table 5-6. These are intended as guidelines and may be modified within limits. An example illustrating this chromatographic procedure is presented in Fig. 5-4.

[6] The pump must be capable of maintaining a constant volume output over a wide range of flow rates. Two pumps which are satisfactory for this purpose are the Accu-Flo Pump (Beckman Instruments, Palo Alto, Calif.) and the Mini pump (Milton Roy Co., Philadelphia, Pa.). It is convenient to equip the pump with a gauge so that operating pressures can be monitored.

Table 5-6. Conditions for operation of Dowex 50-X2 columns

Column size	0.9 × 160 cm	1.9 × 160 cm	3.8 × 160 cm
Volume of buffer contained in each cylinder at start of gradient (liters)	0.3	3.5	10
Flow rate (ml/hr)	15	60	220
Fraction volume (ml)	2	10	20

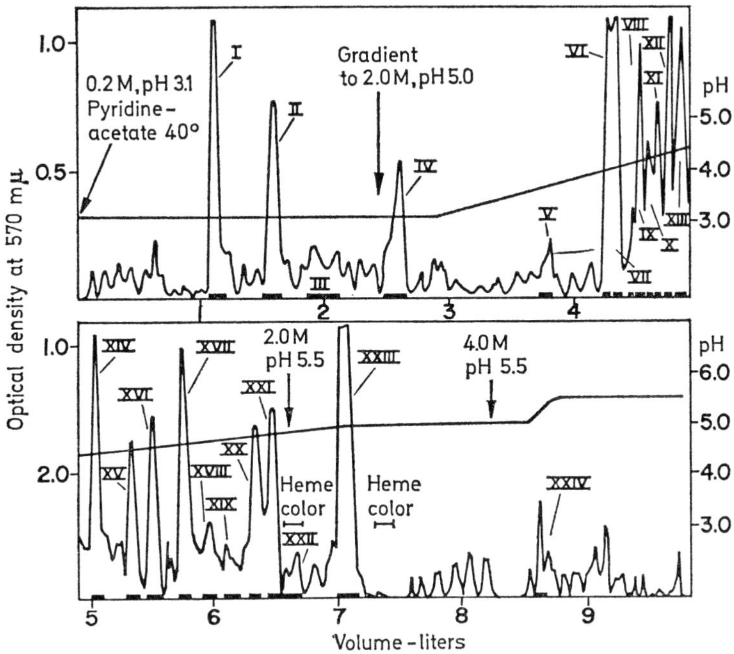

Fig. 5-4. Elution pattern of peptides derived from the tryptic digest of human cytochrome *c*. Thirty-one μmoles were chromatographed on a 2 × 165 cm column of Dowex 50-X2. The solid bars on the abscissa indicate the fractions pooled from which peptides were isolated. (Reprinted from Ref. [822] by courtesy of The Journal of Biological Chemistry)

d) Examination of Column Fractions

The progress of the fractionation procedure is followed by analyzing the column effluent for nonhydrinreactive material after alkaline hydrolysis [547]. Aliquots from alternate fractions are pipetted into clean polypropylene tubes. One ml of 2.5 N NaOH (freshly prepared) is added to each tube, and the rack containing the unstoppered tubes is placed into a boiling water bath for 2 hrs. After the hydrolysis period the tubes are removed and cooled to room temperature. Acetic acid (1.0 ml of 30% by volume) is added to each tube and the contents thoroughly mixed. Next, 2.0 ml of ninhydrin reagent are added to the neutralized hydrolysate and the remainder of the analysis is carried out as described in Section III.C.1. With the peptide elution pattern as a guide, the appropriate fractions are combined and concentrated to dryness on a rotary evaporator at 40°. The residue is taken up in a small volume of water and re-evaporated to remove residual pyridine acetate. The peptide(s) is dissolved in an appropriate solvent (e.g., aqueous solutions of pyridine, trimethylamine, acetic acid, or formic acid) and transferred to a labeled vial for further study. The homogeneity of each fraction is assessed by the two-dimensional separation technique described by INGRAM [609] and further purification carried out as indicated.

e) Comments

The amount of sample (total µmoles) which can be chromatographed on any given size column will depend primarily on the complexity of the peptide mixture. For example, 100 µmole quantities of various chymotryptic digests of cytochrome *c* [220, 907, 926] have been separated on 3.8 × 150 cm columns of Dowex 50-X2. Digests of lysozyme (25 µmoles) [199] and subtilisin BPN′ (60 µmoles) [811] were chromatographed on 1.9 × 150 cm columns. The 0.9 × 150 cm column will usually handle between 8 to 10 µmoles of an enzymic digest. As a first approximation, peptides are expected to elute from an ion-exchange column on the basis of their net charge. This is not always observed, since non-ionic interactions with the resin matrix are also influencing factors. Peptides containing hydrophobic amino acids, particularly the aromatic residues, exhibit a greater degree of retention as a result of the hydrophobic interaction of these side chains with the polystyrene backbone. Behavior of this type can be illustrated by considering the following example. Tryptic peptide T-2 from subtilisin BPN′ [661] was digested with chymotrypsin and the following points of cleavage were noted:

```
       C-2             C-1              C-3
←────────────→ ←────────────→ ←────────────────→
Ala-Pro-Ala-Leu-His-Ser-Gln-Gly-Tyr-Thr-Gly-Ser-Asn-Val-Lys
←────────────────────────────→
            C-4
```

The chymotryptic peptides are numbered in order of their elution from AG 50W-X2 (Fig. 5-5). The neutral tetrapeptide C-1 did not emerge until the gradient had taken effect; this retardation is probably attributable to the presence of the tyrosyl residue since neutral non-hydrophobic peptides usually elute before the gradient is established. Furthermore, the basic aromatic peptide C-4 was retained to a much greater extent than the basic hexapeptide C-3, which lacks an aromatic residue. The occurrence of leucine in both C-2 and C-4 also contributes to their hydrophobic character. Pyridine in the eluting buffers tends to disrupt the non-ionic peptideresin interactions. This point was nicely illustrated by KIMMEL et al. [690] who compared the chromatographic behavior of tryptic peptides derived from oxidized papain on Dowex 50-X2 with use of pyridine-acetic acid and sodium acetate systems. Marked differences were noted between the two elution patterns even though the experiments were conducted under essentially identical conditions with regard to pH, cation concentration, and temperature. In general, peptides were less retarded when columns were eluted with the pyridine buffers, but this effect was most notable with peptides containing tyrosine, phenylalanine, and tryptophan.

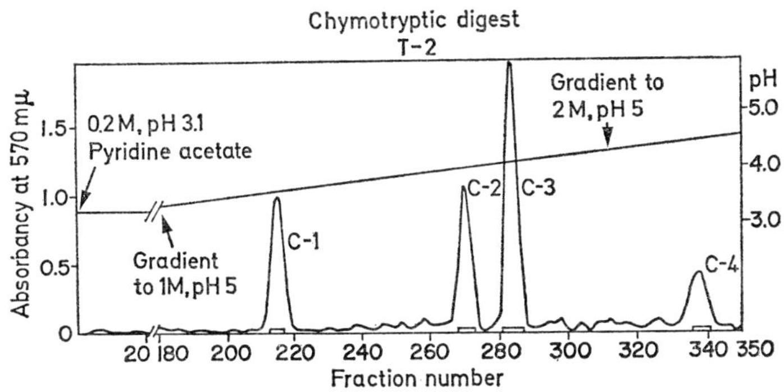

Fig. 5-5. Elution pattern of chymotryptic peptides derived from 11 μmoles of Peptide T-2. Chromatography was performed at 40° on a 0.9 × 160 cm column of Dowex 50-X2. (Reprinted from Ref. [661] by courtesy of The Journal of Biological Chemistry)

2. Dowex 1

This resin is a strongly basic anion exchanger with a cross-linked polystyrene network similar to that found in Dowex 50. The basic properties derive from the quaternary ammonium function attached to the polymer backbone; specifically, Dowex 1 is the trimethylbenzyl ammonium derivative.

a) Developers

The improved gradient system of SCHROEDER and ROBBERSON [1112] for ion exchange chromatography of peptides on Dowex 1 will be described. This elution schedule has a decided advantage over the previously used developers [1111] in that it produces a gradual, slightly sigmoid pH gradient. Thus, the abrupt pH change which results from the addition of acetic acid to the starting buffer is avoided. Five developers are required. The following quantities are for the preparation of 4 l of the pH 9.4 starting buffer: N-ethyl morpholine, 60 ml; α-picoline, 80 ml; pyridine, 40 ml; glacial acetic acid, 0.5 ml. Buffers of pH 8.4 and 6.5 are prepared in a similar manner except for the addition of increasing amounts of acetic acid to achieve the desired pH; approximately 3 and 37 ml of acid, respectively, are needed. Aqueous solutions of 0.5 N and 2 N acetic acid are also required. All reagents except the acetic acid must be distilled before use. Redistilled N-ethyl morpholine is best stored in the cold. It is advisable to use freshly prepared buffers, since they are somewhat unstable and decompose to give non-volatile products on standing. Also, it is important to prepare the alkaline buffers with CO_2 free water so that bubbles will not form in the column as the pH drops during the gradient. This is best accomplished by thoroughly flushing the water to be used for the preparation of buffers with nitrogen before adding the reagents.

b) Preparation of the Resin and Chromatographic Column

The analytical grade of the anion-exchange resin Dowex 1-X2 is obtained from Bio-Rad Laboratories (Richmond, California) as AG 1-X2 (200 to 400 mesh) in the chloride form. Fines are removed by suspending the resin in approximately ten volumes of distilled water, allowing the resin to settle for about 30 min, and siphoning off the supernatant. This step is repeated until the supernatant is essentially clear. The following method of regeneration and equilibration of the resin is patterned after SCHROEDER et al. [1111] and applies to the processing of 1 lb of AG 1-X2. The resin is washed on a Büchner funnel with 2 l of water at 60°, and at room temperature with 1 l of carbonate-free 0.5 N NaOH, 3 l of H_2O, 0.5 l of 1 N HCl, and 3 l of H_2O. The resin may be stored wet as the chloride in a dark bottle. Before pouring the column, the required amount of resin is placed in a Büchner funnel, and the following solvents are allowed to flow through under gravity: 2 volumes of H_2O, 2 volumes of carbonate-free 0.5 N NaOH, 5 volumes of H_2O, 2 volumes of 1 N acetic acid, 2 volumes of H_2O, and 5 volumes of the pH 9.4 equilibrating buffer.

Table 5-7. Conditions for chromatography on Dowex 1 [1112]

Column size	0.6 × 60 cm	1 × 100 cm
Equilibrating volume (ml)	60	100
Volume of buffer (pH 9.4) in mixing flask (ml)	40	135
Fraction size (ml)	1	3[a]
Volume of developer (ml)		
pH 9.4 buffer	10	40
pH 8.4 buffer	30	120
pH 6.5 buffer	40	160
0.5 N acetic acid	60	240
2 N acetic acid	100	400
Flow rate (ml/hr)	60[a]	60[a]
Re-equilibrating volume (ml)	75	150

[a] These values differ from those given in the original table and have been kindly provided by Dr. Walter A. Schroeder (personal communication). The flow rates are attained with a constant volume pump.

The chromatographic columns are of the same design as employed for Dowex 50 separations. The column is clamped in a vertical position and water at 40° C is pumped through the jacket from a constant temperature bath. A small amount of the pH 9.4 buffer is allowed to pass through the column, and the outlet is closed in time to retain a column of buffer of approximately 10 cm. The resin is suspended in the pH 9.4 buffer in the proportion of 1 part of resin to 3 parts of buffer, warmed to 40°, and degassed. The slurry is poured into the column and a gentle vacuum is applied to the top of the column while the resin settles. When settling is complete, excess buffer is removed except for about a 10-cm layer above the resin, and the procedure just described is repeated until a resin bed of the desired height is obtained. The column is equilibrated by washing with the pH 9.4 starting buffer (Table 5-7).

c) Chromatographic Procedure

The buffer is removed from above the resin bed, and the sample, which has been dissolved in a small volume of pH 9.4 buffer and adjusted to pH 10.4 with 0.2 N

NaOH, is carefully applied to the column. After the sample is rinsed into the resin, pH 9.4 buffer is added to fill the column. The gradient device employs a constant volume mixing flask connected to a buffer reservoir by a short section of polyethylene tubing (Fig. 5-6). It is convenient to use the same type of pump that was

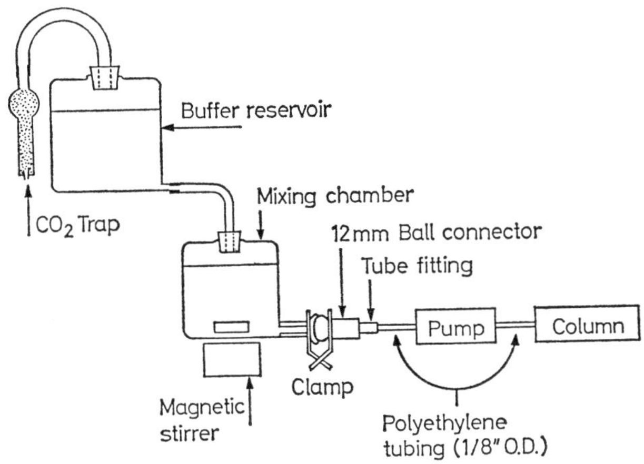

Fig. 5-6. Constant volume gradient apparatus for Dowex 1 chromatography. The 12 mm ball connector (Part No. 312937) and the tube fitting (Part No. 313285) are obtained from Beckman Instruments, Palo Alto, Calif.

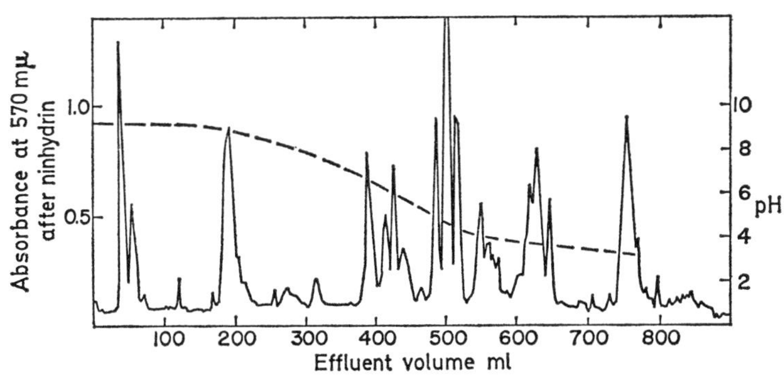

Fig. 5-7. Separation of tryptic peptides obtained from approximately 7 μmoles of aminoethylated-γ-chain of bovine fetal hemoglobin (1 × 100-cm column). The dashed line indicates the pH of the effluent fractions. (Reprinted from Ref. [1112] by courtesy of Analytical Chemistry)

recommended for use with the Dowex 50 columns, but the application of a hydrostatic head has been suitable for obtaining the desired flow rate in certain instances. The volume of buffer placed in the mixing flask, as well as the volume of other developers used during the run, will depend upon the size of the column (Table 5-7). A CO_2 trap (Ascarite, A. H. Thomas Co.) is installed at the air intake of the buffer reservoir, and the elution is begun with pH 9.4 buffer in both the mixing chamber and reservoir. After the required volume of starting buffer has flowed through the column, the pH 9.4 buffer in the reservoir is replaced with the appropriate volume of pH 8.4 buffer which in turn is replaced by the pH 6.5 buffer and so on until the elution is completed with the 2 N acetic acid developer. The column effluent fractions are examined as described for Dowex 50 (Section IV.A.1.d.). A peptide elution pattern illustrating this method of separation is shown in Fig. 5-7. It is possible to re-equilibrate the resin in the column by washing with the pH 9.4 buffer until the pH of the effluent matches that of the influent. Under certain conditions, however, this may not be desirable. When an enzymic digest of a whole protein is fractionated on Dowex 1-X2, certain peptides may not be eluted by the developers, and a regeneration cycle would be required to clean the resin adequately.

d) Comments

Peptides containing histidine, tyrosine, phenylalanine, and tryptophan are eluted from Dowex 1 more slowly than would be predicted on the basis of charge. Generally the order of emergence of peptides from this anion exchanger is approximately the reverse of that noted with Dowex 50. Whereas the method of elution from Dowex 50 employs a gradient of increasing ionic strength and pH, peptides are eluted from Dowex 1 by continuously increasing the acetic acid concentration. With certain peptide mixtures, it may be desirable to modify the pH gradient by changing the volumes of the different developers. For example, if several basic peptides are being separated, resolution may be improved by increasing the volume of pH 9.4 buffer in the mixing flask to provide a more gradual change in pH in the alkaline range.

3. Phosphocellulose

a) Chromatographic Procedure

Phosphorylated cellulose (Whatman Chromedia, P-11) with a capacity of 7 meq. per gram is obtained from H. Reeve Angel (Bridewell, New Jersey). The cellulose is successively washed on a Büchner funnel with 4 volumes of 1 N NaOH, distilled H_2O until the filtrate is below pH 9, 4 volumes of 1 N HCl, and distilled H_2O until the pH of the filtrate is above 4. The resin may be stored as an aqueous suspension in the cold room until ready for use. The conditions to be described are basically those employed by Canfield and Anfinsen for the separation of the chymotryptic peptides of egg white lysozyme [201]. All chromatographic steps are carried out in the cold room, and the salt and pH gradient is generated by a multichambered Varigrad [971]. Prior to pouring the column the cellulose is equilibrated in a Büchner funnel with the starting buffer (Table 5-8). The column outlet is closed and buffer is added to fill approximately 1/5 of the tube. A degassed slurry of phosphocellulose (1 volume cellulose to 3 volumes of buffer) is poured down the side of the column (2.4 × 25 cm) and a 1- to 2-cm bed of exchanger is allowed to form under gravity. The bottom outlet is then opened and additional portions of slurry are added to obtain the desired bed height. Final equilibration of the column is achieved by washing with several bed volumes of starting buffer. The sample is dissolved in 10 ml of distilled water, adjusted to pH 3.4 with formic acid, and applied to the column. A four-chambered Varigrad with a capacity of 1100 ml per chamber is used with ammonium acetate buffers (Table 5-8). A constant volume pump maintains the flow rate at 60 ml/hr and 10-ml fractions are collected. When the Varigrad is nearly empty, 500 ml of 0.2 M ammonium acetate are added to Chamber 1 and titrated to the same pH as the solution in the top of the column (pH 4.91). To Chamber 2,

Table 5-8. Elution buffers for phosphocellulose columns [201]

Buffer system	Chamber No.	Volume ml	Acetate, concentration, M	pH
Ammonium acetate	1	1100	0.02	3.95
	2	1100	0.07	4.03
	3	1100	0.13	4.30
	4	1100	0.20	5.06
Pyridine-acetic acid	1	500	0.05	3.91
	2	500	0.10	4.02
	3	500	0.15	4.21
	4	500	0.20	4.42
	5	500	0.25	4.63
	6	500	0.25	4.80
	7	500	0.30	5.01
	8	500	0.45	5.40

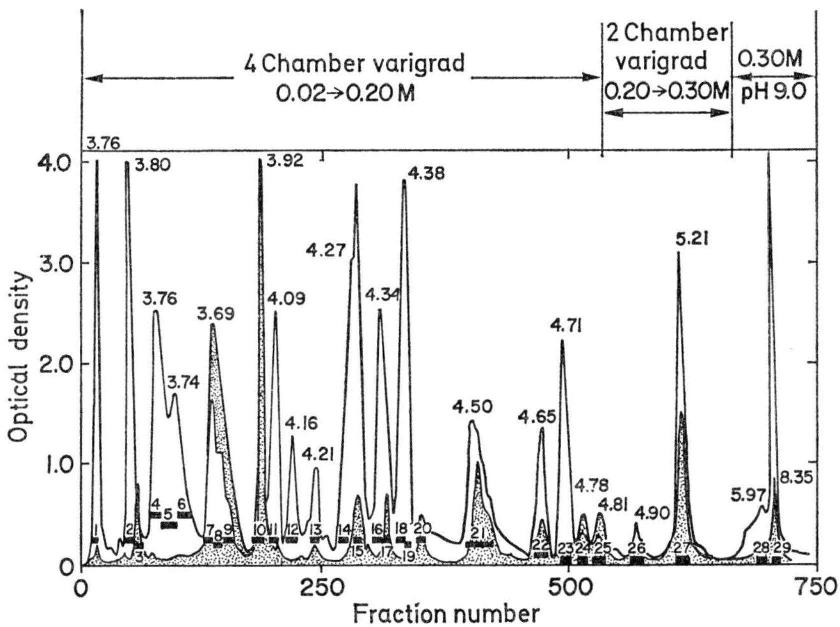

Fig. 5-8. The chromatographic separation of peptides produced by chymotrypsin digestion of carboxymethylated lysozyme. The solid lines represent the optical density at 570 mμ of aliquots subjected to ninhydrin analysis following alkaline hydrolysis. The stippled areas represent the optical density at 280 mμ of the column effluent. The regions taken for analysis are numbered, and the fractions pooled for a given region are symbolized by black bars. The pH of the peak tubes for each pooled fraction is noted. (Reprinted from Ref. [201] by courtesy of The Journal of Biological Chemistry)

500 ml of 0.3 M, pH 6.3 ammonium acetate is added and the instrument is now operated as a two-chambered Varigrad. When these buffers are exhausted, the column is washed with 0.3 M ammonium acetate, pH 9.0. The effluent fractions are examined with respect to pH, 280 mμ absorption, and ninhydrin-reactive material after alkaline hydrolysis. Samples of 200 μl are removed from alternate fractions, transferred to polypropylene tubes, and dried in a vacuum oven at 80° in the presence of P_2O_5 for 24 hrs prior to hydrolysis. The results of the chromatographic separation are presented in Fig. 5-8. The appropriate fractions are pooled and lyophilized. To remove excess ammonium acetate, the lyophilized powder is heated to 50° for 24 hrs over P_2O_5 in a vacuum oven, dissolved in a small volume of water, and lyophilized a second time.

Pyridine-acetic acid buffers have also been used for fractionating peptides on phosphocellulose (Table 5-8). The effluent cannot be monitored at 280 mμ owing to the strong absorption of pyridine, but the pyridine-acetic acid system has two main advantages over ammonium acetate buffers: (a) the pyridine salts are more easily removed by rotary evaporation or lyophilization and (b) aliquots of effluent fractions may be analyzed directly, since the buffer yields a low ninhydrin value in the colorimetric analysis.

b) Comments

Of the derivatized cellulose exchangers, phosphocellulose is best suited for the fractionation of peptide mixtures. It has a high capacity and functions well in the adsorption and elution of peptides in the range of pH 3 to 5; the two pK values for the phosphate group are near pH 2.2 and 6.5, respectively [972]. Carboxymethyl (CM-) cellulose has a pK value of approximately 4 and is therefore only partially ionized in the pH region that favors peptide binding. Consequently, the capacity of this exchanger is markedly reduced, and the titration of the carboxyl groups during an increasing pH gradient can exceed the buffering capacity of the low ionic strength eluting buffer and thus result in an undesirable pH change within the column. Similar limitations also apply to diethylaminoathyl (DEAE-) cellulose. The most promising application of CM- and DEAE-cellulose is for the purification of large peptides, using a salt gradient at a constant pH. From the available results, it appears that peptides are usually eluted in order of their net charge from cellulose exchangers. The interaction of aromatic side chains with the cellulose matrix is not pronounced as is the case with the Dowex resins.

4. Derivatized Dextran Gels

Ion exchangers prepared from cross-linked dextran gels are commercially available as the DEAE-, CM-, and sulfoethyl (SE-) derivatives of Sephadex (Pharmacia Fine Chemicals, Piscataway, New Jersey). Each exchanger may be obtained in two porosity grades which are assigned the numbers of 25 and 50; the prefix A- or C- is used to indicate either an anionic or cationic exchanger. The type 25 has a porosity which corresponds approximately to G-50 Sephadex, whereas the type 50 has a lower degree of cross linkage and is comparable to G-200 Sephadex in porosity. Thus, for peptide fractionation the type 25 exchanger is preferable since exclusion from the interstices of the gel does not occur until the molecular weight of the solute exceeds 10000.

a) Chromatographic Procedure

The cation and anion exchangers are supplied in the sodium and chloride forms respectively and are prepared for use in the following manner. The dry beads are allowed to swell in a large volume of starting buffer at room temperature for 24 to 30 hrs with gentle stirring. Several times during this period, the exchanger should be permitted to settle, the supernatant removed, and fresh buffer added. In this way, equilibration is accomplished while the gel is hydrating. The column is packed in much the same way as described for phosphocellulose and is washed with several bed volumes of starting buffer before the sample is applied. The bed volume of the gel is markedly influenced by pH, ionic strength, and the specific buffer employed. In general, the gel will shrink during the conversion of the exchanger from the charged to the uncharged form. This effect is also noted upon increasing the salt concentration in the eluant. It is desirable to use a starting buffer with an ionic strength of near 0.05 since below this value the volume change associated with increasing salt concentration is quite pronounced. The most common type of gradient is one producing a linear increase in salt concentration; however, both pH and ionic strength may be varied simultaneously, but it is desirable to use a pH gradient only outside of the titration range of the exchanger. Because contraction of the gel bed will occur during the development of the chromatogram, it is advantageous to use a constant volume pump to provide an even flow rate of approximately 6 to 8 ml/cm^2/hr. Alternate effluent fractions are examined for ninhydrin-reactive material after alkaline hydrolysis as described in Section IV.B.1. Cation exchangers are regenerated by successive washings on a Büchner funnel with 0.5 N HCl, H_2O, 0.5 N NaOH, and H_2O to remove excess alkali. For DEAE-Sephadex, the reverse order is recommended.

b) Comments

Derivatized dextran gels can be profitably employed in situations where Dowex columns do not prove satisfactory. For example, extremely basic peptides, particularly those containing aromatic residues, can be so tightly adsorbed to Dowex 50-X2 that sodium hydroxide is required to elute them, whereas these same peptides are likely to pass through Dowex 1-X2 almost unretarded, with the result that there is poor resolution and additional fractionation is necessary. Accordingly, sulfoethyl-Sephadex was used by Konigsberg et al. [716] to further purify peptide fractions obtained by countercurrent distribution of a peptic digest of the β-chain of human hemoglobin. This was of particular value in the case of two peptides (Pβ1, Leu-Ala-His-Lys-Tyr-His; and Pβ18, Tyr-Pro-Trp-Thr-Gln-Arg-Phe) which were completely retained on Dowex 50-X2 and which also failed to separate on Dowex 1-X2. When mixtures containing these peptides were chromatographed on SE-Sephadex C-25 with the usual pyridine acetate gradient employed for Dowex 50-X2 columns, both peptides Pβ1 and Pβ18 were obtained in high yield and in a reasonably homogeneous state. DEAE-Sephadex A-25 proved useful for the fractionation of a mixture of insoluble tryptic peptides derived from the f_2 bacteriophage coat protein [719]. The insoluble fraction was dissolved in 8 M urea and chromatographed on the anion exchanger equilibrated with 0.05 M Tris chloride, pH 8.0, containing 8 M urea. A linear gradient of increasing chloride concentration (0.05 M to 0.5 M Cl^-) was used to develop the chromatogram.

B. Gel Filtration

Two different types of gel filtration media are currently used to fractionate peptide mixtures. These are cross-linked dextrans (Sephadex, Pharmacia Fine Chemicals, Inc., Piscataway, New Jersey) and the polyacrylamide gels (Bio-Gel P, Bio-Rad Laboratories, Richmond, Calif.). Both are prepared in bead form, and for a given type the beads have pores of carefully controlled dimensions which permit resolution based on molecular size. A list of the various types of gels applicable to the separation of peptides is given in Table 5-9 along with the fractionation range for each gel. Gels of a lower degree of cross-linkage capable of accommodating solutes of higher molecular weight have been omitted. Sephadex G-10 is a tightly cross-linked gel of relatively small pore size, whereas Sephadex G-50 contains fewer cross-links and is considerably more porous. The same trend in physical properties holds for the Bio-Gel series.

Table 5-9. Gel filtration media for peptide fractionation

Type	Fractionation range (molecular weight of solute)
Sephadex G-10	up to 700
Sephadex G-15	up to 1500
Sephadex G-25	1000 to 5000
Sephadex G-50	1500 to 30000
Bio-Gel P-2	200 to 2000
Bio-Gel P-4	500 to 4000
Bio-Gel P-6	1000 to 5000
Bio-Gel P-10	5000 to 17000
Bio-Gel P-20	10000 to 30000

A gel filtration column contains two types of water volumes: V_0, the outer or void volume and V_i, the interstitial volume of the gel bead. The total volume of the gel bed, V_t, is described by Eq. (11), where V_g is the volume occupied by the gel matrix.

$$V_t = V_0 + V_i + V_g \,. \tag{11}$$

V_i increases as the porosity of the gel increases. Thus, upon swelling in water, 1 g of G-50 will occupy a volume 3 to 4 times greater than an equivalent amount of G-10. The distribution of the solute between V_0 and V_i may be expressed in terms of a coefficient, K_d. If total exclusion from the interstices of the gel occurs, K_d is equal to zero. If the diffusion of the solute is not restricted by the pore size of the gel, K_d is equal to 1. Therefore, the elution volume, V_e, of a solute with a K_d in the range of 0 to 1 will be [Eq. (12)]:

$$V_e = V_0 + K_d V_i \,. \tag{12}$$

Certain molecules may have K_d values greater than 1, indicating that a process other than sieving occurs. For example, aromatic compounds are retarded on Sephadex to a greater extent than are non-aromatic compounds of the same molecular size. Thus, peptides containing tyrosyl, phenylalanyl, or tryptophanyl side chains do not exhibit normal elution behavior [989]; this adsorption effect is most pronounced with tryptophan. A second property of Sephadex that produces anomalous behavior during the filtration process is charge. The presence of a small number of carboxyl groups in the dextran matrix retards positively charged solutes and is responsible for the partial exclusion of acidic substances from the interstices of the gel. Both of these charge effects are neutralized by increasing the salt concentration in the eluting solvent. Gel-solute interactions frequently prove useful in separating peptides of similar size but different composition. Adsorption effects appear to be minimal with polyacrylamide gels, although hydrogen bonding is possible with the amide bonds of the matrix.

1. Procedure

The choice of the appropriate gel for a particular separation depends on the nature and complexity of the mixture. For the best resolution, a fine particle size is recommended. The gel is prepared by adding the dry beads to a large excess of solvent with constant stirring at room temperature. For the materials listed in Table 5-9, approximately 3 to 5 hrs should be allowed for complete hydration. The amount of gel required for a given column may be estimated from the bed volume (ml/g dry gel) specified for that gel. The gel suspension (3 parts of solvent to 1 part of settled gel) is degassed to remove trapped air and poured down the side of the column which is filled about one-fifth full with equilibrating solvent. A 6- to 8-cm layer of gel is allowed to foım before the outlet of the column is opened. The gel is not allowed to settle completely before the buffer above the settling zone is removed and additional slurry added with stirring. This step is repeated until a continuous gel bed of the required height is obtained. Note that an unsectioned column of gel results from this method of pouring, i.e., separate layers of gel are not observed. Several bed volumes of solvent are passed through the column prior to use. The surface of the gel is protected by the insertion of a filter paper disk. Before starting an experiment, the column packing should be examined for imperfections by passing a solution of colored material (e.g., hemoglobin or Blue Dextran-2000, Pharmacia Fine Chemicals) through the bed and visually inspecting for air bubbles, zone distortion, or channeling. If any of these faults are noted, repacking is indicated. This also permits the determination of the void volume, provided the substance used is totally excluded from the inner portion of the gel particle.

Prior to the application of the sample, the column outlet is closed and the solvent above the gel is removed. The sample is carefully distributed over the surface of the gel and the bottom outlet of the column is opened. If the flow is too rapid, a piece of flexible tubing and a clamp attached to the outlet will control the flow rate. Several small portions of eluant are used to rinse the wall of the column and wash the sample into the gel bed before filling and connecting the column to the eluant reservoir. For optimal separation, the width of the starting zone should be kept at a minimum. The flow rate should be maintained in the range of 6 to 8 ml/cm^2/hr by use of a constant volume pump or a Mariotte flask. Gel columns may be reused without repouring or regenerating.

2. Comments

Gel filtration is an exceedingly mild fractionation procedure which is a useful adjunct to the common ion-exchange methods. Since the sieving phenomenon is relatively insensitive to pH, salt concentration, and eluant, the use of a wide variety of solvents is permitted. This feature is particularly valuable when dealing with peptides that are insoluble in the usual chromatographic developers. For example, the B- and C-chains of aminoethylated α-chymotrypsin, which are exceedingly difficult to separate owing to a strong tendency to aggregate, were easily separated on Sephadex G-50 when hexafluoroacetone was used as the solvent [188]. Aqueous solutions of formic acid containing either 8 M urea or 4 M guanidine hydrochloride have also been useful in fractionating insoluble peptides [813]. The purification of the polypeptide hormone oxytocin has been effectively accomplished by partition chromatography on Sephadex G-25 [1406]. Similar partition columns have proved useful in the fractionation of DNP-peptides [803]. Excellent recoveries are obtained from gel columns, and problems with the irreversible adsorption of solutes have not been reported.

C. Preparative Paper Electrophoresis and Chromatography

The fractionation of peptide mixtures by preparative paper techniques is widely employed in protein sequence studies. The methods are versatile, permitting the processing of small as well as relatively large quantities of material. The separation can involve chromatography or electrophoresis or a combination of both. It is wise to keep the number of steps to a minimum, since yields decrease as the number of manipulations increases. Thus, complex mixtures of peptides are best resolved into simpler mixtures by ion-exchange chromatography, which may be then subject to a final one-step purification on paper.

1. Electrophoresis

Separations can be conducted with either a flat plate or tank type electrophoresis apparatus. The latter piece of equipment is described in detail by Katz et al. [672], and is essentially a modification of the original Michl design [859]. Commercial versions are available from Savant Instruments (Hicksville, N.Y.), Gilson Medical Electronics (Middleton, Wisconsin), and O. Hiller Co. (Madison, Wisconsin). Whatman 3MM is most commonly used and is obtainable as sheets ($18^1/_4 \times 22^1/_2$ inches) or rolls ($18'' \times 300'$). The appropriate buffer system (Table 5-10) and the

Table 5-10. Buffer systems for paper electrophoresis

pH	Components	Ratio (by volume)	Reference
7.9	0.068 M N-ethyl morpholine, 0.025 M acetic acid		[456]
6.5	Pyridine-glacial acetic acid-H_2O[a]	100:4:900	[1053]
4.7	Pyridine-glacial acetic acid-H_2O[b]	1:1:48	[860]
3.6	Pyridine-glacial acetic acid-H_2O	1:10:89	[1053]
1.9	Glacial acetic acid-formic acid (98%)-H_2O	87:25:888	[49]

[a] Dilute with an equal volume of water before use.
[b] The paper is wetted with a 1:10 dilution of this buffer.

time of electrophoresis necessary to separate the peptides in a mixture are determined by trial runs with about 0.05 μmoles of material. It is important that the peptide solution be free of salts in order to achieve satisfactory separations. Small amounts of pyridine acetate or ammonium bicarbonate will not interfere, since these salts volatilize during the drying process. If the sample is dilute, it is convenient to remove all solvent by rotary evaporation at 40° and redissolve at a concentration of about 10 μmoles/ml so that the application of a large volume is unnecessary. The material to be fractionated is applied as a narrow uniform zone across the width of the paper to within 6 cm of each edge; overloading should be avoided since streaking and poor resolution can result. Approximately 0.4 μmole of peptide is applied per cm and usually not more than 10 μmoles are applied to a single sheet. The area of

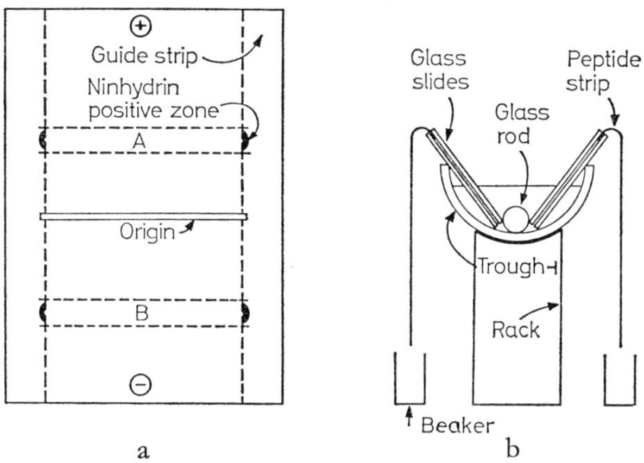

Fig. 5-9. (a) Reconstructed (actual size, $22^1/_2 \times 18^1/_4''$) electrophoretogram. Guide strips have been stained with ninhydrin. Zones *A* and *B* enclosed by the horizontal dashed lines indicate the areas containing peptide material. (b) End view of peptide elution apparatus; the cover is not shown

application should be supported so that it is not in direct contact with a solid surface. Drying is facilitated by directing a stream of unheated air from a hand dryer across the sample zone. The position of the origin with reference to the anode and cathode can be adjusted depending on the electrophoretic properties of the respective peptides so as to provide optimal distance for migration. The paper is then transferred to a flat glass surface and a glass rod is inserted under the sample zone. The paper is wetted with buffer to within about 1 cm of each side of the applied sample. Buffer is allowed to flow evenly and slowly into the sample zone from both sides, thus concentrating and sharpening the peptide band. This is a most important step and must be performed with great care; repeated small applications of buffer are usually necessary. The paper on both sides of the sample is carefully blotted to remove excess buffer and to provide a paper of reasonably uniform wetness. Electrophoresis is carried out at a potential gradient of approximately 50 V/cm for the required period of time. After electrophoresis, papers are clipped to a glass rod and dried in a forced air oven at 40°. To locate the position of the peptides, guide strips which include approximately 6 mm of each end of the sample zone are cut from the paper, sprayed with ninhydrin solution, and heated at 80° for 20 min to develop the stain. If necessary, a narrow strip may also be cut from the center of the paper and stained, thus providing a third point of reference for establishing the peptide zone. Tryptophan peptides are detectable under ultraviolet light as a fluorescent band and may be located directly without staining. The electrophoretogram is reconstructed by attaching the stained strips to the unstained center section with transparent tape at the extreme ends of the paper (Fig. 5-9a). The regions of the paper containing the peptide bands are outlined with a ruler and pencil, labeled, and cut with scissors. Rubber gloves or forceps should be used to handle the paper strips containing the peptide. If the peptide bands are well separated, an additional 6 to 8 mm on each side of the band is included in order to compensate for any irregularities in the shape of the zone. Peptides are eluted from the paper with 20% acetic acid or some other suitable solvent (Fig. 5-9b). The homogeneity of the purified peptide should be ascertained by electrophoresis or chromatography in at least two different systems prior to amino acid analysis.

2. Chromatography

Descending chromatography is performed with full-sized sheets of Whatman 3MM. The origin, guide strips, and lines to indicate folding are drawn on the paper as illustrated in Fig. 5-10. The bottom edge of the paper is serrated with pinking shears so that the solvent will run off evenly should the solvent front progress that far. The paper is folded along lines *a* and *b* as indicated (Fig. 5-10), and the salt-free sample is applied evenly to the origin, which should exceed 6 mm in width. The chromatogram is developed for a period of 16 to 20 hrs with the appropriate solvent (Table 5-11). After chromatography the paper is dried, and the guide strips are removed and stained. The peptide zones are located, cut out, and eluted as described for preparative electrophoresis. Generally, 6 to 10 μmoles of a peptide mixture may be fractionated on a single sheet.

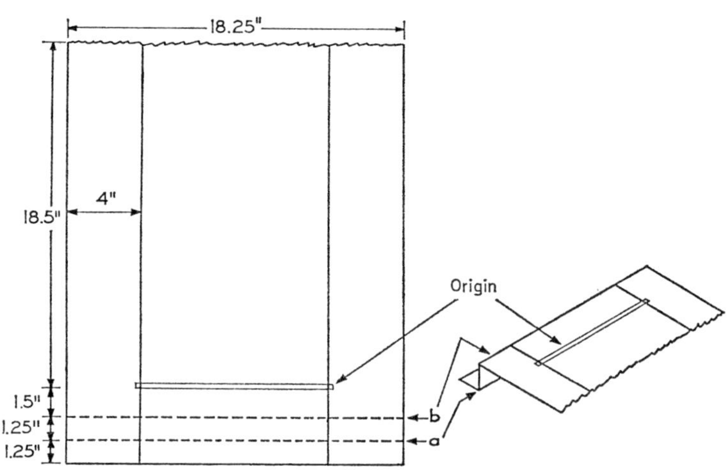

Fig. 5-10. Diagram illustrating preparative paper chromatogram. See text for details

Table 5-11. Solvent systems for paper chromatography of peptides

Components	Ratio (by volume)	Reference
1-Butanol-glacial acetic acid-H_2O	200:30:75	[781]
1-Butanol-glacial acetic acid-H_2O[a]	4:1:5	[672]
1-Butanol-pyridine-glacial acetic acid-H_2O	75:50:15:60	[781]
Pyridine-*iso*-amyl alcohol-H_2O	7:7:6	[1114]
sec-Butyl alcohol-formic acid-H_2O	7.0:0.9:2.1	[454]

[a] The upper phase is used for chromatography and the lower aqueous phase is placed in the bottom of the chromatographic chamber.

3. Assessment of Paper Techniques

Electrophoretic separations require considerably less time than do the chromatographic procedures. Generally, the complete fractionation of a peptide mixture by electrophoresis can be accomplished easily in 1 day, whereas the chromatographic method consumes about twice this amount of time. Recoveries from paper will depend upon the properties of the peptide but usually range between 70 to 90%. Some peptides exhibit very strong adsorption to paper and are obtained in rather low yields. If small amounts (less than 1 μmole) of material are being fractionated, Whatman No. 1 paper may be used, in which case approximately 0.1 μmole of peptide per cm is applied. The main advantages of paper techniques are speed and simplicity. Recently, the development of the diagonal paper electrophoretic technique has permitted the location of disulfide bridges in proteins [179, 180, 368, 1173], the selective purification of methionine peptides [1263], and the preferential isolation of lysine peptides from tryptic digests of trifluoroacetylated [969] and maleylated [192 a] proteins.

Chapter 6

Identification of Specific Amino Acid Residues

ERNESTO SCOFFONE and ANGELO FONTANA

I. Introduction

The determination of amino acid residues in proteins has received new emphasis in recent years because of the increasing significance of sequence studies and because of the importance of evaluating genetic changes and homologies in studies of evolution. Concomitant with the elucidation of the chemical structure of the amino acids, methods were explored for the determination of each of these moieties. During the early period, gravimetric and then colorimetric methods were extensively employed. Only a limited number of amino acids could be estimated by these procedures and, furthermore, the precision of analysis was far from satisfactory. In the same period, microbiological techniques of measurement were introduced which permitted the determination of most amino acids [443].

The complete separation of all common amino acids by a simple procedure and their determination awaited the development of ion exchange column procedures. The automatic amino acid analyzer introduced by SPACKMAN et al. [1198] has been a key instrument in one of the most active fields of all of biochemistry, i.e., the determination of the primary structures of proteins. The analyzer has also been of fundamental help to many workers pursuing research programs involving characterization of the components of physiological fluids. With the present techniques a complete amino acid analysis can be performed in a short time on micro-quantities down to the nmole level with a minimum of effort and provides values of high precision.

The limiting factor in determining the amino acid composition of a protein lies in the fact that the composition of the hydrolyzate of the protein (usually 6 N HCl, 110° C, 20 to 48 hrs) is not always a true reflection of the composition of the parent protein. The two major problems associated with acid hydrolysis in 6 N HCl are the destruction of the labile amino acids (tryptophan, cysteine, cystine, serine, threonine and tyrosine) [778] and the slow hydrolysis of some peptide bonds between bulky, sterically hindered amino acids. With the exception of tryptophan and sulfur amino acids, these problems may be circumvented by extrapolating values obtained from 22 and 72 hrs hydrolyzates. On the other hand alkaline hydrolysis accompanies complete destruction of cysteine and cystine and of some other amino acids and is restricted to the determination of tryptophan, which is rather stable under these conditions.

Determination of the content of a particular amino acid in the intact protein is feasible only in the case when a reactive side chain is implicated. In such instances, with application of appropriate procedures, a quantitative measure can be obtained without the requirement of prior manipulation of the protein sample. Such reactive functional groups include the indole nucleus of tryptophan, the thiol group of cysteine, the disulfide linkage of cystine, the phenolic group of tyrosine, the imidazole ring of histidine and the guanidyl group of arginine.

The limitation inherent in such analytical methods lies in the nature of the protein, since its secondary and tertiary structure allows for different degrees of reactivity of the functional groups; reaction will occur only with those groups which are openly available or unhindered. Therefore for analytical purposes the reactions are carried out in denaturing conditions for the protein, in which loss of organization is accompanied by enhancement in the reactivity of the functional groups.

Besides the analytical interest, the reactivity of functional groups in a protein is a powerful tool in correlating structure and biological function. Selective modification of a particular side-chain group allows conclusions about the participation of an amino acid residue in the active site or about the availability or the degree of burial of such a residue in the protein molecule [1003, 1281, 1386].

In this chapter we shall be concerned primarily with analytical methods for tryptophan and sulfur amino acids and only procedures giving reliable results will be discussed in detail; other methods will be reported more briefly.

II. Presentation of the Results

A. Quantitative Expression

The results of the quantitative determination of single amino acid residues in proteins are best expressed as numbers of residues per molecule of protein. However, in general, an exact determination of the molecular weight of the protein is not always available and therefore the results usually are expressed as "residues/10^5 g of protein" or "grams of amino acid per 100 g of protein". In this way, any ambiguities due to uncertain molecular weights are removed and a direct comparison of results of various workers is possible [156, 1292].

If the weight of solid protein is taken as the basis for the calculation, consideration must be given to the salt, ash and moisture content of the sample. Before analysis, salts are removed by exhaustive dialysis against 0.1 M KCl in order to displace other polyions by K^+ ion and then by dialysis against deionized water to remove the KCl; the protein is then recovered by lyophilization. Analytical determinations are best carried out on the air-equilibrated samples rather than with anhydrous proteins, which may be very hygroscopic. The moisture content of the sample is determined by drying *in vacuo* at 105° C to constant weight [158, 230]. The ash content is determined by combustion in a platinum crucible over an open flame until constant weight is attained.

The total protein nitrogen is indicative of the state of the protein and its accurate estimation is an indication of the purity of the protein sample. The percentage of nitrogen should be 13.5 or more. If the nitrogen content falls below this limit, it is essential to remove non-protein substances. The standard microanalytical procedures of Kjeldahl is the most widely used for the determination of nitrogen. For the description of the standard procedure adopted, the reader is referred to the original literature [144, 230, 970] or to the review by BLOCK [115].

The best and most precise procedure for determination of the protein concentration in solution is the convenient and rapid estimation of the amino acid content by automatic amino acid analysis of a protein hydrolyzate obtained from an aliquot of the solution. By such means the analytical data of the determination of single amino acid residues are referred to the recovery of one or more amino acids in the protein hydrolyzate. When the amino acid composition of the protein is known, its concentration in solution is calculated and the data are directly reported as residues per molecule.

The protein concentration in solution can be also referred either to the nitrogen content of the solution [117] or by using methods which estimate particular groups. Several sensitive and rapid colorimetric analyses are usefully employed, with high precision, when appropriate controls and careful calibration of the methods are made. For this purpose the phenol reagent of FOLIN [364] following the LOWRY method [794] and the biuret procedure of GORNALL [431] are the most extensively used. Earlier methods involved the Sakaguchi reaction for arginine [480, 685] or the coupling of histidine and tyrosine with diazotized sulfanilic acid [480, 561, 1345].

It must be emphasized that the choice of proteins for testing analytical methods of amino acids requires caution since it is clear that comparison of analytical data becomes ambiguous when test proteins of poorly established purity and homogeneity are used. For this reason, the discrepancies in values reported in the literature often may be regarded as arising from the use of improperly defined test proteins, rather than in differing accuracies. Many workers have used casein as a test protein, but this material is by no means of sufficiently constant composition to warrant its use for this purpose [226, 1292].

B. Protein Determination

1. Lowry's Method [794]

Reagents. Reagent A: 2% Na_2CO_3 in 0.1 N NaOH. Reagent B: 0.5% $CuSO_4 \cdot 5\,H_2O$ in 1% sodium or potassium tartrate. Reagent C: alkaline copper solution. Mix 50 ml of reagent A with 1 ml of reagent B and discard after 1 day. Reagent D: diluted Folin reagent. Titrate FOLIN-CIOCALTEU phenol reagent [364] (commercially available, e.g., Fluka AG, Basle, Switzerland) with NaOH to a phenolphthalein end-point. On the basis of this titration dilute the FOLIN reagent (about 2-fold) to make it 1 N in acid.

Procedure. To a sample (25 to 500 γ of protein) in 1 ml solution, 5 ml of reagent C is added. Mix well and allow to stand for 10 min or longer at room temperature. Reagent D (0.5 ml) is added very rapidly and mixed within a second or two. After 30 min or longer, the sample is read spectrophotometrically at 750 nm. If the protein is present in a dilute solution any convenient fraction of the total volume may be employed for analysis. Calculate from a standard curve obtained using a test protein (crystalline bovine albumin).

2. Biuret Method [431]

Reagent. Weigh 1.5 g cupric sulfate ($CuSO_4 \cdot 5\,H_2O$) and 6 g of sodium potassium tartrate ($NaKC_4H_4O_6 \cdot H_2O$); transfer to 1 l volumetric flask and dissolve in about 500 ml of water. Add with constant swirling 300 ml of 10% NaOH. Make to volume with water, mix, and store in a paraffin-lined bottle. This reagent should keep indefinitely but must be discarded if, as a result of contamination or faulty preparation, it shows signs of depositing any black or reddish precipitate.

Procedure. To the sample dissolved in 1 ml solution add 4 ml of the reagent. Allow the solution to stand for 30 min at room temperature and the absorbance at 540 nm is then determined. Calculate the concentration from a calibration curve obtained using a test protein (albumin, casein).

III. Estimation of Tryptophan

Tryptophan is a vital constituent of proteins and constitutes about 1% of their amino acid residues [1179]. Although the amino acid composition of many proteins has been reported, evaluation of the number of tryptophan residues is often omitted. This situation reflects the difficulty associated with quantitative determination of tryptophan residues in a protein, since acid hydrolysis usually results in the extensive destruction of tryptophan [561].

On the other hand, alkaline hydrolysis, though less destructive [1202], allows determination of tryptophan by chromatographic separation of the protein hydrolyzate [1347]. Tryptophan, in the standard procedure for the amino acid analysis, is eluted from the short column of the basic amino acids as a well resolved peak appearing before lysine [393]. Basic hydrolysis unfortunately does not give always satisfactory results. Spies and Chambers [1202] studied factors affecting the stability of tryptophan in 5 N NaOH. They found that tryptophan is more labile when the amino acid is peptide-linked and that losses depend on the protein composition. Oelshlegel et al. [932] hydrolyzed proteins with NaOH in the presence of thioglycol, the recoveries of tryptophan being 89 to 96%. Basic hydrolysis by either $Ba(OH)_2$ or NaOH of proteins has been critically reviewed [705, 1031, 1070, 1203, 1204].

Enzymatic hydrolyses of proteins have also been used to avoid destruction of tryptophan [535]. Proteolytic enzymes such as pronase [927], viokase [1288] or the combination of papain, leucine aminopeptidase and prolidase [535] have been used. Upon completion of enzymatic hydrolysis of the protein, the hydrolyzate is rendered free of proteolytic enzyme by precipitation and centrifugation and the extract directly analyzed with the amino acid analyzer. The method however may not be generally valid, since the possibility of artefacts due to self-digestion of the enzymes used.

Satisfactory alternatives to the hydrolytic procedures have been sought for many years. Methods have been tried to produce colored derivative by exploiting the reactivity of the indole nucleus in the intact protein [393]. Sodium hypochlorite, ferric chloride, cupric sulfate, bromine, sodium nitrite, usually in acid solution, have been found to produce color with tryptophan-containing proteins. Later workers have formed other colored derivatives by reaction with the tryptophan specific reagents 2-hydroxy-5-nitrobenzyl bromide (HNB-Br) [587] or 2-nitrophenylsulfenyl chloride (NPS-Cl) [374]. Still others have used the characteristic absorption at 280 mμ to estimate tryptophan spectrophotometrically or spectrofluorometrically.

Because of uncertainties in each of these methods for tryptophan analysis, a fast accurate general procedure remains to be found [393]. Although the acid hydrolysis with *p*-toluenesulfonic acid of LIU [787] seems to be the most reliable method giving satisfactory results, generally the analytical figures for tryptophan should be checked by different methods.

These methods are here discussed in more detail.

A. Acid Hydrolysis

Although tryptophan is usually destroyed in the usual acid hydrolysis of proteins, MATSUBARA et al. [820] discovered that 2 to 4% thioglycolic acid in the 6 N HCl has a protective effect on the destruction of tryptophan. Recovery of tryptophan to the extent of 80 to 90% for some proteins was reported by these authors. However, when glucose was present in the protein sample, the protective effect of thioglycolic acid was lost and no tryptophan could be recovered from the hydrolyzate. Analogous results have been obtained performing the acid hydrolysis with β-mercaptoethanol as protective agent [393].

LIU and CHANG [786, 787] utilized *p*-toluenesulfonic acid as the catalyst for hydrolysis rather than HCl. This acid was found to be less destructive and when hydrolysis was carried out in 3 N *p*-toluenesulfonic acid with 0.2% 3-(2-amino-ethyl)-indole as a scavenger for tryptophan destruction the recovery of tryptophan was 88 to 98%. This procedure provides a new method for the determination of tryptophan in proteins, giving results comparable with those obtained by other non-hydrolytic methods. In Table 6-1 are reported typical results of the tryptophan analysis of polypeptides and proteins. An advantage of the method lies also in the fact that the hydrolyzate can be placed on the amino acid analyzer column for analysis without prior removal of the solvent, as is required when 6 N HCl is used. In addition, the recovery of all other amino acids is comparable to that obtained from a 6 N HCl hydrolyzate.

Reagents. *p*-Toluenesulfonic acid (Baker Chemical Co., Phillipsburg, New Jersey) can be recrystallized by dissolving in the smallest quantity of water and, after filtering, three volumes of concentrated HCl is added. The precipitate of the acid is filtered off, washed with concentrated HCl and then dried at 45° *in vacuo* over NaOH pellets to remove HCl (mp 104 to 105° C).

3-(2-Aminoethyl)-indole can be prepared from its HCl salt (Eastman Organic Chemicals, Rochester, New York) supplied by the manufacturer. To a solution of 2 g of the salt in 25 ml of water, chloroform is added, and to the mixture 5 ml of 2 N NaOH are added at 0° C with stirring. The chloroform layer is separated, washed with water and dried over $MgSO_4$. Evaporation of the solvent under reduced pressure gives crystalline the free base.

Procedure. Hydrolysis is carried out in heavy-walled ignition tubes (18 × 150 mm) which have been washed with H_2SO_4:HNO_3 (3:1), rinsed with deionized water and oven-dried. The protein (about 1 mg) is hydrolyzed under vacuum at 110° for 22, 48 and 72 hrs, with 1 ml of 3-N-*p*-toluenesulfonic acid containing 0.2% 3-(2-amino-ethyl)-indole.

Table 6-1. Analyses of tryptophan in proteins by acid hydrolysis with 3 N toluenesulfonic acid containing 0.2% 3-(2-aminoethyl)-indole. (Taken from Ref. [787])

Sample	Hours of hydrolysis			Extrapolated Value[a]	
	22	48	72		
	Residues			Residues	Yield (%)
Bovine deoxyribonuclease A	2.80	2.65	2.57	2.92 (3)	97
Bovine trypsinogen	3.74	3.69	3.65	3.81 (4)	95
Sperm whale myoglobin	1.87	1.87	1.89	1.87 (2)	94
Bovine chymotrypsinogen A	7.45	7.19	6.79	7.85 (8)	98
Thrombin	8.02	7.68	7.69	8.20 (9)	91
Streptococcal proteinase	4.29	3.99	3.87	4.40 (5)	88
Zymogen of streptococcal proteinase	4.22	4.08	3.68	4.70 (5)	94
Egg white lysozyme	5.52	5.45	5.16	5.90 (6)	98
Tyrocidine C	1.85	1.70	1,62	1.90 (2)	95
Bradykinin-potentiating pentapeptide, V-3-A	0.92	0.86	0.85	0.95 (1)	95

[a] The theoretical integral residue numbers are given in parentheses.

B. Spectrophotometry

In the absence of prosthetic groups or nucleic acids, the absorption of proteins in the ultraviolet region between 250 to 310 nm is due to the three aromatic residues, namely, tryptophan, tyrosine and phenylalanine [70]. Since phenylalanine has a low molar extinction ($\varepsilon = 195$ at 257.5 nm) and the maxima of absorption of the other two aromatic residues are at higher wavelengths, the absorption of proteins at 280 to 290 nm is due only to tryptophan and tyrosine. Therefore, if the small contribution of phenylalanine is neglected, the absorption of the solution of a protein can be considered as arising from a mixture of the two amino acids. This fact was used by HOLIDAY in 1936 [574] and later by GOODWIN and MORTON [428] as the basis of a spectrophotometric determination of these residues in proteins.

The procedure is to dissolve the protein in 0.1 N NaOH and make up to a convenient volume in order to obtain an absorption of 0.2 to 0.5 near 290 nm. The total molar concentration (a) of tryptophan and tyrosine is obtained [Eq. (1)] from the absorption at 294.4 nm, one of the wavelengths at which the absorption curves of the two amino acids intersect (isosbestic point) and therefore have the same extinction (ε).

$$A_{294 \cdot 4} = a\varepsilon \,. \tag{1}$$

From the extinction at another wavelength, 280 nm, the concentrations of the two components can be calculated. If at 280 nm the molar extinction of tryptophan (ε_1) and of tyrosine (ε_2) is known, the observed absorption A_{280} is given by [Eq. (2)]

$$A_{280} = b\varepsilon_1 + (a - b)\varepsilon_2 \tag{2}$$

(where b is the concentration of tryptophan and a to b the concentration of tyrosine) and b by Eq. (3).

$$b = \frac{A_{280} - a\varepsilon_2}{\varepsilon_1 - \varepsilon_2} \,. \tag{3}$$

After determination of the protein concentration by standard methods (see 6.II), the tryptophan residues/protein (mole/mole) is obtained from the ratio of the two concentrations.

However, the method has not gained general acceptance because of the possibilities of anomalies. Tyrosine residues in proteins sometimes behave abnormally, and incubation at pH 13 is necessary for maximal ionization of the phenolic group [1003, 1281, 1386]. Therefore the method must be tested after exposure of the protein to pH 13 for varying periods of time. On the other hand, SPIES and CHAMBERS [1202] found that a slight destruction of tryptophan could occur in 0.1 N alkaline solution.

The original application of the method to proteins was based on the assumption that the absorption of amino acids was unaffected when linked in a polypeptide chain. Later work has suggested that in preparing the standard solutions, the use of the amino acids in peptide combination is more satisfactory, e.g., glycyl-tryptophan [324], since a shift (1 to 3 nm) in the maxima of absorption toward longer wavelengths occurs compared with that of the free amino acids. BENCZE and SCHMID [73] modified the procedure of GOODWIN and MORTON [428] in order to overcome this problem.

The spectrophotometric method has been used with much success, though difficulties sometimes are encountered when tryptophan is present in the protein at a low level or when the tyrosine content is high. As an example, the A protein component of tryptophan synthetase of *Escherichia coli*, was thought to contain tryptophan [528] until the absence of this amino acid was proved by other methods (N-bromosuccinimide [955], alkaline hydrolysis, microbiological assay).

EDELHOCH [307] further improved the method by performing the absorbance measurements at 288 and 280 nm on the protein dissolved in 6 M guanidine hydrochloride in order to destroy its secondary and tertiary structure. Blocked tryptophanyl and tyrosyl compounds (N-acetyl-tryptophan amide, glycyl-tyrosylglycine) were selected as model substances to represent the absorbance of the two chromophores in proteins. The author stresses that contaminants, organic co-factors and turbidity contribute to absorbance at 280 to 290 nm. A good method for checking for impurities is to examine the absorbance of the protein in neutral 6 M guanidine hydrochloride at 315 nm where tyrosine and tryptophan do not absorb. However, although the method was shown to give values of tryptophan content in very good agreement with those known from sequence studies (trypsinogen, chymotrypsinogen, lysozyme), an imperfect fit of the spectra of the solutions containing the model compounds to those of the proteins was observed. The latter are shifted to the red in all three cases examined.

C. Colorimetry

1. p-Dimethylaminobenzaldehyde

The long known reaction of the indole nucleus of tryptophan in strong acid conditions with aldehydes to give colored products [1320, 1382] was employed by SPIES and CHAMBERS [1202] in a colorimetric procedure for the determination of tryptophan. In this method, the tryprophan-containing compound (peptides or proteins) is dissolved in concentrated sulfuric acid containing *p*-dimethylaminobenzaldehyde (DAB) and a blue color is obtained after a prescribed time by adding sodium nitrite solution.

The nature of the chromophore obtained upon reaction of the tryptophan residue and DAB is so far not completely elucidated. Current evidence does not permit a choice between structure I (derived from 1 mole of tryptophan and 1 mole of DAB) and II (derived from 2 moles of tryptophan and 1 mole of DAB), but II seems more likely since its formation requires oxidation [393].

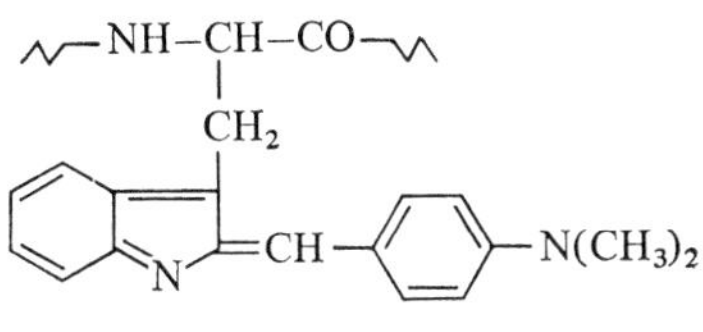

I

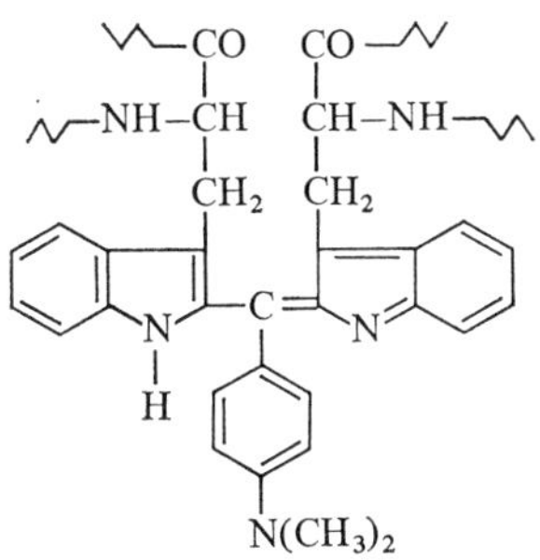

II

The method of SPIES and CHAMBERS has been extensively used and gives reliable results. However, the method was found to suffer some limitations. Variations in the maximum of absorption (from a normal value of 590 to 600 nm to 545 to 560 nm has been verified with different protein samples) [513].

Intense light, oxidizing and reducing agents including H_2S, bisulfite and chloride ions interfere with the method [263, 393, 652]. Interferences have been eliminated to some extent by appropriate addition of silver sulfate to the assay mixture [1202]. Low tryptophan values are obtained due to interactions among sulfur amino acids, serine and threonine with the indole nucleus of tryptophan. The nature of these reactions remains to be established.

FRIEDMAN et al. [391, 392] eliminate interference by cysteine and cystine residues by reducing the protein disulfide bonds and alkylating the native and generated-SH groups with vinyl derivatives. JUNEJA et al. [652] evaluated in detail the interference by sulfur amino acids; they estimated tryptophan content of wool colorimetrically after hydrolysis with 6 N $Ba(OH)_2$ for 5 hrs at 125° C followed by removal of liberated H_2S and barium ion with silver sulfate and sulfuric acid, respectively. Some workers replaced the intact protein method by methods involving basic or enzymatic hydrolysis.

Procedure [1202]. The sample (for lysozyme, 1 mg) to be analyzed is dissolved in 0.5 ml water and added to a mixture of 4 ml of 23.8 N H_2SO_4 and 0.5 ml of DAB solution in 2 N H_2SO_4 (30 mg/ml). After about 15 hrs at 22 to 24° C in the dark, to the reaction mixture is then added 0.1 ml of a freshly prepared 0.045% aqueous solution of $NaNO_2$. After standing further for 30 to 60 min in the dark, the optical density of the solution at 590 nm is measured. The tryptophan content is calculated from a standard curve obtained with tryptophan or with a tryptophan-containing peptide, after correction for a blank value.

Reactions of the indole nucleus of tryptophan with xanthydrol [283, 284] and with glyoxylic acid [1150] also have been employed for analytical purpose, but the procedures have received only limited application.

GAITONDE [405, 406] has found that treating either free or protein-bound tryptophan with ninhydrin in a mixture of formic acid and hydrochloric acid for 10 min at 100° C results in a chromophore with λ_{max} at 390 nm. FRIEDMAN and FINLEY [393] propose for the ninhydrin-tryptophan reaction the structure III.

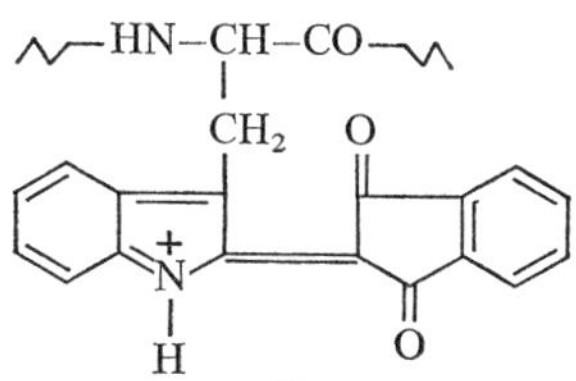

III

The reaction however seems not be unique, since the extinction coefficient of the chromophore is dependent upon the nature of the model compound and reaction conditions.

2. Sulfenyl Halides

Sulfenyl halides have been found to be specific, mild reagents for modification of the tryptophan and cysteine residues of polypeptides and proteins in acidic media [372, 373, 375, 1140]. Tryptophan is attacked at the 2-position of the indole nucleus to give a thioether [Eq. (4)] whereas cysteine is converted to a mixed disulfide Eq. (5).

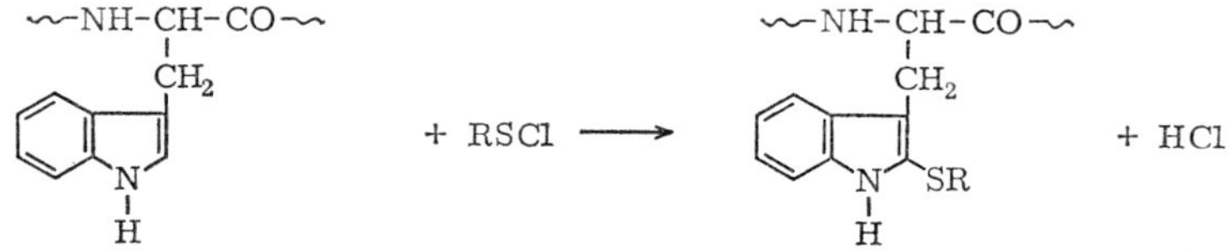

(4)

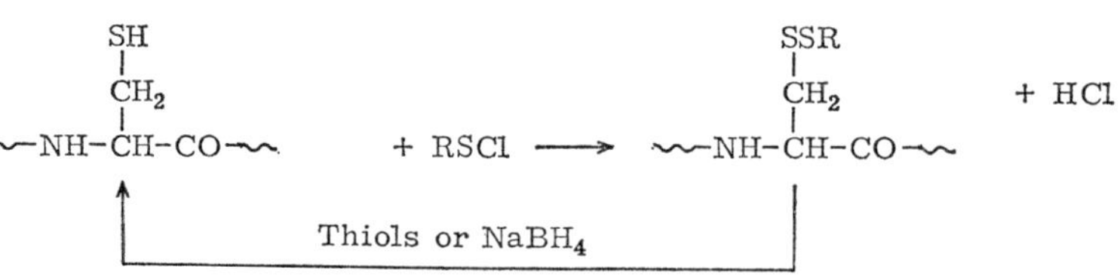

(5)

The unsymmetrical disulfide at the cysteine residue can be easily cleaved by means of reducing agents, such as β-mercaptoethanol, thioglycolic acid or sodium borohydride, or by simply exposing it to alkaline treatment (0.1 N NaOH) [372], thereby allowing specific labeling only of the tryptophan residue. For the analytical application of the reaction to cysteine the reader is referred below (6.IV.C.2.).

In principle, all members of the wide class of sulfenyl halides could be used for the labeling of tryptophan. However, these reagents are highly unstable in aqueous solvents with the exceptions of 2-nitrophenyl compounds, which are more resistant (1 000 times) to hydrolysis in comparison to the other halides.

By using nitrophenylsulfenyl halides a chromophore is generated in the protein which absorbs in the visible part of the spectrum. This fact offers an easy quantitation of the reaction by spectrophotometry. At the present time the most widely used sulfenyl halide is 2-nitrophenylsulfenyl chloride (NPS-Cl) (IV). Other compounds tested are 4-nitrophenylsulfenyl (V) [1140], 2,4-dinitrophenylsulfenyl (VI) [1140], 2-nitro-4-carboxyphenylsulfenyl (VII) [1317] and 2,4-dinitrophenyl-1,5-disulfenyl chloride (VIII) [1318]. The bifunctional reagent VIII has been used for the inter- and intramolecular cross-linking of tryptophan residues in peptides and proteins [1318].

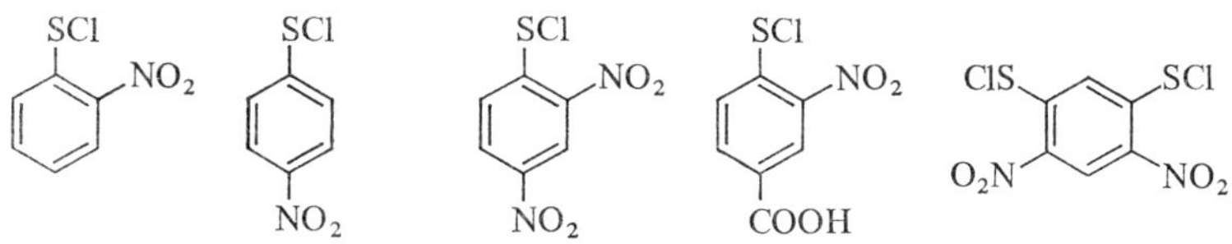

IV V VI VII VIII

The 2-thio-(2-nitrophenyl)-tryptophan residue, obtained by reaction of NPS-Cl with tryptophan-containing proteins absorbs at 365 nm with a molar absorptivity $\varepsilon = 4000\ M^{-1}\ cm^{-1}$ in aqueous acetic acid.

Fig. 6-1 shows the absorption spectrum of chymotrypsin and of its NPS-derivative (eight tryptophan residues modified) and Fig. 6-2 the extent of sulfenylation of the same protein by using increasing amounts of NPS-Cl.

It was shown [1317] that the sulfenyl halide reacts with the indole nucleus of tryptophan much faster when this residue is linked in the protein than in free tryptophan and in small peptides. This enhanced reactivity enables one to perform the sulfenylation reaction in aqueous acetic acid (50%), in spite of the fact that a rapid (2 to 3 min) decomposition of the reagent occurs, provided that the halide is first dissolved in glacial acetic acid. Quantitative modification of tryptophan residues in proteins is achieved readily, by using a low molar ratio of reagent to protein [123, 1140, 1317].

The accuracy of the procedure has been tested on a number of polypeptides and proteins, including glucagon, neocarzinostatin, gramicidin A, ACTH, human growth hormone, trypsin, lysozyme, transferrin, concanavalin, myoglobin, a bacterial proteinase, the A_1-encephalitogenic protein, staphylococcal nuclease and others.

More recently, Santarius and Belitz [1072] proposed a simplified procedure of tryptophan assay using NPS-Cl without isolation of the NPS-labeled protein and at the same time improving the sensitivity of the method.

Procedure [123, 1140]. To a solution of about 1 μmole of the protein in 1 to 2 ml of 50% aqueous acetic acid, 20 to 30 equivalents of NPS-Cl (Fluka AG, Basle, Switzerland) in 0.5 ml of glacial acetic acid are added at room temperature under

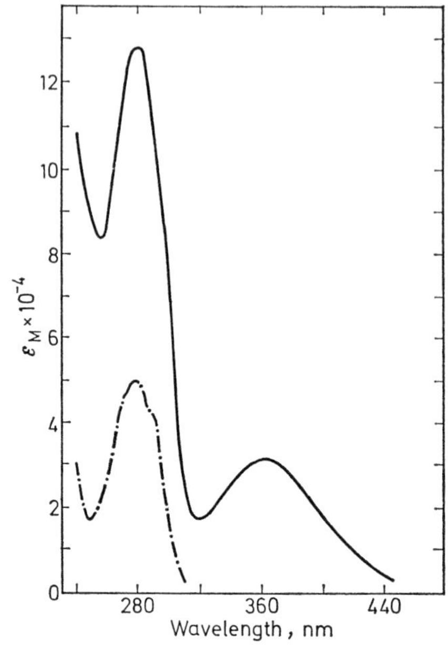

Fig. 6-1. Absorption spectra of α-chymotrypsin (— · — ·) and its NPS-derivative (—) in 80% acetic acid. (Taken from Ref. [1140])

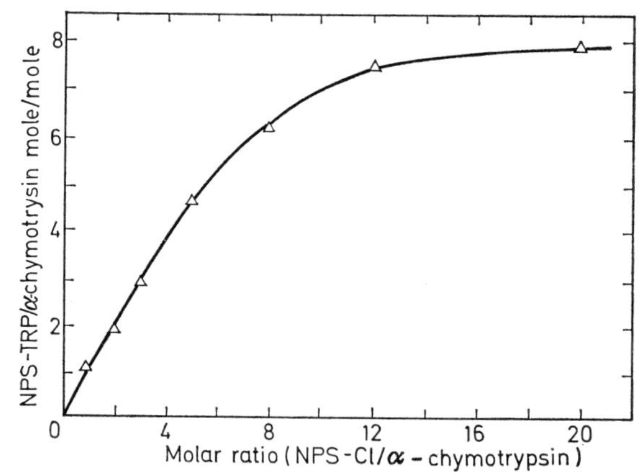

Fig. 6-2. Extent of sulfenylation of α-chymotrypsin terated with increasing amounts of NPS-Cl. The number of modified residues has been determined spectrophotometrically (see text). (Taken from Ref. [1140])

vigorous stirring. When the protein to be analyzed is insoluble in aqueous acetic acid, formic acid can be used [369]. The modified protein is precipitated with acetone-1 N HCl (39:1), separated by centrifugation, and carefully washed several times with acetone and ethyl ether. In some instances, the precipitation and wash procedure is repeated in order to remove any excess reagent. The product is dried *in vacuo* over

P_2O_5. By the procedure reported above, it was shown that practically no products arising from the excess of reagent are present in the samples of the modified proteins. In any case the sample can be further purified by gel filtration on a column of Sephadex G-25 using aqueous acetic acid as the eluent, following known procedure.

If the protein to be analyzed for tryptophan content is supposed to contain cysteine residues, reactive toward sulfenyl halides, a rough spectrophotometric test is made. By dissolving the sample of the modified protein in 0.1 N NaOH, the 2-nitrothiophenol released absorbs at 420 nm and cause an obvious variation in the spectrum of the modified proteins [369].

In order to remove the cysteinyl chromophoric group in a cysteine- and tryptophan-containing protein, an aliquot of the modified protein (20 mg) is dissolved in 2 ml of 8 M urea solution, adjusted to pH 8.6 with 5% methylamine after flushing the solution with a nitrogen stream, treated with 20 μl of β-mercaptoethanol, following the standard procedure for cleaving disulfide bonds in proteins (see 6.V.B.). After 4 hrs at room temperature, the entire solution is applied to a 1 × 40 cm column of Sephadex G-25 which has previously been equilibrated with aqueous acetic acid (0.2 M). After elution, the protein is located in the effluent (spectrophotometry) and recovered by lyophilization. Similar results can be obtained by cleaving the mixed-disulfide bond by dissolving the sample of the modified protein in 0.1 N NaOH and recovering the protein by acid precipitation or by gel filtration [123].

Calculations. In order to estimate the number of 2-thio-(2-nitrophenyl)-tryptophan residues per molecule of protein, the sample of the modified protein (about 0.1 μmole) is dissolved in 80% acetic acid (3 ml) or, for less soluble samples, in 99% formic acid and the concentration of the covalently bound chromophore determined by spectrophotometric assay at 365 nm with the use of an extinction coefficient of 4000 M^{-1} cm^{-1}.

The protein content in the solution can be analyzed in several ways, the most accurate being the use of the amino acid analyzer (see 6.II.).

3. 2-Hydroxy-5-nitrobenzyl Bromide

Koshland and coworkers [62, 586, 724] have used 2-hydroxy-5-nitrobenzyl bromide (HNB-Br) (IX) as a specific reagent for tryptophan. The water soluble reagent X has been prepared by Horton and Tucker [588] by reacting IX with dimethyl sulfide.

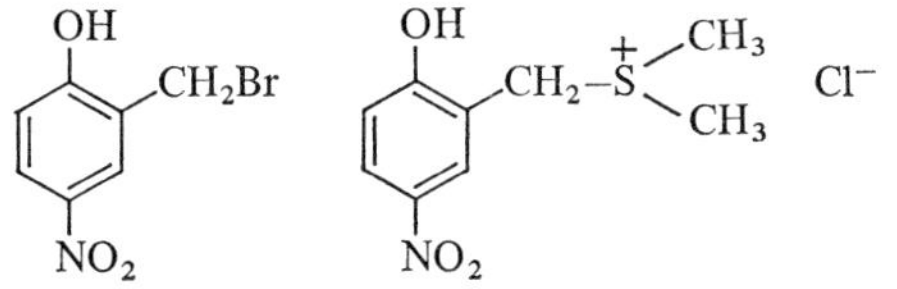

IX X

Upon reaction with tryptophan-containing proteins, a chromophore is generated in the protein which absorbs in the visible part of the spectrum. Reaction of IX with the indole nucleus of tryptophan is complex and has been studied in several laboratories [790, 838, 1190]. The 2-alkylated compound XI has been identified as the

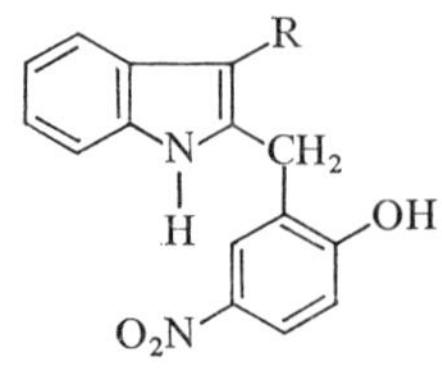

XI

minor product in the reaction of HNB-Br with the model compound skatole (3-methyl-indole), the major products being the cyclic adduct involving the OH group of HNB (XII) and the indolenine (XIII). With tryptophan ethyl ester the cyclic compound XIV has been obtained. The possibility of disubstitution products has been confirmed by SPANDE et al. [1199].

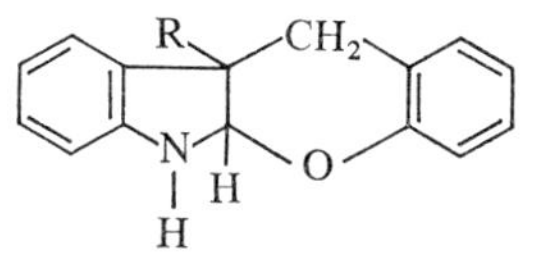

XII

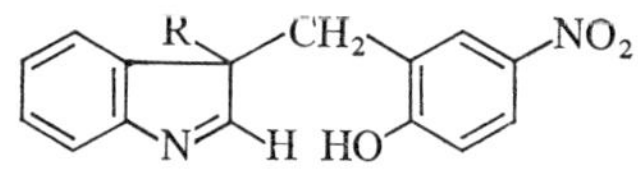

XIII

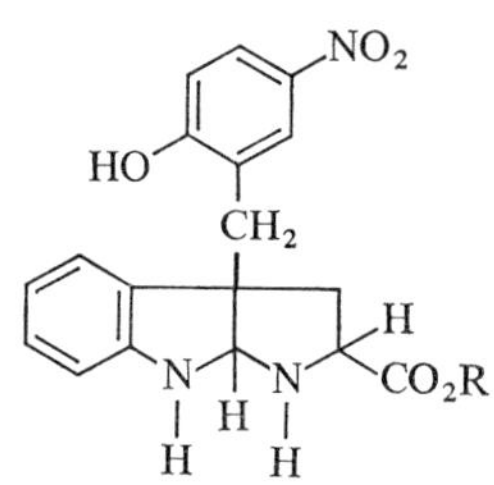

XIV

When the model compound N-acetyl-tryptophan methyl ester was reacted with two equivalents of HNB-Br a mixture of three products was obtained, the major component (37%), isolated after acetylation with acetic anhydride-pyridine, being XV.

This fact agrees with the known polyalkylation of the indole ring by alkylating agents [651] and explains the excessive incorporation of chromophore into the protein when using a high concentration of the reagent [60, 292].

The Koshland's analytical method has been shown to be fairly accurate [62] but appears to have some disadvantages.

The reaction of HNB-Br takes place also with cysteine and a procedure of blocking this residue by carboxymethylation is needed prior to treatment of the SH-containing protein with the reagent.

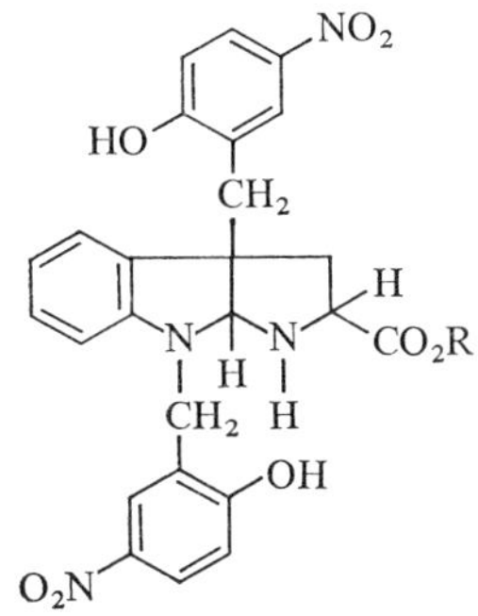

XV

Specific analyses can be performed to determine the extent to which cysteine residues might have reacted with HNB-Br. The S-HNB-cysteine peak emerges from the short column of the amino acid analyzer in the same region as histidine under standard conditions of analysis.

Finally, a quantitative reaction is achieved only under conditions of unfolding of the protein by dissolving it in 10 M urea or at best by cleavage of the S-S bridges. The use of 5 M guanidine hydrochloride as an unfolding reagent causes incorporation of the yellow label not only into tryptophan-containing proteins, but also into ribonuclease, a tryptophan- and cysteine-free protein [62].

Assuming that the chromophore covalently bound at the tryptophan residue has the same molar extinction as the model compound 2-hydroxy-5-nitrobenzyl alcohol ($\varepsilon = 18450$ at 410 nm in alkali solution), the reaction is easily quantitated by spectrophotometry. After reaction with HNB-Br, the absorption of the HNB-protein, free from excess reagent, in alkaline solution allows calculation of the number of tryptophan residues modified by HNB-Br.

Procedure [586, 587]. The protein (about 1 μmole) to be analyzed is incubated for 16 to 20 hrs at 37° C in 1 ml of 10 M urea which had been adjusted to pH 2.7 with concentrated HCl. The urea solution is then cooled at room temperature and approximately 5 mg of HNBB in 0.1 ml of dry acetone are added with vigorous stirring to the protein solution. The mixture usually remains clear but occasionally a slight precipitate forms (usually 2-hydroxy-5-nitrobenzyl alcohol) which is removed by centrifugation. The labeled protein is separated from excess reagent on a column (23 × 1.1 cm) of Sephadex G-25 which has previously been equilibrated with 0.18 M acetic acid (pH 2.7) or with 10 M urea (pH 2.7). The 10 M urea is used when necessary to keep the modified protein in solution.

The protein in the pooled fraction is precipitated by addition of 50% trichloroacetic acid to a final concentration of 5%. If urea is used instead of acetic acid, 5-fold dilution with water is necessary before trichloroacetic acid addition. The precipitate is collected by centrifugation, washed twice with 5 ml of ethanol-HCl (2 ml of concentrated HCl to 98 ml of 95% ethanol), and finally dissolved in 1 ml of 11.2 M HCl. A portion of the HCl solution (usually 0.1 ml) is adjusted to pH 12 by addition of 2.5 M NaOH and diluted if necessary to a final volume of 2.5 ml. The concentration of the reacted tryptophan in this solution is determined by spectrophotometric assay at 410 nm by using an extinction coefficient of 18450 $M^{-1}\ cm^{-1}$. The remaining aliquot can be analyzed for protein content (see 6.II.).

D. N-Bromosuccinimide

The use of N-bromosuccinimide (NBS) as a specific reagent for the modification of tryptophan in proteins has led to a rapid and convenient spectrophotometric method for determining the tryptophan content of a protein [956, 1010, 1200, 1386]. The procedure is based on the large decrease in absorbance at 280 nm which accompanies the oxidation of tryptophan to an oxindole chromophore (Fig. 6-3).

The oxidation is conducted in an acidic medium, usually an acetate or acetate-formate buffer, by using controlled amounts of NBS. Side reactions usually encountered with NBS and proteins, such as oxidation of methionine, cysteine, cystine, histidine and tyrosine, do not interfere with the tryptophan determination. Experi-

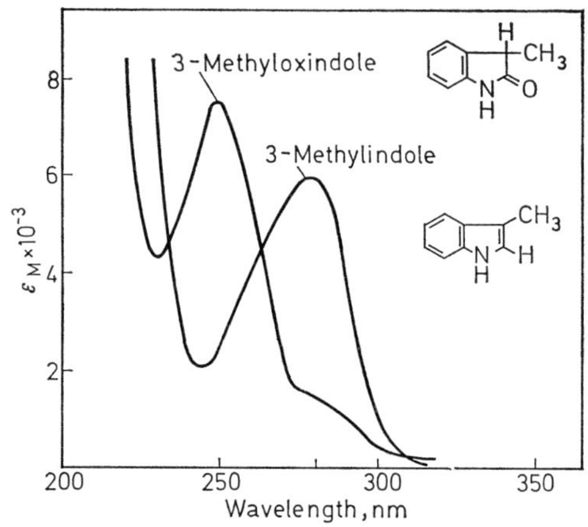

Fig. 6-3. Ultraviolet absorption spectra of a representative indole-oxindole pair, 3-methylindole (skatole) and 2-hydroxy-3-methyl-indole. (Taken from Ref. [1200])

ments with model compounds indicate that the reactivity of tryptophan is much higher in respect to the other amino acid residues. The oxidation of residues other than tryptophan is revealed by increased consumption of NBS by the protein. Usually, 2 to 4 equivalents of reagent are needed for the complete oxidation of tryptophan, but with some proteins excess up to 20 equivalents may be needed [1200]. Some proteins may possess tryptophan residues reacting very slowly or not at all under the usual conditions of assay; in such case the reaction has to be carried out in 8 M urea.

The modest material requirements and the ease with which the determination can be carried out, make the assay for tryptophan with NBS a useful alternative to the other available methods.

Procedure [1200]. The reaction of NBS (recrystallized from glacial acetic acid or water) with tryptophan-containing proteins can be conducted in 0.05 to 0.1 M acetate buffer, pH 4.0 or 8 M urea adjusted to pH 4.0 with acetic acid. Formate-acetate buffers or 70% acetic acid may also be employed. The determination is carried out by placing 2 ml of a protein solution in a cuvette provided with a small stirring bar. The bar can be left in the cuvette as long as it does not obstruct the light path. The protein concentration is selected so as to give an absorption of 1.5 to 2.0 at 280 nm (usually 1 mg/ml). The reaction is followed *in situ* with a recording ultraviolet spectrophotometer, noting the disappearance of the indole absorption at 280 nm. The spectrum of the protein solution is recorded over the range 250 to 400 nm in order to determine the initial optical density at 280 nm. NBS in aqueous solution (about 10^{-2} M) is then added in 5 or 10 μl quantity to the cuvette with stirring. After several minutes, the spectrum of the protein solution is again recorded to measure the absorption at 280 nm. The addition is continued in a stepwise fashion until further additions of NBS do not further decrease the optical density at 280 nm. The minimum absorption is recorded and corrected for the volume increase due to the added NBS solution (usually 100 to 200 μl).

The optical density decrease at 280 nm can be related to the optical density of tryptophan in the sample by the empirical factor 1.31, which corrects for the absorption at 280 nm of the oxidation product of tryptophan, oxindole (Fig. 6-3). The amount of tryptophan in the solution is calculated using an extinction coefficient for tryptophan at 280 mμ of 5500 M^{-1} cm^{-1} [1200].

E. Other Methods

A tentative procedure for determining the tryptophan content in a protein was proposed by PREVIERO et al. [995]. The method involves formylation of the 1-position of the indole nucleus of tryptophan by dissolving the protein in formic acid-HCl solution [Eq. (6)].

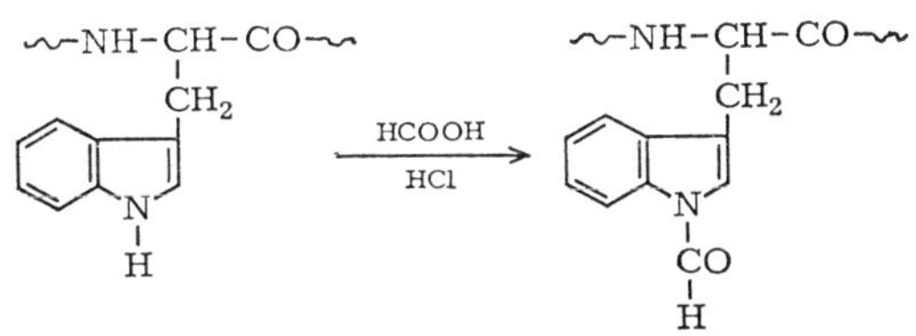

(6)

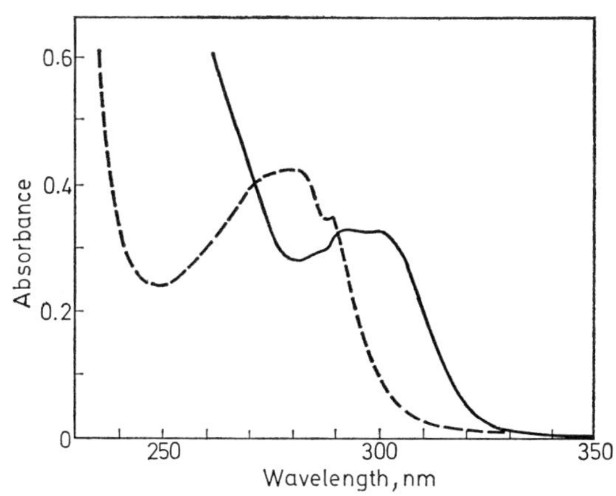

Fig. 6-4. Absorption spectrum of lysozyme (– – – –) and of lysozyme dissolved in a HCl-HCOOH mixture (———). (Taken from Ref. [995])

The high yield (over 90%) of specific formylation of tryptophan, concomitant with a shift of absorption to longer wavelengths (298 nm) (Fig. 6-4) allows a spectrophotometric determination of tryptophan in proteins. The method was shown to be fairly accurate with several test proteins [995].

Tryptophan residues are oxidized to N[1]-formylkynurenine (XVI) by ozone [994, 996] and by proflavin-sensitized photooxidation [403] and then to kynurenine (XVII) by acid catalyzed deformylation [Eq. (7)].

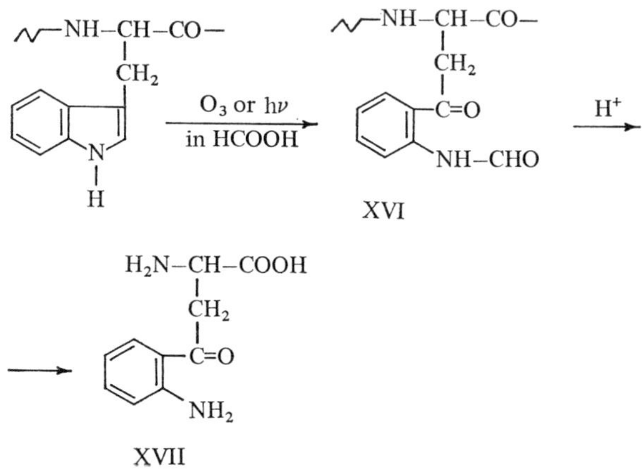

(7)

Since kynurenine is stable to acid hydrolysis, the tryptophan content of a protein may be related to the recovery of kynurenine on the amino acid analyzer, after ozonization or proflavine-sensitized photooxidation of the protein sample in 99% formic acid followed by acid hydrolysis.

In addition, kynurenine has an absorption maximum near 360 nm; this fact permits direct estimation of the kynurenine content in the intact protein by spectrophotometry.

IV. Estimation of Sulfhydryl Groups

The SH group has undoubtedly received more attention than any other reactive group in the protein molecule for analytical purposes as well for structure-activity relationship studies. This is due primarily to the relatively large number of chemical methods available for modifying this highly reactive group. Further, since SH groups play a prominent role in enzymic catalysis, extensive investigation in this area has been carried out.

A large number of methods for the estimation of thiols are available to biochemist, but no single method has emerged which is superior to all others. Since values are influenced by such factors as the reagent, pH, buffer, availability of the SH group to the reagent and therefore on the extent of denaturation of the protein molecule, analytical results must be compared by different methods.

In addition to the references cited here, the reader may consult several comprehensive reviews dealing with the chemistry of sulfur in proteins having well documented analytical sections [74, 153, 213, 214, 376, 480, 758].

A. Mercaptide Formation

Reactions of heavy metals with SH group leads to the formation of mercaptide derivatives [Eq. (8)].

$$\text{Protein-SH} + \text{Me}^{+} \longrightarrow \text{Protein-SMe} + \text{H}^{+}\,. \tag{8}$$

The marked affinity of certain metal cations for RS^- results from their tendency to accept a pair of electrons with the resultant formation of undissociated sulfides. Silver, mercury, copper, cadmium, zinc, lead, arsenic and antimony ions form mercaptides.

The reagents which have been used most successfully for the estimation of -SH groups and in particular of cysteine residues in proteins are those which form highly undissociated mercaptides, namely, silver salts, mercury salts, and organic mercury derivatives of the type RHgX. For an excellent discussion of the interaction of metallic ions with proteins, peptides and amino acids see the review of GURD and WILCOX [458].

The most widely employed method for following these reactions analytically has been titration of the mercaptan with graded amounts of the metal ion and the measurement of the excess by electrometric procedures.

The name *amperometric titration* is given to the special case when the voltage applied across the reference electrode and indicator electrode is constant, and the current increments are plotted against the volume of titrating reagent added. Under well-defined conditions only a small residual current flows through the cell until all the -SH groups are blocked. The end point of the titration is then marked by the appearance of the diffusion current due to excess titrating reagent. A plot of current *versus* the volume of added titrant gives two straight lines which intersect at the equivalence point.

A detailed account of these techniques can be found in the review article of CECIL and MCPHEE [214], of LEACH [758] and in the standard books on the subject by KOLTHOFF and LINGANE [709] and by MEITES [857].

There are two types of electrodes in common use, namely the rotating platinum electrode and the dropping mercury electrode.

The platinum electrode is made from a short piece of platinum wire sealed into a piece of thick-walled glass tubing which is filled with mercury. This electrode has to be rotated at speeds of about 800 r.p.m.

The rotating platinum electrode however has the disadvantage of becoming erratic or unresponsive as the surface of the electrode becomes poisoned by the sulfur compounds being titrated. The electrode must be carefully cleaned prior to each titration.

The dropping mercury electrode has the unique advantage of providing a continuously renewed electrode surface of constant characteristics. However, this method is about ten times less sensitive than that using the rotating platinum electrode.

Mercaptide formation can be also followed by potentiometry. Potentiometric titration [713] is performed by measuring the potential of a reversible electrode under equilibrium conditions. The potential is proportional to the logarithm of the concentration of the ion to which the electrode is reversible. A metal thiol electrode, consisting of a mercaptide layer on the surface of the electrode, is employed, since it is reversible to thiol as well as to metal ion.

1. Silver

The reaction of silver with mercaptans has been most frequently followed by amperometric titration with the rotating platinum electrode. The procedure is sensitive (approximately 10^{-5} M -SH) and can be carried out in neutral, aqueous solution.

Unfortunately, the stoichiometry of the reaction [Eq. (9)] on which this method is based cannot always be relied upon, since it has been shown that the silver mercaptides of simple thiols have a strong tendency to bind additional silver ion [190, 214, 711, 1171]. If this occurs to any extent under the conditions used for estimation of the SH groups in proteins, a positive error would result. This can be minimized by varying the pH of the reaction or by using complexing agents for the heavy metal, which forms an electro-reducible complex with the ion, so that all the heavy metal reagent is present either in the complex form or as the mercaptide. Examples of this procedure are the use of ammonia [711] or tris-(hydroxymethyl)aminomethane buffer [78] with silver ion. Thiols were titrated with $Ag(NH_3)_2^+$ and the reaction was followed with a rotating platinum electrode, using ammonia and ammonium nitrate in 95% alcohol [712] as the supporting electrolyte. The protein is titrated until there is a slight excess of the titrating ion, shown by the increase in the current passing through the solution. The method was improved by replacing the ammonia with tris(hydroxymethyl)aminomethane buffer as the complexing agent [78].

$$RSH + Ag^+ \longrightarrow RSAg + H^+ \tag{9}$$

Recently, Gruen and Harrap [441, 502] used a specific-ion electrode [782, 1017] which is highly selective to Ag^+ as an alternative means of observing the end point in the titration of thiols with Ag^+. The potential at the Ag/S specific-ion electrode depends only on the equilibrium activity of free Ag^+ (the only interfering species known is Hg^{++}), whereas the current at the rotating platinum electrode depends on the dynamic reduction of any reducible species and may therefore not be as selective as the specific ion electrodes.

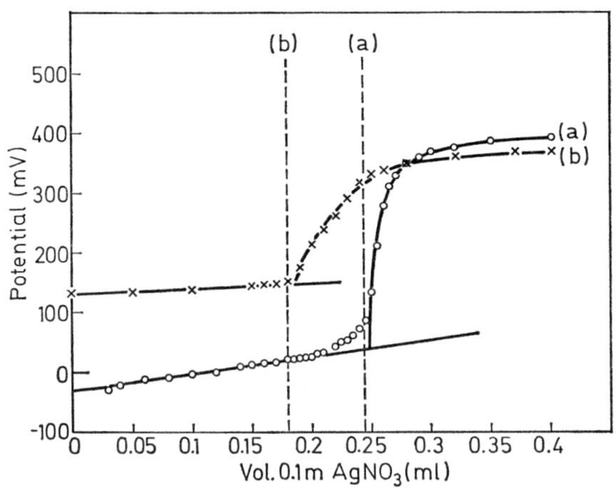

Fig. 6-5. *a* Addition of 0.1 M $AgNO_3$ to 0.2 ml 0.123 M β-mercaptoethanol in 20 ml H_2O. *b* Addition of 1.0 M $AgNO_3$ to 0.2 ml 0.088 M glutathione in 20 ml H_2O. Vertical dashed lines indicate stoichiometric end points. (Taken from Ref. [441])

As Ag^+ was added to the thiol there was very little change in potential until the end point was almost reached (Fig. 6-5). Beyond the end point, the potential rose rapidly due to the excess Ag^+. The end point was taken as the intersection between the linear baseline and the curve of potential *versus* excess Ag^+. Typical titration curves are shown in Fig. 6-5. The method was shown to be accurate ($\pm$ 2%) and found applicable also to the determination of -SH groups in proteins.

2. Mercury Compounds

The titration of thiols with Hg^{2+} has the disadvantage that ambiguities can arise from the divalention. While simple thiol compounds form mercaptides of the type $(RS)_2Hg$, proteins give half-mercaptides (Protein-SHgX). Simple thiols only were estimated accurately using Hg^{2+}. Alkylmercury derivatives of type RHgX have been used with more success. There are numerous mercury derivatives available of differing molecular size, reactivity, solubility and specificity. Examples include compounds such as methylmercuri-iodide, mercurated alkylamides or phenylmercuri-compounds.

All operations involving the use of metallic mercury, HgX_2 or RHgX, whether at the preparative or analytical level, should be carried out in a well-ventilated area, preferably under a fume hood [593, 757]. These precautions apply equally well to work with all highly reactice -SH reagents particularly the volatile ones. All of these reagents are toxic on inhalation and have a vesicant action on the skin at high concentrations. None should be pipetted by mouth.

Organic mercurials of the type RHgX have the outstanding advantage of undergoing only a monovalent reaction with -SH groups, since one of the valences of mercury already is covalently bonded with carbon [Eq. (10)].

$$R'SH + RHgX \longrightarrow R'SHgR + H^+ + X^- \tag{10}$$

Although methylmercuri-iodide reacts more rapidly in alkaline solution than at pH 7, reaction still occurs sufficiently rapidly in neutral and acid solutions. Being a covalent compound the reagents has only limited solubility in water. Alternatively the more water soluble MeHgCl or $MeHgNO_3$ may be used [758].

Methylmercuri-nitrate, which is very soluble, is best prepared in solution from methylmercuri-iodide by a double decomposition reaction with silver nitrate [832].

The problem of detecting the end point in the reaction between -SH groups and organic mercury compounds has in general been followed by amperometric methods. A number of other mercury compounds has been used as mercaptide forming agents in special studies.

Colored mercurials have been used extensively in biochemical work [189]. Horowitz and Klotz [585, 704] have used a colored mercurial, 4-(*p*-dimethylaminobenzeneazo)phenylmercuriacetate (XVIII), which has an absorption peak at 458 nm.

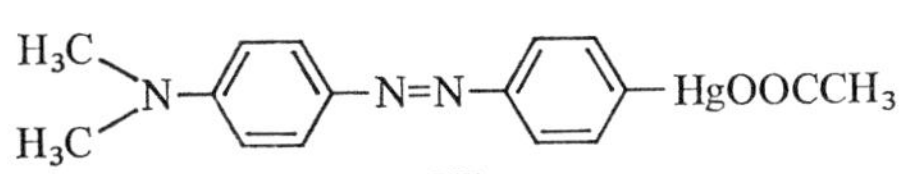

XVIII

Among these colored reagents, 1-(4-chloromercuriphenylazo)-naphthol-2 (mercury orange) (XIX) has found an application as a histochemical reagent for its red staining and for being very specific for -SH groups [82].

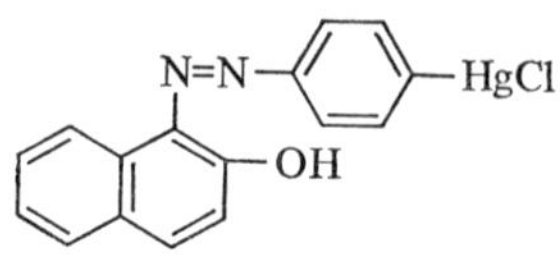

XIX

The colored mercurinitrophenol compounds [840, 936] 2-chloromercuri-4-nitrophenol (XX), 2-chloromercuri-4,6-dinitrophenol (XXI) and 4-chloromercuri-2-nitrophenol (XXII) have been used as chromophoric probes for thiol groups in proteins and other biologically interesting thiols. They bind specifically at the -SH groups,

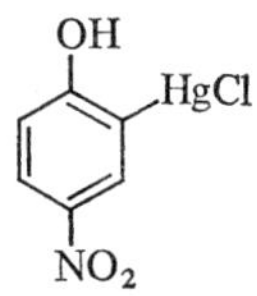

XX

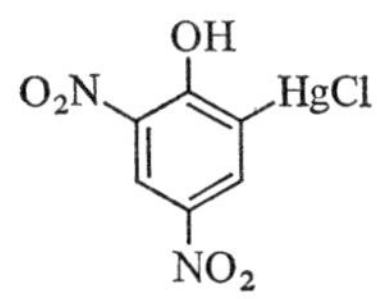

XXI

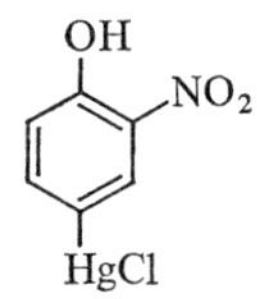

XXII

and the binding reaction induces pK changes and spectral changes of the nitrophenol moiety at certain pH values. The reagents give some indication of the microenvironment at their site of reaction.

3. p-Chloromercuribenzoic Acid

Several organic mercurials undergo an alteration in their spectral characteristics when they react with thiols. However, thus far, *p*-chloromercuribenzoic acid (PMB)

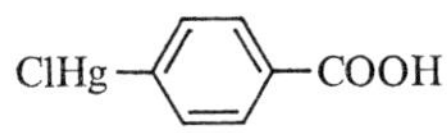

XXIII

(XXIII), [153, 1242] is the only mercurial which gives rise to an adequate increase in absorption in a useful spectral region as the result of mercaptide formation (Fig. 6-6).

PMB exhibits an absorption maximum at 233 nm with a molar absorptivity of 1.69×10^4. On formation of a mercaptide, the molar absorptivity increases to 22×10^4. However, this change is small in comparison with the maximal difference in absorbance which occurs between PMB and its mercaptide in the region from 250 to 255 nm and which directly reflects the amount of reagent reacted [153].

The method of Boyer is the only one which measures directly a property of the newly formed Hg-S bond, i.e., its UV absorption in the region of 250 nm. The technique used is to titrate the mercurial with the -SH compound until there is no further change in the differential absorption. Instead of a single measurement in an excess of PMB, the assay is carried out as a spectrophotometric titration.

Since measurements must be made in a region where proteins absorb, precautions must be taken to ensure corrections for the protein content. But, since practically no spectral changes may be observed in the presence of proteins lacking -SH groups, the specificity of the reaction is evident.

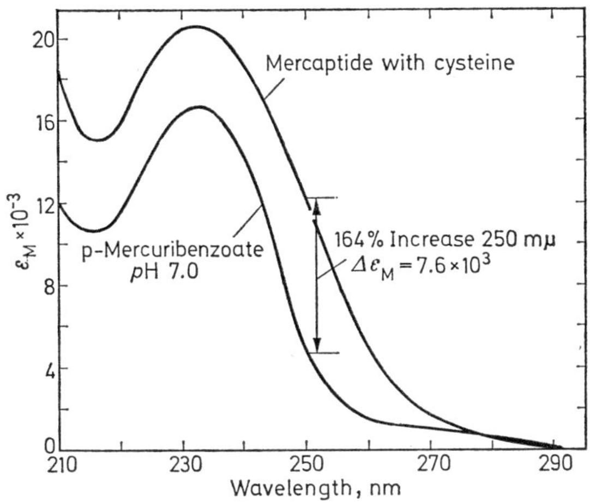

Fig. 6-6. The absorption spectrum of *p*-chloromercuribenzonate and its mercaptide with cysteine in 0.05 phosphate, pH 7.0. (Taken from Ref. [153])

The use of detergents [814, 929, 949] and urea [777, 1252] was reported to permit measurements of less reactive SH groups.

Neuman et al. [913] has modified slightly the analytical method first described by Boyer [153] and this experimental procedure is herewith reported.

Reagent. It is difficult to obtain the reagent in pure form, so that solutions of PMB must be standardized spectrophotometrically either against standard solutions of cysteine hydrochloride or mercaptosuccinic acid [941]. The molar extinction of PMB at 233 nm is 16900 in water solution, pH 7. A suitable concentration of the reagent is $5 - 7 \times 10^{-4}$ M [941].

Procedure [913]. Aliquots of a solution of PMB, adjusted to pH 7.0, is added to the protein solution in 0.1 M phosphate buffer, pH 7.0. The change in absorbance at 250 nm is measured after each addition. The observed absorbance is corrected for dilution and plotted against the volume of reagent added. A blank, in which there is no protein, is treated in an identical manner. The SH content of the protein solution is obtained from the equivalents of the added reagent at the end point of the titration (no further increase in absorption at 250 nm).

B. Alkylating Agents

Alkylating agents (iodoacetate, bromoacetate) [382, 419, 491, 527] have been extensively used in protein chemistry as inhibitors of SH enzymes [153, 691, 799, 941]. Further important use of the alkylation reaction of SH groups lies in the stabilization of these residues in proteins prior to enzymic or chemical degradation (see Chapter 5). The reaction has been also applied for the estimation of the thiol content. There is a wide choice of SH-alkylating reagents and they may be divided into two classes: (1) those carrying a halogen which reacts with the elimination of halogen acid, (2) those with an active double bond to which the thiol groups add.

1. Haloacetates

The more extensively studied and the most useful reagents of type (1) are iodoacetamide or the more soluble iodoacetic acid [Eq. (11)].

$$\text{Protein-SH} + \text{I-CH}_2\text{COOH} \longrightarrow \text{Protein-S-CH}_2\text{COOH} + \text{HI}\,. \qquad (11)$$

The reactions of thiols with active halogens compounds are bimolecular, nucleophilic substitution reactions, in which the nucleophilic reagent is the highly reactive mercaptide ion [68, 153]. However, attack on the amino, phenolic and imidazole groups can occur, but alkylation of SH proceeds more rapidly and side reactions can be minimized by avoiding high concentrations of reagent and strongly alkaline conditions.

The reaction of iodoacetate with proteins is measured by ion exchange chromatographic analysis of S-carboxymethyl-cysteine after acid hydrolysis of the protein (recoveries over 90%) [241, 876] (see also 6.V.B.). When iodoacetamide is employed for the alkylation step, the S-carboxamidomethyl-cysteine is converted by acid hydrolysis to the S-carboxymethyl-derivative.

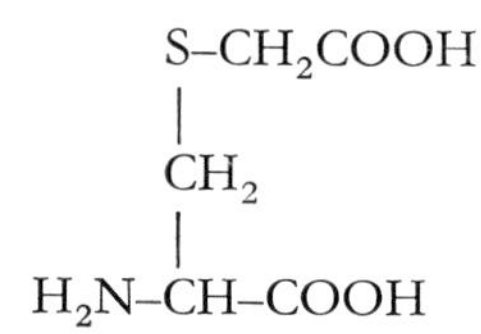

XXIII.a

If precautions are not taken to exclude oxygen during acid hydrolysis, destruction of S-carboxymethyl-cysteine amounts to nearly 10%; much higher values approaching 50% may be observed in certain cases. The deaeration step is therefore extremely important and the use of a water pump is inadequate. Even the method of removing HCl may introduce artifact peaks in the chromatograms and it is important to observe the recommended conditions in each detail.

ANFINSEN and HABER [45] used iodoacetate-1-C^{14} for the coupling reaction and the S-carboxymethyl-cysteine content in the modified protein was determined directly by radioactivity measurements, after separation of excess reagent through Sephadex G-25.

Alternatively, the direct titration of the halogen acid formed by the reaction allows the determination of the thiol content [76]. This procedure is very sensitive, but is feasible only with simple model thiols.

For a detailed discussion on the carboxymethylation of proteins, the reader is referred to the excellent review papers of HIRS [545] and GURD [459].

2. Addition to Double Bonds

Thiols add to double bonds and the reaction has many features in common with the addition of halogen acids [833]. N-Ethyl-maleimide (NEM) reacts rapidly and quantitatively with glutathione and thiolactic acid at pH 7.4 [Eq. (12)] [389, 390, 815]. This reaction has found useful applications in the quantitative estimation of SH groups in proteins [21, 277, 444, 764, 1032].

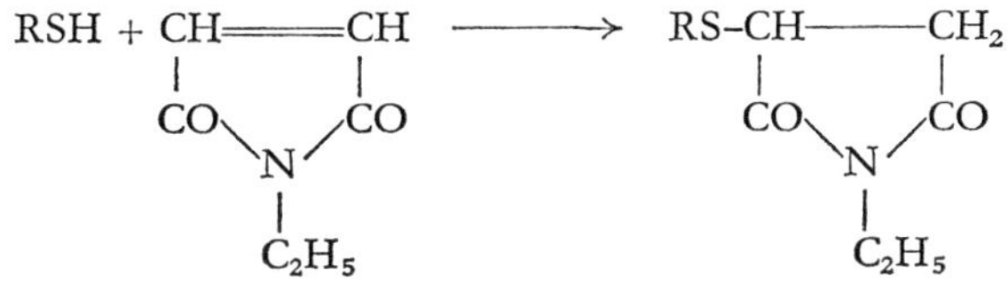

(12)

The extent of reaction was measured by titration with iodine [815], or, better, by following the decrease in absorption at 300 nm that occurs when NEM combines with cysteine residues [21, 277, 444, 764, 1032], as shown in Fig. 6-7. Alternatively,

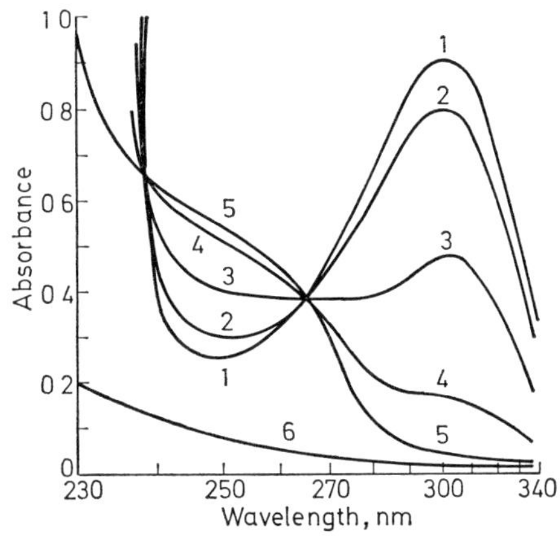

Fig. 6-7. The absorption spectrum of N-ethylmaleimide (1.5×10^{-3} M) in phosphate buffer (0.1 M, pH 6.0) in the presence of increasing amounts of cysteine. (*1*) Without cysteine; (*2*) 0.15×10^{-3} M cysteine; (*3*) 0.75×10^{-3} M cysteine; (*4*) 1.27×10^{-3} M cysteine; (*5*) 1.50×10^{-3} M cysteine; (*6*) 1.50×10^{-3} M cysteine alone. (Taken from Ref. [21])

the amount of acid hydrolysis product, S-succinyl-cysteine (XXIV) has been quantitatively determined [457, 1182, 1201, 1291]. Excess NEM is required in order to obtain a stoichiometric reaction.

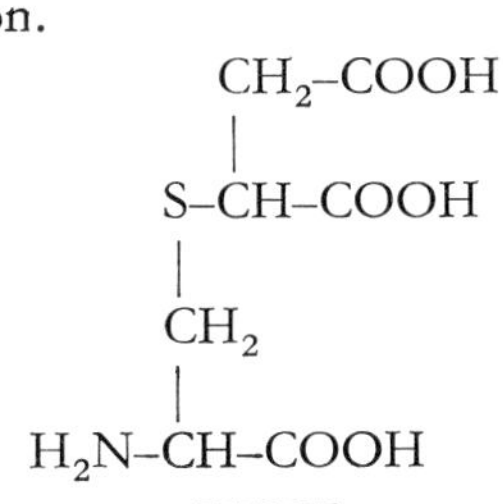

XXIV

NEM is of low reactivity in acid solution and unstable in alkali, so the reaction is carried out by adding the thiol compound at pH 6 to 7. The determination should be repeated using various amounts of -SH compound and allowing the reaction to proceed for various lengths of time. The reduction in optical density of the NEM is converted into a reduction in concentration of the reagent by referring to the calibration curve. The sensitivity is high; 1 μmole of cysteine/ml yields an absorbancy change of 0.60.

A disadvantage of the NEM reaction for analytical purpose in the intact protein is that a quantitative reaction is not always easily achieved; proteins react more slowly with NEM than do simple thiols such as cysteine. Indeed it was found that, at pH 7, cysteine reacts rapidly but the enzyme myokinase [444] required 30 min for complete reaction and bovine serum albumin, 55 min [1032]. Therefore a pre-treatment of the protein sample with urea solution is suggested in order to expose the less reactive SH groups [1409].

The reaction of NEM toward SH groups in proteins is not specific, but reaction occurs also with imidazole and α-amino groups [172, 1183]. These side reactions however are not of great significance unter the experimental conditions for thiol group determinations [172]. On the other hand GUIDOTTI and KONIGSBERG [457], after treating carbon monoxyhemoglobin with NEM, noted a discrepancy between the decrease in absorbance at 300 nm and the appearance of S-succinyl-cysteine in the hydrolyzate. Their data suggested that the amino terminal valine of the α-chain was alkylated.

The reaction has also found useful application for the chromatographic determination of SH-containing compounds. The product of the reaction of NEM with thiols has been found to give an intense red color in alkaline solution [75].

Mention should be made of a number of other substituted maleimides which have proven useful since they introduce a chromogenic substituent into biologically important substances, particularly proteins. Reagents as N-(2,4-dinitrophenyl)-maleimide (XXV) allow modification of reactive -SH groups in the protein and

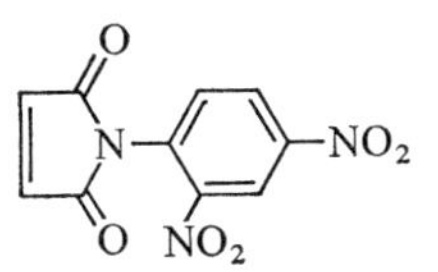

XXV

ready detection of the appropriate peptide after proteolytic degradation and separation [234, 418, 1001].

Acrylonitrile is another effective alkylating agent for thiols [1143, 1342]. The reaction is quantitative and the resulting cyanoethyl derivative is formed [Eq. (13)].

$$\mathrm{RSH} + \mathrm{CH_2{=}CH{-}CN} \longrightarrow \mathrm{RS{-}CH_2{-}CH_2{-}CN}\,. \tag{13}$$

Alkylation of protein sulfhydryl groups with acrylonitrile is extensively used in protein chemistry for protein modification studies as well as for analytical purposes. Upon acid hydrolysis, the S-cyanoethyl derivative of cysteine is converted to the S-carboxyethyl derivative (XXVI) which is estimated by automatic amino acid analysis.

$$\begin{array}{l}\mathrm{H_2N{-}CHCH_2{-}S{-}CH_2CH_2{-}COOH}\\ \qquad\;\; |\\ \qquad\mathrm{COOH}\end{array}$$

XXVI

4-Vinylpyridine reacts specifically with thiol groups of proteins at pH 7.5 [Eq. (14)], cysteine being converted to S-β-(2-pyridylethyl)-cysteine [392], which can be estimated by ion exchange chromatography.

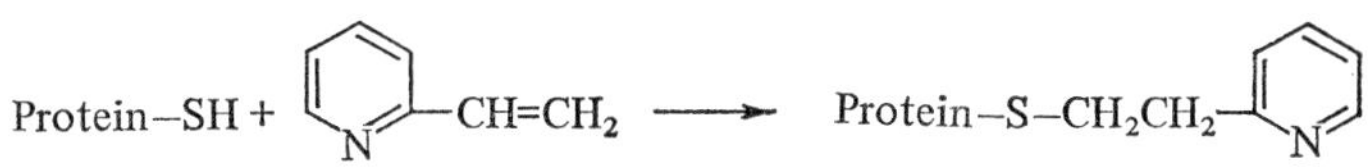

(14)

C. Colorimetry

1. Ellman Reagent

The use of 5,5′-dithiobis-(2-nitrobenzoic acid) (DTNB) for the estimation of sulfhydryl groups was introduced by Ellman [328] and the procedure is now extensively used because of its ease and accuracy. The method involves reaction of thiol groups with the reagent at pH 7 to 8 to produce one mole of 4-nitro-3-carboxy-thiophenol ($\varepsilon = 13600$ at 412 nm) [Eq. (15)].

$$\text{Protein–S}^- + \text{R–S–S–R} \longrightarrow \text{Protein–S–S–R} + \text{RS}^- \quad (15)$$

$$\text{Protein–S–S–R} + \text{Protein–SH} \longrightarrow \text{Protein–S–S–Protein} + \text{RS}^- \quad (16)$$

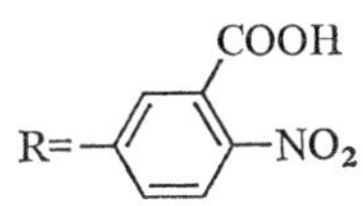

Reaction (15) could be followed by reaction (16), leading to intramolecular and intermolecular disulfide formation. However, whether reactions (15) or (15) and (16) occur, the same stoichiometry applies, namely 1 mole of nitrothiophenol per mole of protein SH group.

The Ellman's reagent has gained popularity rapidly, particularly in the field of protein chemistry, so that it may now be considered as one of the standard reagents for thiol groups.

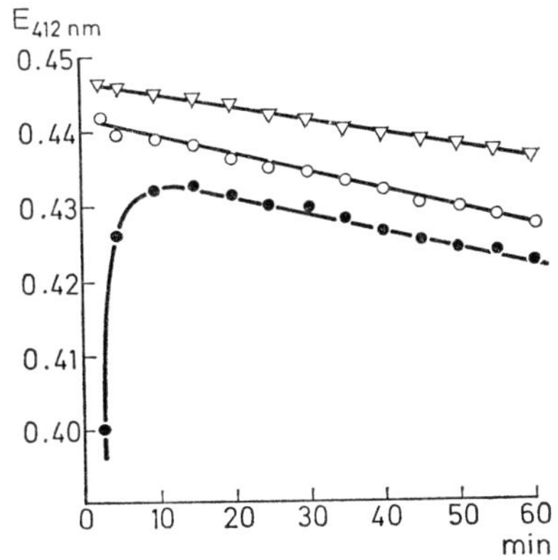

Fig. 6-8. Time dependent absorption of the Ellman's reagent with glutathione (△——△), cysteine (○——○), and mercaptosuccinic acid (●——●). (Taken from Ref. [1349])

The color obtained in the Ellman assay requires metal-free reagents; it is modified by metal ions directly bound to the proteins (ferric, calcium) [805, 1097]. Therefore 10^{-4} M EDTA is generally included in the reagent solution. In addition, the reaction is better performed under denaturing conditions (urea, sodium dodecyl sulfate) to expose buried SH groups [256, 917, 1227].

With DTNB, a solution of 0.01 μmole of sulfhydryl/ml gives an absorbance of 0.136 (1 cm light path) at 412 nm. Simple thiols, e.g., cysteine, give complete color development within 2 min. The color is not stable but slowly fades due to autoxidation (Fig. 6-8). This fact is delayed by the inclusion of EDTA in the reaction medium [73a, 1267].

Reagents. A choice of buffers is available, depending on the specific analytical requirements. Some of the most frequently used are listed below. The DTNB is dissolved (10^{-2} M) in the buffer and the solution flushed with nitrogen.

0.1 M phosphate, pH 7.9, 10^{-4} EDTA, 6 M urea,
0.1 M Tris buffer, pH 8.0, 4 to 6 M urea,
0.1 M Tris maleate buffer, pH 7.0,
0.1 M phosphate, pH 7.9, 5 M guanidine hydrochloride,
0.1 M phosphate, pH 7.9, 1% sodium dodecylsulfate.

Procedure. The protein sample (0.1 ml) is added to 2.9 ml of the reagent solution and the absorbancy at 412 nm is followed until no further increase is observed (about 30 to 60 min). The total SH content in the sample is calculated assuming $\varepsilon = 13600$ for the reagent. For more reliable results standardization of the method is recommended using a cysteine solution, standardized by iodometric titration. After determination of the protein concentration by standard methods (see 6.II.), the SH-groups/protein (mole/mole) is obtained from the ratio of the two concentrations.

2. Dithiodipyridine Derivatives

It has been shown that 2,2′-dithiodipyridine and 4,4′-dithiodipyridine are excellent reagents for the determination of –SH groups, because of a shift in their absorption upon reaction with a thiol [435]. The reaction of these disulfides with thiols occurs as in Eq. (17).

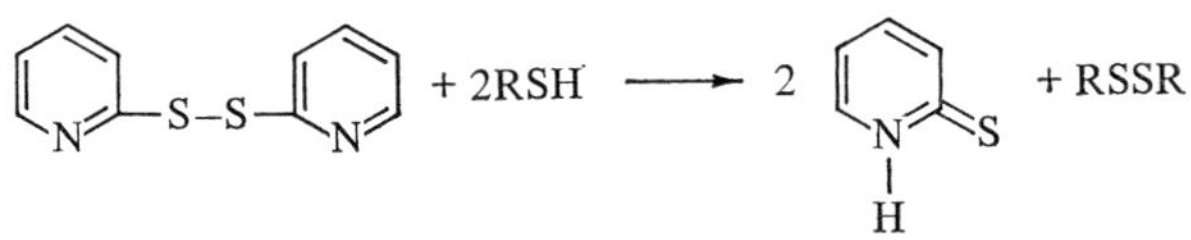

(17)

XXVII

The pyridinethiol formed is virtually exclusively in the thiopyridone form (XXVII) (2-thiopyridone has λ_{max} 343 nm and $\varepsilon = 7060$ and 4-thiopyridone λ_{max} 324 nm and $\varepsilon = 19800$). This fact has the double advantage of causing a shift in the absorption and making the reaction essentially irreversible. These reagents are useful alternatives to the Ellman's reagent when other substances are present which absorb in the region near 400 nm (e.g., heme-containing proteins).

3. Sulfenyl Halides

The selective reaction of sulfenyl halides [372, 373, 1318] with the thiol group of cysteine to form a mixed disulfide can usefully be employed for the estimation of the SH groups of a protein (for the estimation of tryptophan using this reaction see Section III.B.3.).

Taking advantage of the ready cleavage in alkaline solution of alkyl-aryl disulfide with formation of arylthiophenol [951], a method was suggested [374, 1318] involving labeling of the SH-containing protein with *p*-nitrophenylsulfenyl chloride (p-NPS-Cl) in acetic or formic acid. The *p*-nitrothiophenol moiety, covalently bound to the protein by a disulfide linkage, is then quantitatively released by exposing the sample of p-NPS-protein to alkaline media (0.1 N NaOH) (for a discussion on the mechanism of cleavage by alkali of aryl alkyl-disulfides cf. FONTANA and SCOFFONE [373]). The highly colored chromophore ($\varepsilon = 13600\ M^{-1}\ cm^{-1}$ at 412 nm in 0.1 N NaOH) was shown to be quantitatively released from the S-p-NPS-derivatives, if deaerated solutions are employed.

The method was tested with model compounds (e.g., N-acetyl-S — NPS-cysteine, S-NPS-glutathione) as well as with reduced ribonuclease. In this case the SH groups per molecule of enzyme were found to be 7.8 (theory 8 residues) [372].

When a protein containing both tryptophan and cysteine has to be analyzed for SH content, the procedure involving the reaction of p-NPS-Cl with SH groups can still be employed since the 2-thio-(4-nitrophenyl)-tryptophan residue (absorption peak near 323 nm) [1318] shows almost no absorption at the λ_{max} (412 nm) of *p*-nitrothiophenol [1209]. In any case, higher wavelengths can be selected for absorption measurements, i.e., λ 450 nm, at which wavelength the thiol still strongly absorbs ($\varepsilon = 8500$) [374, 1318].

D. Comments

We have above discussed the analytical techniques available for the identification and detection of thiols. The large number of methods which have been published over the years is perhaps the best indication of the technical difficulties that have been encountered. No single method has emerged which is superior to all others.

An instructive paper was recently published by WENCK and coworkers [1349] in which the authors made a comparative study of the most usual physico-chemical methods for the quantitative determination of –SH groups using compounds of different structure. Amperometric titration with $AgNO_3$, Ellman and Boyer's (PMB) method as well as potentiometric measurement were compared with respect to their applicability, reliability and susceptibility to interferences. The potentiometric determination was performed by adding to the –SH compound solution an excess of iodoacetate and following the liberation of iodide ion with the Ag/AgI electrode.

In Table 6-2 are reported the results of the analyses for the –SH content of a series of –SH containing compounds using the four techniques mentioned above. Amperometric titration with $AgNO_3$ yields correct results only for certain types of compounds. Often silver does not combine in a 1:1 stoichiometry. Thiol compounds which contain an additional amino or imidazole group bind additional silver. The Ellman and Boyer's reagent gives reliable results. Both methods require calibration with a substance of known –SH content. Potentiometric measurement with the Ag/AgI electrode is a suitable method for the determination of absolute –SH content, giving reproducible results for all the compounds tested.

Table 6-2. Comparative analysis for SH-content. The results are presented as percentage, calculated on the basis of weighed amount of material taken as 100%. (Taken from Ref. [1349])

Compound	Potentiometry	Amperometry	Boyer	Ellman
Glutathione	99.3	98.0	(100.0)[a]	(100.0)[a]
Mercaptosuccinic Acid	99.3	96.6	99.0	100.0
Cysteine · HCl	99.0	139.0	101.0	99.0
Cysteamine · HCl	92.4	126.0	92.5	93.8
Cysteinamide · HCl	85.5	117.0	85.6	83.5
N-Acetyl-cysteine	95.5	92.6	96.3	98.6
N-Acetyl-cysteinamide	87.0	92.6	90.0	92.4
Homocysteine	97.0	112.0	98.0	100.0
4-SH-Methylimidazole · HCl	94.6	145.0	96.0	94.8
4-SH-Ethylimidazole · HCl	99.2	133.0	99.5	101.5
1-CH_3-5-(2′-SH-Ethyl)- -imidazole · HCl	94.3	129.0	91.0	93.3
Cyclo-cysteinyl-glycine	97.0	96.5	96.0	99.3
N-Acetyl-cysteinyl-histidyl- -aspartic acid	78.7	97.0	78.5	75.3
N-Acetyl-cysteinyl-γ-amino- butyryl-histidyl-γ-amino- butyryl-aspartic acid	88.7	103.8	88.5	87.4
0-Aminothiophenol · HCl	91.7	100.0	88.0	96.0
2-SH-Imidazole	97.2	142,5	—[b]	—[c]
1-CH_3-2-SH-4(5)-Imidazole- carbonate methyl ester	94.4	—[d]	—[e]	—[e]
Ergothioneine	101.0	140.0	—[a]	—[c]
Glutathione + Imidazole	—[b]	98.0	—[b]	—[b]

[a] Taken as standard. [b] Not determined. [c] No quantitative reaction.
[d] Not possible to calculate. [e] Not soluble.

V. Estimation of Disulfides

The disulfide bonds are most conveniently estimated by prior conversion to thiols followed by their quantitative determination. Thus, the cystine content of a protein is determined indirectly from the amount of sulfhydryl groups formed upon reduction. If the cysteine content is known, the determination of cystine can be found by difference.

A direct assay for disulfide groups makes use of their quenching effect on the fluorescence of fluorescein mercury acetate (XXVIII) [659]. In alkaline solution,

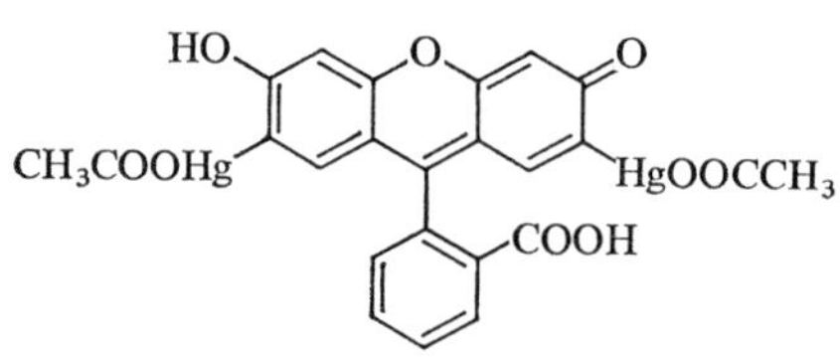

XXVIII

thiols quench to only 5% of the extent of disulfides and their contribution can easily be eliminated, e.g., by alkylation with iodoacetate. The method is very sensitive (10^{-7} M) and gives values for the disulfide content of some proteins which agree well with the known values.

We will now consider the methods available for splitting disulfide bonds.

A. Reaction with Sulfite

Disulfides are cleaved by sulfite to S-sulfoacids and sulfite [Eq. (18)] [235, 523, 524, 797]. The reaction was extensively

$$R\text{–}S\text{–}S\text{–}R + SO_3^{2-} \longrightarrow RS^- + RS\text{–}SO_3^- \tag{18}$$

studied [214, 841, 1239, 1240] and found to be reversible, so that the concentration of sulfite must be kept high.

Since the thiols formed (one SH group per S–S) during the reaction have to be determined in the presence of a large excess of sulfite, alkylating reagents cannot be used because reaction takes place rapidly with sulfite too. Potentiometric titration with silver nitrate [214] or with mercuric chloride [212] was found to be a useful procedure, as is amperometric titration with silver nitrate in an ammonia buffer [711] or mercuric chloride in borate buffer [1241] (for a review on sulfitolysis of disulfide bonds cf. COLE [240]).

By mild oxidation (oxygen, sodium tetrathionate, iodosobenzoate) [240] RS^- is reconverted to R–S–S–R so that the reaction of sulfite progresses to the quantitative conversion to $RS\text{–}SO_3^-$. The same complete reaction may be achieved carrying out the sulfitolysis step in the presence of Cu^{2+} ion [710, 1250]. In this case, concomitant to the reduction of cupric ion all the disulfides are converted to the S-sulfonate.

The reaction with sulfite is used in some instances in the primary structure determination of proteins (e.g., with trypsinogen and chymotrypsinogen [966]), in order to avoid the insoluble derivatives of the proteins obtained upon reduction and carboxymethylation. However a limitation in the use of the S-sulfoacids of proteins in sequence studies lies in the great instability of these derivatives in acidic as well as in alkaline media.

Reduction of cystine residues of proteins with dithiothreitol followed by treatment with an excess of sodium tetrathionate is the basis of the method of INGLIS and LIU [603, 788] for the determination of the half-cystine residues in proteins as S-sulfocysteine.

Opening of disulfide bonds in proteins by phosphorothionate was achieved [913, 915, 916] in a way similar to the action of sulfite. It was shown in a recent study [914] that the reaction is a nucleophilic heterolytic scission which proceeds according to the following [Eq. (19)].

$$R\text{–}S\text{–}S\text{–}R + S\text{–}PO_3^{3-} \longrightarrow R\text{–}S\text{–}S\text{–}PO_3^{2-} + R\text{–}S^- \tag{19}$$

Cyanide will also react with disulfides in a similar way to sulfite [Eq. (20)] [404].

$$RSSR + CN^- \longrightarrow RSCN + RS^- \tag{20}$$

If the disulfides contain free amino groups, as with cystine, cyclic amidine derivatives XXIX are formed [424, 1093].

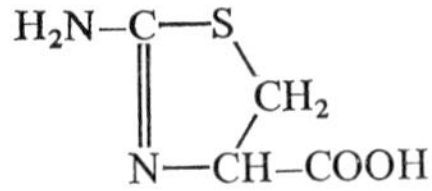

XXIX

B. Reductive Cleavage

Reduction of disulfide bridges has some advantage with respect to the sulfite reaction since two SH groups per mole of disulfide are produced. The sulfhydryl groups formed by reduction can be estimated by using the methods described earlier (cf. 6.IV.).

Older methods of reductive cleavage (zinc dust [1392], magnesium [1084] and sodium amalgam [711]), are no longer employed. Electrolytic reduction, although promising, has been rarely used [287].

The traditional method of reduction of protein disulfide groups is by treatment with high concentrations of thiols. Reagents that have been used for the reductive cleavage of disulfide bonds are cysteine, reduced glutathione, thioglycolic acid, β-mercaptoethanol [1277], β-mercaptoethylamine [814] and thioglycolic acid [663]. The reaction may be represented by the simple Eq. (21)

$$RSSR + 2\,R^1SH \rightleftarrows 2\,RSH + R^1SSR^1 \qquad (21)$$

from which it is obvious that in order to drive the reaction in the desired direction a considerable excess of thiol has to be added [352].

The reagent of choice for the reduction was β-mercaptoethanol until recently when dithiothreitol (DTT) (XXX) [237, 720] became commercially available. The use of DTT as a reducing agent for disulfides was first described by Cleland, and the reagent bears his name.

The reaction of DDT with disulfides occurs as follows [Eqs. (22) and (23)].

$$\begin{array}{c} RSSR \\ + \\ HS{-}CH_2{-}CH(OH){-}CH(OH){-}CH_2{-}SH \\ \text{XXX} \end{array} \longrightarrow \begin{array}{c} RSH \\ + \\ RS{-}S{-}CH_2{-}CH(OH){-}CH(OH){-}CH_2{-}SH \\ \text{XXXI} \end{array} \qquad (22)$$

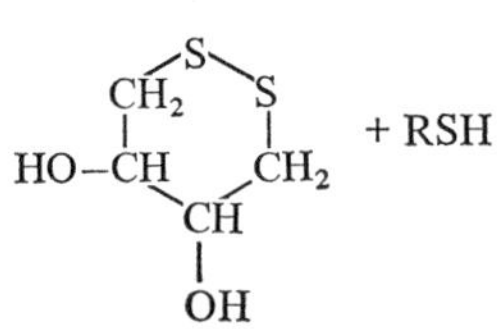

(23)

XXXII

The overall reaction proceeds nearly to completion because the formation of a 6-membered ring (XXXII) containing a disulfide bridge is energetically favored over the mixed disulfide.

The oxidation reduction potential of DTT at pH 7 and 25° is −0.33 V compared to −0.22 V for cysteine. These two values allow one to calculate an overall equilibrium constant of 10^4 for the reduction of cystine by DTT.

This reagent can be used at a much lower concentration than β-mercaptoethanol by virtue of its lower oxidation-reduction potential and its resistance to air oxidation. An additional advantage of DTT is its relative lack of the characteristic unpleasant thiol odor.

Recent studies have indicated that tri-n-butylphosphine (Bu_3P) is also a reductant for the disulfide bonds [Eq. (24)] [801, 1251].

$$RSSR + Bu_3P + H_2O \longrightarrow 2\,RSH + Bu_3PO \tag{24}$$

The cleavage of S–S bonds is a necessary pretreatment for proteins to be subjected to primary structure studies in order to facilitate enzymic hydrolysis of the protein chain. The new SH-groups formed upon reduction may reoxidize and must be stabilized by alkylation (iodoacetic acid or its amide [45], acrylonitrile [1342], bromoethylamine [783], or ethyleneimine [544]).

For sequence studies, reduction with sodium borohydride at pH 9 to 10 [544, 876] is no longer used since the reagent was found to produce cleavage of the peptide bond. By using thioglycolic acid as a reducing agent, there is a danger of "thiolation" of amino groups when impure thioglycolic acid containing reactive polythioglycolides and cyclic thiolactones is used [1365].

A general procedure of reduction of S–S bonds in a protein (ribonuclease) and their estimation by the carboxymethylation procedure was reported by ANFINSEN and HABER [45], WHITE [1365] and CRESTFIELD et al. [254]. The latter procedure is given below.

Procedure [254]. To 5 to 100 mg of protein in a 12 ml screw-cap vial (Kimble No. 609102) maintained under a nitrogen atmosphere [253], add 3.61 g of deionized crystalline urea [78], 0.30 ml of EDTA solution (50 mg of disodium EDTA/ml), 30 ml of Tris buffer; pH 8.6 (5.33 g of Tris and 9 ml of 1 N HCl diluted to 30 ml with water), and finally 0.1 ml of β-mercaptoethanol. The solution is made up to 7.5 ml with water and a solution 8 M in urea and 0.2% in EDTA is used to fill the vial completely. A disk of polyethylene is slid over the top in order to exclude air and is held in place with the screw cap. After 4 hrs at 22 to 25° C, the contents of the vial are transferred to a 25 ml beaker maintained under nitrogen. A freshly prepared solution of 0.268 g of iodoacetic acid in 1 ml of 1 N NaOH is added to the reaction mixture. The iodoacetate added is slightly less on a molar basis than the amount of β-mercaptoethanol; the SH groups of cysteine residues react most rapidly. Three equivalents of Tris per equivalent of iodoacetate keep the solution more alkaline than pH 8.3. Fifteen min after the addition of the iodoacetate, the sample may be applied to a 4 × 40 cm column of Sephadex G-75, medium particle size, equilibrated with 50% acetic acid. The sample is driven in and rinsed under air pressure to produce a column flow rate of 20 ml/hr. Fractions of 8 ml are collected. The protein in the effluent is located spectrophotometrically at 280 nm and then recovered by lyophilization. A sample of the carboxymethyl-protein is hydrolyzed (HCl, 6 N, 110° C, 24 hrs) and the content of S-carboxymethyl-cysteine determined by automatic amino acid analysis [241, 459, 545].

C. Oxidative Cleavage

The oxidation of a protein with performic acid quantitatively converts cysteine and cystine residues to the corresponding sulfoacids and methionine to its sulfone [1098, 1276, 1284]. These oxidation products are stable to acid hydrolysis and can be separated by ion exchange chromatography after hydrolysis of the oxidized protein [543, 1098], permitting the determination of the total amount of cysteine and cystine. The yield of cysteic acid obtained from cysteine and cystine generally exceeds 90% of the theoretical value [545, 876, 1250].

The performic acid reagent is made by mixing five volumes of 30% hydrogen peroxide and 95 volumes of 99% formic acid and the closed container is allowed to stand at 25° for 120 min. The titer of peracid reaches a maximum at about this time and decreases slowly thereafter. Only freshly prepared reagent should be used [545]. At the end of the reaction, the performic acid is removed by lyophilization after addition of water, by evaporation *in vacuo*, or by precipitation of the protein with trichloroacetic acid or an organic solvent. A sample of the oxidized protein is subjected to acid hydrolysis and then analyzed for cysteic acid content by automatic amino acid analysis.

The reaction was first applied by Sanger [1066] in his sequential work on insulin and by Hirs on ribonuclease [543]. A disadvantage presented by the method lies in the complete destruction of tryptophan [1284].

Another possibility for achieving the oxidative cleavage of disulfide bonds involves ozonization of the protein in 99% formic acid [996]. The yield of cyteic acid by this method, using free cystine, was 98%. The procedure is not selective for S–S bridges and other amino acids (methionine, tryptophan and tyrosine) are modified.

Photochemical oxidation [649] in the presence of cresol red or crystal violet as sensitizers is also a useful technique for the quantitative conversion of cysteine to cysteic acid.

VI. Estimation of Other Amino Acids

Little use is being made of methods for the direct determination of tyrosine, histidine, arginine in the intact protein in as much as the determination of these amino acids can be accomplished easily and accurately by automatic amino acid analysis. Selective reagents for these amino acid residues, however, are usefully employed to obtain an indication of the status of these residues in the protein [1281].

The tyrosine residue can be determined by the direct spectrophotometric method of Goodwin and Morton [307, 428] (see Section III.A.). Not all the phenolic groups can be titrated with alkali at the same pH in the native protein as in the free amino acid [251, 1281]. Differences in reactivity of the hydroxyl group in the native and denaturated protein are often observed [1281].

N-Acetylimidazole [1028], cyanuric fluoride [743] and tetranitromethane [1026, 1027, 1190] have been applied as selective reagents for tyrosine residues in proteins. These reagents are now extensively used in protein chemistry and find utility as diagnostic reagents for free and buried tyrosine residues.

Tyrosine and histidine react with freshly diazotized sulfanilic acid in sodium carbonate solution (Pauly reaction). When the reaction is performed in the Weiss-Szobolew modification [480, 561, 1345], the color value obtained is equal to the sum of the contributions of histidine and tyrosine [561]. If the tyrosine content is known, the determination of histidine can be found by difference. Diazo-1-H-tetrazole [582, 1192] also has been used as a histidine- and tyrosine-modifying reagent.

Arginine can be estimated by the well known SAKAGUCHI reaction [1055]. The guanidyl group reacts with α-naphthol and sodium hypochlorite in alkaline solution producing a red color. Since the color is unstable in the presence of hypochlorite, urea is added to destroy excess hypochlorite after the reaction [685, 1336]. Since arginine linked in the protein molecule occasionally does not react completely with the Sakaguchi reagent, prior acid hydrolysis of the protein sample has been suggested [629].

VII. Estimation of Amino and Amide Groups

Since the free amino groups of a protein can be readily calculated from the recovery of lysine by amino acid analysis of a protein hydrolyzate, its direct determination in the intact protein is no longer of importance. On the other hand, since the amide groups of glutamine and asparagine are hydrolyzed by acid or alkaline hydrolysis and the recovery of ammonia on the analyzer generally is not quantitative, the determination of the amide content in the intact protein is necessary.

A. Amino Groups

The total amino nitrogen has been determined by the ninhydrin reaction [928], formol titration [1042] and the nitrous acid method of VAN SLYKE [1308, 1309]. The latter was shown to be the most reliable procedure. Upon reaction of the primary amino groups with nitrous acid, formed by mixing sodium nitrite and cold acetic acid [Eq. (25)], nitrogen is evolved, which is measured by volumetric or manometric techniques. Accounts can be found in the original papers [1308, 1309] or, for a detailed description of the method, in the review article of HAMILTON [481].

$$\text{Protein-HN}_2 + \text{HONO} \longrightarrow \text{Protein-OH} + \text{N}_2 + \text{H}_2\text{O} \qquad (25)$$

The 1-fluoro-2,4-dinitrobenzene (FDNP) reagent of SANGER [1065] has the advantage that the dinitrophenyl-N bond is stable to acid hydrolysis so that the extent of reaction can be determined by colorimetric analysis of the Nε-DNP-lysine after chromatographic separation [381, 990].

More recently the use of 2,4,6-trinitrobenzenesulfonic acid (TNBS) has been suggested [727, 942] as a reagent for the determination of the free amino groups in peptides and proteins. It was shown that TNBS reacts preferentially and quantitatively with $-NH_2$ under mild conditions to give the corresponding trinitrophenyl derivatives. No reaction occurs with imidazole nitrogen of histidine, the guanidinium group of arginine, or the hydroxyl group of tyrosine, serine and threonine. The trinitrophenyl derivative of amino groups can be quantitated spectrophotometrically. A detailed account of this technique may be found in the review article of FIELDS [356].

B. Amide Groups

The amide groups of glutamine and asparagine release ammonia after acid or alkaline hydrolysis which can be determined by automatic amino acid analysis [1198]. However, the values obtained are only approximate since ammonia derived from destruction of other amino acids (in particular serine and threonine) and extraneous sources interferes. Therefore it is necessary to reach a compromise between the acid concentration, time and temperature of hydrolysis [1184], and to measure the ammonia produced as a function of the time of hydrolysis. In this way the secondary deamination arising from partial destruction of amino acids can be evaluated and the ammonia content determined by extrapolating the ammonia produced to zero time.

The estimation of ammonia is most readily carried out by a modification of the CONWAY microdiffusion technique [247]. An aliquot of the hydrolyzate is mixed with an alkaline buffer, and the ammonia which diffuses into a dilute acid solution is determined by the photometric ninhydrin method [550]. Alternatively, ammonia may be determined by distillation from an alkaline buffer in a micro-Kjeldahl apparatus and titration of the distillate [756, 1019].

Hydrolysis is usually performed in 2 N HCl at 100° C for 3 to 7 hrs [227]. Alkaline conditions (1 N NaOH at 25° C for 20 hrs) also can be employed [1217]. 12 N HCl at 37° C for 10 days has been used with success [430, 1019]. Diluted acid (5.5 N HCl) and higher temperature (100° C) lead to high values for the amide content.

The hydrolysis of a protein by enzymatic procedures offers the advantage that glutaminyl and asparaginyl residues are preserved so that their direct determination is possible by the use of chromatographic [1198] or electrophoretic [202, 886] separation. HILL and SCHMIDT [535] employed a two-step procedure using papain and then a mixture of leucine aminopeptidase and prolidase [267]. Several proteins of known composition were investigated and the values so obtained compared favorably with those obtained by acid hydrolysis. A single enzyme preparation, Viokase, a mixture of pancreatic enzymes, has been tried [1288]. Complete hydrolysis was achieved for peptides but for proteins the extent of cleavage was only 50 to 80%.

Enzymic procedures are currently used in the determination of the presence of asparagine and glutamine in small peptides. A digest of the peptide with leucine aminopeptidase can be analyzed by paper electrophoresis [1227] or by the conventional quantitative amino acid analysis by column chromatography. However, since serine is eluted together with asparagine and glutamine, a modification of the temperature of column chromatography is need in order to lead to elution of serine in a different position [1198].

The enzymatic determination of glutamine and asparagine has been reviewed in great detail by TOWER et al. [1289]. The chemical method of CHIBNALL et al. [227] for the estimation of amide groups should be mentioned, though intrinsic difficulties limit its extensive use. The carboxyl groups of aspartic and glutamic acids as well the carboxyl terminal residue are converted to esters by treatment of the protein with methanolic HCl at 25° C for 24 hrs, while asparagine and glutamine remain unaffected. Reduction with $LiAlH_4$ in tetrahydrofuran converts the esterified groups to the corresponding alcohols of the aspartyl and glutamyl residues, and after acid hydrolysis γ-hydroxy-α-amino-butyric acid and δ-hydroxy-α-amino-valeric acid are obtained, respectively, while asparagine and glutamine yield aspartic and glutamic acid; the carboxyl terminal residues give β-aminoalcohols. After separation of these components, the relative amounts are estimated by quantitative procedures. An excellent review on this subject with a detailed experimental section is available [1379].

VIII. Detection of Amino Acids and Peptides in Paper Chromatograms

The column chromatographic methods of amino acid analysis give highly reproducible and quantitative results [2]. There are, however, several procedures which necessitate accurate analysis, but wherein quantitation of the order of $\pm$ 10% is satisfactory. Under such circumstances the more rapid methods of paper and thin layer chromatography and electrophoresis may be used.

These techniques have found extensive application for the qualitative and semiquantitative determination of the amino acid composition of peptide and protein hydrolyzates, for end group determination following treatment with exopeptidases, and for following the course of enzymatic hydrolysis. These techniques have also been of invaluable aid in connection with the phenylisothiocyanate (PTH) sequential degradation procedures. The PTH amino acids released during sequential degradation are hydrolyzed prior to analysis (see Chapter 8).

Paper chromatography is usually carried out by the descending technique on Whatman No. 3 or 3 MM sheets. The procedure provides qualitative identification of amino acids found in peptide hydrolyzates down to the nmole level and quantitative results on samples containing 5 to 50 nmoles. One dimensional high voltage electrophoresis of amino acids permits the separation of 10 to 20 samples in a period of about 2 hrs [83, 295]. Thin layer chromatography is an extremely simple and inexpensive technique that permits very rapid and efficient analysis. A great variety of different chromatographic material is available. Precoated thin layers on glass plates, glass fiber, plastic, or alumina are available, based on silica, cellulose, polyamide, or aluminum oxide [923, 953].

Peptides are also separated with great success by the aforementioned techniques. Mixture of peptides result from partial degradation of proteins. Separation of the individual components of the complex peptide mixtures resulting from enzymatic digest of a protein molecule and establishment of their amino acid sequence forms the basis for elucidating the primary structure of a protein [672]. A combination of preparative paper chromatography and electrophoresis has often been used successfully to separate such mixtures [83]. Thin layer chromatography may also be combined with thin layer electrophoresis [578].

The development of procedures for peptide mapping has brought into widespread use a technique that is capable of giving highly valuable information on the primary structure of protein molecules [83, 609, 672]. As generally employed, it consists in

Table 6-3. Amino acid detection with relatively specific reagents. (Taken in part from Ref. [170])

Amino acid	Reagent	Reference
Tryptophan	*p*-Dimethylaminobenzaldehyde + HCl (Ehrlich reagent)	[262, 947, 1178]
	Cinnamaldehyde + HCl	[500, 638, 1373]
	Xanthydrol	[283]
	Formaldehyde reagent	[1002, 1209]
	Diazotized sulfanilic acid	[340]
	Diazotized *p*-nitroaniline	[340]
	Diazotized benzidine	[238]
	Diazotized ethyl-α-naphthylamine	[262]
	o-Phthalaldehyde	[958]
Cysteine	Sodium nitroprusside	[1285, 1381]
	Iodoplatinate	[1285, 1381]
	Iodine-azide	[51, 218]
	Tetrazolium salts	[191]
Cystine	Sodium nitroprusside/NaCN	[1381]
	Iodoplatinate	[1381]
	Iodine-azide	[51, 218]
Methionine	Iodoplatinate	[1381]
	Iodine-azide	[51, 218]
	Potassium permanganate	[262]
Arginine	α-Naphthol/NaOBr or NaOCl (Sakaguchi reaction)	[1055–1057]
	Na nitroprusside/potassium ferricyanide	[1035]
Histidine	Diazotized sulfanilic acid or sulfanilamide (Pauly reagent)	[168, 386, 964, 965]
	Fast Blue B salt (diazo-reagent)	[308]
	Diazotized *p*-anisidine	[1071]
	o-Phthalaldehyde	[958]
Tyrosine	α-Nitroso-β-naphthol/HNO_3	[3, 410]
	Diazotized sulfanilic acid (Pauly reagent)	[168, 386, 964, 965]
	Tungstomolybdophosphoric acid (Folin-Ciocalteau reagent)	[364,737]
	Millon reagent	[297, 863]

subjecting an enzymatic digest of a protein to chromatography, followed by high voltage electrophoresis in the second dimension [83].

For elution of the peptides the appropriate area of the paper is cut out, eluted with 10% acetic acid and the peptide recovered by evaporation or lyophilization.

Water should not be used with silica layers since it dissolves the gypsum which may interfere later. 1-Butanol and acetic acid have been found to be satisfactory.

A number of reagents have been employed for detecting amino acids and peptides on chromatograms.

For color development, dipping of the chromatogram in a solution of the appropriate reagent is recommended. Of course, a solvent must be chosen for the reagent in which the substances to be detected are insoluble. For spraying chromatograms, a glass atomizer may be used. Care must be taken not to overload the paper with the spray reagent since the spots tend to diffuse. This is probably true when aqueous solvents are employed.

A multiple procedure of spot detection can be done by successive color reaction (ninhydrin, Ehrlich reagent, Sakaguchi, or diazo reagents; see below) leading to the possibility of revealing substances of different nature on the same chromatogram. With care, stains may be applied to only one side of the paper. The application should be in a small burst of spray, with drying in between, so that the liquid does not soak through. For example, in this way a Pauly stain (see below) may be produced on one side, a ninhydrin-collidine stain on the other, and finally the paper may be dipped in Ehrlich's reagent. Thus most of the amino acids or peptides may be identified on the same chromatogram by their position and characteristic color reactions. If more stains are needed, duplicate sheets must be prepared.

At present there is no known reagent specific for amino acids alone. Some non-specific reagents for nitrogen-containing compounds and several which are specific for one or a few amino acids are available (Table 6-3) [170].

The same amino acid specific reagents may also be used to stain peptides in chromatograms. The popular ninhydrin reaction is not always sufficiently sensitive with higher peptides; it fails utterly with cyclic peptides unless these contain free amino groups in side chains. More generally available and more sensitive (limit of detection about 0.1 μg) is the N-halogenation of Reindel and Hoppe (see further) [1020].

The subject of chromatographic techniques is adequately covered in many books [118, 1208] and review papers [170, 923, 954]. Thus, only a description of few selected procedures will be given.

A. Non-Specific Reagents

1. Ninhydrin

Ninhydrin is still the most popular reagent for amino acids. Although the color reaction of indane-1,2,3-trione hydrate with amino acids has been known since 1910 there is still uncertainty about the course of the reaction [194].

Whereas the normal method for color development yields yellow (proline and hydroxyproline) or violet (all other α-amino acids) products, more specific colors may be obtained by addition of bases such as collidine.

Treatment with complex-forming cations (Cu, Cd, Ca) modifies the color obtained with ninhydrin bathochromically towards red and considerably increases the color stability.

Ninhydrin-Acetic Acid [345]: 500 mg of ninhydrin, 5 ml of glacial acetic acid and 100 ml of 1-butanol. The chromatogram is sprayed and kept at 80° C for 30 min or at 110° C for 10 min.

Ninhydrin-Collidine (2,4,6-Trimethylpyridine) [874]: a solution is made from 600 ml of absolute ethanol, 200 ml of glacial acetic acid, 80 ml collidine and 1 g of ninhydrin. The paper is dipped in this solution and allowed to dry in an oven for 5 to 10 min at 80° C. The solution produces unpleasant and toxic fumes and should be used only in a well-ventilated room. A wide spectrum of colors and shades, ranging from pink, yellow and green to deep blue and purple, is produced by different peptides. The colors fade or change with time at room temperature. Accordingly, the stained papers must be stored in a freezer. Similarly, the solution itself should be stored in the cold.

Ninhydrin-Cadmium [522]: a stock solution is made up in the following proportions: water, 200 ml; acetic acid, 40 ml; cadmium acetate, 2 g. A 24-ml aliquot of this stock solution is added to 200 ml of acetone followed by 2 g of ninhydrin. The papers are dipped in this reagent and placed in a glass or other suitable sealed enclosure containing a small vessel of concentrated sulfuric acid. Color development is allowed to proceed for 18 hrs at 25° C, or of appropriately shorter periods of time at higher temperatures. The resulting cadmium complex of the ninhydrin color is much more stable than the usual dye and the background paper blanks are exceptionally low.

Ninhydrin-Cupric [676]: cupric nitrate has also been used as stabilizer of the ninhydrin spots. After visualization with the usual ninhydrin reagent, the chromatogram is sprayed with a mixture of 100 ml of 96% ethanol, 0.2 ml of 10% nitric acid and 1 ml of saturated cupric nitrate and then placed in a chamber containing 25% ammonium hydroxide. The red copper complex obtained in this way is stable only in the absence of free hydrogen ions and powerful complexing agents.

Moreover, the chromatogram must be protected from moisture since the Cu-ninhydrin complex dissociates reversibly between pH 7 and 9 and irreversibly at pH values above 9.

2. Chlorine-Toluidine

The reagent is specific for substances containing the grouping -NH-CO- and generally for nitrogen-containing compounds which can be converted into chloroamines. The reagent is especially useful for peptides (limit of detection about 0.1 μg); free amino acids require relatively high concentrations.

Procedure of Reindel and Hoppe [1020]: About equal volumes (20 to 50 ml) of 1.5% $KMnO_4$ solution and 10% HCl are placed in a tank of suitable size which contains a grating of glass rods, 2 to 3 cm above the bottom. The chromatographic plate to be developed is placed on the grating and the tank is covered with a large glass plate. After 15 to 20 min reaction time, the plate is aerated for 2 to 3 min in a well ventilated fume cupboard; the smell of chlorine should have completely disappeared before subsequent treatment.

If a chlorine cylinder is available, it is simpler to fill an appropriate vessel with chlorine gas and to place the plate for 5 to 10 min in this chlorine atmosphere. Paper chromatograms are sprayed with 2% acetic acid to remove excess chlorine. The spray reagent is prepared by dissolving 1 g of potassium iodide and 0.16 g of *o*-toluidine in 500 ml of 6% acetic acid. A corner of the chromatogram is first sprayed cautiously to establish if chlorine has been completely removed (no blue color should appear). Only when this condition is fullfilled can full spraying be carried out.

Procedure of Greig and Leaback [446]: The plate is lightly sprayed with 2% aqueous sodium hypochlorite solution and then, after 1 to 1.5 hrs at room temperature, sprayed with a solution prepared by mixing equal volumes of saturated *o*-toluidine solution in 2% acetic acid and 0.85% aqueous potassium iodide.

3. Isatin

The chromatogram is sprayed with 0.2% isatin in acetone and then heated for 10 min in a water-saturated oven at 70° C [1177]. The amino acids yield various colors upon reaction: proline and hydroxyproline give blue colors; aspartic and glutamic acid, dark purplish blue; serine, threonine, and tyrosine light, brown.

4. Indicator Dyes

Advantage may be taken of the buffering capacity of acidic, basic, or amphoteric substances by treating the chromatogram with a dilute solution of an indicator dye. After spraying with 0.1% solution of the sodium salt of bromothymol blue and drying, the chromatogram is dipped in 0.2% acetic acid [950]. The sheet turns yellow except for the blue spots indicating amino acids or peptides. A 0.04% solution of bromocresol green in ethanol [1171] as a spraying reagent was also employed.

5. Iodine

Iodine is a general reagent for detection of organic substances giving brown spots. The chromatogram is introduced into a closed vessel on the floor of which some crystals of iodine have been placed. In a modified procedure, the chromatogram is placed in a dense atmosphere of iodine vapor for 5 min or sprayed with an iodine solution (0.5% in chloroform). Excess iodine evaporates on standing in the air. The spots turn blue on spraying with a starch solution (1% in water) [66].

B. Specific Reagents

1. Tryptophan

Ehrlich Reagent [262, 947, 1178]: The chromatogram is dipped or sprayed with a solution made from 90 ml of acetone, 10 ml of concentrated HCl, and 1 g of *p*-dimethylaminobenzaldehyde. Peptides containing tryptophan develop a blue or purple color after a few minutes at room temperature. Since HCl bleaches ninhydrin colors, this reagent may be used to detect tryptophan on chromatograms previously stained with a ninhydrin reagent.

1-Dimethylaminocinnamaldehyde [500] is also conveniently used for indole detection and in particular for tryptophan compounds. To prepare the spray reagent, 2 g of the aldehyde are dissolved in a mixture of 100 ml of 6 N HCl and 100 ml of ethanol; the solution is stored in the dark in the refrigerator. Before use the spray reagent is diluted four times and then the chromatograms are heated at 100° C; pink to violet spots appear. The reagent is unsuitable when ammonia-containing developing solvents have been used since the background becomes colored. This can be mitigated by heating the chromatogram for 10 min at 105° C before spraying.

Xanthydrol [283]: The chromatograms are sprayed with a solution of 0.1 g of xanthydrol in 10 ml of 36% HCl and 10 ml of ethanol and then heated at 110° C until the spots acquire optimum color intensity.

Formaldehyde-HCl [1002, 1209]: The spray reagent is prepared by mixing 10 ml of 35% formaldehyde solution with 10 ml of 25% HCl and 20 ml of ethanol. The sprayed chromatograms are heated 5 min at 100° C. The fluorescent colors (yellow-orange-greenish) in long-wave UV light may be enhanced by exposure to the vapors of aqua regia.

2. Sulfur Amino Acids

Since cysteine in paper chromatograms is oxidized to cystine after application to the filter paper [1285], the methods of detection of cystine are valuable also for cysteine-derivatives.

Sodium Nitroprusside [1285]. *Reagent 1*: Dissolve sodium nitroprusside (1.5 g) in 5 ml of 2 N H_2SO_4. Then 95 ml of methanol and 10 ml of 28% ammonia are added. The solution is filtered and stored in a refrigerator.

Reagent 2: Two g of NaCN are dissolved in 5 ml of water and diluted to 100 ml with methanol. *Tests:* For cysteine use reagent 1; for cystine dip into reagent 1, dry slightly, and, while still damp, dip into reagent 2; for both cysteine and cystine prepare reagents at double strength and treat with an equal mixture of 1 and 2.

Sodium Azide : Iodine Reaction [51, 218]: The dry chromatograms are sprayed with 0.05 N iodine in 50% ethanol containing 1.5% of sodium azide. The spots are best seen under a UV light [1378]. A freshly prepared solution of 0.01 M iodine in 0.5 M KI plus 0.5 M NaN_3 can also be employed [4]. The sensitivity of the reaction is equivalent to 0.5 μg of methionine.

Platinic Iodide [1381]: Add in the following order 4 ml of 0.002 M $PtCl_6^{2-}$, 0.25 ml of 1 N KI, 0.4 ml of 2 N HCl, and 76 ml of acetone. The chromatograms are developed by dipping in the reagent. Cystine, cysteine, methionine as well other reducing agents give a white spot on a red-purple background. If phenol, lutidine, etc. are the developing solvents, all traces must be removed with ether/acetone, petroleum ether, etc.

Palladous Chloride in 0.1 N HCl may be used in place of chloroplatinic acid.

3. Arginine

Sakaguchi Reaction [1055–1057]: The chromatograms are sprayed with a 0.1% solution of α-naphthol in 1 N NaOH. After drying, the paper is sprayed with NaOCl solution prepared from an equal mixture of ethanol and commercial NaOCl. Arginine and arginine-containing compounds appear as red spots. Another more sensitive procedure (0.2 μg of arginine) can also be employed [4]. Dissolve 0.01% α-naphthol in ethanol containing 5% urea. Add KOH to a total concentration of 5% just before spraying. After air drying for a few minutes, spray slightly with 0.7 ml of bromine in 100 ml of 5% KOH.

A modification of the Sakaguchi reaction consists in dipping the chromatogram into 0.1% 8-hydroxyquinoline in acetone. After the chromatogram has dried, it is dipped into a solution of 0.02 ml of bromine in 100 ml of 0.5 N NaOH. Arginine and other guanido compounds give orange-red spots [637].

Phenanthrene-1,2-quinone was found by Gottlieb and Sussman [432] to give strongly fluorescent derivatives with arginine and arginine peptides in alkaline media. The reaction was applicable for detection purpose in paper chromatograms.

4. Histidine

The imidazole group of histidine as well the phenol group of tyrosine are readily detected on paper chromatograms by means of Pauly's reaction [386, 964, 965].

Sulfanilic Acid. To a fresh mixture of two volumes of sulfanilic acid (9 g of the compound in 90 ml of concentrated HCl/l of water) plus one volume of 5% (w/v) $NaNO_2$, add one volume of 20% (w/v) NaOH. Spray the chromatogram and after air drying spray further with 10% Na_2CO_3 solution. Histidine appears as a bright yellow spot with a red margin.

Several modifications of the procedure have been suggested [118]. The use of sulfanilamide [149], *p*-bromoaniline [116] and *p*-anisidine [1071] has also been reported.

5. Tyrosine

This amino acid reacts with the Pauly reagent and with the Folin-Ciocalteu reagent (see 6.II.A.).

α-Nitroso-β-Naphthol [3, 410]. Spray the chromatogram with 0.1% α-nitroso-β-naphthol in 95% ethanol. After drying in warm air, spray with 10% aqueous HNO_3 and heat at 100° C for 3 min. The reaction is given by phenols and 3- and 5-substituted phenols, but not by *ortho*-substituted or 3,5-disubstituted phenols.

Chapter 7

Amino Acid Composition by Column Chromatography

P. E. HARE

I. Introduction

Twenty-five years ago a determination of the amino acid composition of a protein hydrolysate by column chromatography took about 10 days to complete, used three separate columns, and required the hydrolysis of about 10 mg of protein [878].

In 1974, with commercially available equipment, a complete amino acid analysis of a peptide or protein can be routinely determined in less than an hour using a single column and with less than a microgram of sample hydrolysate. Amino acids at the picomole (10^{-12} mole) level are being analyzed by column chromatography coupled with some of the newer detection techniques that will be discussed later in this chapter.

A number of reviews of the spectacular advances made in the quantitation of amino acids by column techniques has been published over the years and should make interesting reading, particularly for those of us who take for granted the current state-of-the-art [485, 883, 884].

After a discussion on the interrelationships of resolution, speed, and sensitivity in amino acid analysis by column techniques, there is a brief consideration of some of the innovative commercial instrumentation currently available. Next, a section on the modification of older equipment is included as an aid to those who have older amino acid analyzers and may wish to update them to take advantage of recent developments, including the fluorescent detection of amino acids. Following this is a section for the do-it-yourselfer who, by inclination and/or lack of funds, prefers to piece together an instrument from available components that is inexpensive but capable of the high sensitivities available in the most sophisticated commercial instrumentation. The last part of the chapter is concerned with (1) problems of contamination, (2) the preparation of buffers and reagents, (3) the preparation of samples for analysis, (4) racemization and the determination of D and L amino acids, and (5) sequencing and the amino acid analysis of peptides and PTH amino acids.

II. Principles

A. Resolution, Speed, and Sensitivity in Column Chromatography

The complete amino acid analysis of a protein is accomplished by hydrolyzing the protein to a mixture of around 20 free amino acids and then separating this amino acid mixture into individual amino acids and subsequently determining the concentration of each amino acid.

MOORE and STEIN [878] performed the first amino acid analysis by column techniques on glass columns packed with starch and eluted with various alcohol-aqueous HCl solvent systems. The separation of the different amino acids was a function of their interaction with the starch in a particular solvent system. Because of overlaps it was necessary to use three columns. When sulfonated polystyrene resins [879] replaced starch for the column packing and aqueous sodium citrate buffers of increasing pH were used for the eluting solvents, a complete analysis could be performed on a single column 0.9×150 cm [880]. The separation of amino acids was effected by the ion exchange interaction between the anionic sulfonic acid groups of the resin and the cationic form of the amino acids in the buffer solutions.

Since the introduction of ion exchange resins and citrate buffers for amino acid analysis, virtually all column chromatographic techniques have used these materials for the separation of mixtures of amino acids. It is a tribute to the careful and thorough work of MOORE and STEIN over 20 years ago that this system has become the standard of comparison for all other methods of amino acid analysis.

B. Resin Effects

Improvements in resolving power and analysis speed have come about through a better understanding of the effect of particle size on resolution and flow rate [481, 485]. Smaller size resin particles provide a larger surface area for a given volume of resin and give better resolution but require higher pressures to maintain a given flow rate. Early column chromatography of amino acids on starch and on ion exchange resins used gravity feed and relatively large particle diameters to get reasonable flow rates. With the introduction of a pump to deliver a constant buffer flow rate at higher pressures, it became possible to use a finer particle resin to achieve higher resolution and to increase the flow rate so that a complete amino acid analysis could be completed in 1 day [882]. HAMILTON's hydraulic procedure [481] for obtaining resin particles, with a homogeneous particle size distribution within a range of 1 to 2 μ of any given diameter, was critical for the development of the first completely automated instrument [1198] as well as for subsequent developments.

In fact, the recent interest and expansion in liquid chromatography, leading to the development of high performance liquid chromatography (HPLC), is due in large part to the recognition that a homogeneous particle size of column packing material leads to better resolution; and the smaller the particle diameter the better the resolution. The capability of operating at pressures of 1000 p.s.i. and above has made possible the rapid separation of many organic mixtures that could not be resolved by other methods. It seems probably that other materials will be found that are even more suitable than ion exchange resins for the rapid separation of amino acids by column chromatography.

In addition to resin particle size and flow rate, there are a number of other factors that influence the resolution of amino acids eluted from ion exchange columns. One of these is the degree of cross-linking of the resin or the amount of divinylbenzene copolymerized with the styrene [482, 485]. Although most resins used in amino acid work today are nominally 8% cross-linked, successful separations have been made on 4% [880] and 12% [978]. HAMILTON [482] has shown how different degrees of cross-linking affect amino acid resolution and concluded that mixing resin batches of different degrees of cross-linking could result in relatively poor resolution of some amino acids.

A properly packed column of the finest resin available, used with the optimum flow rate, will not insure a satisfactory separation of amino acids from the column. On the other hand, a poorly packed column with inferior resin will never yield satisfactory resolution, regardless of how carefully other variables are controlled.

C. Buffers and Temperature Effects

The composition, ionic strength, and pH of the buffers and the temperature of the column are interrelated factors that can be varied over a considerable range to achieve optimum resolution or other desired effects. Different amino acids are influenced to different degrees by changes in the column temperature and by changes in the buffers. HAMILTON, in his classic paper [482], shows the resolution of the basic amino acids as a function of pH, temperature, sodium concentration, and buffer composition. LONG and GEIGER [791] have also considered the effect of these variables on the resolution of the common amino acids and, in addition, have included the effect of alcohol added to some of the buffers.

Most of the papers published on new techniques in amino acid analysis report some variant in the temperature and/or buffer composition to achieve a faster analysis or to improve the resolution. It is interesting to note that, in spite of all the variations that have been suggested, most of the commercial instruments today use buffer systems almost identical to those used in the first automatic amino acid analyzer in 1958 [1198]. This system made use of three buffers and two columns for the complete separation of the common amino acids. The acidic and neutral amino acids were separated on a 0.9 × 150 cm column using 0.2 N sodium citrate buffers of pH 3.25 and pH 4.25. A third buffer, 0.35 N sodium citrate of pH 5.28, resolved the basic amino acids from a 0.9 × 15 cm column. Duplicate aliquots of sample were needed for the two columns.

The earlier ion exchange system of MOORE and STEIN [880] was a single-column system in which all the amino acids were eluted and resolved from a single column; but the ubiquitous presence of ammonia in the laboratory atmosphere, and in the buffers used for elution, caused problems for the single-column system that could be readily solved by using a two-column system. Ammonia in the acidic buffers is added to the column during the elution of the neutral and acidic amino acids. During the elution of the basic amino acids this excess ammonia from the buffers is eluted as a broad peak or plateau and may interfere with the detection of the basic amino acids if its concentration is too high. This was not as serious a problem in the fraction collector system as it was in the automatic amino acid analyzers, in which the effluent was being monitored continuously. In a two-column system the run is terminated on the long column before the elution of ammonia, while on the small column the ammonia in the buffer is continuously eluting and is an integral part of the baseline. The basic amino acids, including any ammonia formed by the hydrolysis of glutamine and asparagine or by deamination, are resolved and eluted from the column with no interference from the ammonia in the buffer.

D. Column and Extra Column Effects

The dimensions of the ion exchange column directly affect the resolution, speed, and sensitivity of amino acid analysis [482, 485]. Decreasing the length of the ion exchange resin bed and keeping the flow rate constant decrease the resolution but increase the analysis speed and sensitivity. By using improved spherical resins of smaller diameter and increasing the flow rate, Spackman showed it was possible to complete an analysis in a quarter of the time required by the first automatic analyzer [1195, 1197]. The resolution in this accelerated run, while not as good as in the longer original run, was more than adequate for hydrolysate samples. Sensitivity is increased in a shorter column, because the peak spreading varies with the square root of the column length [485].

Much effort has been extended in attempting to increase the speed of analysis [341, 590]. By using still finer resin particles and smaller volume columns it is possible to complete an amino acid analysis in less than 1 hr. It seems safe to predict that with the capability of some of the modern instruments to operate at pressures in excess of 3000 p.s.i. it will be possible to maintain adequate resolution and still be able to complete an analysis in less than 15 min.

The resolution of individual amino acids from a column may subsequently be lost by (1) remixing during the addition of the reagent used for detection, or (2) because of its prolonged dwell time in the heating coil or in the flow cell used for making the final optical measurement. Clearly the effluent from the column can be maintained on the way to the colorimeter by passing it through tubing with a small enough bore, or by segmenting the flow with gas [341] to prevent remixing. This is a particularly serious problem with small-volume columns.

Sample injection onto the column may result in poor resolution of some of the amino acids if the sample pH is too high or too low. The sample pH should be in the range of 1.5 to 2.5. Too large a volume of sample (greater than 10% of column volume), high concentrations of salts, soluble proteins from enzyme treatments, or particulate contamination in the samples are other factors that can lead to a loss of resolution. Careful sample preparation is essential in order to maintain resolution and for long-term reproducibility. Resin settling produces a dead space at the top of the column and can be a serious problem for resolution with continued use of a column, but the addition of some resin to the column usually restores the resolution. Most commercial instruments have a device for eliminating any dead space by a single adjustment at the top of the column.

E. Sensitivity — Column Effects

Decreasing the column diameter directly increases the sensitivity. This method of increasing sensitivity has been termed column amplification [483, 485] and is one way of increasing sensitivity without increasing the size of the buffer change peaks and the ammonia plateau. Because there is less buffer and reagent volume needed for elution, there is a significant economic advantage in small-bore columns. The smallest bore columns that have been used for ion exchange amino acid analysis to date are the 0.5 mm bore columns of Kirsten and Kirsten [698]. An 0.5 mm diameter column has only about 1/300 of the cross-sectional area of the standard 9 mm diameter columns, and the results indicate that the microbore column is about 300 times more sensitive. The extremely slow flow rates required (0.1 ml/hr) in this system, achieved by means of a special pump, and the necessity of using relatively small sample injection volumes have discouraged significant commercial development of this system, although modifications to an existing commercial microanalyzer have been described [447]. A compromise small-bore column size of 1.75 to 2.8 mm is used in a few commercial amino acid analyzers, but the vast majority still use the standard 9 mm bore columns.

It may not be generally recognized that the resin itself can contribute to sensitivity differences. A poorly graded resin will tend to give shorter and wider peaks, while a well-graded resin will give higher and narrower peaks, indicating that the amino acids are eluting from the column in narrower bands. This is the reason why, with better graded and finer particle sizes, it is possible to use relatively small-volume columns to achieve good resolution, fast analysis speed, and high sensitivity.

F. Sensitivity — Detectors

The sensitivity of any instrument depends directly on the method used for detection. When a fraction collector was used to collect discrete fractions from the column, a measured volume of ninhydrin reagent [820] was added to each tube. The tubes were subsequently heated in boiling water to produce the characteristic purple color with most amino acids, or yellow with proline and hydroxyproline. In the first automatic amino acid analyzer [1198], the ninhydrin reagent was modified and mixed continuously with the column effluent. The mixture was then kept in a boiling water bath for 15 min by using a 95-ft coil or teflon tubing of 0.7 mm bore (22 gauge). The solution then passed into the cuvettes or flow cell of the detector, where variations in absorbance of the solution were measured by photocells monitoring light of two path lengths (2.2 mm and 0.7 mm). The resulting series of peaks represents the concentration of the various amino acids eluted from the column.

Increasing the path length of the flow cell directly increases the sensitivity. Jones and Weiss [647] described a flow cell with a 20 mm path length, and more recently a flow cell with a 60 mm path length and only a 10 µl volume has been described [387, 497, 498]. This gives an increase of sensitivity of nearly 30 over the original design. Increasing the sensitivity of detection by increasing the path length of the flow cell has been termed optical amplification [483, 485]. Unlike column amplification, increasing the optical path length increases any baseline fluctuation due to (1) poor mixing, (2) the buffer change peaks, or (3) the ammonia plateau problem. The use of long path length flow cells has led to the recognition of problems contributing to baseline fluctuations that were not even noticeable with shorter path length cells. If the volume of an optical flow cell is too great, resolution will suffer. It is necessary, therefore, to keep the volume of the flow cell as small as possible while maintaining as long a path length as possible for higher sensitivity.

A third method to increase sensitivity has been termed electrical amplification [483, 485] and makes use of range expansion with zero displacement on the recorder. If a 10 mv signal is fed from the colorimeter to a 1 mv full-scale recorder, it is possible with a zero displacement device to make 10 mv the baseline position at one edge of the chart paper and 9 mv the position at the other edge of the chart. The effect is to get a 10 X expansion of the signal, but it is clear that peaks whose optical density is greater than 0.1 will be "off scale" and not be usable. A convenient solution is to have a two-pen recorder, with one pen on expanded scale and the other with normal range for larger peaks. Electrical amplification, like optical amplification, increases the sensitivity of the baseline artifacts along with the desired increase in sensitivity for the amino acid peaks.

By optimizing the column parameters and the flow cell length, and by using recorder range expansion it is possible to get acceptable chromatograms of 10 picomole levels of amino acids and to detect less than 1 picomole levels with some confidence, using ninhydrin for detection. The limiting factor for a further increase of sensitivity is the presence of contaminants in the buffers that cause artifacts during buffer changes and baseline shifts that interfere with the detection of the amino acid peaks. Because of the serious baseline shifts that occur during the elution of the basic amino acids in the single-column procedure, a two-column system with only a single buffer change is better suited for ultramicro analysis than is the single-column system, in spite of the fact that two sample aliquots are needed for a two-column system. It should be possible to eliminate, or significantly reduce, baseline artifacts; but until this is accomplished the limiting factor in ultramicro amino acid analysis is buffer contamination and not the ninhydrin sensitivity.

G. Sensitivity of Fluorescence Systems

Although several reagents have been suggested [114, 412, 485, 802, 1043] over the years as alternatives to ninhydrin for the detection of amino acids from ion exchange column, none has made as much of an impact as the recent discovery of fluorescamine (Fluram-Roche Diagnostics) by Udenfriend and his associates [1300, 1301]. Stein et al. [1220] and Georgiadis and Coffey [409] have described systems using fluorescamine to monitor the effluent from ion exchange columns. Fluorescamine reacts with primary amines almost instantly at pH 9 at room temperature to form a product that is highly fluorescent (excitation—390 nm, fluorescence—475 nm). Unfortunately proline and hydroxyproline do not react directly with fluorescamine but must be oxidized to primary amines by N-chlorosuccinimide [1341] (or other suitable reactant) prior to reaction with fluorescamine. A system utilizing N-chlorosuccinimide has been described [354, 355]. Fluorescamine itself hydrolyzes in aqueous solutions (and also alcohols) to form nonfluorescent products so that it must be kept in a nonaqueous, nonalcoholic organic solvent. Acetone, acetonitrile, and dioxane have been used successfully. Because it is necessary to buffer the column effluent to pH 9 for optimum sensitivity, a separate borate buffer reagent (potassium borate or lithium borate is more soluble than sodium borate) is mixed prior to or simultaneously with the fluorescamine. The fluorescence of the stream is monitored by an appropriate device such as the Fluoromonitor (Fig. 7-1) from the American Instrument Company. Fluorescence is proportional to the concentration of the amino acid and is also a function of the primary lamp intensity, which means that a brighter lamp produces a higher intensity of fluorescence.

It is difficult to compare the sensitivity of fluorescamine with that of ninhydrin. Both systems have been used extensively in our laboratory over the past 2 years, and it is our conclusion that, with the present state of the art, fluorescamine is not significantly more sensitive than ninhydrin for the detection of amino acids from an ion exchange column [499]. Both systems are limited, not by their inherent sensitivities but by chromatographic artifacts which are both ninhydrin- and fluorescamine-sensitive. It is significant that fluorescamine is relatively insensitive to ammonia compared to ninhydrin; and on a single-column system it shows little or no ammonia plateau. However, there are other artifacts that show up more in the fluorescent

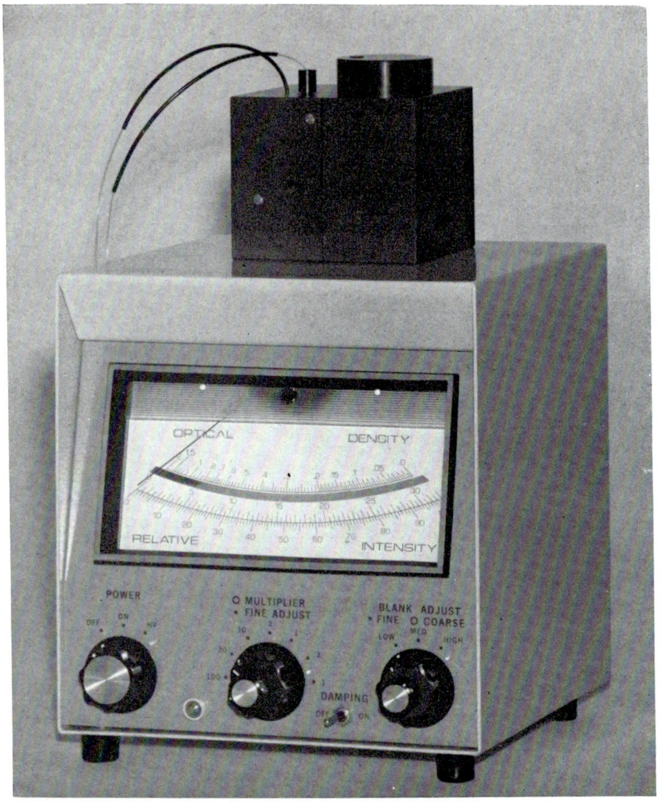

Fig. 7-1. The AMINCO Fluoromonitor, a filter fluorometer designed specifically to monitor the fluorescence in a flowthrough system. Unit is supplied with a quartz flowcell, a 4-watt mercury lamp, and a photomultiplier detector (American Instrument Company)

system than with ninhydrin. Earlier papers on the use of fluorescamine have implied, or even stated, that fluorescamine is up to two orders of magnitude more sensitive than ninhydrin [354, 409, 1301]. This is from a comparison of a sample analyzed on a standard-sized column with ninhydrin and a narrow-bore column with fluorescamine. The column used for fluorescence also was shorter so that the column amplification factor was a significant factor in the sensitivity difference. In another study, HARE [499], using the same ion exchange column and comparing the ninhydrin and fluorescamine detection systems on separate runs, found that the ninhydrin system, with a 6 mm optical path flow cell had about equivalent sensitivity.

A fluorescent system using *o*-phthalaldehyde (OPA) and mercaptoethanol was described by ROTH [1043, 1044] and appears to have some advantages over fluorescamine. It is stable in aqueous solutions and, consequently, can be combined into a single reagent with the borate buffer. Like fluorescamine it does not react directly with proline or hydroxyproline. OPA reaction products with amino acids are excited and fluoresce at wavelengths similar enough to those of fluorescamine reaction products so that the same filters on the detector may be used for both systems. It has been found by a direct comparison of the two systems that OPA is between 5 and 10 times more sensitive than fluorescamine. Because it is possible to use OPA in an aqueous solution there are fewer problems in mixing with the column effluent than with either ninhydrin or fluorescamine. The result is that the baseline is significantly more stable with OPA than with either ninhydrin or fluorescamine. The OPA system, at present, has not been modified to detect proline or hydroxyproline; and until this capability is developed very few protein hydrolysates will be run with this system.

H. Sensitivity of Systems Other than Ion Exchange

The recent advances in high performance liquid chromatography (HPLC) suggest some potential applications to amino acid analysis. It is possible that amino acids may be rapidly separated by HPLC using silica gel columns and typical TLC solvent systems to elute the amino acids. Column effluents could be monitored by ninhydrin or fluorescent systems in much the same way as are present amino acid analyzers.

The development of systems for the rapid separation of PTH amino acids [433, 983] may make it desirable to make PTH or other suitable derivatives (e.g. dansyl) of protein hydrolysates and separate the derivatives by HPLC and monitor the effluent for UV absorption of fluorescence. It would not be necessary to add a reagent to the effluent stream.

Whenever a derivative step is involved, as it is in GLC techniques for amino acids or in the formation of PTH and dansyl amino acid derivatives, there is always the concern for reproducibility of the derivatization step. Generally it is preferable to do no more than necessary in sample preparation, as each step introduces the possibility of contamination and/or loss of the sample. There are, however, cases where this is necessary; unless there is a significant improvement in sensitivity, speed, reliability, or cost of analysis it does not seem that techniques requiring the formation of amino acid derivatives will be widely used for determining protein hydrolysate compositions.

With continuing improvement in column chromatographic techniques it will be interesting to see what develops in the next few years in instrumentation for amino acid analysis. It seems likely that for the foreseeable future column chromatographic techniques for amino acid analysis, although 25 years old, will remain the standard method of comparison.

III. Instrumentation

A. Commercial Amino Acid Analyzers

Table 7-1 is taken from the International Chromatography Guide for 1974 to 1975, which lists 17 companies that offer amino acid analyzers. Most of the instru-

Table 7-1. Companies currently supplying amino acid analyzers as listed in the 1974—75 International Chromatography Guide (Journal of Chromatographic Science, Vol. 12, G 5, 1974)

American Instrument Company, Inc.
Travenol Laboratories, Inc.
8030 Georgia Ave.
Silver Spring, MD 20910
301 589-1727

J. T. Baker Chemical Co.
222 Red School Lane
Phillipsburg, NJ 08865
201 859-2151
Foreign Headquarters:
J. T. Baker
608 Gross Gerau
West Germany

Beckman Instruments, Inc.
Spinco Division
1117 California Ave.
Palo Alto, CA 95304
415 326-1970

Carlo Erba S.p.A.
Scientific Instruments Div.
P.O. Box 4342
20100 Milano,
Italy
9568013/16/33/59

Cenco Instruments
2600 S. Kostner Ave.
Chicago, IL 60623
312 227-8300

Durrum Instrument Corp.
3950 Fabian Way
Palo Alto, CA 94303
415 321-6302

Glenco Scientific, Inc.
2802 White Oak Drive
Houston, TX 77007
713 861-9123

Japan Electron Optics Laboratory Co., Ltd.
1418 Nakagami
Akishima
Tokyo, Japan
0425-43-1111
Foreign Headquarters:
JEOLCO (U.S.A.) Inc.
Analytical Instruments
477 Riverside Ave.
Medford, MA 02155
617 396-8620

LKB—Produkter AB
Fredforsstigen 22-24
Fack, S-16125
Bromma, Stockholm
Sweden
08/980040
Foreign Headquarters:
LKB Instruments Inc.
12221 Parklawn Dr.
Rockville, MD 20852

Optica S.A.S.
21 Via Gargano
Milano, Italy
564908

Phoenix Precision Instruments
The Virtis Co.
Rt. 208
Gardiner, NY 12525
914 255-5000

Rank Precision Industries Ltd.
Analytical Division
Westwood Industrial Estate
Margate, Kent, England
0843 24261
Foreign Headquarters:
Rank Precision Industries, Inc.
411 Jarvis Ave.
Des Plaines, IL 60018
312 297-7720

Sanda, Incorporated
Gypsy & School House Lanes
Philadelphia, PA 19144
215 849-8100

Schoeffel Instrument Corporation
24 Booker St.
Westwood, NJ 07675
201 664-7263
Foreign Headquarters:
Schoeffel Instrument GmbH
2351 Trappenkamp
Celsiusstraße 5, Postfach 119
West Germany

Systems & Components Ltd.
Process Instruments Division
The Broadway, Market Lavington
Devizes, Wilts.
England
Lavington 3226

Table 7-1 (continued)

Technicon Industrial Systems A Division of Technicon Instruments Corp. 511 Benedict Ave. Tarrytown, NY 10591 914 631-8000 Foreign Headquarters: Technicon International Division S.A. Geneva, Switzerland	**Jose Henao Velez & Cia. Ltda.** Correra 4a 8-63 P3 Cali, Valle, Colombia 831924/841057

ments use an ion exchange column for the separation of the amino acids. Most also use ninhydrin for the detection reagent; but at least one, the Aminanalyzer (American Instrument Company), is built around a fluorometric detector and may be used with either fluorescamine or *o*-phthalaldehyde as the fluorescent reagent. The instruments of at least two of the companies (Durrum and American Instrument) are capable of delivering buffers at pressures in excess of 3000 p.s.i., making them capable of high performance liquid chromatography applications. A brief discussion follows on the instruments that are known to make use of smallbore ion exchange columns and/or other innovative features.

The Technicon TSM amino acid analyzer is a two-column system using moderately small-volume columns. It is designed to operate at pressures less than 1000 p.s.i. and has a 40-sample capacity. Run times for protein hydrolysates are between 1 and 2 hrs with a sensitivity of approximately 1 O.D. (full scale) for 100 nmoles of most amino acids. This instrument has a number of innovative features and was discussed in considerable detail in the earlier edition of this book [342].

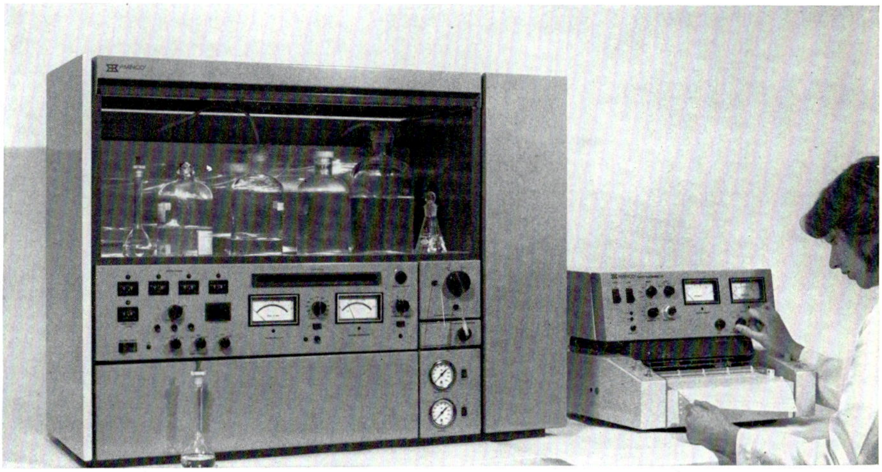

Fig. 7-2. AMINCO Aminalyzer, the first amino acid analyzer designed specifically around a fluorescence detector. Uses a narrow-bore ion-exchange column, high pressure (500 p. s. i.) displacement pumps, and a high pressure injection valve

Rank Precision Industries advertises an instrument capable of detecting 500 picomoles using a single column and a gradient elution with two buffers. Sixty samples can be loaded for consecutive operation.

Aminco's instrument (Fig. 7-2) is built around a ratio fluorometer and is capable of operation up to 5000 p.s.i. It features a small-bore column and is capable of subnanomole sensitivity.

Beckman's Spinco Division was one of the first in the amino acid analyzer field with a commercial version of the SPACKMAN et al. instrument [1198]. Spinco has recently introduced its Model 121M Microbore Amino Acid Analyzer, which features a number of changes from the earlier Model 120 series. It has a 2.8 × 500 mm column and a short column for two-column operation, or optionally it can use a single-column procedure. It is designed for use at pressures less than 1000 p.s.i. and uses specially designed displacement pumps for low flowrates, to deliver the buffers

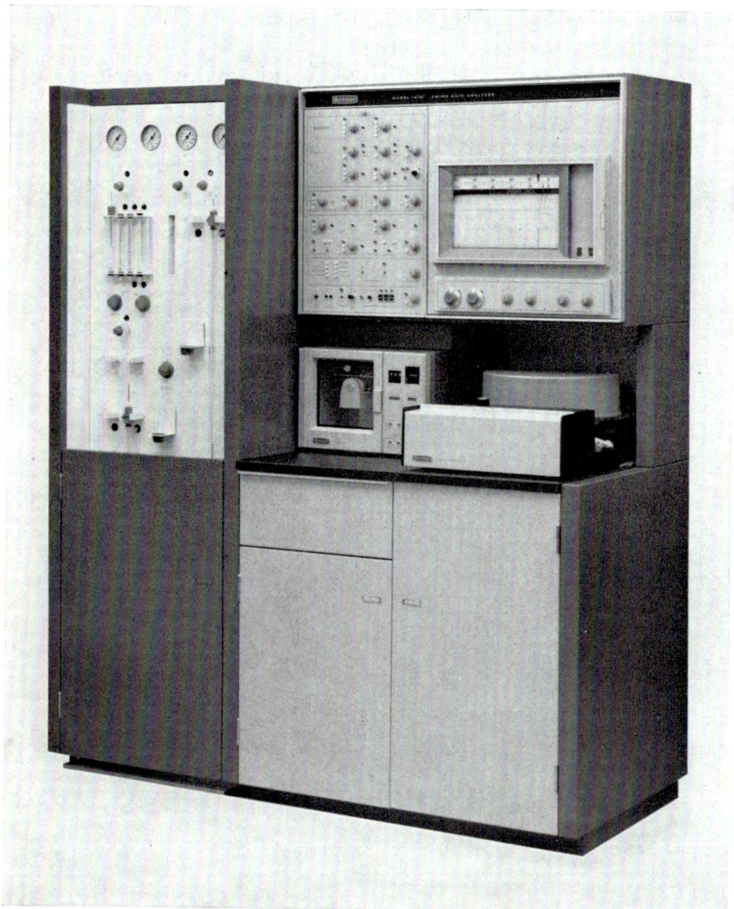

Fig. 7-3. Beckman Model 121-M Amino Acid Analyzer. Designed for pressures less than 1000 p. s. i. Uses single or dual narrow-bore ion-exchange columns. Instrument shown with the 80-loop automatic sample injector

to the columns, and ninhydrin reagent to the mixer and reaction coil. It is capable of completing a protein hydrolysate in about 1.5 to 2 hrs on the two-column system. An automatic injector with 80 sample loops can be used to run 80 consecutive samples on a single column or 40 double sample aliquots on the two-column system. Results of the analyses are recorded and simultaneously processed by the System AA Computing Integrator, which identifies and integrates each peak area and calculates its concentration or percentage from a calibration constant for that particular amino acid.

The Durrum Model D-500 Amino Acid Analyzer (Fig. 7-3) is at present the most sophisticated commercial instrument in the field of amino acid chromatography. It is capable of operation at pressures in excess of 3000 p.s.i. The pump is a combination of eight syringe pumps (Fig. 7-4) that can be used to deliver eight different

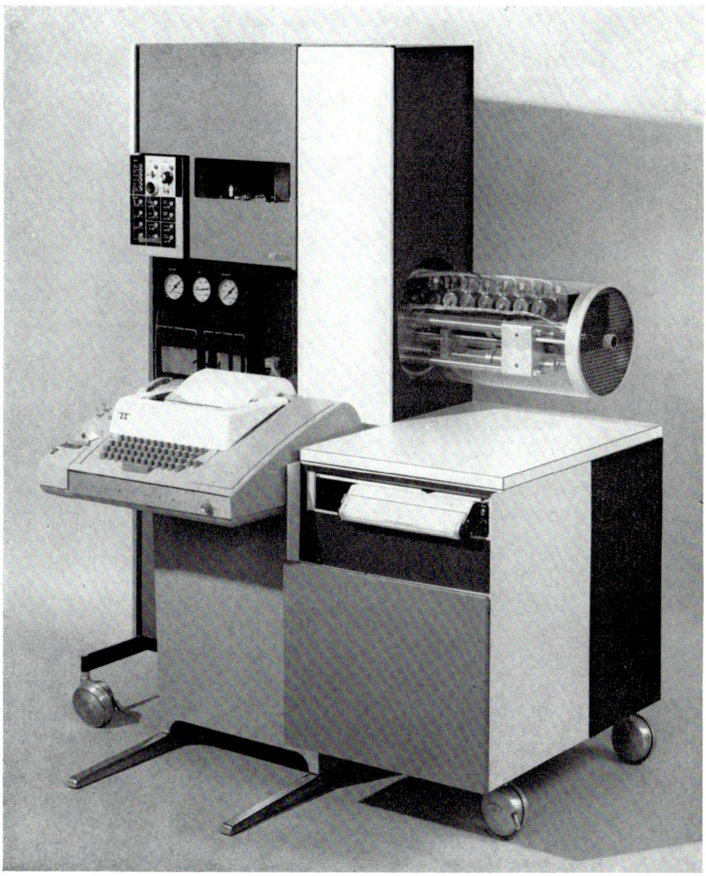

Fig. 7-4. Durrum Model D-500 Amino Acid Analyzer. Built-in computer controls, sample injection, buffer changes, and column regeneration, and also identifies and calculates the concentration of each amino acid in each sample. The high pressure pump assembly is located in the clear cylinder on the right and consists of eight syringe pumps and solenoid control valves for buffers and reagents

buffers, solvents, or up to an eight-step gradient solvent system for high performance liquid chromatography. As currently used for amino acid analysis, it uses two pumps for the ninhydrin reagent and the other six for buffers and regenerating reagents. It uses a smaller bore column (1.75 mm) than most other commercial instruments and is capable of a 1 O.D. (full chart peak height) sensitivity with only 10 nmoles of most of the amino acids. The chart from a 500-picomole run is shown in Fig. 7-5. This is far from its maximum sensitivity. It can do a protein hydrolysate in less than an hour (45 min). The multiple-sample injector is almost unlimited in capacity, because each sample is placed in a cartridge; and the cartridges are placed in a gravity feed holder located in a refrigerator compartment above the injector mechanism.

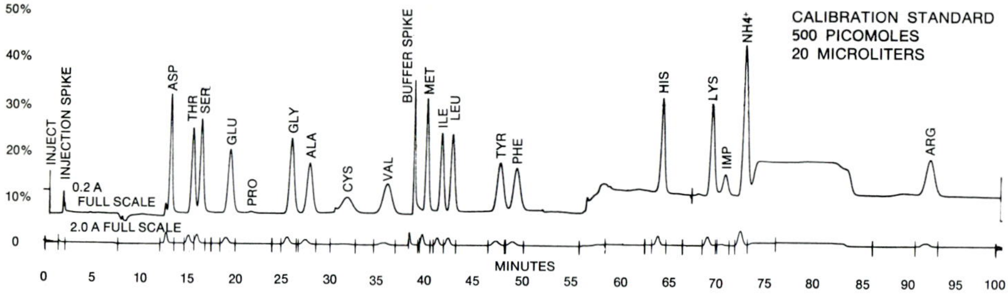

Fig. 7-5. Chart from Durrum Model D-500 from 500 picomoles of each amino acid. Vertical lines on the lower tracing indicate start and finish of integration of each peak

=

5/10/74 500 PICOMOLE SENSITIVITY
PROC= 1 RUN= 7

PEAK	MIN-SEC		NAME	QUANT	TYPE	AREA	FACTOR	BASE
1	1	33	INJ	1311.2	0	125	143	28
2	12	52	ASP	55.1	0	1286	34987	28
3	15	6	THR	51.1	1	1230	36081	28
4	16	1	SER	53.8	2	1351	37689	28
5	19	10	GLU	49.5	0	1267	38356	30
6	25	37	GLY	51.3	0	1441	42098	29
7	27	35	ALA	50.1	0	1215	36387	29
8	31	35	CYS	63.6	0	885	20858	29
9	35	48	VAL	54.3	0	1188	32837	29
10	38	35	BSP	1408.7	0	941	1002	29
11	39	45	MET	56.5	0	1396	37083	29
12	41	18	ILE	56.5	1	1220	32363	29
13	42	25	LEU	56.5	2	1419	37662	29
14	47	29	TYR	61.7	0	1553	37748	29
15	49	11	PHE	62.3	0	1499	36118	29
16	64	24	HIS	46.4	0	1319	42672	53
17	69	37	LYS	51.0	0	1504	44247	49
18	71	7	IMP	1329.6	0	437	493	49
19	73	2	NH4	196.5	2	2711	20691	49
20	92	40	ARG	49.7	0	1251	37740	50

100 0
ERR= 0

Fig. 7-6. Computer printout for 500 picomole analysis shown in Fig. 7-5. Each detected peak is numbered consecutively and identified by retention time. Peaks 1, 10, and 18 are artifacts caused by injection (INJ), buffer change (BSP), and an impurity (IMP) in the buffer

The heart of the instrument is the PDP-8 computer. The computer controls the operation of the instrument such as the temperature changes and buffer cycles and also computes the results upon completion of a run (Fig. 7-6). New instructions for varying the buffer change times can be readily entered. The ninhydrin system is unique in that the reaction coil temperature is higher (120 to 130° C); consequently, the residence time is only 1 or 2 min for nearly maximum reaction. This permits the use of less tubing between the detector and the mixer and improves resolution significantly.

One of the many innovations introduced in the D-500 was in the design of the colorimeter for reducing bareline artifacts due to poor mixing, air bubbles, or even particulates. It uses a reference wavelength of 690 nm, where the absorbance of the ninhydrin reaction products of amino acids is negligible. The ratio of 570 nm (actually it uses 590 nm) to 690 nm and 440 to 690 nm is used to monitor the column effluent and leads to a significant improvement over the use of 570 and 440 nm alone. Noise not related to the amino acid color development will absorb at 690 nm as well as at 570 and 440 nm. By using the 690 nm as a reference wavelength, this noise, in effect, can be canceled out. This innovative concept has been incorporated more recently by other instrument manufacturers such as Beckman in the Model 121 M.

Most of the other commercial systems use 9 mm bore columns and thus do not take advantage of the column amplification factor for increasing sensitivity as previously discussed. Many are capable of high sensitivity by combining optical and electrical amplification at the detector and recorder. The trend in the newer commercial instruments is toward a versatile instrument with automatic multiple sample capability coupled with automatic data reduction.

B. Modification of Standard Amino Acid Analyzer Equipment

Substituting a 2.8 × 600 mm narrow-bore column in place of the standard 0.9 × 150 cm column on an amino acid analyzer will increase the sensitivity of the instrument by approximately a factor of 20. However, unless other changes and modifications are made, the resolution will be so poor that the increased sensitivity will be of little value. The standard 22-gauge teflon tubing leading from the column through the mixer and reaction bath to the colorimeter must be replaced by narrower bore teflon tubing (preferably 32-gauge). The paper by Liao et al. [775] describes in detail the necessary modifications to convert a standard amino acid analyzer to a narrow-bore column system. The necessary components are commercially available, and little or no specialized machining is necessary. Liao et al. suggest shortening the reaction coil from 100 to 50 feet. Although the residence time is only 4 min, over 90% of the longer reaction coil color is achieved.

Liao et al. also suggest diluting the ninhydrin reagent with water (2:1) and pumping the ninhydrin and buffer at equal flowrates to achieve better mixing and baseline stability. The sensitivity of the ninhydrin is not significantly affected by dilution with water in the ratio of two volumes of ninhydrin reagent to one volume of water.

Modification of the pumps involved eliminating excessive dead space in the outlet end of the pump to minimize buffer mixing during buffer changes.

Photometers need not be modified provided their internal volumes are 100 ml or less. Of course, it is obvious that a considerable increase in sensitivity is possible if the standard photometers are replaced with flow cells with longer optical path length. If the baseline noise is sufficiently low it may be possible to increase the sensitivity by 10 times or more by longer optical paths. Recorder expansion is an alternative method or may be used in addition to optical amplification.

It is possible to replace the ninhydrin system completely by using a fluorescence detector [1301]. Felix and Terkelsen [354, 355] have described a narrow-bore column system that uses three reagents: fluorescamine, N-chlorosuccinimide, and a borate buffer. The N-chlorosuccinimide converts hydroxyproline and proline to primary amines that will react with fluorescamine. Three separate pumps are used to meter in the three reagents. Since heating is not required to develop the fluorophor, the system does not require a long reaction coil and heater. This helps to improve the resolution of the amino acid separation.

It remains to be seen if, or to what extent, fluorescence methods will replace ninhydrin methods for the amino acid analysis of protein hydrolysates. The simultaneous dual-wavelength monitoring of the ninhydrin reaction products gives qualitative as well as quantitative information. Retention time, coupled with the value for the 440 to 570 nm ratio, is better for peak identification purposes than retention time alone. On the other hand, since the peak identification in protein hydrolysates is relatively straightforward, based only on retention time, it may be that fluorescent systems will become popular if they can be proved to be simpler or to be significantly more sensitive. So far the difficulties involved in the fluorescent detection of proline and hydroxyproline have led to a detection system that is in some respects more complex (3 reagents and 3 pumps) than the ninhydrin system (1 reagent, 1 pump, 1 reaction bath). In addition, direct comparisons from the same ion exchange column with currently available detectors of the fluorescamine and ninhydrin system do not show significant differences in sensitivity between the two systems [499].

Sensitivity at present is limited by other factors such as buffer contaminants and not by the sensitivity of the reagents (ninhydrin or fluorescent) or the detectors. The ultimate measure of sensitivity must be the minimum detectable concentration of amino acid or other compound with some generally accepted value such as 2 to 1 for the signal-to-noise ratio.

Of the hundreds of modifications suggested for existing equipment, procedures, or techniques in amino acid analysis, only a relatively few have significantly improved the performance of existing amino acid analyzers for the determination of the amino acid composition of protein hydroloysates. Future significant modifications to existing instruments will reflect contributions in three areas: (1) optimization of the column parameters, (2) advances in the technology of column packings, and (3) the development of new and more sensitive detection systems.

C. Construction of Simplified Instrumentation for Amino Acid Analysis

There are some who prefer to build an instrument from component parts rather than to buy a new instrument or to modify an existing one. An advantage in this approach is that the instrument initially may be constructed as a simpler manual version and then subsequently automated to various degrees.

Details of construction for the first automatic amino acid analyzer [1198] indicate the general approach that has been used in virtually all subsequent instrumentation. A pump is used to deliver the buffers to an ion exchange column, and another pump delivers reagent to a mixer, where the column effluent and reagent are mixed. The mixture then flows through a reaction coil to a colorimeter, where changes in absorbance are monitored by a recorder. Buffer changes are generally made by a buffer-selector valve and automatic timer. Samples which were originally pipetted directly onto the resin surface are more often injected directly onto the column while maintaining the normal operating pressure and flowrate.

Most of the instrumental procedures described since the SPACKMAN et al. paper [1198] have used the same basic approach, using pumps of one kind or another, with modifications being made for a single-column gradient elution system [978], a single-column step-wise buffer system with narrower bore columns [482], and many variations on the two-column and single-column procedures. Some recently described systems utilizing narrow-bore columns with ninhydrin [775] and with fluorescamine [354, 355, 409, 1220] could be built from components, most of which are readily available from commercial sources.

The author described a system [493] that uses nitrogen gas pressure to drive the buffers and reagents in a narrow-bore ion exchange column system. This system, in its simplest form, does not require any buffer change timers or pumps. A single reservoir of pH 3.25 buffer (tubing or cylinder capable of withstanding 500 p.s.i. pressure) is attached to a three-way valve (Chromatronics or Altex) to teflon tubing (100 ft $\times$ 20-gauge, 4 ft $\times$ 32 gauge) and then to a second three-way valve. The second valve is connected to the ion exchange column. A ninhydrin reservoir (Saran tubing or stainless steel reservoir) or three fluorescamine reservoirs (teflon tubing or stainless steel) are connected to a mixer with the column effluent. A reaction coil (32 gauge, 10 feet) is immersed in a heating bath for ninhydrin or left at room temperature for fluorescamine. The reaction coil is connected to the flow cell of a colorimeter or fluorimeter. A recorder is used to monitor the results.

Two regulators for nitrogen or helium capable of delivering up to 400 or 500 p.s.i. are needed to control the flow rates of the buffers and ninhydrin. For fluorescamine, two additional regulators are desirable but not essential. Fig. 7-6 shows a diagram of the simple system.

The buffer changes are made by adding with a syringe the appropriate volumes of pH 3.25, pH 4.25, pH 5.2, and pH 10.1 buffers (or other buffers) into the 100-feet length of 20-gauge teflon tubing between the two three-way valves. The sample is drawn into the 4-feet length of 32-gauge tubing. This can be done with the three-way valves in the by-pass position with pH 3.25 buffer flowing through the system. Any desired number of discrete buffer changes can be made, including a sodium hydroxide wash. When the valves are moved to inject position the sample is injected into the column, followed by the series of buffers previously added to the teflon tubing.

To change the buffer sequence or times it is only necessary to fill the tubing with different buffers or with smaller or larger volumes of the various buffers. The 20-gauge tubing prevents undue mixing of buffers and makes it possible to achieve discrete buffer changes. A two-column system requires another buffer reservoir and regulator.

It is desirable but not essential to add flow meters to monitor the flow rates and back pressure filters to help maintain uniform flow and to achieve better mixing.

Various detectors have been used, but the flow cell volume must be kept low (50 µl or less). Fig. 7-7 shows a flow cell that has been described [387, 497, 498] with

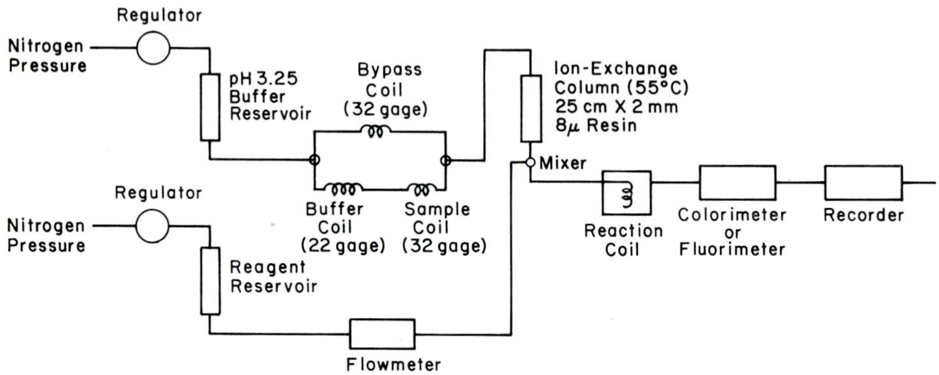

Fig. 7-7. Flow diagram of simple gas-driven constant pressure amino acid analyzer. The buffer coil contains a total of approximately 10 mls volume and is filled with the appropriate volumes of pH 3.25, 4.25, 5.2, and 10.1 for a single run. The sample is added to the sample coil and injected into the column followed by the series of buffers. Finally, pH 3.25 buffer from the reservoir follows the pH 10 buffer, and the system is ready for the addition of another sample and series of buffers

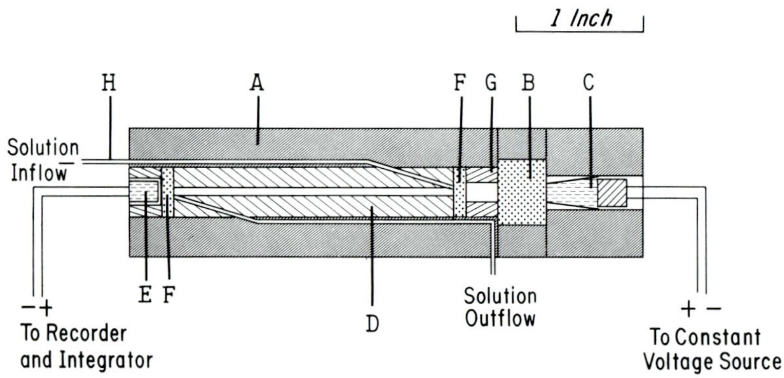

Fig. 7-8. Flow through cell for pathlengths up to 60 mm. *A* Delrin housing block. *B* Interference filter. *C* Lens-end type lamp. *D* Teflon rod with 0.5 mm diameter flow channel and 0.25 mm diagonal channels for entrance and exit of fluid stream. *E* Photoconductor cell. *F* Quartz windows held against teflon cell by threaded removable inserts *G* in housing block *A*

up to a 60 mm optical pathlength and a bore of about 0.5 mm. Enough light energy can be transmitted by using the lens-end type bulb (Chicago Miniature CM-3). In combination with a photoconductor cell (Clairex 905L) it may provide up to 100 mv output. In the Geophysical Laboratory system four colorimeters are used in series: two 2 mm flow cells at 440 nm and 570 nm, and two 60 mm flow cells at

440 nm and 570 nm. These are monitored on two 2-pen recorders and provide a system that covers a range of concentrations from less than 0.05 nmoles to over 100 nmoles. With recorder range expansion, picomole levels of amino acids are readily detected. Fluorescence detection systems can be readily adapted to the gas pressure system. For analysis of proline and hydroxyproline, three additional buffer reservoirs are used for N-chlorosuccinimide, borate buffer, and fluorescamine reagent.

By adding separate buffer reservoirs and on-off valves for each buffer, it is possible to have an instrument that is better adapted to multiple automatic sample injection (Fig. 7-8), although it is no faster or more sensitive. In this system the buffer reservoirs (under identical pressures) are controlled by on-off valves and a timer that provide the desired sequence and sample injection. The instrument in our laboratory uses a modified (100 μl sample volumes) 20-port Chromatronix sample injector. Depending on the complexity of the samples and the resolution desired, it analyzes from 10 to 20 samples per day. An integrator (CSI 100 A) is used to provide areas and retention times to simplify data handling. More sophisticated data reduction systems, such as the Spectra Physics System AA Computing Integrator, are available and can be readily added to any system to have a complete on-line data system. Automatic data handling becomes more essential as analysis times decrease. Manual calculations of chromatograms now take as much or more time than the actual instrumental analysis time!

IV. Procedures and Techniques

A. Preparation

Regardless of which instrumental technique is used to determine the final amino acid concentrations in a protein hydrolysate, the preparation of the sample for analysis is of utmost importance and often will make the difference between meaningful and worthless results. This is particularly true for samples in the submicrogram range. It is difficult but straightforward to dilute down a standard mixture of amino acids and demonstrate that it is possible to analyze reproducibly a nmole or less of each amino acid. It is considerably more difficult to carry a fraction of a microgram of protein or peptide through the necessary hydrolysis and preparation steps for analysis and to determine the amino acid composition without encountering serious losses or contamination. Organic geochemical work with carefully collected and controlled lunar samples from the Apollo program has demonstrated that some unique problems exist in the recovery of amino acids at low levels [486, 498]. In one experiment, when aliquots of an acidic solution containing 0.1 nmole of each amino acid were placed in glass vials, it was shown to be virtually impossible to obtain a quantitative recovery of the amino acids methionine and cystine [498]. This experiment did not simulate hydrolysis but only recovery of amino acids after various storage and drying procedures. In another experiment, amino acids added to pulverized chert [486] and lunar material [408] resulted in substantial losses for many amino acids.

B. Contamination Problems

Contamination is generally assumed to cause erroneous results by contributing a spectrum of amino acids not present in the original sample. The finding of substantial levels of amino acids in human fingerprints [484, 492, 944] has led to a careful evaluation of the various steps in sample preparation as well as reagent contamination. Significant levels of amino acids have been found in reagent-grade hydrochloric acid [487, 492, 1389] and distilled water [487], two materials used extensively in the preparation of buffers for amino acid analysis and also for sample hydrolysis.

The source of much of the contamination appears to be air borne particulate material and is difficult to control. Ammonia, as a constituent or contaminate in air, is virtually ubiquitous and may reach extreme proportions in tobacco smoke. Dust has been found to contain amino acids upon hydrolysis [492, 1013]. It is doubtful that any reagents or materials are completely free or that contamination can be entirely eliminated. It is important to be aware of potential contamination problems and to control contamination to a realistic level.

Besides contributing unwanted amino acids to the system, contamination may also lead to a loss of amino acids, particularly at subnanomole concentration levels. In an early study [529] it was found that glassware cleaned with the usual sulfuric acid-dichromate cleaning solution still had traces of dichromate after ten rinses with water. This could lead to serious losses when submicrogram levels of protein and peptides are hydrolyzed in dichromate-cleaned tubes.

Several studies [486, 487, 496] have suggested that heating the glassware above 500° C is an adequate procedure for cleaning that eliminates most if not all acid-containing contaminants. A method of cleaning glassware, recently suggested by HAMILTON [481], involves a preliminary wash and rinse followed by a 30-min wash in sodium bicarbonate solution (50 g/l). After a final rinse the glassware is wrapped in aluminum foil and baked in a muffle furnace for at least 4 hrs at 625° C. The wash in sodium bicarbonate was designed to combine sodium ions with all negatively charged sites on the glass surface.

Contamination is usually monitored by running a reagent blank, including the same volumes of all reagents and using the identical procedures used in the sample preparation steps. Besides running a reagent blank, particularly when working at low concentrations of amino acids, a diluted (100 to 1) standard amino acid solution should also be carried through all of the sample preparation steps. The recovery of the amino acids from the diluted standard, and the presence or absence of significant amino acid levels in the reagent blank, should be of help in the interpretation of possible contamination in sample processing.

C. Preparation of Reagents

1. Clean Water

The quality of distilled or deionized water is usually measured by its specific conductivity. This index of "purity" does not indicate the presence of nonionized organic matter, including proteinaceous material. What may be adequately "pure" water for one application may be unacceptable for certain other applications. The final test of acceptability must be the results obtained in an analysis when it is used in the buffers and reagents of the analytical system.

It is difficult to produce distilled or deionized water free of amino-acid-containing material, and even more difficult to keep the quality of the water at a high level during storage [59]. As water from the storage tank is withdrawn, laboratory air with air borne contaminants is drawn into the storage tank to replace the volume of water used. The use of particulate and ammonia filters help but do not eliminate the problem. The use of an ammonia-free and filtered inert gas, such as helium or nitrogen kept at a low positive pressure above the water in the storage tank, may be a satisfactory solution for the storage of high-quality water.

A distilling system for producing ultrapure water has been described [532] and consists of a continuously operating double still and a sealed reservoir. Oxygen (or air) is introduced into the water vapor during the primary distillation step and the mixture passed over a glowing platinum wire to a secondary still. The reservoir is designed to recycle the stored water back to the secondary still. Special precautions against contamination must be made at the reservoir outlet where the purified water is withdrawn for use.

Hamilton [487] has found most standard distillation systems and deionizing systems unreliable for producing amino-acid-free water. He describes one commercial deionizing system (Model 1-R recycling system from Continental Water Corporation, El Paso, TX 79926) that has been successful in reducing amino acid contamination levels to less than 1 μg/l (after acid hydrolysis).

Standard distillation and deionizing systems may be adequate to produce water of sufficient purity for most applications to amino acid analysis. No system can be taken for granted but must be checked out by careful analysis of reagent blanks. Following are a few suggestions for using water from standard distilling or deionizing systems:

1. Use freshly distilled water rather than water that has been stored for even a few days.

2. Let the still operate for a time to purge the system if it has not been operating continuously.

3. Avoid the use of plastic tubing unless it has been checked out [59].

4. Evaporate a liter of deionized (or distilled) water to dryness and check the residue for amino acids before and after hydrolysis. If the deionized water leaves a brown colored residue it will probably contribute to baseline noise and buffer change spikes during analysis. It should be distilled to remove this residue (residue bleed from the resins used for deionizing).

5. Prohibit smoking in the laboratory room where the still is located, because tobacco smoke can significantly increase the ammonia content of the water after it has been distilled.

6. Turn off any exhaust hoods in the room to avoid bringing dirty air from other rooms in the laboratory.

2. Hydrochloric Acid

Next to water, hydrochloric acid is probably the most frequently used reagent in sample and buffer preparation for amino acid analysis. Several commercial brands of reagent-grade hydrochloric acid have been found to contain significant quantities of free amino acids ranging up to nearly 10 nmoles/ml [487]. With milligram quantities of proteins and peptides available for hydrolysis, commercial grade hydrochloric acid diluted to six normal with water is adequate for hydrolysis procedures. For microgram and submicrogram quantities of proteins and peptides the quantity of amino acids present in the hydrochloric acid become significant and may equal, or even exceed, the amino acid levels present in the sample. It has been shown [1389] that distillation can reduce the amino acid levels by more than two orders of magnitude. Several studies have used distilled water and gaseous HCl (Matheson, electronic grade) to make up hydrochloric acid. An advantage to this method is that small quantities may be readily made when needed, and the problem of long-term storage of the hydrochloric acid is eliminated.

Regardless of the source of the hydrochloric acid, whether it is diluted from commercial brands, triply distilled, or made by passing hydrogen chloride gas into a good grade of water, it must yield a hydrolysis blank that contains a significantly lower level of amino acids than that found in the sample itself. It is rather embarrassing to discover that the supposed amino acid composition of a particular sample is in reality the amino acid composition of the hydrochloric acid used in the hydrolysis!

D. Preparation of Buffers and Reagents

1. Buffers

The buffers currently used in most column systems for amino acid analysis are nearly identical in composition to the buffers used in the first automatic amino acid analyzer [1198]. Sodium citrate is the most frequently used buffer salt, because it is an effective buffer from below pH 3 to above pH 6. The citrate buffers can most easily be made by using reagent-grade sodium citrate dihydrate (M.W. 294.1, Pierce Chemical Company, Rockford, Ill.) and dissolving 19.6 g for each liter of buffer desired for the neutral and acid buffers, or 34.4 g/l for the basic buffer. Redistilled 6 N hydrochloric acid is used to adjust the pH to the desired value and the final dilution made to 1 l. For single column elution, a basic buffer of higher ionic strength of higher pH is needed. Adding sodium chloride to the citrate buffer to make the final sodium ion concentration 1 N and a final pH of 6.4 is one possibility. Many variants have been suggested. A borate buffer of 0.2 N sodium ion concentration and pH 10 is also satisfactory [498]. For gradient elution buffers see Eveleigh and Winter [342]. Lithium citrate buffers have been used and are effective for resolving glutamine and asparagine. For recipes for sodium and lithium citrate buffers used in standard procedures see Benson and Patterson [86].

The addition of thiodiglycol to prevent methionine oxidation during chromatography may add contaminants to the buffers unless a specially purified grade is used (Pierce Chemical Company). This is also true of other additives designed for antibacterial action. Filtering is usually recommended, but again, filters are a possible source of contamination, particularly ammonia.

To remove ammonia contamination, the buffers may be processed through a column of cation exchange resin that has been washed with 2 N sodium hydroxide and water [978]. An alternative is to take a slightly alkaline solution of the sodium citrate and boil it before adding the hydrochloric acid for the final pH adjustment.

Still another way to remove ammonia is to install a large capacity in line ammonia filter (of action exchange resin) between the acidic buffers and the pump intake, or to install a small ammonia filter between the pump and the ion exchange column through which the acidic buffers flow and are stripped of ammonia. The basic buffer bypasses this small column and thus leaves the ammonia from the early buffers on the small column, which is regenerated each run cycle.

Many prefer to buy buffers ready-made as concentrates. Pierce Chemical Company (Rockford, Ill.), Beckman Instruments (Palo Alto, Calif.), Durrum Instruments (Palo Alto, Calif.), and Hamilton Company (Reno, Nevada) all offer buffer concentrates that are generally 10X concentrated. To use, 100 mls of buffer concentrate is diluted to 1 l with clean water, and the buffer is ready to use. These buffer preparations are adequate for most amino acid analyzers but are not entirely free from contamination. Improper storage of the buffer concentrate may lead to contamination by ammonia or other undesirable constituents.

2. Ninhydrin Reagent

The original ninhydrin formulation for the automatic amino acid analyzer system [1198] used 2% ninhydrin in a mixture of methyl cellosolve (3 parts), and 4 M sodium acetate buffer (1 part). 0.4 g/l of stannous chloride was used as a reducing agent to form the necessary component, hydrindantin, from some of the ninhydrin. More recently JAMES [635] has reported the use of titanous chloride as an improved substitute for stannous chloride. MOORE [881] suggested a new reagent formulation in which 1.25 g of hydrindantin per 2 l was added directly to the reagent, consisting of 2% ninhydrin in a mixture of dimethyl sulfoxide (3 parts) and 4 M acetate (1 part).

The needed reagents can be obtained separately or in a prepared reagent form. If the reagent is to be made in the laboratory, precautions must be made to insure an oxygen-free environment by flushing the reagent with an inert gas. In the pressure-drive amino acid analyzer described earlier the ninhydrin reagent is kept in a stainless

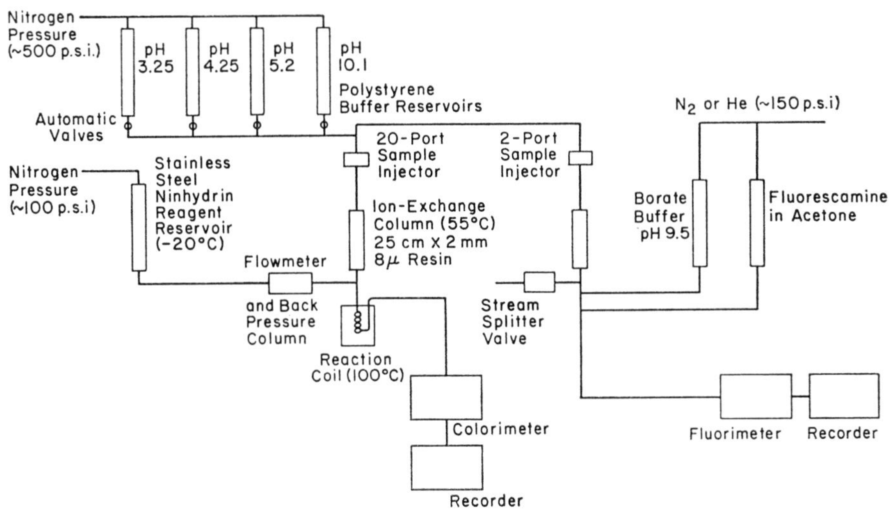

Fig. 7-9a. Diagram of gas-driven constant pressure system for automatic consecutive sample analysis with a second ion-exchange column for simultaneous analysis of separately injected samples by fluorescent techniques. Also shown on the effluent from the second column is a stream-splitting valve allowing the collection of most of the effluent and the simultaneous detection of part of the effluent by fluorescent techniques

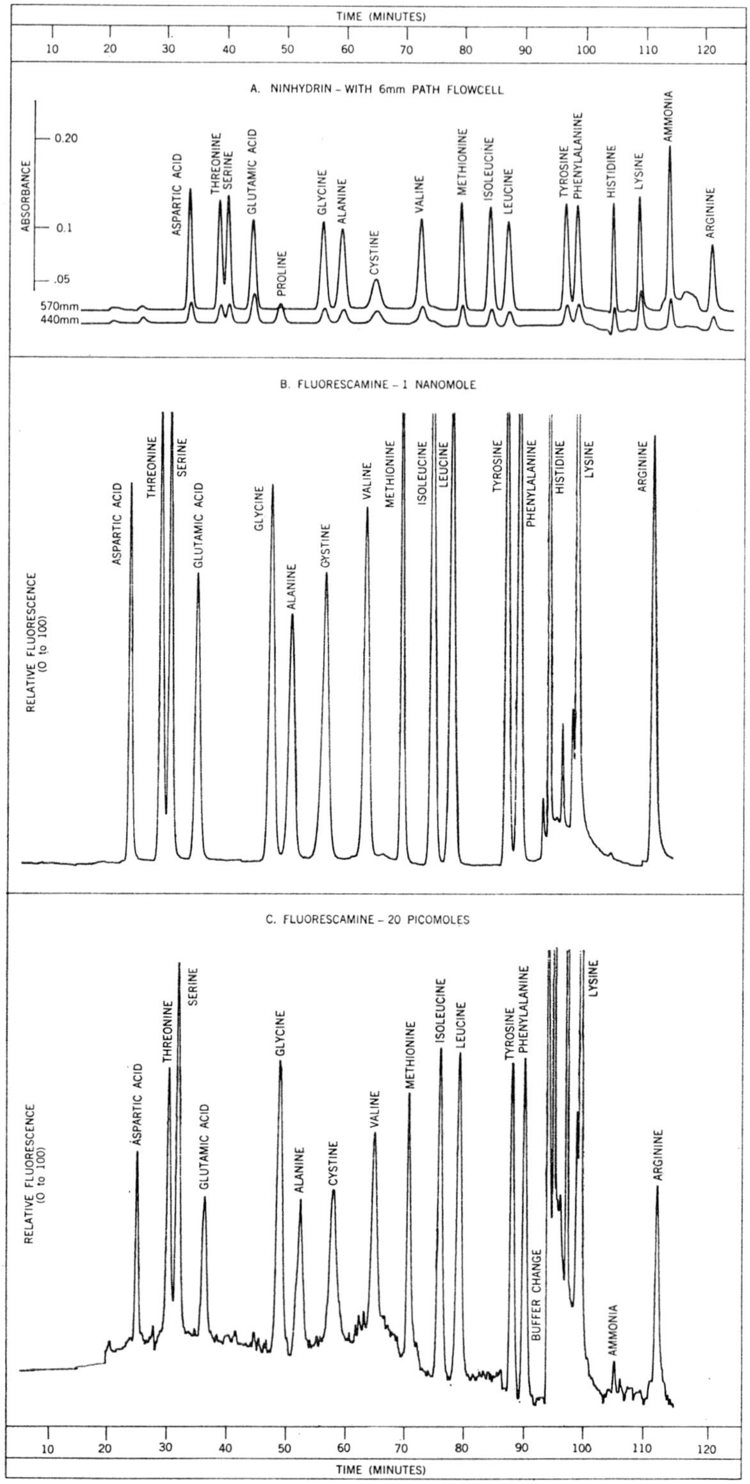
TIME (MINUTES)
10 20 30 40 50 60 70 80 90 100 110 120
A. NINHYDRIN – WITH 6mm PATH FLOWCELL
ABSORBANCE
0.20
0.1
.05
570mm
440mm
ASPARTIC ACID
THREONINE
SERINE
GLUTAMIC ACID
PROLINE
GLYCINE
ALANINE
CYSTINE
VALINE
METHIONINE
ISOLEUCINE
LEUCINE
TYROSINE
PHENYLALANINE
HISTIDINE
LYSINE
AMMONIA
ARGININE
B. FLUORESCAMINE – 1 NANOMOLE
RELATIVE FLUORESCENCE (0 to 100)
ASPARTIC ACID
THREONINE
SERINE
GLUTAMIC ACID
GLYCINE
ALANINE
CYSTINE
VALINE
METHIONINE
ISOLEUCINE
LEUCINE
TYROSINE
PHENYLALANINE
HISTIDINE
LYSINE
ARGININE
C. FLUORESCAMINE – 20 PICOMOLES
RELATIVE FLUORESCENCE (0 to 100)
ASPARTIC ACID
THREONINE
SERINE
GLUTAMIC ACID
GLYCINE
ALANINE
CYSTINE
VALINE
METHIONINE
ISOLEUCINE
LEUCINE
TYROSINE
PHENYLALANINE
BUFFER CHANGE
LYSINE
AMMONIA
ARGININE
10 20 30 40 50 60 70 80 90 100 110 120
TIME (MINUTES)

Fig. 7-9 b

steel reservoir in a freezer at −20° C. It is stable almost indefinitely as long as oxygen is kept out of the system. A number of modifications have been suggested for the ninhydrin reagent. LIAO et al. [775] suggested dilution of the DMSO reagent (2 volumes) with water (1 volume) and indicated that the lower viscocity of the diluted reagent improved mixing.

3. Fluorescamine Reagents and Buffer

Because fluorescamine is not stable in aqueous solution, it is necessary to use two reagents. 200 mg of fluorescamine (Fluram-Roche Diagnostics) are dissolved in a liter of acetone or other suitable organic solvent. The second reagent is made by dissolving 30 g of boric acid in a liter of water and adding potassium or lithium hydroxide to a pH of 9.5 to 10.0. The borate buffer is needed to bring the column effluent to a pH of about 9 for optimum sensitivity. The sodium borate buffer originally suggested is considerably less soluble than either the lithium or potassium borate.

If proline and hydroxyproline are to be determined, a third reagent is needed: 30 mg/l of solution of N-chlorosuccinimide (Eastman No. P6578). This reagent is used only during the time hydroxyproline and proline are being eluted from the column. To achieve a better baseline, water, or some other buffer, may be pumped through the system when the N-chlorosuccinimide is not being pumped through. In this way the total flow rate through the fluorescence detection system remains more nearly constant. There is only a slight decrease in sensitivity due to the dilution with water or buffer.

4. Phthalaldehyde Reagent

The fluorogenic reagent described by ROTH [1043] is stable in aqueous solutions and can be combined in a single reagent with the borate buffer. Boric acid (30 g) is dissolved in 1 l of water and the pH adjusted to 10 by the addition of potassium hydroxide. Phthalaldehyde (Eastman No. 8154) is dissolved in ethanol (500 mg in 10 ml), and 1 ml of mercaptoethanol is added. The mixture is then added to the liter of borate buffer. Proline and hydroxyproline do not react. Cystine and lysine have low-color yields, but the reagent is even more sensitive than fluorescamine for the other amino acids, peptides, and proteins. It is also sensitive to ammonia, particularly if a short, heated reaction coil is used.

E. Sample Preparation

1. Acid Hydrolysis — HCl

The standard procedure is to hydrolyze a purified protein or peptide in a large excess of 6 N hydrochloric acid for 20 to 24 hrs at 110° C. Hydrolysis losses of amino acids are minimized if oxygen is excluded by flushing the hydrolysis tube

---

Fig. 7-9b. Ninhydrin and fluorescamine detection of amino acids with system shown in Fig. 9a. (A) One nmole of each amino acid with ninhydrin using 6 mm path flowcells and no recorder expansion. Lag in time compared to (B) is due to the 10-min reaction coil time for ninhydrin. (B) One nmole of each amino acid with fluorescamine. Note absence of reaction with proline and low ammonia yield. (C) 20 picomoles of each amino acid with fluorescamine. Buffer contaminants especially in the basic region limit sensitivity. Both ninhydrin and fluorescamine detection systems are capable of significantly higher sensitivities

with nitrogen or by evacuation. A combination of nitrogen flushing and evacuation is even more effective. Some amino acids such as aspartic acid, threonine, and serine, are readily released within a few hours from peptide linkage, while others such as valine and isoleucine are not completely released from some proteins even after several days of hydrolysis at 110° C. The most serious losses occur to tryptophan, serine, threonine, tyrosine, and cystine, with the amount of destruction of the various amino acids dependent on the nature of the protein or peptide itself and on the time and temperature of the hydrolysis.

The hydrolysis of sample aliquots for various lengths of time makes it possible to correct for the losses of certain amino acids, as well as to correct for the incomplete release of other amino acids in 24 hrs. A study of serine and threonine destruction during the acid hydrolysis of bovine submaxillary mucin showed that the destruction of serine followed a first-order reaction, while threonine followed a zero-order reaction [294]. One of the problems in using this approach is that several days are required to complete the series if the hydrolysis is carried out at 110° C. At higher temperatures it is possible to complete a series of timed hydrolyses in a much shorter time, and, by extrapolation to zero time similar corrections for losses may be made. For example, a series of collagen samples were hydrolyzed at 155° C for intervals ranging from 10 min to 8 hrs. The data, when extrapolated back to zero time, gave the same amino acid composition as did a series at 110° C for a week. It was found that 25 min at 155° C was approximately equivalent to 24 hrs at 110° C. Roach and Gehrke [1030] have suggested 4 hrs at 145° C as equivalent to thc usual 24 hrs at 110° C. Higher temperatures develop higher pressures during hydrolysis; and because of this, small tubes (13 mm diameter or less) should be used.

Hydrochloric acid is used primarily because it can be readily evaporated, but it is important how this is accomplished. Evaporation of the hydrolysates in a vacuum desiccator over sodium hydroxide can lead to significant losses of glutamic acid and serine [597]. A rotary evaporator is recommended for volumes of more than a few mls. For volumes of few mls or less a convenient method of evaporating the excess HCl is to use a stream of clean, dry nitrogen and a heated aluminum block (80 to 90° C). The stream of nitrogen is directed above the surface while the vial is heated. The evaporation of 1 ml takes less than 5 min, and good recoveries of amino acids are obtained.

An alternate method for evaporation of the HCl is to neutralize the acid with NaOH to a final pH of 2. This requires considerable dilution of the sample but avoids losses due to drying. To avoid resin shrinkage the sodium concentration of the sample solution added to the column should be 0.5 N or lower. Therefore, 1 ml of 6 N HCl hydrolysate should be diluted with the equivalent of 0.5 N NaOH to about 12 mls. If 12 mg of protein were originally taken, then the final concentration would be 1 mg/ml or 100 μg in 100 μl. This sample would be too concentrated for some narrow-bore columns. An original sample weight of 1 mg of protein hydrolyzed and diluted to 12 mls would be equivalent to over 8 μg in 100 μl, which is adequate for many instruments.

2. Acid Hydrolysis for Tryptophan — Methanesulfonic Acid

Tryptophan is labile and is not generally recovered in acid hydrolysates. Special procedures are required for the determination of tryptophan. Usually alkaline hydrolysis is performed, but it has been found that if 2 to 4% mercaptoacetic acid is added to the 6 N HCl that up to 90% of the tryptophan may be recovered if carbohydrate is absent [821]. Another procedure [407] suggest the addition of 5% thioglycolic acid. LIU and CHANG [786] described the use of *p*-toluenesulfonic acid and 2.0% tryptamine for hydrolyzing proteins, which gave better than 90% tryptophan recovery from carbohydrate-free proteins. Methanesulfonic acid has been found to be even more effective [884]. Hydrolysis is carried out in 1 ml of 4 N methanesulfonic acid containing 0.2% tryptamine for 24 hrs at 115° C. The hydrolysate is neutralized with 1 ml of 3.5 N NaOH and diluted to 5 mls. If more than 5% carbohydrate is present there are appreciable losses of tryptophan.

3. Alkaline Hydrolysis for Tryptophan

The preferred method of recovering tryptophan from proteins containing carbohydrate is by alkaline hydrolysis with sodium hydroxide. HUGLI and MOORE [591] report recovery of nearly 100% of tryptophan by hydrolysis in 4.2 N NaOH. Starch is added as an antioxidant, and polypropylene test tube liners are used to prevent etching of the glass test tubes. They suggest using 0.6 ml of 4.2 N NaOH with 25 mg of starch for 16 hrs or longer at 110° C. The hydrolysate is then neutralized down to pH 4.25 with HCl, diluted, and analyzed for tryptophan, using a 0.9 $\times$ 8 cm column and eluting with 0.21 M sodium citrate buffer at pH 5.4. Tryptophan is resolved well ahead of the basic amino acids.

This procedure is not suitable for most of the other amino acids, because of the presence of artifacts and the extensive destruction of several amino acids. If accurate values for tryptophan are desired, separate sample aliquots should be heated for progressively longer periods of time, as tryptophan is released slowly from some proteins.

4. Enzymatic Hydrolysis

It is also possible to effect essentially complete hydrolysis of proteins by the use of proteolytic enzymes. One advantage is that the amino acids labile to acid and alkaline hydrolysis are not altered appreciably during enzymatic hydrolysis. Asparagine and glutamine, which are converted to aspartic acid, glutamic acid, and ammonia by acid hydrolysis, are preserved as the amides in the enzymatic methods. A disadvantage of enzymatic hydrolysis is that no one proteolytic enzyme is capable of hydrolyzing all the different types of peptide bonds found in proteins; therefore, a mixture of enzymes must be used [114, 533]. This specificity of different enzymes for particular peptide linkages is, of course, most useful in obtaining overlapping peptides from proteins by partial enzymatic hydrolysis. The enzymic approach for total hydrolysis seems capable of being a valuable supplement to other methods of hydrolysis. BLACKBURN [114] outlines procedures for enzymatic hydrolysis that uses micromole amounts of protein and isolates the free amino acids by dialysis or picric acid precipitation. Possible contamination of enzymes with free amino acids must be considered as well as incomplete hydrolysis of the sample.

F. Racemization and the Determination of D and L Amino Acids

Extensive racemization occurs during alkaline hydrolysis of peptide-bound amino acids. Free amino acids are racemized relatively slowly unter the same conditions.

Racemization also occurs during acid hydrolysis but not nearly as rapidly as in alkaline hydrolysis. In 6 N HCl the rate of racemization for peptide-bound and free amino acids is usually not greatly different, although MANNING [807] has shown that some amino acid residues may be completely racemized during acid hydrolysis, depending on the nature of the linkage in the peptide chain. MANNING and MOORE [806] have developed an elegant method for the determination of the D and L isomers of amino acids using an amino acid analyzer. The N-carboxyanhydride of L-leucine (or L-glutamic acid for the basic amino acids) reacts with D and L amino acids in the cold at pH 10 within 2 min to form diastereoisomeric dipeptides. The solution is acidified to pH 2, and the dipeptide diastereoisomers are resolved on an amino acid analyzer with modified buffer programs. With this method it is possible to detect as little as one part of D-amino acid in the presence of 1000 parts of the L-isomer. Proteins and amino acids in shells, bones, and teeth have been found to hydrolyze and racemize under natural conditions so that in progressively older fossils the degree of racemization for various amino acids generally increases [494, 495, 1340].

In the Edman degradation method used in peptide sequencing, the PTH-amino acids contain a thiazoline ring which, on hydrolysis to the free amino acid, leads to extensive racemization of the amino acid. The configuration of a particular amino acid residue in a peptide can be determined by a subtractive Edman method coupled with the Manning-Moore technique.

G. Amino Acid Analysis and Sequencing

The sensitivity and speed of ion exchange amino acid analysis has increased to such an extent that it is now possible for this technique to be of much greater use in sequencing studies. The PTH amino acids from the Edman degradation cycle can be identified in several ways. Thin layer chromatography [320, 1039] gas chromatography [984], and high-speed liquid chromatography [433, 983, 1427] are all useful for the rapid determination of the PTH amino acid. Another method is to hydrolyze the derivative back to the amino acid and use ion exchange methods to determine the identity and concentration [1181]. The subtractive Edman method is useful for checking the results of other methods by determining the amino acid composition of the peptide before and after the PTH amino acid has been split off [770]. The methods of sample preparation for the residue peptide hydrolysis are the same as described above, and if enough material is available both acidic and alkaline hydrolysates should be obtained.

The hydrolysis of the PTH amino acids (back hydrolysis) presents some special problems. SMITHIES et al. [1181] have developed a procedure that uses hydroiodic acid (57% HI) for 16 to 20 hrs at 127° C and also 0.2 N sodium hydroxide/dithionite for 2.5 to 3.5 hrs at 127° C. The HI hydrolysis destroys methionine. Serine and cystine both yield alanine. Hydrolysis in $NaOH/Na_2S_2O_4$, however, destroys serine and cystine making it possible to distinguish them from alanine, which is not destroyed. The alkaline hydrolysis gives quantitative yields of methionine and tryptophan. Experiments at higher temperatures (up to 160° C) for shorter times (10 min to 1 hr) with the HI hydrolysis procedure may make it possible to speed up the hydrolysis step so that the identification of an amino acid residue could be made within an hour or so of its being isolated from the sequence. Hydrolysis of small aliquots of the residue peptide after each Edman cycle could likewise be speeded up to make it possible to determine its amino acid composition as a check on the direct identification of the PTH amino acid.

V. Conclusion

The improvement in sensitivity and speed in the ion exchange chromatography of amino acids makes it possible to determine picomole concentrations of amino acids in proteins and peptides. The problem of reagent contamination at present limits the sensitivity and makes careful reagent control and sample preperation essential in obtaining meaningful results on picomole concentration levels of amino acids.

Future efforts to improve sensitivity still further will be directed toward identification and control or reagent contaminants. Improvements in ion exchange resins or perhaps the development of other column packings will result in 10 to 15 min analysis times. It seems doubtful that the use of column chromatography for the determination of the amino acid composition of proteins and peptides will ever become obsolete in spite of many predictions to that effect!

Chapter 8

Sequence Determination

Pehr Edman and Agnes Henschen

I. Introduction

Anyone entering the field of protein structure determination will be faced with a bewildering choice between techniques for sequence determination. To sort the useful from the less useful, or even unworkable, could be difficult indeed. However, well-trodden tracks exist. These will be described here in some detail together with traps not to be fallen into, sidetracks not to be taken, and spooks not to be feared.

A certain bias cannot be avoided in giving preference and prominence to certain techniques. This is so because the authors would not like to recommend a procedure unless they themselves are thoroughly familiar with it. However, alternative procedures will be mentioned and sometimes critically discussed. Certain procedures will be described, not because they have been worked out in sufficient detail for an evaluation, but because they represent novel approaches which may in time lead to superior sequence techniques.

Sequence determination by mass spectroscopy has not been treated by the authors since it would fall outside their competence.

II. Isothiocyanate Degradation

This procedure was first described some 25 years ago by one of the authors [309, 311]. In the following years it has been used extensively, in one or another of its modifications, to establish amino acid sequences of peptides and proteins.

A. Reaction Mechanism

The following series of reactions forms the basis of the phenylisothiocyanate degradation [102, 312, 313]:

$$C_6H_5\cdot NCS + NH_2\cdot CHR\cdot CO\cdot NH\cdot X \longrightarrow \underset{\text{I}}{C_6H_5\cdot NH\cdot CS\cdot NH\cdot CHR\cdot CO\cdot NH\cdot X} \quad (1)$$

(2)

II.

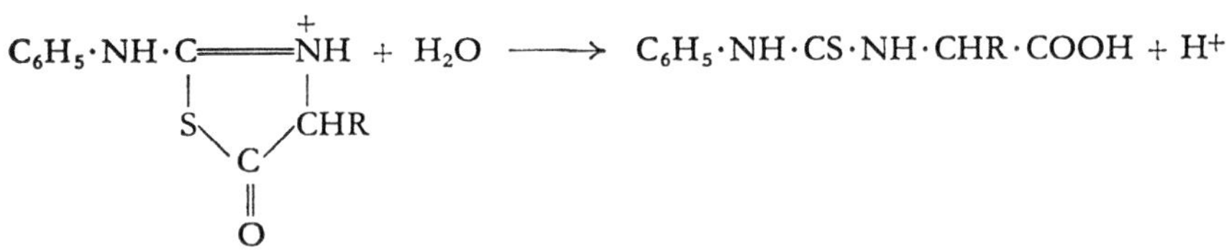

(3a)

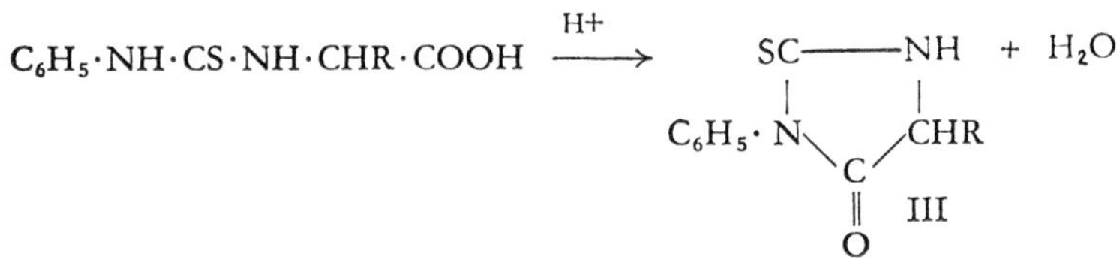

(3b)

The first reaction (1) is the formation of a phenylthiocarbamyl[1] derivative of the peptide (I) by coupling the free α-amino group with phenylisothiocyanate. This reaction will be referred to as the *coupling* reaction.

The next reaction (2) involves the cleavage of the PTC-peptide at the peptide bond nearest to the PTC substituent. This requires a strongly acid medium and leads to the formation of a 2-anilino-5-thiazolinone derivative (in the ammonium form) (II) and a peptide with one amino acid less than the original. This is termed the *cleavage* reaction.

The shortened peptide has a free α-amino group, and may therefore be subjected to a new cycle of the degradation.

The thiazolinone cleaved off is a derivative of the N-terminal amino acid, and could therefore in principle be used for its identification. In practice this is not possible because of the instability of this type of derivatives. However, the thiazolinone may be converted to the isomeric and stable 3-phenyl-2-thiohydantoin (III). This *conversion* reaction (3) takes place in an aqueous acid medium and consists in fact of two reactions, i.e. the hydrolysis of the thiazolinone to the PTC-amino acid (3a) and the cyclization of the latter to the corresponding PTH-amino acid (3b). The first reaction, however, is much faster than the second, and for practical purposes they may be treated as one reaction. It may be added that there exists also a direct route, which does not involve hydrolysis, from the thiazolinone to the phenylthiohydantoin [313]. This isomerization reaction, however, has not been studied in detail.

Needless to say, a full understanding of the nature of these reactions is essential for an intelligent use of them in sequence work. A detailed discussion is left to the sections dealing with the experimental conditions of the degradation. However, even at this stage the importance of separating the cleavage and the conversion reactions is emphasized. Since they are both acid catalysed it may seem logical to combine them. Yet, this does not take into account the fact that the conversion reaction requires an aqueous acid medium and is slow, whereas the cleavage reaction does not require water and is fast. A combination of the two reactions will therefore result in an unnecessary and prolonged exposure of the peptide to hydrolytic condi-

[1] Abbreviations used are: DFP, diisopropyl fluorophosphonate; PTC, phenylthiocarbamyl; PTH, 3-phenyl-2-thiohydantoin; GLC, gas-liquid chromatography; TLC, thin-layer chromatography; X (in formulas), rest of peptide.

tions. The better procedure is to extract the thiazolinone from the reaction mixture after cleavage, and to convert it in a separate operation.

It is illuminating to compare the mechanism of isothiocyanate degradation with those of some other reactions which have been proposed for the sequential degradation of peptides. In the degradation scheme of LEONIS and LEVY [759, 760] a dithiocarbamate of the peptide is initially formed, which under the influence of acid splits off the N-terminal amino acid in the form of a 2-thio-5-thiazolinone derivative (IV):

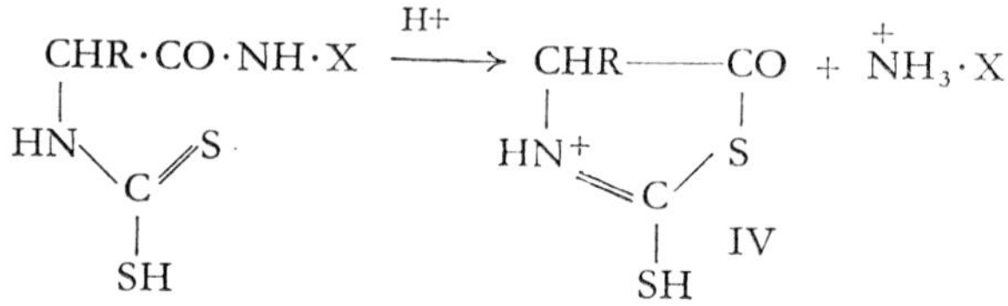

(4)

Similarly, in the procedure of KENNER and KHORANA [682] an N-thioncarbalkoxy peptide in acid splits off a 2-alkoxy-5-thiazolinone derivative (V):

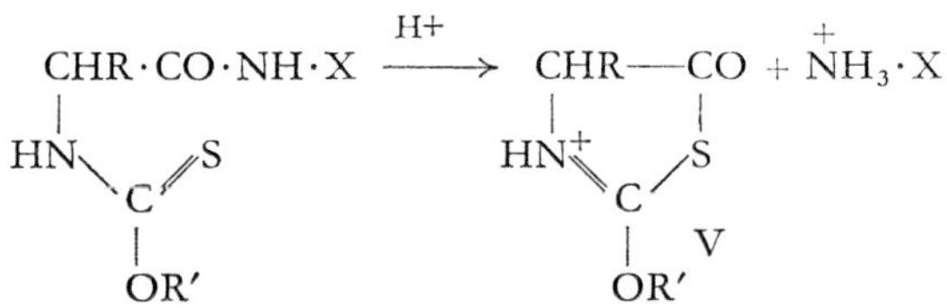

(5)

In the scheme of ELMORE and TOSELAND [330, 331] a 2-acylamido-5-thiazolinone (VI) is formed by exposing an N-acylthiocarbamyl peptide to anhydrous acid (6):

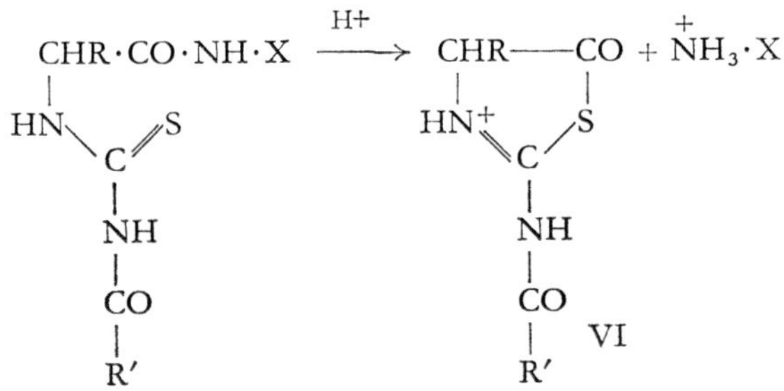

(6)

BARRETT [64] has described a reaction (7) where N-thiobenzoylated peptides are easily cleaved in anhydrous acid forming 2-phenyl-5-thiazolinone derivatives (VII):

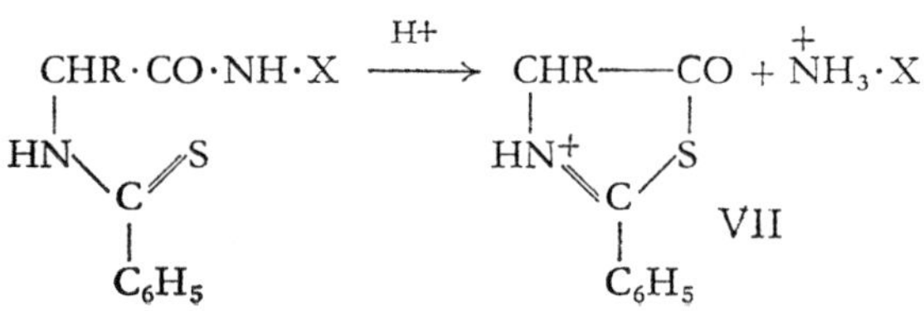

(7)

Variants of this procedure have been described by PREVIERO and PECHERE [998], MROSS and DOOLITTLE [888], PRAGER et al. [992] and MROSS and OHMS [890].

In all these reactions the cleavage of the peptide bond is concomitant to the formation of a thiazolinone derivative. The explanation for the extreme facility of these reactions obviously lies here. On the other hand, the phenylisocyanate degradation of ABDERHALDEN and BROCKMANN [1] is unrelated to the reactions discussed, since obviously no thiazolinone can be formed here. The absence of a reactive thioketonic group makes it a very sluggish reaction.

B. Preparation of Phenylthiohydantoins

The synthesis of a PTH-amino acid [167, 310, 318, 600, 605, 769b, 1034] is usually carried out in two steps. First, the amino acid is coupled with phenylisothiocyanate in a weakly alkaline medium to produce the PTC-amino acid. The latter is then converted to the corresponding PTH-amino acid in acid medium, usually at elevated temperature.

1. Preparation of PTC-Amino Acids

The first reaction, i.e. the formation of the PTC-amino acid, is straightforward and the same for all amino acids. Phenylisothiocyanate reacts with the amino acid only in its amino form. The reaction rate is therefore dependent on the *pK* of the amino group, i.e. the higher the *pK* the lower the reaction rate. This applies particularly to lysine where the ε-amino group ($pK = 10.53$) is noticeably slower in reacting. To ensure a complete reaction it is advisable to follow the coupling by the alkali consumption at a pH held constant at say 8.6. In practice one finds that the alkali consumption never ceases completely, but falls off to a constant rate due to the slow hydrolysis of the phenylisothiocyanate. After completion of the reaction excess reagents and by-products are removed by extraction, and subsequently the PTC-amino acid is precipitated from the aqueous solution by the addition of acid. No attempt is made to obtain an analytically pure preparation at this stage, since gradual conversion of the PTC-amino acid to the PTH-amino acid cannot be avoided.

The phenylsiothiocyanate seems to react exclusively with primary and secondary (prolines) amino groups. Other functions, e.g. alcoholic, phenolic, imidazolic and guanidino groups, do not appear to react. At any rate these groups are unmodified in the end products, i.e. the PTH derivatives.

The amino acid (10 mmoles) is dissolved or suspended in pyridine-water (1:1; 50 ml) in a beaker. The pH is then adjusted to 8.6 (glass electrode) with 2 N NaOH, and the volume added is noted. The temperature is brought to 40° C, preferably on an electric hot plate with provisions for magnetic stirring and phenylisothiocyanate (2.4 ml) is added. The solution is stirred and the pH is kept steady by the addition of 2 N NaOH from a burette. Usually the alkali consumption becomes slow and steady in less than 30 min, indicating the completion of the reaction. The total alkali consumption, including that of the initial pH adjustment, should be equivalent to the number of amino groups. Excess reagents and by-products are extracted with benzene (3×1 vol.). The remaining aqueous solution is cooled in ice and an amount of 2 N HCl equivalent to the total alkali consumption is added gradually and with stirring. By a further small addition of acid the pH is brought to about 1. The PTC-amino acid usually precipitates as a mass, which solidifies on storage in the cold for a few hours. The supernatant is discarded, and the preparation is then ready for conversion to the PTH-amino acid.

Some PTC derivatives do not precipitate or precipitate incompletely on acidification. *PTC-ariginine* and *PTC-histidine* may be precipitated isoelectrically at pH 7.5 and 3.5, respectively. The aqueous solution of *PTC-cysteic acid* is evaporated to dryness *in vacuo.*

In the case of *PTC-serine* and *PTC-threonine*, the aqueous solution remaining after benzene extraction is made 0.5 N in HCl and the conversion is carried out in this medium as described below.

2. Conversion of PTC-Amino Acids

The conversion is carried out in aqueous acid usually at elevated temperature. Most PTH-amino acids are perfectly stable under these conditions and remain unaffected even if the conditions are more vigorous than is strictly necessary to bring about a complete conversion. However, certain PTH-amino acids tend to decompose on prolonged exposure to acid and heat. The chemical nature of these changes is discussed in the section "Properties of Phenylthiohydantoins", and here is only given the standard procedure applicable to most PTH-amino acids followed by special procedures for some PTH-amino acids.

The PTC-amino acid preparation is suspended or dissolved in 1 N HCl (25 ml). The temperature is quickly brought to 80° C, kept there for 10 min, and then dropped to ambient. The reaction mixture is repeatedly extracted with ethyl acetate (3 $\times$ 1 vol.) and the extracts are combined and evaporated to dryness. The residue is taken up in a small volume of hot glacial acetic acid and made to crystallize by the gradual addition of water. Recrystallizations are made from ethanol-water (in the case of PTH-aspartic and PTH-glutamic acid ethanol should be avoided because of the risk of esterifying the carboxyl group).

PTH-arginine, *PTH-histidine* and *PTH-cysteic acid* are not extracted by ethyl acetate, The aqueous phase is evaporated to dryness *in vacuo*, water added, and the process repeated until excess HCl has been removed. *PTH-arginine* and *PTH-histidine* may then be recrystallized as their hydrochlorides from water and ethanol, respectively. The dry *PTH-cysteic acid* preparation is dissolved in a small volume of water. The solution is brought to pH 4.2 with 1 N KOH and evaporated to dryness *in vacuo*. The dry product is extracted with acetone-water (4:1), which leaves the inorganic salts behind. Most of the solvent is evaporated *in vacuo*, and to the resulting viscous solution is added acetone to the point of incipient crystallization, which is completed in the cold. Recrystallizations are made from water-ethanol. The resulting preparation is the potassium salt of *PTH-cysteic* acid.

The conversion of *PTC-serine* to *PTH-serine* is carried out in 0.5 N HCl at room temperature for 24 hrs. The precipitated product is filtered off and dried *in vacuo*. The dry product is extracted exhaustively with absolute ethanol and any insoluble residue filtered off and discarded. The ethanolic solution is concentrated at room temperature until incipient crystallization, which is completed by the addition of heptane. Recrystallization is made from ethanol. The procedure for *PTH-threonine* is the same as for *PTH-serine*, but the product is considerably less soluble in ethanol, and crystallization is therefore made from hot ethanol.

PTH-asparagine and *PTH-glutamine* are prepared from the corresponding PTC-derivatives by crystallization from boiling glacial acetic acid.

The over-all yield of PTH-amino acid calculated on the amino acid is generally about 80%. The asymmetric center at C-5 of the thiohydantoin ring is completely or almost completely racemized during synthesis, and the configuration at the α-carbon of the starting material is therefore of little consequence.

Apart from elementary analysis and melting point determination the identity and purity of a PTH-amino acid preparation may be established by other criteria, e.g. chromatography (cf. "Identification of Amino Acids") and ultraviolet absorption (Table 8-1).

A wide range of PTH-amino acids are now commercially available.

C. Properties of Phenylthiohydantoins

PTH-amino acids are generally easily crystallizable compounds with high melting points. Their chemical stability is with few exceptions excellent. The less stable compounds are those with an OH or an SH group on the β-carbon of the side chain. These compounds show to a varying degree a tendency for β-elimination. So, for example, PTH-threonine tends to undergo the following reaction [310]:

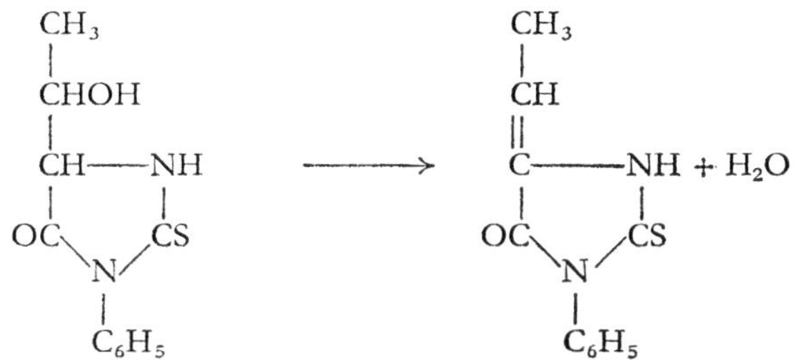

(8)

This tendency is stronger in PTH-serine. Here the formation of a highly reactive methylene group at C-5 of the thiohydantoin ring leads to further reactions, e.g. polymerization. In PTH-cystine and PTH-cysteine the tendency for β-elimination is so strong that the derivatives are not useful for identification purposes. On the other hand, the S-alkylated PTH derivatives of cysteine are more stable as is also PTH-cysteic acid. The tendency for β-elimination becomes important in the identification of the susceptible compounds, and this problem is discussed in the section on "Conversion". It seems likely that a tendency for β-elimination is also present in the corresponding thiazolinones.

The PTH derivatives of asparagine and glutamine are unstable to the extent that the amide group is hydrolyzed by acids and alkalis. In sequence determination these PTH-amino acids are found to be contaminated by the corresponding acids. PTH-tryptophan is also susceptible to the action of strong acids, presumably due to decomposition of the indole nucleus. However, the acid conditions of the degradation procedure do not seem sufficiently vigorous to affect appreciably PTH-tryptophan.

Solutions of PTH-amino acids show a tendency for photodecomposition, which becomes apparent after prolonged exposure to daylight or ultraviolet light. Solutions should therefore be stored away from light.

The PTH-amino acids are, with few exceptions, poorly soluble in water, and easily soluble in organic solvents. The exceptions are PTH-arginine, PTH-histidine and PTH-cysteic acid. This is important in the conversion procedure (cf. "Conversion"). After conversion the PTH-amino acids are extracted from the aqueous acid medium with ethyl acetate, leaving only the PTH derivatives of arginine, histidine and cysteic acid behind in the aqueous phase.

The PTH-amino acids may be hydrolyzed by acids or alkalis to regenerate the amino acid. However, this requires vigorous conditions. Certain labile amino acids are destroyed in the process. The identification of the amino acid after hydrolysis of the PTH has been used in sequence work and this approach is discussed in the section "Identification of Amino Acids".

All PTH-amino acids show a strong absorption in the ultraviolet with a maximum around 268 mμ and a minimum around 245 mμ (Fig. 8-1). The molar absorptivity at 269 mμ is usually around 16000 (Table 8-1)[2]. The ratio A_{245}/A_{269} lies around 0.4. This ratio is a useful index in assessing the purity of a PTH-amino acid preparation

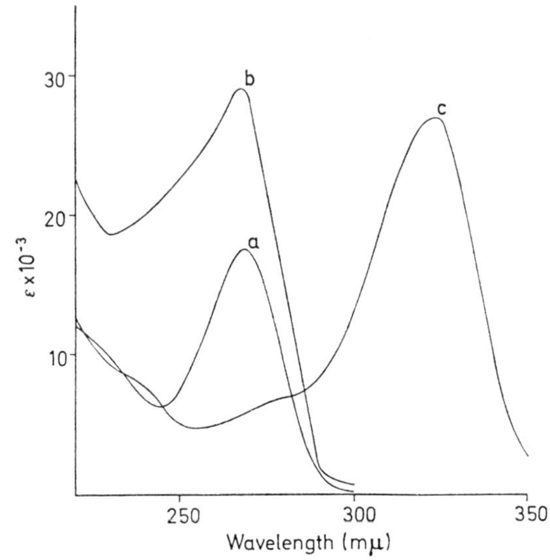

Fig. 8-1. Ultraviolet absorption spectra of PTH-leucine (*a*), PTH-N$^{\varepsilon}$-PTC-lysine (*b*) and PTH-dehydrothreonine (*c*) in ethanol

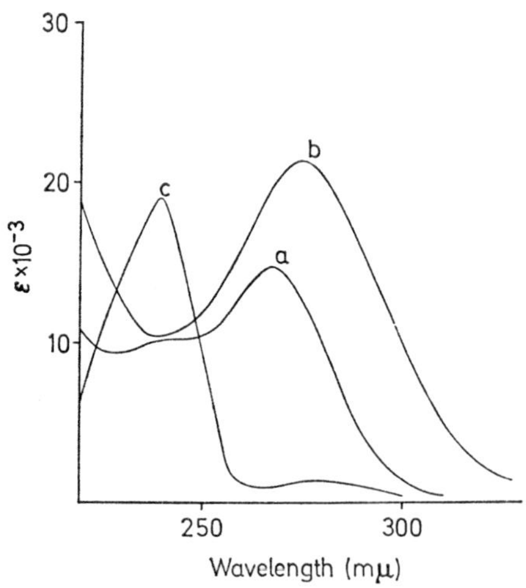

Fig. 8-2. Ultraviolet absorption spectra of phenylthiourea (*a*), *s*-diphenylthiourea (*b*) and phenylcarbamyl leucine (*c*) in ethanol

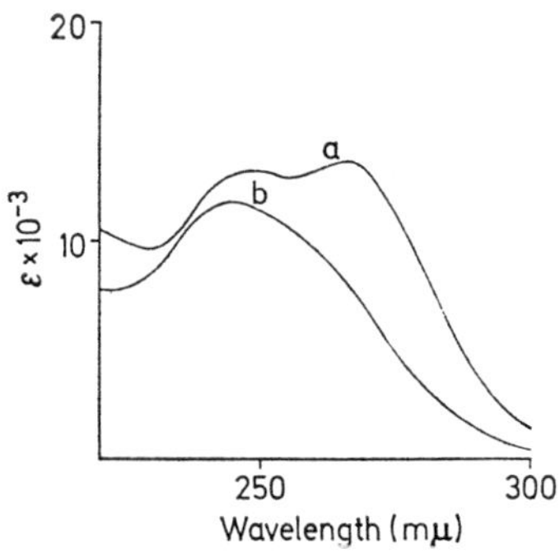

Fig. 8-3. Ultraviolet absorption PTC-glycine in ethanol (*a*) and in 0.1 N hydrochloric acid (*b*)

[2] This wavelength is used throughout for quantitation of PTH-amino acids by spectrophotometry although it does not always coincide exactly with the wavelength of maximum absorption.

Table 8-1. Molar absorptivities and melting points of phenylthiohydantoins. Light absorptions are measured in ethanolic solution

PTH-Amino acid (m.w.)	ε_{245}[a]	ε_{269}[a]	$\frac{\varepsilon_{245}}{\varepsilon_{269}}$	M. p.[b] °C
PTH-alanine (206.3)	6,300	16,000	0.39	185
PTH-arginine-HCl (327.8)	6,000	15,900	0.38	189
PTH-asparagine (249.3)	6,500	17,200	0.38	235—237
PTH-aspartic acid (250.3)	6,500	16,100	0.40	229
PTH-cysteic acid, K salt (324.4)	5,900	16,600	0.36	193—197 (d.)
PTH-glutamic acid (264.3)	6,200	15,900	0.39	166—167
PTH-glutamine (263.3)	6,500	17,000	0.38	201—211 (d.)
PTH-glycine (192.2)	6,300	14,900	0.42	245—248 (d.)
PTH-histidine-HCl (308.8)	6,400	15,500	0.41	200—206 (d.)
PTH-isoleucine (248.3)	6,500	17,000	0.38	173—175
PTH-leucine (248.3)	6,000	16,700	0.36	178
PTH-N$^{\varepsilon}$-PTC-lysine (398.5)	21,000	29,000	0.72	162—164
PTH-methionine (266.4)	6,900	17,100	0.40	132
PTH-methionine sulfone (298.4)	6,400	15,300	0.42	210
PTH-phenylalanine (282.4)	5,700	15,500	0.37	187
PTH-proline (232.3)	9,600	14,300	0.67	179
PTH-S-allyl cysteine (278.3)	7,300	17,600	0.41	134
PTH-S-carboxymethyl cysteine (296.5)	7,100	16,600	0.43	164
PTH-S-ethyl cysteine (266.3)	6,600	16,300	0.40	128
PTH-S-methyl cysteine (252.3)	6,500	16,400	0.40	171
PTH-serine (222.3)	5,500	15,500	0.35	176—178 (d.)
PTH-threonine (236.3)	7.200	15,600	0.46	194—198 (d.)
PTH-tryptophan (321.4)	8,000	19,700	0.40	177
PTH-tyrosine (298.4)	6,600	15,600	0.42	216
PTH-valine (234.3)	6,200	16,500	0.38	206

[a] The data are mostly from reference [1166].
[b] The data are mostly from references [310, 318, 600, 605].

since impurities tend to increase the absorption at 245 mμ more than at 269 mμ. PTH-N$^{\varepsilon}$-PTC-lysine is exceptional in that it has a much higher molar absorptivity at 269 mμ, and also a higher ratio A_{245}/A_{269} (Table 8-1). This is due to the additional absorption caused by the N$^{\varepsilon}$-PTC-group. The prolines also show a high ratio. The β-elimination product of PTH-threonine, i.e. PTH-dehydrothreonine, has, because of the conjugation of the newly formed double bond with the thiohydantoin ring, an entirely different spectrum with a maximum at 323 mμ (Fig. 8-1). PTH-dehydroserine has a similar spectrum, but the compound is unstable, and is rapidly transformed into other products with less characteristic absorption. The strong ultraviolet absorption of PTH-amino acids may be used for their location in chromatograms and in quantitation.

The individual PTH-amino acids have characteristic infrared spectra, which may be used for their identification as shown by Ramachandran et al. [1011].

The ease of racemization at the asymmetric C-5 of the thiohydantoin ring is a tendency shared by other hydantoins [151, 261]. The optical rotary dispersion of phenylthiohydantoins have been reported on by Djerassi et al. [286].

D. Identification of Amino Acids

Two different principles are commonly used in the identification of the amino acid residue removed in a degradation cycle. The *direct* method relies on the positive identification of the residue split off. The *indirect* method depends for the identification on a difference in analysis of the peptide before and after a degradation cycle.

PTH-amino acids may be separated and identified by chromatography. For this purpose paper chromatography [746, 1165], thin-layer chromatography [222] and partition chromatography, either in liquid-liquid [1046] or gas-liquid [985] systems, have been proposed. Paper, thin-layer and gas chromatography techniques have been most commonly used, and have been worked out in all detail. These techniques are less well suited for the identification of some PTH-amino acids with strongly ionized groups in the side chain, i.e. PTH-arginine, PTH-histidine and PTH-cysteic acid, but here spot reations and paper electrophoretic techniques may be used.

The PTH-amino acids may also be identified by mass spectrometry [1225].

1. Thin-Layer Chromatography

The systems described here are not necessarily superior to others that have been proposed. However, they have stood up to the test of time, they permit unambiguous identification of all amino acids normally encountered in proteins and some others in addition, and, not least important, they invariably give satisfactory results in the hands of a competent technician.

The room where the chromatography is performed should be thermostated at 20° C $\pm$ 2. Chromatography is ascending, and is carried out in a glass tank with inner dimensions 22 $\times$ 14 cm and 22 cm in height. The glass lid should be tight fitting with one or two holes bored in it, to allow the introduction of the solvent into the tank without lifting the lid. Glass troughs to hold the developing solvent sit on the bottom of the tank. A sheet of filter paper covers the walls of the tank on the inside (systems E and H), and is held in place by loops of stainless steel wire.

The thin layer consists of silica gel to which have been added a fluorescent dye and calcium sulphate as a binder (Camag, Muttenz, Switzerland, type DF-5). If the iodine-azide reaction will be used later to locate the PTH-amino acids, starch is also incorporated in the layer. This may be done routinely, since the starch does not interfere in any way. A slurry of silica gel (50 g) in 0.5% aqueous starch solution (110 ml) is spread in a 0.3 mm thick layer on glass plates, 20 $\times$ 20 $\times$ 0.4 cm, using an applicator (e.g. Camag). The plate is allowed to dry in a horizontal position overnight at room temperature. Lines are then scored in the layer to separate the sample lanes. The lines are drawn parallel to each other and to the direction in which the layer has been spread. The lines are 0.9 cm apart, beginning 2.5 cm from what will be the lower edge of the plate during chromatography, and ending 7.5 cm from the top edge of the plate. Where the lines end, a transverse line is scored across the plate. The lanes ensure that the chromatogram is not distorted should the migration of the solvent not be even over the whole plate. The plates are suitably stored over P_2O_5 in a Perspex box.

Three solvent systems are used:

a) Xylene/Formamide (System D) [319]

The plate is dipped into a solution of formamide in acetone (3:7) for 1 min, and then placed horizontally to dry in the air for a few minutes. The formamide-acetone stock solution should be protected from evaporation, but after some use it tends to become concentrated, and a fresh solution should then be made up. The references and samples in ethylene chloride solution are applied at the start of the lanes, avoiding the two extreme lanes, where chromatography is often disturbed. When many samples are to be handled, it is convenient to use an applicator manifold (Fig. 8-6.) which allows ten samples to be applied simultaneously. The plate is placed in the trough at the bottom of the rank, xylene is added to the trough, and chromatography is started without prior equilibration. It is allowed to proceed until the solvent front in all lanes has reached the line scored across the plate. This usually takes about 40 min. The plate is then dried in an ordinary oven at about 100° C or in a vacuum oven at about 50° C for 15 min. The formamide should be evaporated as it absorbs the ultraviolet light used in locating spots.

b) n-Butyl Acetate/Propionic Acid/Formamide (System E) [319]

The layer is impregnated with formamide as described for system D. The butyl acetate (60 ml) is equilibrated with a mixture of water (1.0 ml), propionic acid (1.8 ml) and formamide (8 ml). Only the upper phase is used. The filter paper lining of the tank is moistened with the solvent mixture. The plate is placed in the empty trough, and allowed to equilibrate for $^1/_2$ hr, and chromatography is then started by filling the trough with the solvent mixture through the hole in the lid. The chromatography takes about 35 min.

c) Ethylene Chloride/Acetic Acid (System H)[3]

To ethylene chloride (60 ml) is added glacial acetic acid (14 ml). The paper lining of the tank is moistened with this mixture, which is also filled in the trough. No equilibration is required. The development takes about 50 min.

The following specifications or purification procedures are recommended for the solvents used in TLC and paper chromatography:

Xylene is shaken three times with conc. H_2SO_4 (0.1 vol.) for $^1/_2$ hr and with cooling to below 30° C. The organic phase is washed with H_2O, dried over pellets of KOH overnight, and finally distilled.

n-Butyl acetate is distilled, and the fraction boiling at 126–127° C collected.

Propionic acid is refluxed for 4 hrs with CrO_3 (10 g/750 ml), distilled on a Widmer column, and the fraction boiling at 140–141° C collected.

n-Heptane is shaken three times with concentrated H_2SO_4 (0.1 vol.), and then washed successively with H_2O, 10% aqueous NaOH and H_2O. It is finally dried over anhydrous Na_2SO_4 and distilled. B.p. 98° C.

Ethylene chloride and formic acid are *purissimum* grades (Fluka A.G., Buchs, Switzerland).

Acetic acid is a *purissimum* (distilled from CrO_3) grade (Fluka A.G.).

Formamide is a *purissimum* grade (Fluka A.G.). It is important for successful chromatography that the formamide is pure. Some reagent grade formamide preparations have been found to be heavily contaminated with ammomiun formate, and are then unsuitable.

The samples and the PTH-amino acid references are dissolved in ethylene chloride. Many other solvents tend to dissolve out the formamide at the application point, and thus disturb the chromatogram. However, some PTH-amino acids are poorly soluble in ethylene chloride, notably the PTH derivatives of aspartic acid, asparagine, glutamic acid and glutamine, and it may therefore be necessary briefly to warm the solution before use. The solutions are best stored in the cold (−15° C) and away from light. They are then stable for 1 to 2 months, but after that time foreign spots may appear in the chromatograms. It is convenient to use three reference mixture solutions consisting of the PTH derivatives of (I) *asparagine*, *serine methionine sulfone*, *glycine*, *alanine*, *S-methyl cysteine*, *valine* and *proline*, (II) *glutamine*,

---

[3] P. Edman and I. Kluh, unpublished.

threonine, *tyrosine*, *lysine*, *methionine*, *S-allyl cysteine* and *leucine*, and (III) *aspartic acid*, *glutamic acid*, *tryptophan*, *S-ethyl cysteine*, *phenylalanine* and *isoleucine*, respectively. With this arrangement the relevant reference spots are well separated from each other. A suitable concentration of each PTH-amino acid for samples and reference solutions is 5 μmoles/ml, and of these solutions 2 μl (or 10 nmole) is applied to the plate. *PTH*-lysine produces a more intense spot, and *PTH-proline* a weaker spot, and are therefore preferably used in half and double this concentration, respectively.

The spots may be located by several independent methods. One method [352b] depends on the fact that divalent sulfur compounds catalyse the reaction between free iodine and azide ions. The plate, which has been pretreated with starch, is sprayed with a solution of iodine-potassium iodide and sodium azide. The layer takes on a blue color, and against this background appear bleached areas where the PTH-amino acids are located. The iodine-azide reaction is very sensitive, and it is possible to detect less than 0.5 nmole of PTH.

The spray reagent is prepared by mixing equal volumes of (a) an aqueous solution 0.01 M in I_2 and 0.5 M in KI, and (b) an aqueous solution 0.5 M in NaN_3. The reagent is applied in a fine spray. The spraying should be done in a hood with a good draft because of the toxic nature of the reagents. To prevent the plate from drying before the spots are sufficiently bleached it is placed under a Perspex cover together with a dish of hot water or a hot solution of dilute ammonia. This helps to bring out the spot of PTH-proline, which otherwise is slow to appear.

The iodine-azide reaction has properties that make it less than ideal for its purpose. Firstly, it is not sufficiently specific, since any compound that reacts with iodine will give a positive reaction. Secondly, the intensity of a spot is not proportional to the amount of PTH. Given enough time to develop, even extremely small amounts of PTH (or impurities) will give rise to positive reaction.

An alternative method for locating the spots is a reaction with ninhydrin [1038]. The PTH derivatives of the common amino acids give a more or less intense color reaction. Many produce characteristic colors, and this helps in the identification. About 5 nmole of PTH is required.

The thin layer plate is heated in an oven at 110° C for 15 min. It is then sprayed with a solution of ninhydrin (100 mg) and collidine (5 ml) made up to 100 ml with absolute ethanol. The plate is again heated at 110° C. The PTH spots react at different rates and give different color intensities.

The authors rely chiefly on another way of locating PTH spots. This method is based on the strong ultraviolet absorption of PTH derivatives. A fluorescent indicator with excitation in the ultraviolet and emission in the visible is incorporated in the layer as already described.

The ultraviolet light source is a battery of tube-shaped low pressure mercury lamps (Phillips Electrical Ltd., type TUV 15 W) which have been arranged in parallel to form a square. When the plate is held with the layer towards the light source, and viewed from the other side, the PTH-amino acids appear as dark spots against a luminous background. The contrast is considerably increased if the plate is viewed through a Wratten No. 45 filter. A permanent record of the chromatogram may be obtained by photographing the plate through the same filter. Kodak Micro-file film, Type 5453, is suitable for this purpose. Even better results are obtained when the ultraviolet light is prefiltered as is the case with a commercial transilluminator (Ultra-violet Products Inc., San Gabriel, Calif., U.S.A.).

The method is sensitive to about 1 nmole of PTH in a spot. It has the advantage over the iodine-azide reaction that the intensity (and area) of the spot is roughly proportional to the amount of PTH. The specificity is of course low, since any material absorbing light at 254 mμ would give a positive reaction. However, in practice, false reactions are rarely seen. The iodine-azide or ninhydrin reaction may be applied to the plate afterwards. However, it is then advisable to disregard faintly positive reactions with iodine-azide unless they tally with the findings of the fluorescence extinction method.

It should be added that some PTH-amino acids already show a characteristic color reaction when the plate is heated during drying. Thus, *PTH-glycine* produces a brownish-red spot, and *PTH-tyrosine* and *PTH-tryptophan* may be recognized by a lemon-yellow coloration. Some other PTH-amino acids also give rise to colors under these circumstances, but the reactions are too faint to be of use in the identification.

A chart showing the positions of the PTH-amino acids in the various systems is shown in Fig. 8-4. To this the following comments may be added. Generally, the D system resolves the non-polar PTH-amino acids, and the E and H systems resolve the polar PTH-amino acids. However, neither of the latter systems alone will allow the identification of all polar PTH-amino acids, but will do so when used together. The normal practice is to use only the D and H systems, and to resort to the E system only when one wants to differentiate between *PTH-tyrosine* and *PTH-glutamic* acid, or between *PTH-methionine sulfone* and *PTH-aspartic* acid. The PTH derivative of *S-carboxymethyl cysteine* (not shown in the chart) falls between PTH-aspartic and PTH-glutamic acid in system H, and coincides with PTH-aspartic acid in system E, and is difficult to differentiate from these PTH-amino acids. Here the resolution may be increased to a sufficient degree in the following way. Chromatography is performed in the H system, the plate is allowed to dry in the air, and is then irrigated in the same system. The present authors prefer other S-alkyl, e.g. S-methyl, S-ethyl or S-allyl, derivatives of cysteine in degradation work, since *PTH-S-methyl cysteine*, *PTH-S-ethyl cysteine* or *PTH-S-allyl cysteine* are more conveniently identified (Fig. 8-4). In the D system *PTH-leucine* and *PTH-isoleucine* are close together, but their relative positions are very reproducible, and may be relied on in the differentiation. However, it may be necessary to rerun the chromatogram, e.g. if the solvent front is skewed. In the rerun, one places the unknown in the middle and the appropriate reference mixtures (II and III) in the neighboring lanes. A similar rerun may also be indicated for other less well resolved pairs of PTH-amino acids.

Samples obtained in a degradation may show spots due to compounds other than the expected PTH-derivatives. *PTH-serine*, due to partial decomposition during conversion (cf. "Conversion"), is always contaminated by other products, which are seen as multiple spots between PTH-serine and the front in solvent systems E and H (because of the lower yield of PTH-serine the sample should be applied in double strength). The positive identification of serine requires that a spot of PTH-serine is clearly present in the chromatogram. *PTH-threonine* is always accompanied by some PTH-dehydrothreonine, which may be seen as a faint spot between PTH-alanine and PTH-methionine in system D. The spots of *PTH-asparagine* and *PTH-glutamine* are always accompanied by strong spots of PTH-aspartic and PTH-glutamic acid, respectively. This helps greatly in the identification, since the latter are better separated than the former (Fig. 8-4). A spot at the origin becomes more apparent as the degradation progresses. It is particularly noticeable in system D, and is due to the accumulation of background material. *Phenylthiourea* and *s-diphenylthiourea* normally are not seen in the chromatograms, since they are removed in the extractions following coupling, but may appear if the extractions have not been thorough enough. Phenylthiourea remains at the origin in system D, in system E it appears between PTH-threonine and PTH-glycine, and in system H it coincides with PTH-

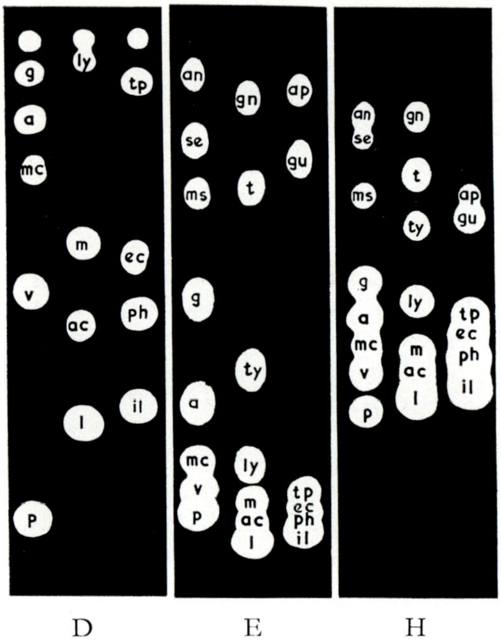

Fig. 8-4. Chart showing the TLC of PTH-amino acids in systems D, E, and H, (see text). PTH derivatives of asparagine (*an*), serine (*se*), methionine sulfone (*ms*), glycine (*g*), alanine (*a*), S-methyl cysteine (*mc*), valine (*v*), proline (*p*), glutamine (*gn*), threonine (*t*), tyrosine (*ty*), lysine (*ly*), methionine (*m*), S-allyl cysteine (*ac*), leucine (*l*), aspartic acid (*ap*), glutamic acid (*gu*), tryptophan (*tp*), S-ethyl cysteine (*ec*), phenylalanine (*ph*) and isoleucine (*il*)

glycine. Diphenylthiourea appears in system D between PTH-methionine and PTH-valine, and in systems E and H among the fastest moving PTH-amino acids. It gradually decomposes and disappears from the thin layer plate on heating at 110° C.

Other systems for the identification of PTH-amino acids using TLC have been published by, among others, Cherbuliez et al. [222], Brenner et al. [171], Pataki [954 a], Wang and Wang [1328], Jeppsson and Sjöquist [636] and Kulbe [735]. A micro-TLC procedure, with the claimed detection limit of about 0.1 nmole, has recently been described by Summers et al. [1244]. The sensitivity of this technique is approaching that of the dansyl technique. It is difficult, and perhaps pointless, to give preference to any particular system for the TLC of PTH-amino acids. More important is probably that the operator is fully conversant with the system he has chosen.

2. Identification of PTH-Arginine, PTH-Histidine and PTH-Cysteic Acid

These PTH-amino acids may be identified by chromatography, but their positions on the chromatogram are not reproducible. However, better reproducibility has recently been claimed for a new chromatography system [602]. In our laboratory PTH-arginine and PTH-histidine are identified by color reactions, and PTH-cysteic acid by its electrophoretic mobility.

PTH-arginine is identified by a spot reaction on filter paper using a modified Sakaguchi reaction [3].

Two spray reagents are used: Reagent A consists of a solution of α-naphthol (5 mg) and urea (2.5 g) in ethanol (50 ml), to which is added immediately before use a solution of NaOH (3 g) in water (5 ml). Reagent B is a solution of Br_2 (0.70 ml) in ice cold aqueous NaOH (5 g NaOH in 100 ml). This solution should be freshly prepared. The sample (approximately 0.1 μmole) is deposited as a small spot on a strip of Whatman No. 1 filter paper. The aqueous phase of the preceding degradation cycle (if this does not contain PTH-arginine) serves as a blank. The strip is sprayed with reagent A until moist, allowed to dry in the air, and then sprayed with reagent B. A red color signifies a positive reaction.

PTH-histidine is identified by a spot reaction with a modified Pauly reagent [306]:

Three solutions are used: Reagent A is a solution of sulphanilic acid (0.45 g) and HCl (0.5 ml conc. HCl) in water (50 ml). Reagent B is a solution of $NaNO_2$ (1.25 g) in water (25 ml). This reagent should be freshly prepared. Reagent C is a solution of Na_2CO_3 (5 g) in water (50 ml). All reagents are chilled in an ice bath. Reagents A and B are mixed immediately before use. The sample (approximately 0.1 μmole) solution is applied as a small spot on a strip of Whatman No. 1 filter paper and dried. It is then sprayed first with the mixed A and B reagents and, after drying, with the C reagent. An orange color marks a positive reaction. A blank should be used.

It should be noted that a short peptide containing arginine or histidine may simulate PTH-arginine or PTH-histidine, respectively, and thus give rise to false identifications.

PTH-arginine, PTH-histidine and *PTH-cysteic acid* may be identified by their electrophoretic mobilities on filter paper[4].

A conventional apparatus for paper electrophoresis is used. The buffer is a 0.025 M potassium phosphate buffer, pH 6.20, the paper is Whatman No. 1, and the voltage approximately 10 V/cm. Reference PTH-amino acids are run alongside the sample. After 2 to 3 hrs the current is discontinued, the paper dried, and the spots located in the usual manner. At pH 6.20 PTH-arginine moves cathodically, PTH-histidine remains stationary (except for the electroendosmotic displacement), and PTH-cysteic acid moves anodically.

3. Quantitative Paper Chromatography

In the identification of PTH-amino acids the paper chromatographic procedures have been superseded by the TLC techniques. For quantitative determinations, however, paper chromatography still plays an important role. In sequence work, the quantitation of PTH-amino acids poses no problem, since normally only one PTH-amino acid is produced at a time, and this may be determined by direct spectrophotometry. On the other hand, the quantitative N-terminal determination of a mixture of proteins, or of a single protein with several different N-terminal amino acids, necessitates the separation of the PTH-amino acids prior to quantitation. This may be done by paper chromatography, since the separated PTH-amino acids can be eluted

[4] In an earlier publication [317] is described the use of thin-layer electrophoresis on silica gel for this purpose. This techniques has later been found insufficiently reproducible to be recommended.

from the paper, and determined spectrophotometrically. The technique described [319, 921b] has been used successfully in the authors' laboratory for many years. It was originally described, perhaps somewhat too modestly, as semiquantitative. In fact, the results for most PTH-amino acids are accurate to $\pm$ 5%, and the method may therefore qualify as quantitative. The technique is in large part similar to or identical with that already described for the TLC of PTH-amino acids, and an abbreviated account should therefore be adequate.

The chromatography room should be thermostated at 20 $\pm$ 2° C. Chromatography is descending, and is carried out in a glass tank, 20 $\times$ 30 cm and 60 cm in height. The glass lid should be well fitting with one or two holes bored in it to allow the introduction of the solvent without lifting the lid. A sheet of filter paper covers the inside walls of the tank, and is held in position by loops of stainless steel wire. The filter paper used in the chromatography is Whatman No. 1.

Three solvent systems are used:

a) Xylene/Formamide (System D)

The system and the procedure are the same as described for TLC. It is convenient to use for the formamide solution a small measuring cylinder, into which the rolled paper is dipped. The soaked paper is immediately placed between sheets of filter paper to remove excess liquid, and suspended in the air for a few minutes to allow the acetone to evaporate. After completion of the chromatography the paper is suspended and dried in a stream of warm air (e.g. from a hair dryer) until it feels dry to the fingers.

b) n-Butyl Acetate/Propionic Acid/Formamide (System E)

The system and the procedure are the same as described for TLC of PTH-amino acids, and the impregnation of the paper with formamide and the final drying of the paper are the same as described for system D.

c) n-Heptane/Ethylene Chloride/Formic Acid (System F)

The upper phase obtained after mixing *n*-heptane (1 vol.), 75% aqueous formic acid (2 vol.) and ethylene chloride (2 vol.) is used for developing the chromatogram, and the lower phase is used to soak the filter paper lining of the tank. An equilibration time of $^3/_4$ hr is required. It is often found, when a new tank is set up, that the first runs are unsatisfactory with a tendency for phase separation on the paper. This trouble usually disappears after the tank has been in use for a day or two. The liquid accumulating in the bottom of the tank should not be removed until necessary, since this seems to upset the equilibration. The final drying of the paper is the same as described for system D.

To locate the PTH spots a fluorescent screen is used [319]. The screen is made of a sheet of translucent material, and one side is covered with a fluorescent material which is excited by ultraviolet light, and emits visible light. The paper is stretched over the fluorescent surface of the screen, and viewed through the screen, while illuminated by an ultraviolet light source. The spots appear as dark areas on an illuminated background. The light source described in the preceding section on TLC is also suitable here. The fluorescent indicator recommended has its secondary emission in the red, and outside the range of visible light emitted by the light source. A considerable increase in contrast may therefore be had by viewing the chromatogram through red colored glasses. The screen is made in the following way.

A sheet of polyethylene, 40 × 30 × 0.15 cm, is coated on one side with a layer of fluorescent cadmium borate (AB Lumalampan, Stockholm, Sweden). Finely cut polyethylene is dissolved in boiling heptane (0.25 g/100 ml), and a fine suspension of cadmium borate in this medium is prepared. The suspension, while still hot, is sprayed on to the polyethylene sheet. After drying in the air the polyethylene sheet is placed on a glass plate with the coated surface upwards, and then left in an oven at 100° C to allow the coating to sinter. The spraying and sintering processes are repeated until an even opaque layer of fluorescent material is obtained.

The purification of solvents is described in the section "Thin-layer Chromatography".

A preliminary qualitative chromatography is made to establish which PTH-amino acids are present in the mixture.

In the quantitative determination 0.03 to 0.06 μmoles of PTH are applied to the paper. The solution of the sample to be determined is applied to the paper alongside an accurately made solution of the reference PTH-amino acid. The reference solution is applied in two different concentration levels, one above and one below the expected concentration of the sample. An adjacent empty lane serves as a blank. All solutions are made up in ethylene chloride. They are applied to the paper 3 cm apart with constriction pipettes. Certain PTH-amino acids are poorly soluble in ethylene chloride, and it may be necessary to use a larger volume, and apply several times with drying in between.

After completion of the chromatography and drying of the paper the spots are marked out in pencil with the aid of the fluorescent screen and cut out with a margin of approximately 3 mm. An area corresponding in size and level is cut out from the blank lane. The cut out pieces, after being divided in smaller pieces with scissors, are extracted with 1.5 ml of 70% aqueous ethanol overnight at room temperature and in the dark.

The absorbances of the extracts are read at 269 mμ in micro quartz cells (1 cm light path) with a capacity of 1 ml. Two corrections are made, one for background absorption due to the paper and one for losses of PTH during chromatography. The first is made by subtracting the absorbance of the paper blank from the other absorbances. The second correction is made by multiplying the sample absorbance with a correction factor. This factor is the ratio between the expected and the found absorbances of the reference samples, the expected absorbances being calculated from the molar absorptivities in Table 8-1. This correction factor varies depending on the PTH-amino acid, or more precisely on its position in the chromatogram. In system F the chromatographic losses are almost negligible, but may in system D be considerable, amounting to 25 to 35% for the spots closest to the solvent front. The amount of PTH-amino acid in the sample is calculated from the corrected sample absorbance using the values in Table 8-1.

Table 8-2 shows the R_f-values of the PTH-amino acids in the three systems. Most of the PTH derivatives are sufficiently well separated in one or the other system to allow elution without contamination from adjacent PTH-amino acids. However, in certain combinations of PTH-amino acids there is a partial overlap, and the estimation is then less accurate.

PTH-arginine, *PTH-histidine* and *PTH-cysteic acid* are estimated after electrophoresis on paper (cf. "Identification of PTH-Arginine, PTH-Histidine and PTH-Cysteic Acid").

The combination of chromatography and ultraviolet spectrophotometry for the estimation of PTH-amino acids occurs in many variants. In this area SJÖQUIST [1165—1167] has been particularly prolific. He has shown that most PTH-amino acids may be separated on a partition column, and estimated with a considerable degree of accuracy [1166]. The incorporation of a light absorption flow cell in the effluent line and the use of a recording spectrophotometer make the estimation particularly rapid and convenient. In conjunction with a microprocedure for the quantitative synthesis of PTH-amino acids from amino acids [1168] the method may also be used for amino acid analysis. SJÖQUIST has also described several paper chromatographic systems suitable for quantitation of mixtures of PTH-amino acids [1167].

Table 8-2. R_f-values of phenylthiohydantoins in paper chromatography systems D, E, and F

PTH-amino acid	R_f-value in system		
	D	E	F
PTH-asparagine	0	0.31	0.80
PTH-serine	0	0.47	0.14
PTH-glutamine	0	0.41	0.15
PTH-aspartic acid	0	0.38	0.16
PTH-glutamic acid	0	0.56	0.27
PTH-threonine	0	0.55	0.34
PTH-tyrosine	0	0.75	0.41
PTH-glycine	0.09	0.63	0.62
PTH-alanine	0.18	0.77	0.78
PTH-N^ε-PTC-lysine	0.05	0.85	0.81
PTH-tryptophan	0.12	0.86	0.82
PTH-methionine	0.45	0.89	0.88
PTH-valine	0.53	0.87	0.89
PTH-proline	0.83	0.90	0.89
PTH-phenylalanine	0.57	0.90	0.91
PTH-isoleucine	0.65	0.92	0.91
PTH-leucine	0.67	0.92	0.92

The sensitivity in detecting PTH-amino acids may be increased by suitable radioactive labeling. CHERBULIEZ et al. [223] and LAVER [754] have utilized S^{35}–labeled phenylisothiocyanate to prepare labeled PTH-amino acids. It is then possible to locate and estimate in a paper chromatogram as little as 0.1 nmole of PTH-amino acid [754]. Tritiated phenylisothiocyanate may also be used in quantitation procedures [750].

4. Gas-Liquid Chromatography

In recent years GLC has been used extensively in the identification and quantitation of PTH-amino acids. PISANO et al. [985] were the first to demonstrate the feasability of GLC for this purpose. Difficulties were, however, experienced with the PTH derivatives of serine, threonine, asparagine, glutamine and the basic amino acids. These difficulties have been reduced by silylation of the PTH-amino acids [453, 501, 979 to 981], but PTH-arginine can still not be identified by GLC, and the yields for the PTH derivatives of serine, threonine, asparagine, glutamine, histidine and lysine tend to be low [980, 984].

Many different stationary phases have been tried, but so far no single phase has permitted the resolution of all PTH-amino acids. However, this may be achieved with certain blends of phases [980, 981, 984].

A few isothiocyanates other than phenylisothiocyanate have been tried in the search for an improved GLC identification. Most commonly methylisothiocyanate has been used [343, 981, 1302, 1335], but the advantage seems marginal. Pentafluorophenylisothiocyanate has also been tried [763], but insufficient data are available for evaluation.

The GLC method has been claimed to be superior to the TLC method in several respects, e.g. resolution, sensitivity and speed. However, this may be questioned. For a complete resolution of all commonly occurring PTH-amino acids more than one set of conditions is needed with both types of methods. Furthermore, the amount of PTH-amino acid required for a single chromatography is the same in both cases, i.e. 2 to 10 nmole. One single gas chromatography of a sample takes 40 to 60 min [cf. 980, 981, 984]. In the same time 10 samples and 10 reference mixtures can be run on one thin layer plate (cf. Fig. 8-6). The TLC technique therefore has not only the advantage of simplicity, but also that of efficiency. The latter point is decisive when a large output from a sequenator is to be processed.

5. Mass Spectrometry

Mass spectrometry of PTH-amino acids was introduced by Stepanov et al. [1225] and Weygand [1358]. The PTH derivatives of all amino acids occurring in proteins may be identified by mass spectrometry as shown by Hagenmaier et al. [472], Fales et al. [350], Sun and Lovins [1245, 1246] and Brewer et al. [175]. Several isothiocyanates other than phenylisothiocyanate have been tried in connection with mass spectrometry. Methylthiohydantoin derivatives [347, 1024, 1245, 1246] and *p*-bromophenylthiohydantoin derivatives [1294, 1295, 1358, 1359] have been successfully used in sequence analysis. Some deuterium and/or halogen substituted phenylthiohydantoin derivatives have been used as internal standards in quantitative determination [350, 1294, 1359].

The temperature of the mass spectrometer probe is high enough to yield a thiohydantoin from a thiocarbamyl peptide [327, 348] or a thiazolinone [349, 793]. This thermal conversion has been used to identify N-terminal amino acids in peptides [327, 348], as well as thiazolinones produced in sequence determination [349, 793]. It is noteworthy that the spectra obtained in this way are virtually identical with those obtained with pure thiohydantoins.

Mass spectrometry may be combined with GLC of the PTH-amino acids [472].

6. Hydrolysis

The amino acid may be regenerated from the PTH-amino acid by hydrolysis, and identified as such. This was in fact the procedure used in the first account of the phenylisothiocyanate degradation [311]. The PTH-amino acid was hydrolyzed in 0.25 N barium hydroxide at 140° C for 48 hrs. In most cases the parent amino acid was recovered, but expectedly PTH-arginine gave rise to ornithine, PTH-asparagine produced aspartic acid (PTH-glutamine, PTH-serine and PTH-threonine were not available), and the yields were generally low. LEVY [767] advocated the use of acid hydrolysis although the PTH derivatives of serine, threonine and tryptophan (and presumably those of asparagine and glutamine) did not produce the parent amino acid. AFRICA and CARPENTER [9] have studied the hydrolysis more thoroughly, and have found that the best results are obtained by alkaline hydrolysis with rigorous exclusion of oxygen. Under these conditions the yields were generally high and often quantitative. However, the general shortcoming of the hydrolytic procedure was again apparent in that the PTH derivatives of serine, threonine, arginine, asparagine and glutamine did not produce the parent amino acids. SMITHIES et al. [1181] and INGLIS et al. [604] claim somewhat better results when hydroiodic acid is used for back hydrolysis, but the principal difficulties remain.

7. General Comments

In the *indirect* method the amino acid residue is identified by a difference in analytical composition of the peptide before and after a degradation cycle. Two techniques are commonly in use.

The first is due to HIRS et al. [549]. Here quantitative amino acid analyses [1198] are performed on the peptide before and after a degradation cycle. The reduction of a particular amino acid by one mole indicates the residue split off (in practice the reduction is often considerably less than one mole). The technique was used extensively in the now classical elucidation of the first enzyme structure [549, 1184, 1185], and has since been widely used in structural studies. Its chief, and perhaps unique, merit lies in the fact that the experimental findings are easily documented. This is done by presenting the results of the amino acid analyses.

The second technique was devised by GRAY and HARTLEY [439]. Its essential feature is that only the N-terminal amino acid is determined before and after a degradation cycle. The new N-terminal amino acid created by a degradation cycle is the one next in sequence. This has the advantage over the preceding technique that the identification is less ambiguous. In the technique of HIRS et al. [549], it is not infrequently seen that more than one amino acid is reduced after a degradation cycle. The uncertainty to which this leads is avoided by the identification of the new N-terminal acid. For identification the sensitive dansyl technique is used, which permits sequence determination on very small amounts of peptide. This technique is not described here, since it is treated in another chapter of this book (cf. "End Group Determinations"). For the following discussion, however, it is important to know that the identification by the dansyl technique requires complete hydrolysis of the peptide.

It should be clear from what has been said earlier in this chapter that the present authors favor the direct identification method, and it only remains to give the reasons.

The chief criticism of the indirect method derives from the fact that it requires a complete hydrolysis of the peptide. Therefore the difference in composition of the peptide before and after a degradation cycle relates only to the *hydrolysate* of the peptide, but not necessarily to the peptide itself. Thus, an amino acid residue may be so modified by the hydrolysis that it cannot be identified by the indirect method. This applies to asparagine, glutamine and tryptophan. Here one is forced to rely on even more indirect and circumstantial evidence for the identification, such as the electrophoretic migration rate of the peptide before and after removal of the amino acid (asparagine and glutamine), or the action of proteolytic enzymes (tryptophan). Furthermore, many proteins contain modified amino acid residues, e.g. by a covalently bound carbohydrate moiety. If these groups are removed by hydrolysis in concentrated acids, and they usually are, they cannot be located by the indirect method. The awareness of their presence may not even be aroused, because there is nothing in the method to indicate their presence. The weaknesses enumerated here are inherent in the indirect method, and cannot be removed by any improvement in technique.

The value of a sequence technique depends largely on the length of sequence which may be determined. The important factor here is a high repetitive yield, i.e. the yield of amino acid calculated from one degradation cycle to the next. Losses of only a few per cent severely limit the length of the degradation (cf. "Sequential Degradation"). The indirect method requires that in each degradation cycle an aliquot of the peptide is removed for analysis. This removal in effect reduces the repetitive yield, and therefore prevents extended degradation.

The rapid automated technique for sequence determination calls also for rapid identification of amino acids. Here the indirect method cannot be used for reasons which should be obvious.

The indirect method was conceived at a time when reliable techniques for the positive identification of amino acid residues removed in a degradation were not available. Today several techniques for this purpose exist. They embrace all amino acids normally occurring in proteins, permit an unambiguous identification, and require little experimental time and labor. On the other hand, the indirect techniques are based on an analytically less sound principle, are not comprehensive, are more cumbersome, and are not suited for automated sequence techniques. For these reasons the indirect method should no longer be used.

E. Sequential Degradation

The value of any sequence method depends largely on the length of the sequence that can be elucidated by it. A large, if not the larger, part of the time spent in determining the primary structure of a protein is taken up by the preparation of (a) fragments sufficiently small to be determined by the available sequence techniques, and (b) fragments furnishing overlapping sequences. Obviously, the magnitude of this arduous task would be substantially reduced, if the sequence method were capable of determining larger structures. Ideally, it should be possible to dispense with the fragmentation altogether. However, present techniques for various reasons fall short of this ideal, although the isothiocyanate degradation is closest to it.

The chemical yield becomes a major factor in the determination of long sequences, which may be illustrated by a simple calculation. Let it be arbitrarily assumed that a sequence determination ceases to give useful information when the over-all yield, i.e. the yield over n degradation cycles, has fallen to 30%. It may then be calculated that an average repetitive yield, i.e. from one cycle to the next, of 97, 98 and 99% would allow 40, 60 and 120 cycles, respectively. It is then obvious that even minor side reactions become significant.

It has already been mentioned (cf. "Reaction Mechanism") that each step in the degradation consists of three operations, i.e. *coupling*, *cleavage* and *conversion*. Only the two first-mentioned reactions are of importance for the repetitive yield, and the associated side reactions and their elimination will be discussed in some detail.

1. Coupling [120, 311, 316]

The reaction is:

$$C_6H_5 \cdot NCS + NH_2 \cdot CHR \cdot CO \cdot NH \cdot X \longrightarrow C_6H_5 \cdot NH \cdot CS \cdot NH \cdot CHR \cdot CO \cdot NH \cdot X \quad (9)$$

Amino groups react with phenylisothiocyanate only in their *amino* form, and the reaction rate is therefore dependent on the pK of the amino group. The higher pK of the ε-amino group of lysine must therefore also be taken into consideration. A reaction pH of 9 and a temperature of 40° C give a convenient rate of reaction. Because hydrogen ions are produced during the reaction a buffer is required, and this should preferably be volatile for easy removal after completion of the reaction. The amine used here, N-dimethylallylamine, is suitable because, (a) it is a tertiary amine and therefore unreactive to phenylisothiocyanate, (b) it has $pK \sim 9$, and (c) it is volatile. To raise the solubility of the non-polar phenylisothiocyanate it is necessary to use a reaction medium with a high content of an organic solvent miscible with water. A mixture of pyridine and water (1:1) is suitable, since it is also a good solvent for peptides and proteins.

Several side reactions have to be considered. Most important of these is the oxidative desulphuration of the phenylthiocarbamyl group observed by ILSE and EDMAN [600]. The oxygen dissolved in the medium is sufficient to bring about this reaction. The resulting phenylcarbamyl group acts as a blockage to the degradation, since obviously no thiazolinone can be formed. This side reaction may be eliminated by performing the coupling reaction in an inert atmosphere, e.g. nitrogen [600]. In the manual degradation it is not practicable to exclude oxygen completely. This is probably the main reason for the lower repetitive yields compared to the sequenator procedure.

The amino group may be made unreactive by various blocking reagents. Important are aldehydes, which tend to form Schiff bases. Traces of aldehydes occur universally, and precautions should be taken to remove them from solvents and reagents. This applies particularly to the sequenator procedure.

It seems likely that heavy metal ions could interfere with the degradation. However, this is hypothetical, and based only on the well known fact that thiourea derivatives are strong chelating agents.

Reactions of phenylisothiocyanate with functional groups other than primary amino groups and the secondary amino groups of the prolines have not been observed by the present authors. Yet, it has been suggested that the imidazole group of histidine and the phenolic group of tyrosine [715] become thiocarbamylated during the coupling. If so, these reactions are reversed at a later stage, since the derivatives do not appear among the products of the degradation or in the synthesis of PTH-amino acids (cf. "Preparation of Phenylthiohydantoins"). An interesting observation, referred to by KONIGSBERG [715] is the reaction of phenylisothiocyanate with free sulfhydryl groups of cysteine residues, a reaction which subsequently leads to scission of the peptide chain at the cysteine residues. In preparations used for degradation the sulfhydryl groups have normally been blocked beforehand, and this reaction should therefore not occur.

A side reaction which does not involve the peptide is the hydrolysis of phenylisothiocyanate:

$$C_6H_5 \cdot NCS + 2\,H_2O \xrightarrow{OH^-} C_6H_5 \cdot NH_2 + CO_2 + H_2S \qquad (10)$$

$$C_6H_5 \cdot NH_2 + C_6H_5 \cdot NCS \longrightarrow \underset{\text{VIII}}{C_6H_5 \cdot NH \cdot CS \cdot NH \cdot C_6H_5} \qquad (11)$$

The *s*-diphenylthiourea (VIII) may simulate a PTH-amino acid in the chromatographic identification (cf. "Thin-layer Chromatography"), and should be removed (together with phenylthiourea formed from ammonia). This is done in the extractions following coupling.

The reaction is carried out in a thick-walled centrifuge tube (inner diam. 1.4 cm) with a capacity of 9 ml, and provided with a C 14 stopper. The weight of the dry tube is determined. The preparation to be degraded remains in this tube during all subsequent operations.

The peptide or protein preparation (0.1 to 5.0 μmoles) is weighed into the tube and dissolved or finely suspended in pyridine-water (1:1; 1 ml) made 0.4 M with respect to N-dimethylallylamine. The pH is adjusted to 9 with dilute aqueous trifluoroacetic acid. Phenylisothiocyanate (50 μl) is added, the tube flushed with N_2, stoppered, and left at 40° C for 1 hr with occasional shakings to keep the solution saturated with the isothiocyanate. The solution is then repeatedly extracted with benzene (5 × 2 ml) using centrifugation to separate the phases. The aqueous phase is subsequently freeze-dried. The sticky residue is repeatedly extracted with ethyl acetate (3 × 0.5 ml) leaving behind a powdery material. The residual ethyl acetate is removed first in a gentle stream of N_2 and finally *in vacuo*. The dry preparation is weighed at this stage in each degradation cycle.

To this description the following comments are added. The amount of material used in the degradation depends on the nature of the preparation and on the purpose of the degradation. If the peptide is short, and the intention is to carry the degradation through to the C-terminal end, it is advisable to use a large amount. The manipulative losses increase as the peptides become shorter, mainly due to an increased solubility of the peptide in the organic solvents used for washing. A warning of increasing losses is given by a more rapid drop in the weight of the PTC-peptide. The main losses in the coupling operation occur during the extraction with ethyl acetate. In the last degradation cycles one may leave out the ethyl acetate extractions, but should then expect spots caused by impurities, e.g. diphenylthiourea, to appear in the chromatograms. On the other hand, if it is only a matter of identifying the N-terminal amino acid in a protein, 0.1 μmole is ample. For an accurate quantitative estimation somewhat more, 0.2 to 0.3 μmoles, are required. In preparations with several different N-terminal amino acids, the quantitative determination involves separation by paper chromatography (cf. "Quantitative Paper Chromatography"), and the material requirements are higher, say 0.3 μmoles of the component occurring in the lowest concentration.

The solubility of the preparation in the coupling medium usually poses no problem. If the preparation is soluble in aqueous media, its solubility in the coupling medium is sufficient for the reaction with phenylisothiocyanate to be quantitative. Some proteins form a fine suspension rather than a clear solution in the coupling medium, but this does not seem to interfere with the reaction. Where it is necessary to handle larger amounts of protein, e.g. proteins of high molecular weight, it is advisable to scale up all volumes, as a very concentrated protein solution is difficult to extract properly. It is sometimes seen that a protein becomes insoluble after a few degradation cycles, and forms a gummy material. This material is difficult or impossible to handle in the degradation procedure. It may be mentioned already here, that this problem is much less troublesome in the sequenator procedure, because the insoluble protein is still reactive when spread in a thin film.

2. Cleavage [120, 314, 316]

The reaction taking place is:

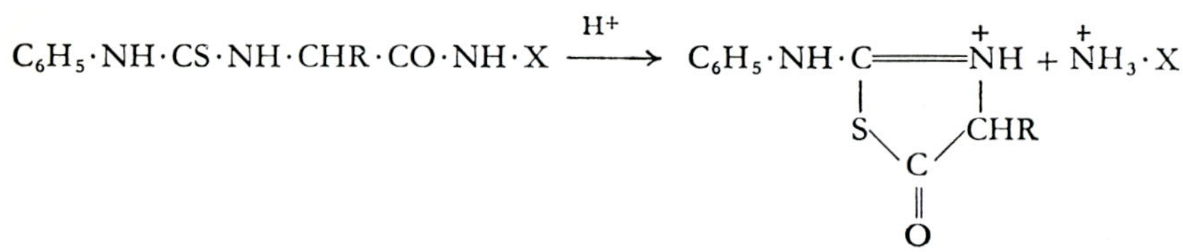

(12)

The two most important facts about this reaction is that it is non-hydrolytic, and that it is rapid. The cleavage may thus be brought about by the action of anhydrous acid for a short time. The peptide bonds are therefore not exposed to hydrolytic conditions. The rate of cleavage varies somewhat depending on the amino acid residue. PTC-leucylglycine in acetic acid saturated with hydrogen chloride is cleaved in less than 3 min at 37° C [315]. Some other amino acid residues react more slowly, but kinetic data are not available. It may be surmised that the relative rate constants for the various amino acids residues are similar to those observed in the better studied conversion reaction [600].

If suitable conditions are used for the cleavage, few side reactions are observed. However, if the conditions are too severe, e.g. glacial acetic acid saturated with hydrogen chloride at 100° C for 1 hr [549], side reactions become prominent. The reactions described involve specific amino acid side chains. Thus, it is reported [1185] that, when the degradation exposes a glutamine residue in N-terminal position, this cyclizes to form a pyroglutamic acid residue (IX):

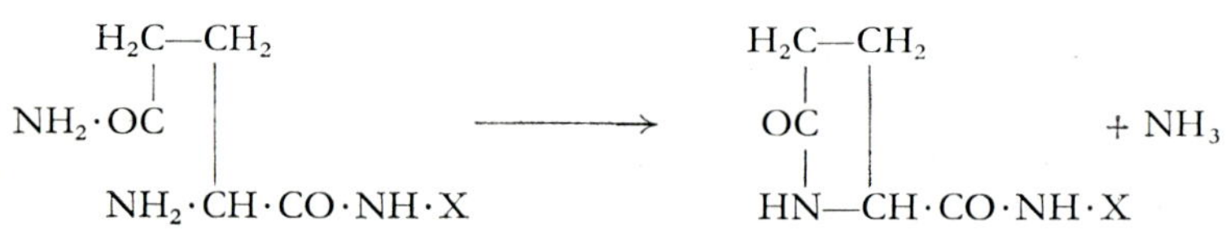

(13)

IX

As the α-amino group is now blocked, the degradation comes to a halt. Reports conflict [448, 715, 1185] whether this reaction takes place under the milder cleavage conditions suggested by Konigsberg and Hill [717], i.e. anhydrous trifluoroacetic acid at 25° C for 1 hr. Under the conditions recommended here, anhydrous trifluoroacetic acid at 40° C for 15 min, this reaction has not been observed. It is relevant that the side reaction follows the cleavage reaction. Measures should therefore be taken to protect the shortened peptide from the conditions causing pyroglutamyl formation.

This is done by removing the trifluoroacetic acid and by storing the shortened peptide in the cold until the next cycle is started.

Under the conditions of HIRS et al. [549], serine and threonine residues become acetylated at their hydroxyl groups, and when the adjacent amino group is released in the course of the degradation, an O → N acetyl migration takes place, blocking the α-amino group [1185]. An $\alpha \rightarrow \beta$ peptide shift at aspartic acid and asparagine residues has also been reported to take place under the above conditions [1185]. It should be stressed, that all these reactions are caused by too severe conditions, and are not unavoidable concomitants of the cleavage reaction. Nevertheless, they serve to emphasize the importance of not exposing the peptide unnecessarily to the cleavage medium.

In the sequenator procedure the protein goes through a great number of degradation cycles; and side reactions which are not noticed in shorter degradations become apparent [317]. What is seen is the cumulative effect of minor side reactions. It is difficult to establish with certainty the nature of these reactions, since in each degradation cycle they may only implicate a few nmoles of the protein. The mechanisms proposed are therefore no more than tentative.

The first reaction to be mentioned is probably not a side reaction in a strict sense. It makes itself noticeable in a gradually increasing overlap between consecutive cycles in the degradation. This overlap is not reduced, if the time allowed for cleavage is increased. On the other hand, if the thiazolinone is extracted, and the cleavage operation repeated, the overlap is substantially reduced or disappears altogether. The suggested explanation [317] is that the end point of the cleavage reaction is an equilibrium rather than a quantitative reaction, although the equilibrium is close to a quantitative reaction. This interpretation is supported by observations made on model substances[5].

The other side reaction noticeable in extended degradations is the appearance in the chromatogram of an increasing general background of other PTH-amino acids. This effect is apparent only if large structures, say with more than 60 amino acid residues, are being degraded, and is the more pronounced the larger the structure. The cause of the background is not known with certainty. It may be significant that the dominating amino acids in the background are also those which occur most frequently in the structure. A small degree of non-specific cleavage along the peptide chain would produce this result. A hydrolytic action of the acid seems unlikely, since precautions have been taken to perform the cleavage reaction under anhydrous conditions. An alternative explanation is acidolysis caused by the organic fluoro acid used in the cleavage. A precedent is the observation that anhydrous trifluoroacetic acid may cause acidolysis of peptide bonds [722].

The amide groups of asparagine and glutamine residues are not appreciably hydrolysed even after many degradation cycles. However, a small degree of hydrolysis would be difficult to detect, since it would be obscured by the partial hydrolysis of the amide group that occurs during the subsequent conversion reaction (cf. "Conversion"). At any rate, no problem arises in the identification of asparagine and glutamine residues. The situation is not as clear in regard to tryptophan residues. The repeated exposure to anhydrous acid expectedly would lead to some destruction of the indole nucleus, and this has also been reported in the phenylisothiocyanate degradation of

[5] P. EDMAN, unpublished.

gramicidin C [1073]. On the other hand, when apomyoglobin was degraded by the sequenator procedure, a tryptophan in position 14 [317] could be identified in the normal way (Fig. 8-11). The stability of tryptophan residues during extended degradations needs further investigation.

Another side reaction, involving histidine residues, has been reported [121, 889, 1108]. It is claimed that when the degradation reaches a histidine residue, not only this residue but also the one following it in the sequence is released. Both appear as PTH derivatives in the identification procedure. These observations have recently been confirmed and extended [1279]. However, the abnormal behavior was only observed when the volatile N-dimethylallylamine buffer was used in the coupling, but not with non-volatile buffers like N-methylmorpholine or Quadrol. The present authors can only state that they have never encountered this side reaction. Examples are the three histidine residues in position 24, 36 and 48, respectively, of the apomyoglobin structure [317], where the histidine residues behaved normally (Fig. 8-11).

As has been pointed out on several occasions, the cleavage reaction takes place under anhydrous acid conditions. The medium first proposed, nitromethane saturated with hydrogen chloride [311], is suitable for short peptides, but not for larger peptides because of their poor solubility. A better solvent is glacial acetic acid saturated with hydrogen chloride, in which the cleavage reaction proceeds with equal ease [315]. A medium where also proteins are soluble, trifluoroacetic acid, was originally suggested by Elmore and Toseland [332], and first used in the isothiocyanate degradation by one of the present authors [314]. Konigsberg and Hill [717] have studied more thoroughly the use of trifluoroacetic acid for this purpose. They recommended anhydrous trifluoroacetic acid at 25° C for 1 hr as a standard set of conditions applying generally to cleavage operations and causing little or no side reactions. The sequenator procedure [317] called for a medium with a lower vapor pressure, and for this reason heptafluorobutyric acid was introduced. The lower homologs of the organic perfluoro acids are in many ways ideal media for the cleavage reaction. They are (a) excellent solvents for peptides and proteins [613], (b) sufficiently strong acids to bring about a rapid cleavage, (c) sufficiently volatile to allow easy removal after completion of the reaction, and (d) easy to prepare in a high degree of purity. On the debit side is tendency to cause acidolysis of peptide bonds (cf. above). From their known properties [1360] one would expect them to cause O-acylation of serine and threonine residues. This may well occur, but, if so, does not lead to the expected complication, i.e. an O $\rightarrow$ N acyl migration and consequent blockage of the amino group, when released in the course of the degradation [1185].

The dry PTC preparation is dissolved in trifluoroacetic acid (50 to 100 µl), the tube stoppered and placed in a water bath at 40 °C for 15 min. The shortened peptide is precipitated with ethylene chloride (3 ml). The precipitate, which is often difficult to see because of a refractive index close to that of ethylene chloride, is spun down in a centrifuge, and the ethylene chloride drawn off with a Pasteur pipette. To the sticky precipitate is added a fresh portion of ethylene chloride (1 ml), and the precipitate is then thoroughly macerated with a glass rod. The ethylene chloride extracts, which contain the thiazolinone, are combined and transferred to a glass-stoppered centrifuge tube (inner diameter 1 cm, capacity 5 ml). The tube containing the precipitated peptide is placed without delay in a precooled desiccator together with a dish containing pellets of KOH. The desiccator is evacuated on the oil pump and put in the cold room over night. This preparation is used for the next coupling. The combined ethylene chloride extracts are evaporated to dryness in a stream of N_2, and the residue is then ready for conversion.

Particular instances may warrant small variations in this procedure. Thus, if larger amounts of protein are handled, it may be necessary to increase the volume of trifluoroacetic acid. When short peptides are being degraded, considerable losses may occur in the ethylene chloride extraction. These may be reduced by (a) evaporating off most of the trifluoroacetic acid subsequent to the cleavage, and (b) reducing the volume of ethylene chloride used in the extraction. Another way of reducing the losses has been suggested by PETERSON et al. [973a]. They used peroxide-free ethyl ether instead of ethylene chloride for the extraction.

The rapid and careful removal of excess of trifluoroacetic acid from the residual peptide and the storage of the latter in the cold serve to prevent side reactions, e.g. pyroglutamyl formation from an N-terminal glutamine residue (cf. above).

3. Conversion [120, 314, 317, 323]

This comprises, strictly speaking, two reactions:

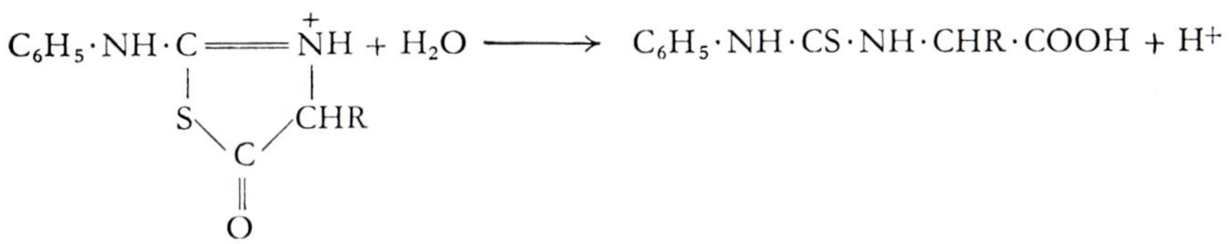

(14)

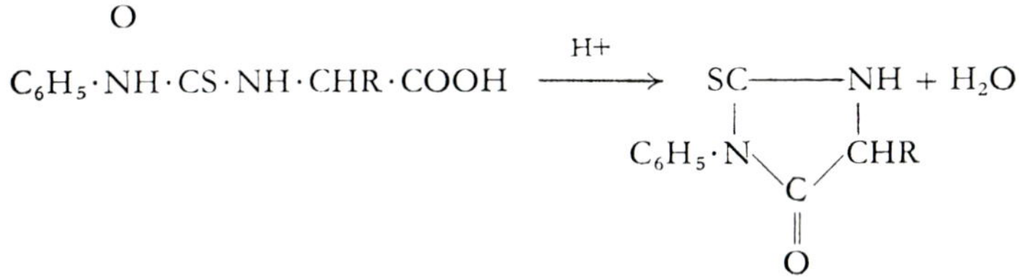

(15)

However, the first reaction is much faster than the second, and for practical purposes they may be treated as one reaction.

The conversion is carried out in an aqueous medium. The kinetics of the cyclization reaction has been studied in some detail [323]. The reaction follows first order kinetics, and, expectedly, the rate increases with the hydrogen ion concentration. The rate varies depending on the amino acid. Thus, the rate constants ($k_1 \times \text{min}^{-1}$) at pH 1.0 and 80° C for the conversion of PTC-valine and PTC-glycine, are 0.13 and 0.03, respectively.

Several side reactions are of importance. As has already been mentioned (cf. "Coupling"), the phenylthiocarbamyl group is easily desulfurized by oxidation, e.g. through the action of oxygen dissolved in the medium. The conversion is therefore carried out in an atmosphere of nitrogen.

Certain PTH-amino acids tend to decompose during the conversion. Most rapidly decomposed is *PTH-serine*, which initially undergoes a β-elimination reaction (cf. "Properties of Phenylthiohydantoins") followed by other reactions, presumably polymerizations. The β-elimination reaction is slower at higher hydrogen ion concentrations [323], and it is therefore possible to select a pH for the conversion reaction which minimizes the tendency for β-elimination. Under the conditions described below the PTH-serine is obtained in about 40% yield. Several of the decomposition products later appear in the chromatograms as multiple spots. *PTH-*

threonine shows much less tendency for β-elimination, and is recovered in near quantitative yield. The decomposition product, PTH-dehydrothreonine, is, unlike that of PTH-serine, rather stable, and reveals its presence by an ultraviolet absorption peak at 323 mμ (Fig. 8-1) and by a spot in the TLC (cf. "Thin-layer Chromatography"). The amide groups of *PTH-asparagine* and *PTH-glutamine* are partially hydrolyzed, and PTH-aspartic and PTH-glutamic acid are formed, respectively. This fact is taken advantage of in the chromatographic differentiation of PTH-asparagine and PTH-glutamine, since the acids are better separated than the amides. For quantitation by paper chromatography (cf. "Quantitative Paper Chromatography") it is preferable to hydrolyze the amides completely to the acids.

In the final extraction of the conversion mixture with ethyl acetate, *PTH-arginine*, *PTH-histidine* and *PTH-cysteic acid* remain in the aqueous phase.

To the dry sample of the thiazolinone is added 1 N HCl (0.2 ml). The tube is flushed with a jet of N_2, care being taken also to stir the solution. The tube is stoppered, heated at 80° C for exactly 10 min, and then cooled under the tap. The solution is extracted repeatedly with ethyl acetate (3 × 1 ml), and the extracts combined. If many samples are to be extracted, it is convenient to use a manifold as shown in Fig. 8-8. The organic phase is evaporated to dryness in a jet of N_2. Here again one way use a manifold (Fig. 8-7). The residue is dissolved in a suitable volume of ethylene chloride (approximately 200 μl/1 μmole of PTH). The aqueous phase is frozen for a later identification (cf. "Identification of PTH-Arginine, PTH-Histidine and PTH-Cysteic Acid").

All solutions of PTHs should be stored at − 15° C and away from light.

4. Quantitation and Identification

The amount of PTH in the sample is determined from the ultraviolet absorption at 269 mμ. The position of the actual absorption maximum varies from one PTH-amino acid to another, and may be as low as 266 mμ. The shape of the absorption curve provides valuable information about the optical purity of the sample and in certain cases about its identity. The absorption measurements should therefore cover the whole range 340 to 220 mμ. This makes a recording ultraviolet spectrophotometer almost indispensable.

Typically (cf. "Properties of Phenylthiohydantoins"), a sample obtained in a degradation should have no absorption at wave lengths longer than 300 mμ, an absorption maximum around 268 mμ, a minimum around 245 mμ and a gradually increasing end absorption at shorter wave lengths. The ratio $\varepsilon_{245}/\varepsilon_{269}$ is a sensitive index of the purity of the product. This ratio should normally be around 0.4 (Table 8-1). The notable exceptions are the PTH derivatives of N^{ε}-PTC-lysine and PTH-proline. Due to decomposition products formed during the conversion (cf. "Conversion") samples of PTH-serine and PTH-threonine show atypical spectra. The absorption curve of PTH-serine samples have a lower absorption at the maximum, and the peak is broader. Samples of PTH-threonine show, in addition to the typical absorption, a smaller peak with a maximum at 323 mμ, which is caused by PTH-dehydrothreonine (Fig. 8-1).

The sample solution in its stoppered tube is heated briefly in hot water to insure complete solution of the sample, and immediately cooled. A measured volume (approximately 10 μl) is added to a light absorption cell (1 cm path length) containing ethanol (3.0 ml). The solution is thoroughly mixed (ethylene chloride is much heavier than ethanol) in the cell with the aid of a Pasteur pipette. The absorbance is read against the same ethanol. The total quantity of PTH in the sample is calculated from the formula:

$$\mu\text{moles PTH} = \frac{a}{b} \times K \times A_{269}$$

where a is the total sample volume, b the volume used for spectrophotometry, K a factor and A_{269} the absorbance at 269 mμ. The magnitude of the factor K depends on the particular PTH-amino acid and is found in Table 8-3.

Table 8-3. Factors (K) for calculating yields of phenylthiohydantoins from ultraviolet absorption at 269 mμ

PTH-amino acid	K	PTH-amino acid	K
PTH-alanine	0.188	PTH-methionine sulfone	0.196
PTH-arginine	0.189	PTH-phenylalanine	0.193
PTH-asparagine	0.174	PTH-proline	0.210
PTH-aspartic acid	0.186	PTH-S-allyl cysteine	0.171
PTH-cysteic acid	0.181	PTH-S-carboxymethyl cysteine	0.181
PTH-glutamic acid	0.189	PTH-S-ethyl cysteine	0.184
PTH-glutamine	0.176	PTH-S-methyl cysteine	0.183
PTH-glycine	0.201	PTH-serine	0.194
PTH-histidine	0.194	PTH-threonine	0.192
PTH-isoleucine	0.177	PTH-tryptophan	0.152
PTH-leucine	0.180	PTH-tyrosine	0.192
PTH-N^{ε}-PTC-lysine	0.104	PTH-valine	0.182
PTH-methionine	0.175		

The identification of PTH-amino acids has already been described in detail (cf. "Thin-layer Chromatography", and "Identification of PTH-Arginine, PTH-Histidine and PTH-Cysteic Acid").

5. Reagents and Solvents

Much emphasis is laid on the purity of reagents and solvents. More effective and simpler purification procedures than those originally described [317] have been developed. However, a procedure may be adequate for a starting material of one provenance but not for that of another. Therefore, we have also described, in addition to the actual purification procedures, criteria of purity. These are also helpful in selecting suitable batches of starting material. This saves labor since, when a good batch is found, it is often possible to delete cumbersome steps in the purification procedure. This applies particularly to ethyl acetate and 1-chlorbutane, where large quantities of otherwise expensive preparations are being used up[6].

The drying of solvents on the molecular sieve is superfluous if the solvents will only be used in manual degradations.

[6] The cost of running a sequenator may vary considerably depending on the local setup. An assessment of our own costs has led us to an estimate of US $ 8 per amino acid residue. This post may be divided into costs of chemicals, $ 1, and technical assistance, $ 7. It should then be noted that technical assistance involves purification of reagents and identification of amino acids by TLC.

Ethanol and n-propanol. The alcohol is stirred magnetically under N_2 for 6 hrs with $NaBH_4$ (1.3 g/l). It is then dried over CaO (100 g/l) for 2 days. The alcohol is decanted, and distilled on a 50 cm Widmer column. N_2 is passed through the alcohol and the still until boiling starts. Several fractions are collected and ultraviolet spectra determined. Store under N_2.

Purity criteria:

Ethanol: Ultraviolet absorption less than 0.3 at 220 μm. Tollens test (dilute 1:1 with H_2O) no more than faintly positive.

n-Propanol: Ultraviolet absorption less than 0.4 at 220 mμ. Tollens test (dilute 1:1 with H_2O) negative. For GLC conditions see below.

Benzene. An analytical grade benzene is purified by fractional freezing. The preparation is slowly frozen at about + 4° C. When the volume of the mother liquor is about 1/10th of the total volume it is drained off and discarded. This process is repeated once, and the preparation is then distilled.

Purity criteria:

Peroxide test negative.

Tollens test negative.

1-n-Chlorbutane. The most important and troublesome impurities make themselves evident by a high ultraviolet absorption in the region above 230 mμ. Commercial preparations vary greatly in this respect, and it is advisable to select one with a low absorption in this region. The charcoal used in the purification should also be checked for its adsorptive efficacy since great differences occur.

A column (6 × 60 cm) is made from activated, granular charcoal. The crude chlorbutane is then passed through at a rate of about 20 ml/min. The ultraviolet spectrum of the effluent is determined at regular intervals, and the process is stopped when the absorption starts to rise at 240 mμ. The chlorbutane is then distilled.

If the test for peroxides is positive the preparation should be passed through a column of activated Al_2O_3 (Aluminium oxide, active, neutral, activity stage I, cat. no. 1077, E. Merck, Darmstadt, Germany). The preparation is finally distilled, dried on the molecular sieve (see below) and stored under N_2.

Purity criteria:

Peroxide test negative.

Tollens test negative.

Ultraviolet absorption less than 0.15 above 230 mμ and less than 0.01 above 240 mμ.

For GLC conditions see below.

Ethyl Acetate. Very harmful impurities are oxidants. These are removed, together with aldehydic impurities, by the procedure described here. However, the oxidants have been seen to reappear during storage, and it is therefore advisable to repeat the test for oxidants immediately before use.

The preparation is passed through a column (3 × 10 cm) of activated Al_2O_3 (Aluminium oxide, active, neutral, activity stage I, cat. no. 1077, E. Merck) and fractions are collected. These are tested for oxidants.

An impurity with a low absorption at 270 mμ is often observed. This impurity may be removed by adsorption on activated charcoal.

The preparation is finally distilled, dried on the molecular sieve (see below) and stored under N_2.

Purity criteria:

Peroxide test negative.

Tollens test negative.

Ultraviolet absorption less than 0.1 above 260 mμ and less than 0.01 above 265 mμ.

For GLC conditions see below.

Ethylene chloride (*purum* grade, Fluka AG, Buchs, Switzerland) is stirred twice with concentrated H_2SO_4 (0.1 vol.), then washed in succession with distilled water, dilute Na_2CO_3 solution and distilled water. It is then shaken in a mechanical shaker overnight with an aqueous solution of $KMnO_4$ (30 g/l). The $KMnO_4$ is removed by repeated washings with distilled water, and the preparation then dried over anhydrous $CaSO_4$, and finally distilled.

n-Hexane is a spectroscopic grade (cat. no. 4372, E. Merck).

Pyridine (L.R. grade, British Drug Houses Ltd., Poole, England) is refluxed over solid KOH for 3 hrs, and then distilled.

N-dimethylallylamine is synthesized as follows [248]: N-dimethylammonium chloride (50 g) is dissolved in water (38 ml) in a thick-walled glass bottle (capacity 250 ml) with a tight fitting glass stopper. Allyl chloride (41 ml) is added, and the mixture cooled in an ice bath. Pellets of NaOH (30 g) are added in portions over 15 min, with the stopper only lightly fitted to allow gas to escape. The reaction is started by gentle shaking of the reaction mixture, and the rate is controlled by cooling. After 45 min the reaction mixture is brought to room temperature, and the reaction completed at this temperature. When the reaction has subsided sufficiently, the stopper is tied down, and the bottle shaken mechanically overnight. The oily amine layer is separated, and to the aqueous phase is added solid KOH (1 g/ml of aqueous phase), whereupon a new layer of amine is formed, which is added to the bulk. The amine is distilled off on a Widmer column. The distillation is interrupted when the temperature rises, and water starts to distill. The distillate is refluxed with phthalic anhydride (5 g/100 ml) for 3 hrs, and then distilled on a short column of glass beads. The free amine should be protected against atmospheric CO_2 during all operations. B.p. 59 to 60° C. (All operations involving allyl chloride should be carried out in the hood and with the use of rubber gloves.)

n-Heptafluorobutyric acid (Minnesota Mining and Manufacturing Co., St. Paul, Minn., U.S.A.) is refluxed with CrO_3 until the color does not appear to change. The acid is distilled off, and again refluxed with solid CrO_3 to ensure that the oxidation has been exhaustive. It is then dried over anhydrous $CaSO_4$, and fractionally distilled on a 50 cm Widmer column. B.p. 120° C.

Trifluoroacetic acid is purified as described for heptafluorobutyric acid.

Phenylisothiocyanate. Common impurities are diphenylthiourea (needle-shaped crystals) and aniline (yellowing of preparation). These are formed by hydrolysis, and the preparation should therefore be protected from moisture. Phenylisothiocyanate is effectively purified by crystallization from absolute ethanol at low temperature. Phenylisothiocyanate (*purissimum* grade, Fluka AG) is mixed with an equal volume of absolute ethanol (for preparation see above) in a glass-stoppered Erlenmeyer flask and placed in the deep freeze (about —25° C). When the solution is cooled down it is seeded with crystals of phenylisothiocyanate. The next day a heavy cake of needle-shaped crystals has formed. The mother liquor is drained off simply by inverting the flask and leaving it for a while in the deep freeze. The crystals are then thawed, and the liquor transferred to a large beaker. Most of the remaining ethanol is then removed in a desiccator *in vacuo.*

The preparation is finally distilled *in vacuo* (40° C/0.05 mm Hg), the first fraction being discarded. It is stored away from daylight in an evacuated desiccator over P_2O_5. (Storage in the cold is unsuitable because of condensation of moisture.)

Purity criterion:

For GLC conditions see below.

N,N,N',N'-tetrakis-(2-hydroxypropyl)-ethylenediamine (Quadrol, Wyandotte Chemicals Corp., Wyandotte, Mich., U.S.A.). 300 g of Quadrol are dissolved in 500 ml absolute ethanol (for purification of ethanol see above) and the solution cooled in an ice bath. Dry HCl is passed through the solution with cooling (below +30° C) until somewhat more than 2 moles have been taken up. At this stage Quadrol dihydrochloride starts to crystallize. The solution is left overnight at —20° C to complete crystallization. Crystals are filtered off on a clean sintered glass funnel with suction, and the mother liquor is removed as completely as possible.

The crystals are then uniformly suspended in 100 ml absolute ethanol. Sixty grams of NaOH (A.R. grade) are dissolved in 600 ml absolute ethanol with stirring. This solution is added gradually and with vigorous stirring to the suspension of Quadrol dihydrochloride, The monohydrochloride goes into solution, and NaCl crystallizes out, and is filtered off on a sintered glass funnel.

The filtered ethanolic solution is cooled in an ice bath, and dry HCl passed through until 50 g HCl has been taken up. The preparation is left to crystallize at —20° C overnight. The following day the crystals are filtered off sharply on a sintered glass filter.

Two moles of NaOH are dissolved in 800 ml absolute ethanol. The crystals of Quadrol dihydrochloride are transferred to a 2 l beaker, and the ethanolic NaOH is added gradually with vigorous stirring. After each addition a sample (about 1 ml) is withdrawn, diluted with 4 ml of distilled water, and the pH of the solution measured with a glass electrode. The alkali addition is continued until pH reaches 11.0. Salts are removed by filtration on a sintered glass funnel.

The ethanolic solution is concentrated in a rotary evaporator, first on the water pump and finally on the oil pump. When the ethanol is removed, salts may again start to crystallize. They may be removed by filtration under suction through a sintered glass filter.

The preparation is stored under vacuum in the cold and in the dark.

Purity criterion:

Tollens test (dilute 1:1 with H_2O) only faintly positive.

Drying on Molecular Sieve. A column (100 × 3 cm) is made of $^1/_{16}''$ pellets of molecular sieve Type 4A (Union Carbide Limited — Chemicals Division, London, England). The solvent is passed through at a rate of 10 to 15 ml/min.

Peroxide Test. To 2 ml of solvent in a test tube is added 1 ml of a 1% aqueous KI solution. The organic solvent is completely evaporated in a stream of N_2 (a weakly positive reaction may otherwise be masked). A blue color on the addition of a starch solution signifies a positive reaction.

Tollens Test. The test is carried out in a darkened room. One ml of a 10% (w/v) aqueous $AgNO_3$ solution and 1 ml of a 10% (w/v) aqueous NaOH solution are mixed in a test tube, and a dilute aqueous NH_3 solution added dropwise until the precipitate is redissolved. To this solution is added 2 ml of the sample (after dilution with H_2O where appropriate). Where two phases are formed the tube is shaken at frequent intervals. In a negative test no discoloration or cloudiness should occur within 1 hr. (Caution: Explosive silver fulminate is formed in the Tollens reaction. Tubes should not be left standing overnight, but should be emptied, washed with distilled water and dilute nitric acid.)

GLC Conditions. The columns used are of glass, 180 cm long and 4 mm inner diameter. The solid support is Chromosorb W, and the carrier gas is N_2. Detection is by flame ionization. Special conditions are as follows:

n-Propanol: Liquid phase is 10% SE-30; isothermal at 50° C.

n-Chlorbutane: Liquid phase is 10% SE-30; temperature program 30 to 60° C with 2°/min.

Ethyl acetate: Liquid phase is 10% Carbowax 1000; isothermal at 40° C.

Phenylisothiocyanate: Liquid phase is 10% SE-30; temperature program 80 to 170° C with 4°/min.

6. Protein Sequenator

This instrument was designed to speed up the process of primary structure determination.

The design and operation of the sequenator have been described in full detail [317]. Since then no major changes have been made, and a description in outline will therefore suffice here.

The operation of the instrument is based on the principle that solutions of reactants and extracting solvents are spread in thin films inside a rotating cyclindrical glass cup. This principle makes it simple to perform the many mechanically diverse operations in the degradation. Thus, a spinning film is ideally suited for extraction by another film of an immiscible solvent sliding over its surface, for centrifugation and also for drying under reduced pressure, because of the large surface and the stabilizing centrifugal force.

The sequenator performs only the coupling and cleavage operations. The conversion reaction is carried out in the conventional way. The omission of the latter operation simplifies the design problem, but does not reduce the speed of the analysis, since the conversion may conveniently be carried out on a large number of samples simultaneously. The evaporation, extraction and application manifolds make the conversion and the TLC identifiation of many samples no more time comsuming than for a single sample.

The general design of the sequenator is shown in Fig. 8-5. The reaction vessel is a cylindrical glass cup (*A*) mounted on the shaft of an electric motor (*B*). The cup spins continuously, and solutions and solvents entering the cup are therefore spread as thin films on the wall of the cup. Reagents and solvents enter through the feed line (*R*) at the bottom of the cup. Extracting solvents climb to the top of the cup, where they are scooped off, and leave through the effluent line (*S*). The cup is enclosed in a bell jar (*Q*), and the system can be evacuated by means of a vacuum pump (*P*). The reaction chamber is thermostated at 50° C.

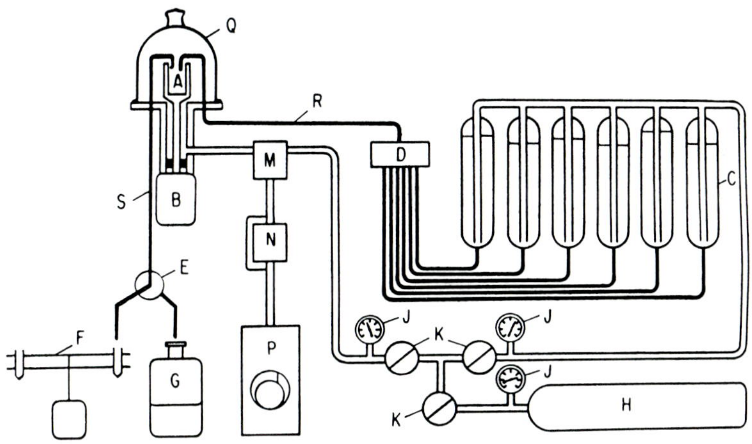

Fig. 8-5. Diagram of sequenator. *A* spinning cup; *B* electric motor; *C* reagent (solvent) reservoir; *D* valve assembly; *E* outlet stopcock assembly; *F* fraction collector; *G* waste container; *H* nitrogen cylinder; *J* pressure gauges; *K* pressure regulators; *M* 3-way valve; *N* 2-way valve with bypass; *P* rotary vacuum pump; *Q* bell jar; *R* feed line; *S* effluent line. Gas lines are doubly contoured, and liquid lines are filled. (Reprinted from Ref. [317], p. 81, by courtesy of Europ. J. Biochem.)

Reagents and solvents are stored in reservoirs (*C*) and are admitted to the cup through an assembly of valves (*D*). The reservoirs are under a constant low pressure of nitrogen supplied by a nitrogen cylinder (*H*) and pressure regulators (*K*). The contents of the bell jar are likewise held at a fixed although lower pressure of nitrogen through a similar arrangement. The pressure differential between the reservoirs and the bell jar is constant, and the volume of reagent or solvent admitted to the cup is therefore determined by the time a valve is kept open.

The effluent line leads, *via* a 3-way stopcock (*E*), either to a fraction collector (*F*) or to a waste container (*G*). The valves in the assembly (*D*) and the gas valves (*M* and *N*) are operated by solenoids and the 3-way outlet stopcock (*E*) and the fraction collector (*F*) by electric motors. All these functions are governed by an electronic programming unit.

A detailed drawing of the reaction chamber is presented in Fig. 8-6.

The design of the sequenator requires modification of certain reagents and solvents as compared to the manual procedure. Thus, the large volume of the bell jar necessitates the replacement of more volatile reagents with reagents of lower vapor pressure in order to reduce losses due to evaporation. N,N,N′,N′,-tetrakis-(2-hydroxypropyl)-ethylenediamine (QUADROL) and *n*-heptafluorobutyric acid have been substituted for N-dimethylallylamine and trifluoroacetic acid, respectively.

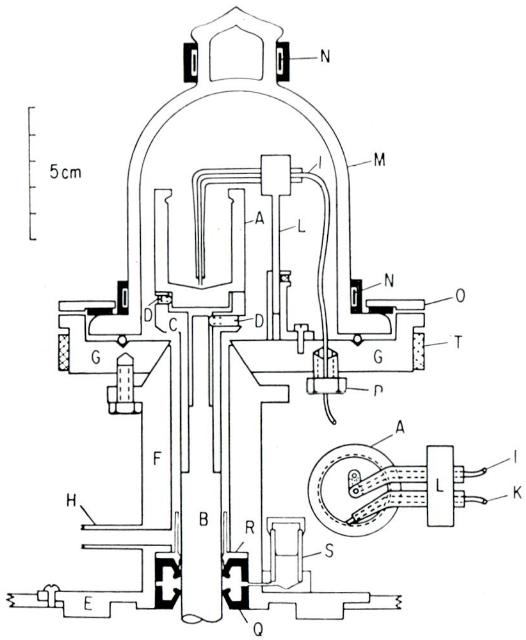

Fig. 8-6. Reaction chamber. *A* glass cup; *B* motor shaft extension; *C* cup support; *D* grub screws; *E* motor support; *F* column; *G* base plate; *H* side tube; *I* feed line; *K* effluent line; *L* adjustable stand; *M* bell jar; *N* electrodes; *O* rubber padded ring; *P* sealing bolt; *Q* oil seal; *R* teflon sleeve; *S* oil reservoir; *T* band heater. A top view of the cup, the feed line and the effluent line is shown in the lower right hand corner. (Reprinted from Ref. [317], p. 82, by courtesy of Europ. J. Biochem.)

Another deviation from the manual procedure is that the cleavage operations is repeated in each cycle. The reason for this is discussed under "Cleavage". The following reagents and solvents are used (cf. Table 8-4):

Reagent 1 is a solution of phenylisothiocyanate in *n*-hexane.

Reagent 2 is a buffer of QUADROL-trifluoroacetic acid (pH 9.0).

Reagent 3 is heptafluorobutyric acid.

Solvent 1 is benzene.

Solvent 2 is ethyl acetate.

Solvent 3 is 1-chlorobutane.

Table 8-4. Operations in a cycle of the sequenator[a]

Reagents 1, 2 and 3 are in reservoirs I, II and III, respectively, and solvents 1, 2 and 3 occupy reservoirs IV, V and VI, respectively. The valves are numbered correspondingly. Other functions are the 3-way vacuum-pressure valve (M), the 2-way vacuum valve (N), the 3-way outlet stopcock (E) and the motor driving the fraction collector (F). A + or a − sign means that the operating solenoid or motor is energized or deenergized, respectively, at the beginning of the stage. The + sign also means that a valve is open, in the case of valve M to the vacuum line. The three positions of the outlet stopcock are indicated by the letters c (collect), v (vacuum) and w (waste). The duration of stages refers to a reaction temperature of 50° C.

Stage	Duration min	(Volume) (ml)	Function I	II	III	IV	V	VI	M	N	E	F
1. Reagent 1	0.15	(0.40)	+								+w	+
2. Reagent 2	1.75	(0.40)	−	+								−
3. Reaction	30.00			−							+v	
4. Restric. vacuum	3.00								+			
5. Vacuum	6.00									+		
6. Delay	0.05								−	−		
7. Solvent 1	5.00	(11.5)				+					+w	
8. Solvent 2	8.00	(23.5)				−	+					
9. Delay	1.00						−					
10. Restric. vacuum	3.00								+		+v	
11. Vacuum	6.00									+		
12. Delay	0.05								−	−		
13. Reagent 3	0.70	(0.23)			+						+w	
14. Reaction	3.00				−						+v	
15. Vacuum	1.50								+	+		
16. Delay	0.05								−	−		
17. Solvent 3	1.75	(5.5)						+			+c	
18. Delay	1.00							−				
19. Restric. vacuum	3.00								+		+v	
20. Vacuum	1.00									+		
21. Delay	0.05								−	−		
22. Reagent 3	0.70	(0.23)			+						+w	
23. Reaction	3.00				−						+v	
24. Vacuum	1.50								+	+		
25. Delay	0.05								−	−		
26. Solvent 3	2.25	(7.0)						+			+w	
27. Delay	1.00							−				
28. Restric. vacuum	3.00								+		+v	
29. Vacuum	6.00									+		
30. Delay	0.05								−	−		

[a] Reprinted from Ref. [317], p. 87, by courtesy of Europ. J. Biochem.

The programming of the degradation cycle requires that it be divided in 30 stages The stages, their duration and associated operations are shown in Table 8-4. The time required for one degradation cycle is 93.6 min, which is equivalent to 15.4 cycles in 24 hrs.

The repetitive yield is about 98%. This high yield makes it possible to carry out much longer degradations with the sequenator than with the manual procedure. With apomyoglobin D of the humpback whale it was possible to perform 60 degradation cycles, and establish the N-terminal structure [317]:

Val.Leu.Ser.Asp.Ala.Glu.Trp.Gln.Leu.Val.Leu.Asn.Ile.Trp.Ala.Lys.Val.Glu.Ala.Asp.Val.
1 2 3 4 5 6 7 8 9 10 11 12 13 14 15 16 17 18 19 20 21

Ala.Gly.His.Gly.Gln.Asp.Ile.Leu.Ile.Arg.Leu.Phe.Lys.Gly.His.Pro.Glu.Thr.Leu.Glu.Lys.
22 23 24 25 26 27 28 29 30 31 32 33 34 35 36 37 38 39 40 41 42

Phe.Asp.Lys.Phe.Lys.His.Leu.Lys.Thr.Glu.Ala.Glu.Met.Lys.Ala.Ser.Glu.Asp.
43 44 45 46 47 48 49 50 51 52 53 54 55 56 57 58 59 60

The identification of the PTH-amino acids in solvent system D are shown in Fig. 8-7. The termination of the degradation was determined by two factors, the cumulative effect of a 2% loss of yield in each cycle, and the gradual appearance in the chromatograms of an increasing general background of other PTH-amino acids, which eventually made the identification impossible. The origin of this background is discussed under "Cleavage". In the degradation of very long peptide chains (more than 60 to 70 residues) the background is a limiting factor. On the other hand, the degradation of very short peptides in the sequenator is subject to other limitations. The difficulty here is the small difference in solubility between the thiazolinone and the short peptide, making a clean extraction of the former impossible (cf. "Cleavage"). Therefore considerable losses of material occur in the last few cycles of the degradation. These losses may be reduced by leaving out the repetition of the cleavage operation. However, a much more effective way of reducing the losses was proposed by Braunitzer and co-workers [162 to 166]. Peptides containing lysine, e.g. tryptic peptides, are first coupled with a highly polar isothiocyanate, i.e. 4-sulphobenzeneisothiocyanate, 3,5-disulphobenzeneisothiocyanate, 4,8-disulfonaphtalene-2-isothiocyanate or, preferably, 4,6,8-trisulfonaphtalene-2-isothiocyanate. The reaction mixture is then transferred to the cup, and the ordinary sequenator procedure is started. All positions in the sequence, except the first and also those containing lysine, yield the usual PTH-amino acids. The hydrophilic sulfonic acid groupings at the lysine ε-amino groups reduce the solubility of the peptide in the organic solvents sufficiently to allow complete sequencing.

The peptide preparation (0.5 to 1 μmole) is dissolved in a mixture of N-dimethylallylamine (40 μl) and water (70 μl). To this solution is added a Braunitzer reagent in sodium salt form (3 to 5 mg), and the solution is kept at 50° C for 60 min. It is then diluted to 0.4 ml with 50% ethanol, and immediately transferred to the sequenator cup. Following a programmed drying period the degradation is started at Stage 1.

The Braunitzer reagents have later been successfully used by, among others, Inman et al. [610]. They also showed that S-aminoethyl cysteine residues reacted similarly to the lysine residues. Other methods of reducing the solubility of the peptides in organic solvents were suggested by Crombie et al. [257] and Foster et al. [377]. Here carboxyl groups were amidated with aminomethanesulfonic acid [251] or 2-amino-1,5-naphthalenedisulphonic acid [311] using a carbodiimide. Again, the introduced sulfonic acid groupings made the peptides less extractable. The amidation of the ω-carboxyl groups of aspartic and glutamic acid residues prevents a normal identification of these residues.

The application of the manual degradation procedure to proteins is often restricted by the fact that proteins become completely insoluble after a few degradation cycles. This limitation is much less severe in the sequenator procedure. It appears that the protein, when spread in a thin layer, retains much of its reactivity even after it has become insoluble.

7. General Comments

Most problems likely to be encountered in the course of a degradation have already been commented upon and it only remains to discuss the scope of the degradation technique.

The repetitive yield in the manual version of the phenylisothiocyanate degradation is 90 to 95%. The loss in yield is probably mostly due to oxidative desulfuration of the phenylthiocarbamyl group. This in its turn is caused by small amounts of oxygen, which are difficult to exclude in the manual technique. Higher losses are to be expected when the degradation approaches the C-terminal end of the peptide. The causes of these losses and ways to minimize them are discussed in the sections "Coupling" and "Cleavage". The degradation of a protein is sometimes prevented by the fact that the preparation becomes insoluble after a few degradation cycles. Sometimes this problem may be overcome by oxidizing the cystine residues to cysteic acid residues [316]. Another limitation in the degradation of proteins is a gradual increase in background amino acids.

In not unfavorable cases one may expect the manual technique to produce 15 to 20 degradation cycles with clearly interpretable results. To perform one degradation cycle requires the best part of a working day. However, it is quite possible for a trained person to run in the same time several parallel degradations.

A blocked N-terminal is not infrequently encountered in peptides and proteins. This may be due to the amino group being acetylated, e.g. cytochrome c [810b],

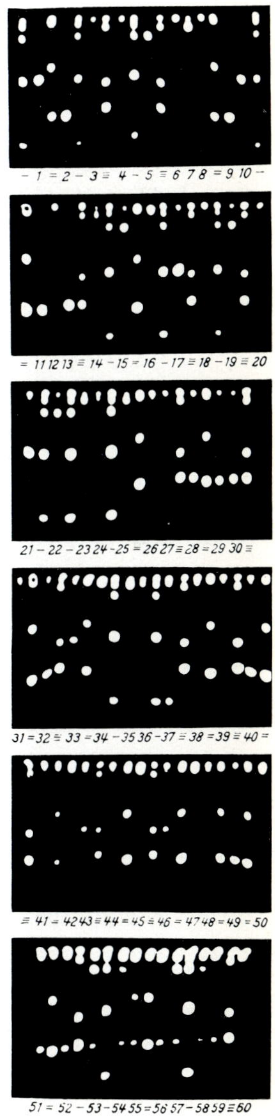

Fig. 8-7. Degradation of apomyoglobin. TLC identification using solvent system D of the PTH-amino acids obtained in first 60 degradation cycles. Reference mixtures are indicated by horizontal bars, and show in order from bottom to top the migration of (—) PTH-proline. PTH-valine, PTH-alanine, PTH-glycine; (=) PTH-leucine, PTH-methionine; (≡) PTH-isoleucine, PTH-phenylalanine and PTH-tryptophan. (Reprinted from Ref. [317], p. 89 by courtesy of Europ. J., Biochem.)

to a Schiff base, e.g. hemoglobin A_{Ic} [577] or to an N-terminal pyroglutamic acid residue, e.g. human fibrino-peptide B [120, 121]. The last may represent the native configuration, but may also be an artifact produced during isolation [1184], since N-terminal glutaminyl residues are notorious for their great tendency to cyclize to pyroglutamyl residues [848] (cf. "Cleavage"). Regardless of the mode of blocking, the isothiocyanate degradation obviously cannot be applied. The way out has been to create a new N-terminal amino acid, preferably close to the blocked group, by the use of some specific cleavage method, e.g. proteolysis. The structure of the fragment with the blocked N-terminal has to be determined by other techniques [121]. A promising approach to the problem of N-terminal pyroglutamic acid has recently been described by DOOLITTLE and ARMENTROUT [289]. They have succeeded in isolating from a strain of *Pseudomonas fluorescens* a new enzyme, pyrrolidonyl peptidase, which is claimed to split off selectively N-terminal pyroglutamic acid residues [1301]. An enzyme with similar specificity has also been reported for *Bacillus subtilis* [1256].

The great tendency for an $\alpha \rightarrow \beta$ peptide shift at aspartyl and asparaginyl peptide bonds is well documented [530, 639, 903, 1185, 1249]. This rearrangement may occur even under mild conditions, e.g. during the isolation of a peptide [474]. A β-aspartyl peptide bond is not cleaved in the isothiocyanate procedure and further degradation is therefore blocked [448, 1337].

A blocking reaction involving N-terminal S-carboxymethyl cysteine residues has recently been described [1186]:

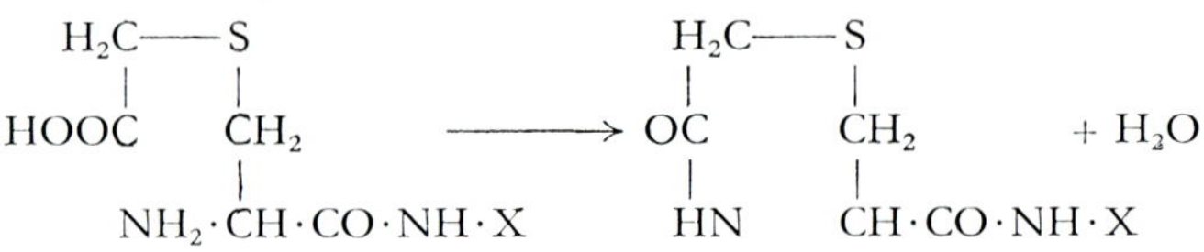

(16)

The reaction takes place even under mildly acid conditions. This may be another reason for using other S-alkyl derivatives of cysteine, e.g. S-methyl or S-ethyl cysteine (cf. "Identification of Amino Acids"), in sequence determinations.

Sequential degradation is normally not carried out on peptides with intact cystine residues. In a few instances where this has been done, it was found that a half-cystine residue does not block the degradation. Expectedly, no PTH derivative is obtained in the corresponding degradation cycle.

Incomplete degradation of proline residues has recently been described [530]. The cause is unknown.

The possibility that an amino acid residue is modified, e.g. carries a prosthetic group, has generally been disregarded in sequence determinations. The wide use of the method of indirect identification (cf. "Identification of Amino Acids") is condusive to this practice. The presence of a modified amino acid residue may, in fact, have been more frequent than is apparent from degradation protocols. The attachment of a carbohydrate moiety through an N-glycosidic linkage to an asparagine residue or through an O-glycosidic linkage to residues of serine and threonine have been established [911b]. Almost nothing is known about the behavior of such modified residues in the isothiocyanate degradation. However, FLETCHER et al. [363a] have shown that an asparagine-carbohydrate complex from ovalbumin with an N-glycosidic linkage between the β-amide group and the carbohydrate forms what appears to be a normal PTH derivative.

The value of quantitative N-terminal determinations in the characterization of proteins has not been realized generally. This is surprising, since evidence of this kind is even more informative than, and often complementary to, ultracentrifugation data. Thus, for a pure protein preparation a quantitative N-terminal determination will give direct information about the minimum molecular weight. This figure, it should be noted, is independent of the quarternary structure of the protein as contrasted to physico-chemical data. In addition, N-terminal analysis may, in the absence of blocked N-terminals, provide information about the number of peptide chains in the smallest unit. It may also reveal, by the occurence of N-terminal amino acids in fractional molar quantities, the cleavage of peptide bonds, e.g. by the accidental action of proteolytic enzymes during the preparation [292b], changes which are not necessarily revealed by physico-chemical measurements. For these reasons it seems appropriate to require, when new protein preparations claimed to be homogeneous are described, that the data should also include a quantitative N-terminal determination.

In many other cases N-terminal determination may be used to great advantage, and a few will be mentioned here. One instance is the characterization of protein mixtures, e.g. plasma proteins [921b], by the quantitative composition of their N-terminal amino acids. Another example is monitoring the progress of a protein purification. The first phase in a structural determination is usually a fragmentation of the structure, and here an N-terminal analysis of the initial digest may serve as a check that fragments are not lost in the later isolation procedure. Also, a quantitative N-terminal determination on an isolated fragment will define its size.

F. Related Procedures

The isothiocyanate degradation exists in numerous modifications. Most of these concern less essential features, and the advantages, if any, would be so marginal that they will not be described here. However, certain modifications, because they are widely in use, or because they represent new approaches, will be mentioned.

The use of organic isothiocyanates other than phenylisothiocyanate has been investigated. Thus, 4-dimethylamino-3,5-dinitrophenylisothiocyanate [1019b] and *p*-azobenzene isothiocyanate [1011b] have been suggested because they give rise to strongly colored thiohydantoins, which are easy to locate on chromatograms and to quantitate. Naphthylisothiocyanate has reactivities similar to phenylisothiocyanate, but naphthylthiohydantoins have a much stronger absorption than phenylthiohydantoins in the ultraviolet [671]. This advantage is at least partly offset by the fact that the absorption maximum is at 222 mμ, a region of the spectrum which is inconvenient to use. Ivanov et al. [623—627] have reported a promising new reagent, 2-*p*-isothiocyanophenyl-3-phenylindone, for the identification of very small amounts of amino acids ($<$ 1 nmole). The sensitivity is claimed to be comparable to that of the dansyl technique [439], and the method has the important advantage of a direct identification. The technique has been used in the sequential degradation of proteins [627]. Several other isothiocyanates have been recommended because of advantages in the gas chromatographic or mass spectrometric identification of the thiohydantoins (cf. "Identification of Amino Acids").

The isothiocyanate degradation of proteins is occasionally limited, at least in the manual version of the technique, by the fact that the protein becomes insoluble and unreactive after only a few degradation cycles. In the paper strip method of FRAENKEL-CONRAT [381] the protein is absorbed on a piece of filter paper, and remains there during all subsequent operations, i.e. couplings, cleavages and conversions. Since the protein is prevented from forming a dense coagulum, it retains at least partly its reactivity and it is therefore possible to carry through several successful degradation cycles. SCHROEDER [1108] has later modified the paper strip method to allow the degradation also of peptides.

In the paper strip method the cleavage and conversion operations are combined. As has already been pointed out (cf. "Reaction Mechanism") the former reaction is rapid and the latter is slow. As a result the peptide is unnecessarily exposed to the hydrolytic and otherwise detrimental (e.g. destruction of tryptophan) action of a strong aqueous acid.

LAURSEN [749—752], LAURSEN et al. [753], PREVIERO et al. [999], DIJKSTRA et al. [285] and SCHELLENBERGER et al. [1083, 1083a] have all reported on the successful isothiocyanate degradation of short peptides covalently bound to resins, so-called solid-phase degradation. The chief advantage of attaching the peptide to a solid support is that all operations in coupling and cleavage may be done by filtration. This simplifies the automation of the procedure [751]. For smaller peptides there is the additional advantage that losses due to extraction are eliminated.

In establishing the linkage between the peptide and the resin the bridge head on the peptide side poses some problems. In one type of procedure the peptide is linked by its previously activated carboxyl groups to the aminopolystyrene resin [285, 749, 751, 752, 1083, 1083b]. Ideally, the bridge head should be exclusively the C-terminal carboxyl group. However, in practice it is difficult to restrict the reaction to this group, since the carboxyl groups of aspartic and glutamic acid residues have similar reactivities. When the degradation passes an attached amino acid residue, this as well as the anchorage point are lost. Another, and unexpected, problem with the modified aspartic acid residues is that they block the degradation, presumably due to an α,β-imide formation [751]. However, in a recent publication the difficulties with the aspartic and glutamic acid residues are claimed to have been overcome through a modification of the carboxyl group activation procedure permitting a selective attachment over the C-terminal carboxyl group [999]. Two as yet unsolved problems with the carboxyl attachment are that a C-terminal arginine does not couple to the resin [752], and that serine residues are lost in the degradation [751].

In an alternative procedure for solid-phase degradation the peptide is linked to the resin *via* its amino groups [752, 753]. The free amino groups of the peptide are coupled to the aminopolystyrene resin through a bifunctional isothiocyanate, *p*-phenyldiisothiocyanate. The procedure is especially suited for tryptic peptides with C-terminal lysine. The α-amino group of the peptide is also bound to the resin, and the N-terminal amino acid can therefore not be identified, but is, on the other hand, split off from the peptide in a normal cleavage procedure.

The method described by STARK [1212] and DOWLING and STARK [293] uses the principle of solid support in reverse, since here the isothiocyanate group is carried by the solid support. The peptide in solution is coupled to the isothiocyanate groups of the resin. In a subsequent cleavage step the shortened peptide is released from the resin leaving the removed amino acid residue attached to the resin. The procedure is then repeated with a fresh batch of resin. In its present stage of development the technique gives rather low repetitive yields. The identification of the amino acid removed depends on the indirect method with its known limitations (cf. "Identification of Amino Acids").

Several other degradation reactions related to the isothiocyanate degradation have already been discussed (cf. "Reaction Mechanism").

III. Other Chemical Degradation Procedures

In all the schemes for sequential degradation discussed so far the essential feature of the mechanism has been the formation of a thiazolinone. However, other schemes with different mechanisms exist. These have generally not been worked out in such detail that they may be applied routinely to sequence determinations, and therefore only the principles will be presented.

A. From N-Terminus

In the method of Holley and Holley [575], the driving force in the cleavage of the peptide is the formation of a lactam. The terminal amino group is first reacted with 4-carbomethoxy-2-nitrofluorobenzene. The nitro group is then catalytically hydrogenated to an amino group (17). The lactamization and the concomitant peptide bond scission (17) produces a 7-carbomethoxy-3,4-dihydro-2-quinoxalone derivative (X) containing the N-terminal amino acid residue. The reaction takes place in aqueous solution under rather mild conditions, 5 hrs at ambient temperature or 15 min at 70° C:

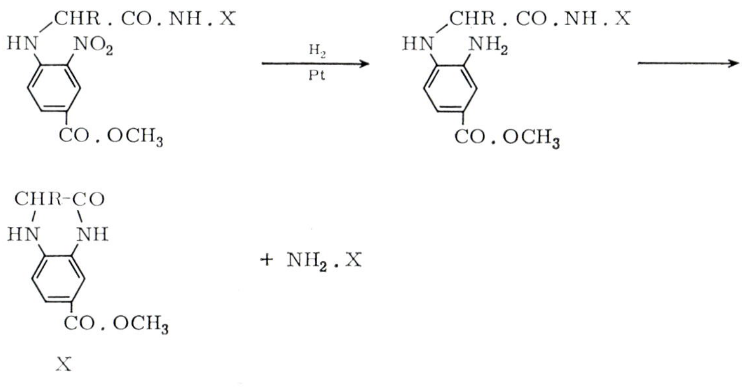

(17)

The procedure has been tried out on short peptides with non-polar amino acid residues. The yield was 84% on the average. A variant of the procedure [608, 653] uses instead the 2,4-dinitrophenyl derivative of the peptide, which is hydrogenated to the 2,4-diaminophenyl derivative, and made to form a lactam in analogous fashion. The general applicability of this type of procedure remains to be demonstrated.

A remarkable degradation reaction has been described by Buckingham, Collman and coworkers [184, 185, 242, 243]. A tetradentate Co(III) complex, β-[Co(trien)-

$OH(H_2O)]^{2+}$ (trien = triethylenetetramine), reacts with a peptide (or an amino acid ester) as follows:

$$\begin{aligned}&\beta\text{-}[Co(trien)OH(H_2O)]^{2+} + NH_2 \cdot CHR \cdot CO \cdot NH \cdot X \longrightarrow \\ &\beta\text{-}[Co(trien)NH_2 \cdot CHR \cdot CO_2]^{2+} + NH_2 \cdot X\end{aligned} \tag{18}$$

The reaction between the cobalt complex and the peptide is stoichiometric and an excess of the former leads to a scission also of the peptide bond next in sequence. The reaction conditions used were rather mild, aqueous solution around pH 7, 65° C and a few hours. The amino acid-cobalt complex split off is strongly colored and may be identified by paper or thin-layer chromatography. Recently, FENN and BRADBURY [353] have made a thorough evaluation of the reaction as a basis for sequence determination. They draw attention to several complicating factors, among them complexing of the reagent with carboxyl groups. Obviously, more work is required.

B. From C-Terminus

The first method for the sequential degradation of peptides by chemical means is due to SCHLUCK and KUMPF [1087]. Their idea, inspired by the studies of JOHNSON and NICOLET [640] and KOMATSU [714] on 1-acyl-2-thiohydantoins, was first to convert the C-terminal amino acid residue into a 2-thiohydantoin substituted in the 1-position by the rest of the peptide (XII) and subsequently to hydrolyze off the acyl group (20). A new C-terminal residue would then be created, on which the procedure could be repeated. The formation of the 2-thiohydantoin derivative was achieved by treating the peptide at 100° C for $^1/_2$ hr with ammonium rhodanide and acetic anhydride. This led to the formation of an acylisothiocyanate (XI) which rearranged spontaneously (19) to the 1-acyl-2-thiohydantoin derivative (XII):

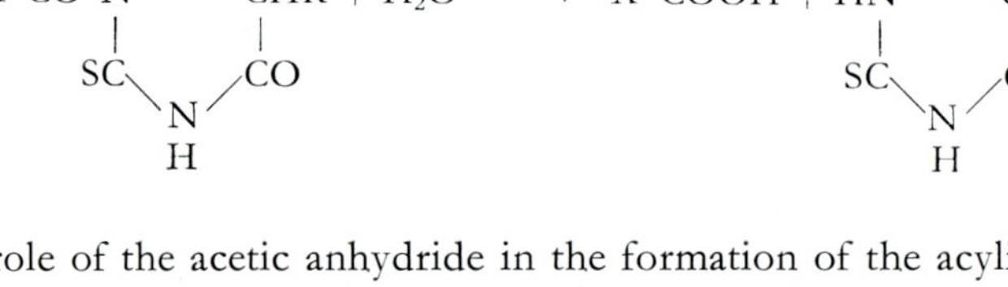

(19)

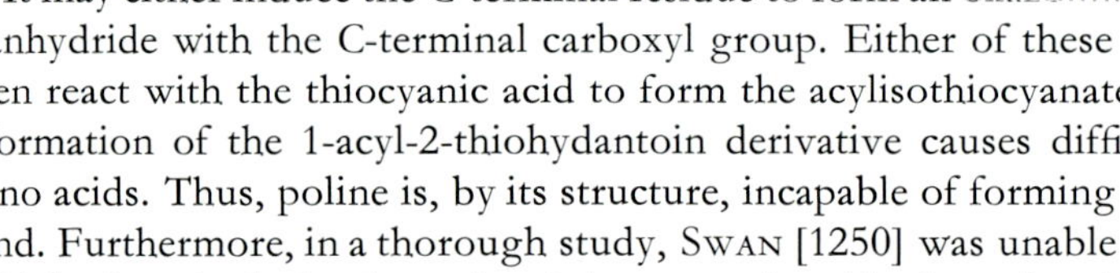

(20)

The exact role of the acetic anhydride in the formation of the acylisothiocyanate is not clear. It may either induce the C-terminal residue to form an oxazolinone, or form a mixed anhydride with the C-terminal carboxyl group. Either of these compounds would then react with the thiocyanic acid to form the acylisothiocyanate.

The formation of the 1-acyl-2-thiohydantoin derivative causes difficulties with some amino acids. Thus, poline is, by its structure, incapable of forming a derivative of this kind. Furthermore, in a thorough study, SWAN [1250] was unable to form the 1-acyl-2-thiohydantoin derivatives of arginine, aspartic acid, glutamic acid, serine and threonine. In N-acylated aspartic and glutamic acids the acetic anhydride induced the formation of cyclic anhydrides, and these were found to be unreactive towards thiocyanic acid. The reason for the failure with the hydroxyamino acids is not clear.

KENNER et al. [681] have suggested a milder way of preparing the acylisothiocyanate. For this purpose they used the reagent diphenyl phosphoroisothiocyanatidate (XIII) which produces the 1-acyl-2-thiohydantoin in excellent yield.

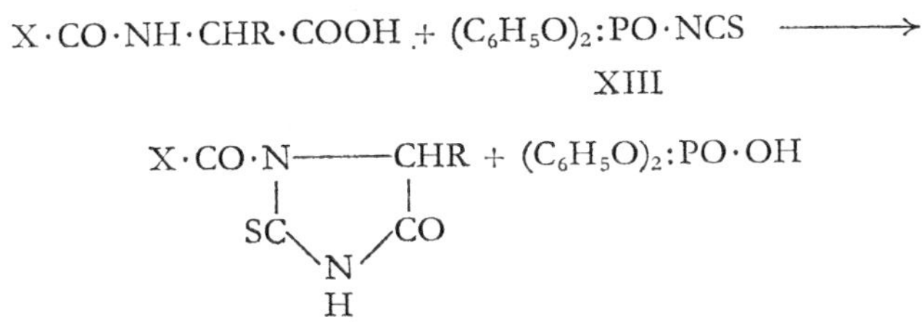

(21)

This reaction has only been applied to small peptides containing amino acid residues with unreactive side chains, and nothing is therefore known about possible side reactions.

The hydrolysis of the 1-acyl-2-thiohydantoin (20) may be done by acids or alkalis. SCHLUCK and KUMPF [1087] used 1 N sodium hydroxide for 3 hrs at room temperature, but WALEY and WATSON [1324] and KJAER and ERIKSEN [702] showed that these conditions were unneccessarily vigorous, 0.01 N sodium hydroxide at room temperature for 1 hr being sufficient to bring about complete hydrolysis.

The application of the reaction, even solely as a C-terminal method, to larger peptides and proteins [60, 1280, 1298, 1324] has produced less than satisfactory results. Perhaps more thorough studies on model peptides are required. Bearing in mind the very reactive nature of acylisothiocyanates, a search for reactions competing with the isomerization reaction (19) could be worthwhile.

A notable extension in the applicability of the method has been reported by STARK [1211]. Milder conditions for the formation of the acylisothiocyanate (XI) as well as for the deacylation step (20), e.g. 0.1 M acethydroxamic acid in 50% aqueous pyridine for 2 hrs at 50° C, have permitted the inclusion of all amino acids except aspartic acid and proline in the scheme. Even with large peptides it was usually possible to carry through a sequential degradation for at least two cycles and quite often for several more. With proteins CROMWELL and STARK [258] preferred to use 12 N hydrochloric acid at room temperature for the cleavage of the acylthiohydantoins (20).

According to recent publications of YAMASHITA [1407] and YAMASHITA and USHIKAWA [1408] improved results are obtained when the peptide is first converted into a mixed peptide-acetic acid anhydride, which is then allowed to react with thiocyanate. The acylthiohydantoins could be cleaved in 0.1 N hydrochloric acid or in the presence of a cat-ion exchange resin. With this method YAMASHITA [1407] was able to split off even aspartic acid and proline residues.

KHORANA [687] has suggested the use of carbodiimides for the removal of C-terminal residues. The peptide is first made to react under very mild conditions with *p*-tolylcarbodiimide ($R'\cdot N:C:N\cdot R'$; $R' = p\text{-}CH_3\cdot C_6H_4$) to form an acylurea (XIV). The latter derivative is then degraded at room temperature with 0.01 N

$$\begin{array}{l} \text{X}\cdot\text{CO}\cdot\text{NH}\cdot\text{CHR} \\ \qquad\qquad\quad | \\ \text{R}'\cdot\text{NH}\cdot\text{CO}\cdot\text{NR}'\cdot\text{CO} \\ \qquad \text{XIV} \end{array} + H_2O \xrightarrow{OH^-} \underset{\text{XV}}{\text{R}'\cdot\text{NH}\cdot\text{CO}\cdot\text{NH}\cdot\text{CHR}\cdot\text{CO}\cdot\text{NH}\cdot\text{R}'} + \text{X}\cdot\text{COOH} \quad (22)$$

$$\xrightarrow{OH^-} \text{R}'\cdot\text{NH}\cdot\text{CO}\cdot\text{NH}\cdot\text{R}' + \text{X}\cdot\text{CO}\cdot\text{NH}\cdot\text{CHR}\cdot\text{COOH} \quad (23)$$

sodium hydroxide in aqueous ethanol (22) when the C-terminal amino acid is split off as an N-tolylcarbamyl toluidide derivative (XV). However, in a quantitatively significant side reaction the acyl urea is hydrolyzed in such a way that the original peptide is reformed (23). Obviously, this interferes with the use of the degradation reaction in a sequential manner.

In the degradation scheme of BAILEY [54, 55] the peptide is first made into an ester, and the ester is then split reductively with a metal hydride, e.g. lithium borohydride (24). The resulting β-hydroxy amide (XVI) is then induced, e.g. by the action of phosphorus oxychloride, to undergo an N $\rightarrow$ O acyl shift (25) to form a β-amino

$$\text{X}\cdot\text{CO}\cdot\text{NH}\cdot\text{CHR}\cdot\text{CO}\cdot\text{OCH}_3 \xrightarrow{\text{LiBH}_4} \underset{\text{XVI}}{\text{X}\cdot\text{CO}\cdot\text{NH}\cdot\text{CHR}\cdot\text{CH}_2\text{OH}} + \text{CH}_3\text{OH} \quad (24)$$

$$\text{X}\cdot\text{CO}\cdot\text{NH}\cdot\text{CHR}\cdot\text{CH}_2\text{OH} \xrightarrow{\text{POCl}_3} \underset{\text{XVII}}{\text{X}\cdot\text{CO}\cdot\text{O}\cdot\text{CH}_2\cdot\text{CHR}\cdot\text{NH}_2} \quad (25)$$

$$\text{X}\cdot\text{CO}\cdot\text{O}\cdot\text{CH}_2\cdot\text{CHR}\cdot\text{NH}_2 \xrightarrow{\text{LiBH}_4} \text{X}\cdot\text{CH}_2\text{OH} + \text{HOCH}_2\cdot\text{CHR}\cdot\text{NH}_2 \quad (26)$$

ester (XVII). The latter may be cleaved reductively (26) to produce a new β-hydroxy amide, on which the degradation cycle may be repeated. The amino alcohol derivatives removed are identified, e.g. by paper chromatography [641]. Little is known about the scope and limitations of the method. Clearly it could not deal with peptides having internal serine and threonine residues, since cleavage of the peptide would occur at these residues.

IV. Enzymatic Degradation Procedures

The use of aminopeptidases and carboxypeptidases for sequence determination may seem attractive because of technical simplicity. However, the simplicity is only apparent, and for theoretical or practical reasons the scope of the approach is limited. Nevertheless, much valuable and perhaps otherwise unobtainable information has been gained in this way. All enzymatic degradation procedures are to a varying degree subject to the same limitations, and these will therefore first be discussed in a general way.

In the chemical methods *all* molecules of a peptide are made to go through a degradation cycle (at least ideally) before a new cycle is started. This is not so in an enzymatic degradation, and here lies its main limitation. The order of the amino acid residues is not determined in a stepwise fashion, but rather from the rate at which the amino acids appear in the digest, i.e. an amino acid appearing faster than another presumably precedes it in sequence. The differences in rate of release tend to become smaller as the degradation proceeds, and ambiguity then ensues. Although the digestion technique in itself is simple, the rate measurements necessitate a supplementary technique for accurate amino acid analysis.

Another limitation lies in the specificity of the enzyme. So, for example, proline and arginine residues in C-terminal positions, are not released by carboxypeptidase A, and the degradation will therefore come to a halt at these residues. Furthermore, the rate at which a peptide bond is split may vary greatly depending on the particular amino acid residue occupying the terminal position. If a rapidly released amino acid follows in sequence a slowly released amino acid, the two amino acids will appear in the digest at similar rates.

A serious practical difficulty is for obvious reasons the contamination of the enzyme preparation with other proteolytic enzymes. This must be guarded against by the use of properly purified enzyme preparations and by the addition of specific inhibitors. Short peptides are less likely to be affected by traces of endopeptidases, but the likelihood of a susceptible peptide bond being present increases with the size of the substrate molecule.

The enzymatic degradation techniques have an advantage over the chemical techniques in that they provide information about the optical configuration of the amino acid released. Only L-forms are released, the occurrence of an amino acid residue of the D-series constituting an absolute blockage to further degradation. This property of the enzymes is made use of in checking synthetic peptides for absence of racemization [568].

Procedures for the preparation and the activity measurement of the enzymes are described in another chapter (cf. "End Group Determinations"), and these descriptions will not be repeated here.

A. From N-Terminus

The aminopeptidases most commonly used in the degradation of peptides from the N-terminal end is the leucine aminopeptidase from hog kidney (EC 3.4.1.1) described by Spackman et al. [1196]. The properties of the enzyme have been reviewed by Smith and Hill [1175]. Leucine aminopeptidase has not been prepared in a chemically homogeneous form. Less pure preparations have been found to display other proteolytic activities, e.g. carboxypeptidase [1184] and various endopeptidase [388, 448] activities and also glutaminase activity [566]. The endopeptidase activity can be considerably suppressed, but not entirely eliminated, by treating the enzyme preparation with DFP and iodoacetic acid [388]. Therefore this type of enzyme preparation cannot be used with confidence in determining N-terminal sequences of proteins and large peptides. On the other hand, shorter peptides, which are less likely to contain bonds susceptible to endopeptidase activity, may be degraded solely in accordance with the specificity of leucine aminopeptidase. However, when possible the sequence deduced should be corroborated by independent evidence. A different method for the preparation of leucine aminopeptidase has been described by Himmelhoch and Peterson [541]. Here the enzyme preparation is claimed to be free of endopeptidase activity.

Leucine aminopeptidase splits the peptide bond adjacent to the free α-amino group of the peptide. This peptide bond must be of the amide type. If the N-terminal residue is linked to a proline residue (imide bond) no cleavage takes place. All other α, α'-peptide bonds are split, but the rate varies greatly depending on the amino acid side chain. Amino acids with large non-polar side chains, e.g. leucine, are released most readily, whereas amino acids with polar side chains and proline are slow to come off [1176b].

The enzyme is a metallo-enzyme and requires for its activity and stability Mg^{2+} or Mn^{2+} ions. It loses its activity outside the pH range 7 to 9. A solution of the enzyme preparation, 0.005 M in Tris (pH 8.0) and 0.005 M in magnesium chloride, may be stored in the frozen state for long periods without loss of activity [536].

ADFP-treated preparation is commercially available. It is supplied as a suspension in 75% saturated ammonium sulfate solution, 0.005 M in Tris (pH 8.0) and in magnesium chloride.

Before use the enzyme preparation is dialyzed against a 0.005 M Tris — 0.005 M $MgCl_2$ buffer (pH 8.5) in the cold. Any insoluble material is centrifuged off. The approximate concentration of the protein solution is determined spectrophotometrically using the formula

$$A_{280} \times 1.2 = \text{mg/ml}$$

where A_{280} is the absorbance at 280 mμ.

The correct amount of enzyme to be used in a digestion depends on several factors, e.g. the nature of the terminal amino acid residue and the amount of peptide, and is found by trial in small scale experiments. However, the range of 3 to 30 enzyme units/μmole of peptide may serve as a guide. For the digestion the following procedure is suggested.

The peptide is dissolved (1 μmole/ml) in 0.05 M Tris — 0.0025 M $MgCl_2$ buffer (pH 8.5) in a stoppered test tube. The solution is incubated at 40° C, and a suitable aliquot of the solution of the activated enzyme is added. Samples (0.1 ml) are withdrawn at the start and at intervals during the next 10 to 20 hrs. A blank without substrate should be included.

The withdrawn sample is transferred to a glass-stoppered centrifuge tube containing the sample dilution buffer (2.1 ml) used in the amino acid analysis. The solution is quickly frozen and stored for later analysis. If necessary, the solution is cleared by centrifugation before analysis.

In calculating the results of the amino acid analysis it should be realized that the analytical procedure does not separate serine, asparagine and glutamine when the ordinary sodium citrate buffers according to Spackman et al. [1198] are used. However, the lithium citrate buffers according to Benson et al. [84] will separate even these three amino acids. It should also be borne in mind that an undigested small peptide may simulate an amino acid in the analysis. This should be suspected if the height of a peak decreases with time, or if it is unduly broadened. The Tris buffer produces a peak in front of lysine on the short column [549].

A plot of concentration *versus* digestion time is made for each amino acid. The relative rates of release indicate the order in which the amino acid residues occur in the sequence, as has already been discussed.

A second aminopeptidase from hog kidney has been isolated from the microsomal fraction by Wachsmuth et al. [1321, 1322]. This enzyme, called aminopeptidase M, is more stable, does not require metal ion activation, and splits the peptide bonds of the various amino acids at a more uniform rate. The enzyme is commercially available.

Finally, it may be mentioned that procedures have been worked out for the sequencing of peptides with dipeptidyl aminopeptidase I [196, 197, 784]. The enzyme releases dipeptides sequentially from the N-terminal end of the peptide. The sequence of the dipeptides is determined, and their order within the peptides is obtained from the release rates of the dipeptides [196]. Alternatively, the N-terminal amino acid of the substrate is first split off chemically. A subsequent digestion with the dipeptidyl aminopeptidase produces overlap peptides, and these can be used in deducing the sequence [784].

B. From C-Terminus

Three different types of carboxypeptidases have been used for the release of C-terminal amino acids. Two of these, carboxypeptidases A and B, are obtained from pancreatic glands [48, 365], and the third type is isolated from a wide range of plant sources [206, 521, 648, 1205, 1293, 1428—1431]. The properties of carboxypeptidases of animal origin have been reviewed by NEURATH [919]. The first plant carboxypeptidase to be described was isolated from the peel of citrus fruits by ZUBER [1428, 1429, 1431] and was called carboxypeptidase C. A similar carboxypeptidase, also termed C, was obtained from orange leaves [1205, 1293]. Other plant carboxypeptidases have been isolated from French bean leaves (phaseolain) [206], *Penicillium ianthinellum* (penicillocarboxypeptidase-S) [648] and yeast (carboxypeptidase Y) [521]. The designations of the carboxypeptidases seem somewhat confusing.

Although all three carboxypeptidases split the peptide bond adjacent to the C-terminal α-carboxyl group, they differ in their specificities. Bovine carboxypeptidase A releases all amino acids except proline and arginine, but the rate differs greatly depending on the amino acid side chain. The amino acids with aromatic or large non-polar side chains are split off faster than those with short or polar side chains. Released at the slowest rate are glycine and the acidic amino acids. Lysine is released slowly. It has been reported that the rate of release of the acidic amino acids may be increased by lowering the pH of the digestion [442, 1169], and conversely that lysine is split off more rapidly at a higher pH [943]. A C-terminal homoserine (resulting from the cleavage of a peptide chain at a methionine residue by cyanogen bromide) is readily released [25].

The specificity of carboxypeptidase B [416b] has not been as well defined. Like carboxypeptidase A it is inert to C-terminal proline. The enzyme releases preferentially arginine and lysine. The activity towards other amino acids is not quite clear. The most highly purified carboxypeptidase B preparations from pork are said to be uncontaminated by carboxypeptidase A activity [366], whereas a pure preparation of the enzyme from beef shows a considerable and apparently intrinsic activity against typical substrates of carboxypeptidase A, e.g. carbobenzoxyglycyl-L-phenylalanine [1384].

The carboxypeptidases of plant origin seem to have a wider specificity than the two enzymes of animal origin [206, 521, 648, 1205, 1293, 1428, 1429, 1431]. Thus, all plant carboxypeptidases mentioned will release proline. The rate of release of the other amino acids does not vary greatly, excepting glycine [206, 521, 648, 1205, 1429, 1431], aspartic acid [206, 521] and glutamic acid [206], which are released only slowly by some of the enzymes.

With carboxypeptidase A it is usually not possible to determine sequences longer than 3 to 4 residues. The limitations here are those inherent in the enzymatic method itself. However, the fact that some amino acids are not released at all (proline and arginine) and others at a very slow rate (glycine and the acidic amino acids) also effectively limits the length of the degradation. The use of carboxypeptidase A in combination with carboxypeptidase B removes the obstacle presented by arginine residues, and speeds up the otherwise slow release of lysine.

The use of carboxypeptidase A and B for the elucidation of large structures (including proteins) has been more successful, and led to fewer false results than the use of leucine aminopeptidase for the same purpose. This is probably due to the fact that extraneous proteolytic activities in carboxypeptidase preparations are better known, and more easily suppressed. A protein substrate in its native conformation is often found to be resistant to the action of carboxypeptidases, whereas the unfolded structure is easily digested. This unfolding may be brought about by breaking the disulphide bonds and converting the cysteine residues to stable derivatives by established methods.

Preparations of bovine carboxypeptidase A (EC 3.4.2.1) and porcine carboxypeptidase B (EC 3.4.2.2) are commercially available. These preparations are contaminated by endopeptidases, mainly chymotrypsin and trypsin, and may not be used as such. The trypsin and chymotrypsin activities are inhibited by treatment with DFP, and preparations so treated are procurable. Also, it should not be relied on that carboxypeptidase A preparations are entirely free of carboxypeptidase B and *vice versa*. Both enzymes require zinc ions for their activity [366, 1301 b].

The crystalline DFP-treated carboxypeptidase A preparation is obtained from the manufacturer as an aqueous suspension to which toluene has been added as a preservative. The preparation may be stored in the cold for long periods without loss of activity. The method of solubilizing the enzyme and the condition of the digestion described here are essentially those of Hirs et al. [549].

Before use the enzyme crystals are washed with distilled water to remove traces of free amino acids. The preparation is then suspended in 2 M NaCl (6 ml), and brought into solution by the addition with stirring of 0.01 N NaOH to pH 10. The pH is immediately reduced to 8.3 by the addition of 1 M sodium phosphate buffer (0.6 ml), pH 8.0. After standing for 2 hrs at 4° C the solution is clarified by centrifugation. The concentration of the enzyme solution is determined spectrophotometrically using the formula

$$A_{278} \times 0.51 = \text{mg/ml}$$

where A_{278} is the absorbance at 278 mμ.

The suitable ratio between enzyme and substrate is found by trial. The range of 0.1 to 0.5 mg enzyme per μmole of substrate serves as guide.

The peptide is dissolved (1 μmole/ml) in 0.1 M Tris buffer at pH 8.3, and the solution is incubated at 40° C in a stoppered test tube. A suitable aliquot of the enzyme is added. A blank without substrate should be included.

Sampling, treatment of samples and analysis are the same as for leucine aminopeptidase digestions (see above).

The commercial DFP-treated carboxypeptidase B preparation is supplied as a frozen solution in 0.1 M sodium chloride. The protein concentration is determined from the formula

$$A_{278} \times 0.47 = \text{mg/ml}$$

where A_{278} is the absorbance at 278 mμ.

A suitable ratio of enzyme to substrate is approximately 5 enzyme units/micromole of substrate. The conditions of the digestion, the sampling, the treatment of the samples and the analysis are the same as for carboxypeptidase A. As has already been pointed out, the two enzymes may be used in conjunction.

A preparation of the carboxypeptidase C of orange peels or leaves, lyophilized from sodium citrate buffer, is now commercially available. Whether the enzymes from peels and leaves are identical is not clear [1205]. The method described here was proposed by TSCHESCHE and KUPFER [1293]. A suitable enzyme to substrate ratio is 500 to 5000 enzyme milliunits/μmole of substrate.

The peptide is dissolved (1 μmole/ml) in 0.05 M sodium citrate buffer at pH 5.3, and the solution is incubated at 30° C. A suitable aliquot of enzyme solution is added.

Sampling, treatment of samples and analysis are essentially the same as for leucine aminopeptidase (see above).

V. Conclusion

The task of determining amino acid sequences is likely to remain with us for quite some time. It is perhaps worth contemplating that the amount of DNA in a mammalian germ cell is equivalent to 2 billion base triplets, which is then the maximum coding capacity for the sequential ordering of amino acid residues. One may contrast this figure with the estimate that the sequential ordering of no more than a few thousand amino acid residues is at present published annually. Perhaps (and hopefully) only a small fraction of the coding capacity is actually used in protein synthesis, the rest being held in reserve.

However, all speculations apart, it is obvious that the amount of structural work waiting to be done is great indeed. It is therefore important that new developments in technique shoul d permit not only accurate but also rapid analysis of protein structures. Other merits of a method, e.g. great sensitivity, are here of secondary importance. In fact, if the advantage is bought at the price of a more unwieldy technique, the method would represent a retrograde step.

Chapter 9

Analysis of Amino Acid Phenylthiohydantoins by Gas Chromatography and High Performance Liquid Chromatography

JOHN J. PISANO

I. General Methods for PTH Identification

Amino acid phenylthiohydantoins (PTHs) are formed in Edman's now classic technique for determining the primary structure of peptides and proteins. Introduced over 20 years ago, it is still the most effective and widely used method for sequence analysis. While X-ray crystallographic and mass spectrometric approaches remain attractive because they do not involve tedious step-wise analyses, their utility is restricted by requirements inherent in the methods. Thus, for the X-ray technique, suitable crystals and adequate stability during X-ray bombardment are unattainable for many proteins and peptides. Even with good crystals, unambiguous distinction of all amino acids is often impossible with obtainable data. Limited volatility is the main deterrent to the wider use of mass spectrometry in peptide analysis. Greater use of the method will follow improvements in chemical techniques for converting peptides to suitable derivatives, but it is unlikely that it will ever be possible to analyze peptides containing more than about 15 residues.

Analysis of PTHs formed in the step-wise Edman degradation was first achieved by paper [146, 1165] and thin-layer chromatographic methods [222, 320]. Although thin-layer chromatography remains an important method of analysis, many investigators have sought alternative methods because the procedure is tedious, difficult to quantitate and limited in resolving power. Identification of some derivatives requires as many as four different solvent systems.

Several valuable indirect procedures have been developed to overcome these difficulties including hydrolysis of the PTHs [9, 1181], the subtractive-Edman [549, 720] and the dansyl-Edman methods [438, 439]. All three procedures circumvent the need to determine the amino acid PTH directly. In the hydrolytic method, the PTH [9] or thiazolinone [1181] is hydrolyzed under carefully controlled conditions and the recovered amino acid is identified and quantitated by conventional means. A major advantage of this method is the lack of interference from contaminants present in the sample or arising from the buffer and solvents. Larger aliquots of the PTH sample may be taken for analysis than with direct methods of PTH analysis which are subject to interference by gross contaminants (Ettore Appella, personal communication). Possible disadvantages include the labor and expense of amino acid analysis, the need for two hydrolytic methods and the possible uncertainty of asparagine and glutamine determinations because they are based on ammonia production [1181].

In the subtractive-Edman method [549, 720] the sequence is deduced by amino acid analysis of an aliquot of the shortened peptide after the cleavage and conversion steps of the Edman method. While this approach will undoubtedly continue to be highly useful it has certain drawbacks. It is limited to small peptides where the loss of a single residue may be determined with confidence. Loss of sample at each step, accumulation of blocked peptides which obscure the analysis, expensive and time-consuming amino acid analysis, and the need to use other methods to identify asparagine, glutamine and tryptophan residues are other limitations of the subtractive method.

In the more sensitive dansyl-Edman procedure, an aliquot of the shortened peptide is taken for identification of the new N-terminal amino acid using the dansyl technique [438, 439]. This valuable method has not worked well with proteins which are insoluble in the sodium bicarbonate-acetone buffer or have sterically hindered N-terminal amino acids. Other limitations include the need for additional methods to determine asparagine, glutamine and tryptophan residues and the difficulty in quantitating dansyl amino acids.

The ideal method for sequence analysis is one which would be applicable to proteins or large fragments and could be automated. Neither the subtractive-Edman nor the dansyl-Edman indirect methods are suited for this purpose.

Direct analysis of the PTHs formed at each step of the Edman procedure is, in principle, the most logical approach in sequencing. A further stimulus to the development of suitable methods for direct PTH analysis was the invention of the protein sequenator for automated Edman degradation [317]. In an attempt to overcome the limitations of the TLC and hydrolytic methods for PTH analysis cited above, gas chromatographic and more recently high performance liquid chromatographic methods [1427] have been developed. Both procedures offer high speed sensitivity, specificity and ease of quantitation.

II. Gas Chromatography

The first demonstration that most amino acid PTHs could be successfully analyzed by gas chromatography (GC) also revealed difficulties with the seryl, threonyl, asparaginyl, glutaminyl, lysyl and arginyl derivatives [985]. These difficulties were overcome [453, 501, 980, 981, 982, 984] by the use of new thermally stable and more polar polysiloxane liquid phases and the powerful silylating reagents, N,O-bis-(trimethylsilyl)acetamide (BSA) [703] or N,N-bis(trimethylsilyl)trifluoroacetamide (BSTFA) [1210] which readily convert the less volatile and unstable PTHs to trimethylsilyl derivatives having excellent chromatographic properties. The very polar argininyl PTH does not form a stable silyl derivative and cannot be analyzed by GC.

A. Equipment

Several manufacturers offer suitable instruments. However, glass columns and on-column injection are recommended because destruction of PTHs occurs in metal injection ports and columns. A hydrogen flame detector is also recommended because of its high sensitivity, thermal stability, wide linear dynamic range and ruggedness. To accommodate the large number of samples produced in the automated Edman degradation, the capacity for the analysis of two samples simultaneously is recommended. This may be accomplished most economically with a double column oven, two hydrogen flame detectors, electrometers and recorders and one temperature programmer. With this instrument package and simultaneous injection, one has the capacity of two chromatographs but saves the cost of a separate column oven and temperature programmer. Busy laboratories may employ two such instruments and take full advantage of the many complementary columns which can be employed.

B. Materials

The support, Chromosorb W, 100 to 120 mesh, stationary phases DC-560 (or the equivalent SP400), OV-210, OV-225, silylating reagents BSA, BSTFA, and dichlorodimethylsilane may be purchased from various supply houses (e.g. Supelco, Inc., Bellefonte, Pa.; Pierce Chemical Co., Rockford, Ill.).

C. Preparation of the Support

The preparation of the support [584] has been modified to include a Na_2CO_3 prewash which slightly improves column efficiency [981]. Chromosorb W, 50 g, is mixed with 500 ml of 0.5 M Na_2CO_3 in a 1 l beaker. After standing overnight fine particles are removed by several decantations using distilled water. Concentrated HCl, 500 ml, is added to the almost neutral support and after 16 to 24 hrs standing with occasional swirling, the fine particles are again removed by several decantations using distilled water. The support is dried at 140° and, while still warm, a 25 g portion is transferred to a 1 l flask. Approximately 200 ml of a 5% solution (v/v) of dichlorodimethylsilane in toluene is added and the mixture degassed by the use of an aspirator. The flask is gently swirled 2 or 3 times during degassing and the 10 to 15 min reaction period. The excess reagent is decanted and the support is rinsed 3 times with anhydrous (reagent grade) toluene. Since the support may contain bound reactive silyl chlorides it should be protected from atmospheric moisture. This is conveniently achieved by keeping the support wet with toluene and carrying out the washings without interruption. Decativation is effected by the addition of 300 ml of anhydrous methanol. After standing 10 to 15 min, the methanol is decanted and the support is rinsed with anhydrous methanol until the rinses are clear. The mixture is then filtered using a course sintered-glass funnel, the support is rinsed with acetone while in the funnel, air-dried and finally dried in an oven at 140°.

Removal of the fine particles produced during the above procedures is recommended for the preparation of efficient columns. After the acid and base washes it may be necessary to decant 25 times with distilled water. Chromosorb W is a fragile and support is easily crushed. Since crushed material contains newly exposed active sites, it should not be mixed with intact support.

The filtration techniques is used to coat the support with stationary phase [584]. The amount of stationary phase used is expressed as its percentage (*w/v*) in solvent.

Thus a 10% DC-560 packing is prepared with 10 g of DC-560 made up to 100 ml with acetone. Usually, 75 ml of solution is added to 5 g of support in a 125 ml filter flask. The mixture is degassed by gentle swirling while under reduced pressure (aspirator), filtered on a 150 ml sintered glass funnel until apparently dry, transferred to a dish and thoroughly dried at 140°. The volume of coating solution is not critical but should be sufficient to allow transfer of the support (with swirling) to the sintered glass funnel.

Of the numerous stationary phases tested, none is capable of separating all the PTHs. However, two phases, DC-560 (or SP400) and XE-60 (or the similar OV-225) are complementary; PTHs unresolved with one phase are resolved with the other.

A superior single column is obtained with a blend of phases referred to as CFC (18) consisting of equal volumes of acetone solutions of 5.5% SP400, 4% OV-210, and 0.5% OV-225.

D. Preparation of Columns and Chromatographic Conditions

Glass columns approximately 4 feet × 2 mm i.d. and glass wool plugs are silylated like the support with dichlorodimethylsilane in toluene. Columns are filled with the reagent and the glass wool is soaked in it for about 15 min. When filling the columns, they are gently tapped to promote even packing. A vibrator should not be used as it may damage the support or cause it to pack too tightly giving prohibitively slow flow rates.

Columns are conditioned using an initial temperature of approximately 50° and a helium flow rate of 150 ml/min. After about 30 min the temperature is raised at the rate of 0.5°/min until it reaches 290°, where it is held for at least 16 hrs or until the baseline rise is less than about 10% full scale when the temperature is increased from 170 to 290°. Helium is superior to nitrogen and argon carrier gasses, giving better resolution and a wider latitude of flow rates without affecting efficiency [981]. Some laboratories may prefer nitrogen because it costs less and still gives satisfactory results. Ultrapure gasses have been used throughout; lower grades of purity have not been tested.

Analyses are usually performed with the injector temperature at 250 to 270°. At lower temperature, e.g., 200°, volatilization of asparaginyl, glutaminyl, tyrosyl, lysyl, histidyl and tryptophanyl PTHs may be incomplete. Temperatures higher than 270° may cause decomposition of glutaminyl, lysyl, and histidyl PTHs. At 300° most PTHs decompose. The hydrogen glame detector bath is held at 300°. Most analyses are performed at column temperatures ranging from 165° to 290°. The optimum temperature should be determined by the investigator. Electrometer controls are usually set so that a full scale deflection corresponds to about 3×10^{-10} A with a 5 mV recorder.

It is not uncommon for the first few samples to tail on a new column. After a few analyses the columns stabilize and may be used for months or even years if not abused by the use of excessive temperatures or dirty samples. The unsuitability of a column is determined by the increase in baseline rise during temperature programming and the decrease in resolution or yield, or the disappearance of peaks. Contaminated columns are usually discarded, but removal of the top inch of packing will occasionally improve the column.

E. Standard Solutions

A suitable solvent for all the PTHs has not been found. Ethyl acetate and ethylene dichloride, used by Edman, will not dissolve sufficient crystalline asparaginyl, glutaminyl and histidinyl PTHs to give the convenient concentration of 1 mg/ml. Methanol is a good solvent but it cannot be used when the standards are silylated (see below). Although reagent grade acetonitrile, N,N-dimethylformamide and pyridine are also good, PTHs are unstable in these solvents. In ethyl acetate, ethylene dichloride and methanol, however, they are stable for months when stored in the dark below 5°. Derivatives obtained from a degradation are much less stable than crystalline standards in any solvent. Upon storage overnight some derivatives (especially seryl PTH) may be completely destroyed, presumably due to contaminants in the sample.

F. Silylation of Phenylthiohydantoins

Because amino acid PTHs differ greatly in their chemical and chromatographic properties, it is useful to divide them into three groups (Table 9-1). Group I derivatives are most volatile and generally give symmetrical peaks. Group II derivatives are least volatile and, with the exception of tryptophan, show the greatest tendency to adsorb to the column packings giving tailing peaks and low responses. Group III derivatives include those (aspartic, glutamic and cysteic acids) which must be converted to the volatile silyl derivatives before analysis and others which when silylated have significantly better chromatographic properties [453, 501, 980, 981, 982, 984]. Silylation of Group I and II derivatives can provide useful confirmatory data. One laboratory [530] prefers to silylate all PTHs routinely and thereby reduce the number of analyses necessary to identify an unknown. However, chromatograms are often much more complex following silylation and occasionally members of Groups I and II may be missed if not analyzed before silylation.

Table 9-1. Grouping of amino acid PTH derivatives according to gas chromatographic behavior

Group I	Group II	Group III
Alanine	Asparagine	Aspartic acid
Glycine	Glutamine	S-Carboxymethylcysteine
Valine	Tyrosine	Cysteic acid
Leucine	Histidine	Glutamic acid
Isoleucine	Tryptophan	Lysine
Methionine		Serine
Proline		Threonine
Phenylalanine		

Group I amino acids are most volatile and generally give symmetrical peaks. Members of Group II are least volatile and with exception of tryptophan show the greatest tendency to adsorb to the column packing giving tailing peaks and low responses. Group III derivatives include those which must be silylated before analysis (aspartic, glutamic, and cysteic acids) and others which, when silylated, have significantly better chromatographic properties.

Silyl derivatives of PTHs have excellent chromatographic properties, and they are easily prepared using BSA or BSTFA. These reagents readily silylate PTHs at sites containing exchangeable protons [501, 982]. However, variable reactivities and stabilities of the derivatives makes it necessary to compromise on the conditions for silylation [980]. Most convenient is "on-column" silylation in which the reagent and then the sample (about equal volumes) are withdrawn into the syringe and injected onto the column. Ethyl acetate, ethylene dichloride, acetonitrile and N,N-dimethylformamide are suitable solvents for the silylation reaction. Aspartyl, glutamyl and lysyl PTHs react well, but the serine and threonine derivatives barely react. Most uniform results are obtained when all the PTHs are reacted for 15 min at 50° in a sealed tube. Although they have high thermal stability, the silyl derivatives are chemically unstable and should be analyzed immediately.

A brief communication on the preparation and GC analysis of trifluoroacetylphenylthiohydantoins did not reveal any advantages of trifluoroacetylation over silylation [1036]. In another approach, the use of the more volatile and organic soluble pentafluoraphenylisothiocyanate (PFPITC) as an alternative to PITC has been investigated but it is too early to determine the eventual utility of this reagent. Alternate use of PFPITC and PITC in a degradation could simplify interpretation of results [763].

G. Chromatography

One of the most widely used columns is 10% DC-560 or SP400 (Fig. 9-1). It is particularly suited for the Groups I and III derivatives; some members of Group II,

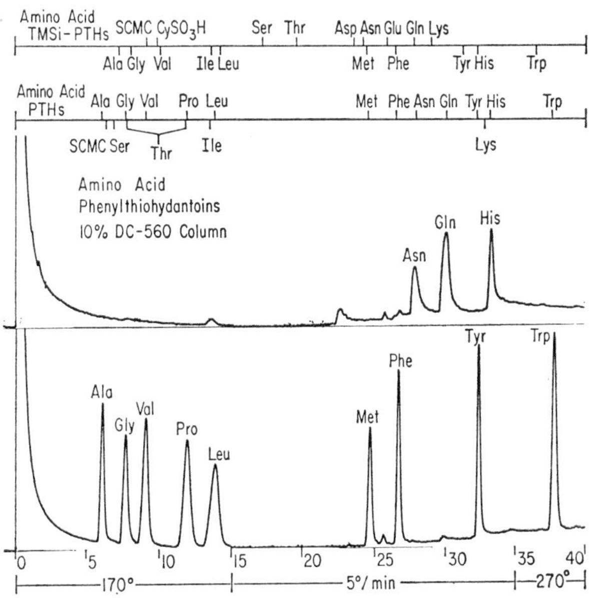

Fig. 9-1. Separation of phenylthiohydantoins on a 4 ft × 2 mm i.d. 10% DC-560 column. Sample sizes: 1 μg each alanyl, glycyl, valyl, prolyl, leucyl, methionyl, phenylalanyl, and tyrosyl PTHs; 2 μg tryptophanyl, asparaginyl, glutaminyl, and histidyl PTHs. Helium flow 65 ml/min. (From Ref. [981])

notably asparaginyl, glutaminyl, and histidinyl PTHs, tail significantly. Group II derivatives separate well and with little tailing on a 2 foot 1% XE-60 column (Fig. 9-2), which also better separates isoleucyl from leucyl PTH. Together these two columns may be used to identify all the PTHs (except argininyl PTHs). An

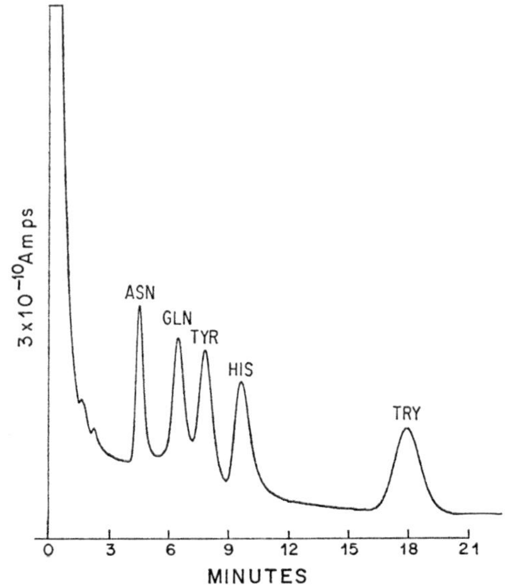

Fig. 9-2. Separation of Group II derivatives. Conditions: column, 2 ft × 2 mm, 1% XE-60 on 100 to 120 mesh Gas Chrom P; column temperature, 265°; argon flow, 50 ml/min; sample size, 3 μg of the asparaginyl, glutaminyl, an histidyl PTHs, 1 μg of the tyrosyl and tryptophanyl phenylthiohydantoins. (From Ref.[980])

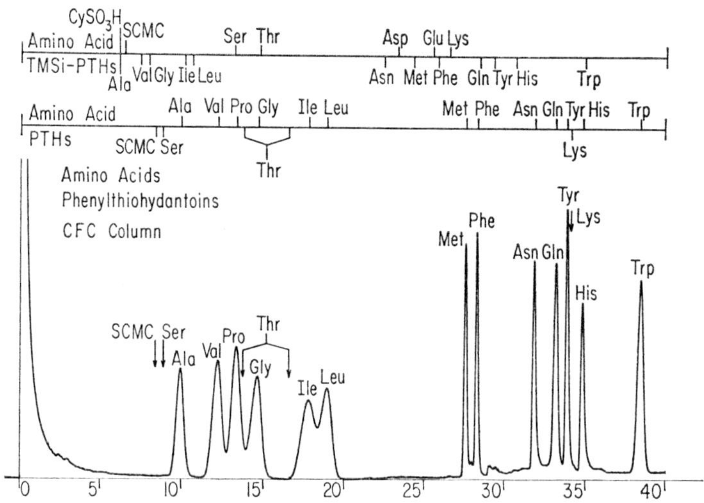

Fig. 9-3. Separation of amino acid phenylthiohydantoins on a 4 ft × 2 mm i.d. CFC column. Sample sizes: 1 μg each alanyl, valyl, prolyl, glycyl, isoleucyl, leucyl, and tyrosyl PTHs; 2 μg each asparaginyl, glutaminyl, histidyl and tryptophanyl PTHs; 0.5 μg each methionyl and phenylalanyl PTHs. Helium flow 105 ml/min. (From Ref. [981])

alternative to the use of two columns is a single column containing the CFC blend. Only the tyrosyl and lysyl PTHs are poorly separated, but their silyl derivatives are completely resolved (Fig. 9-3). The behavior of prolyl PTH on all columns is unique inasmuch as its position, relative to the other derivatives, is temperature dependent. If a newly prepared CFC column does not adequately resolve prolyl PTH from valyl

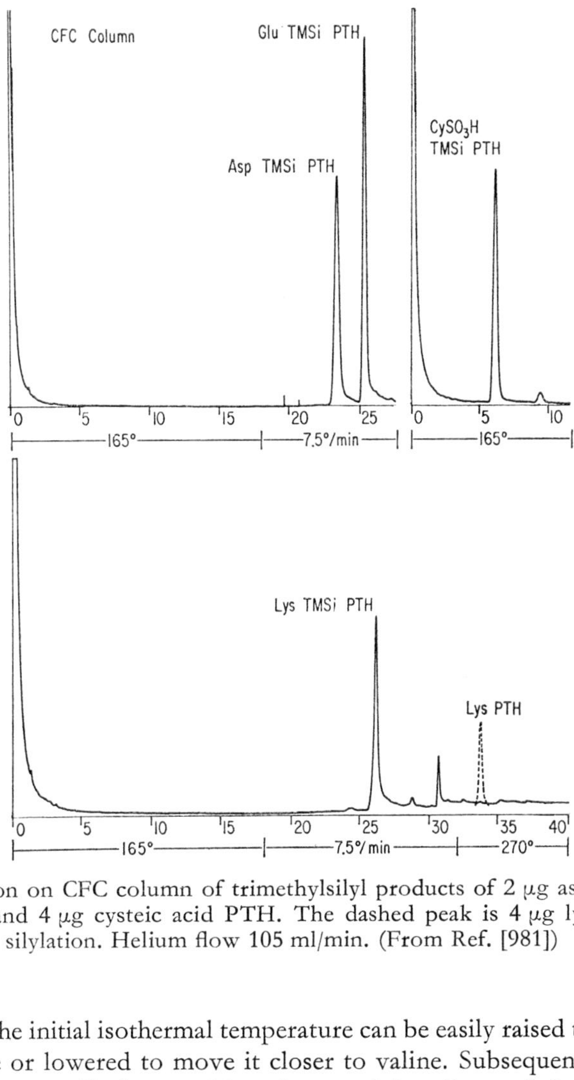

Fig. 9-4. Separation on CFC column of trimethylsilyl products of 2 μg aspartyl, glutamyl, and lysyl PTHs and 4 μg cysteic acid PTH. The dashed peak is 4 μg lysyl PTH before silylation. Helium flow 105 ml/min. (From Ref. [981])

or glycyl PTH, the initial isothermal temperature can be easily raised to move proline closer to glycine or lowered to move it closer to valine. Subsequent adjustment of the helium flow rate will then position the remaining Group I derivatives to any desired time without affecting column efficiency. Silylated Group II thiohydantoins are easily identified on this column (Figs. 9-4 and 9-5).

Although the author has experienced little difficulty in several years of trials, other laboratories have not been able to prepare satisfactory CFC columns routinely and have come to rely instead on the DC-560 or SP400 columns. It is difficult to understand why problems have arisen with the blend, but improper preparation of the support has been suspected in many cases as evidenced by excessive peak tailing. Another cause may be poor quality control in the manufacture of liquid phases, which are usually prepared for other purposes. A useful alternative to the CFC blend is, as stated earlier, the two-column system employing an XE-60 (or OV-225) and a DC-560 (or SP400). These two columns may be housed in the same column oven, and after suitable adjustments in length, operated at the same temperatures. This combination has been used extensively in the author's laboratory.

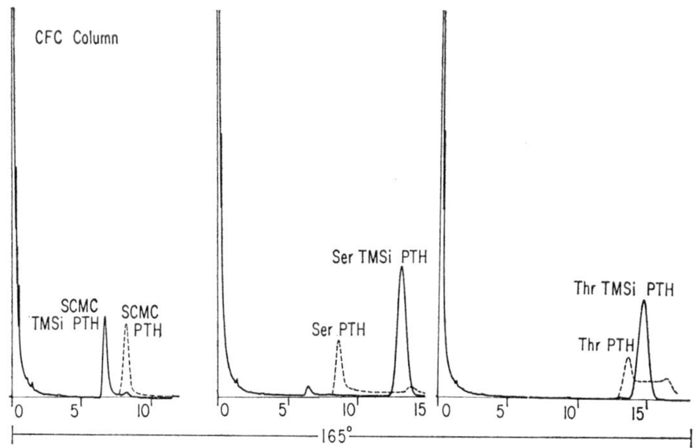

Fig. 9-5. As for Fig. 4: 1 to 2 μg SCMC, seryl, and threonyl PTHs before (dashed peaks) and after silylation

H. Methylthiohydantoins (MTHs)

Several laboratories have considered substituting methylisothiocyanate (MITC) for phenylisothiocyanate used by Edman [31, 50, 285, 343, 1024, 1223, 1224, 1302, 1335]. The MTHs are more volatile and have a wider range of volatilities than the PTHs. This has led to the claim that their identification by GC and mass spectrometry (MS) can be made with greater ease and assurance. Excellent separations (Fig. 9-6) of MTHs and their trimethylsilyl derivatives have been achieved on packed [745, 981] and on capillary columns [343]. Advocates of capillary columns sometimes believe that the calculated higher number of "theoretical plates" attainable with capillary columns means better resolution. However the data often do not support the theory, and a comparison of the results with MTHs is another example. The preparation of capillary columns is difficult and few laboratories choose them when packed columns are applicable.

Although the MTHs are at least as easy to identify and quantitate as the PTHs, the merits of MITC in sequence analysis are unknown if indeed any exist. In the most careful study of MITC [1023], rather severe conditions were recommended for the sequential degradation of polypeptides. Coupling was performed at 60° for 2 to 3 hrs and simultaneous cleavage and conversion at 60° for 1 hr in TFA. These conditions are likely to cause non-specific cleavages of acid-sensitive bonds, the

O-acylation of seryl and threonyl hydroxyl groups (followed by O → N acyl shift when the amino acid becomes N-terminal), cyclization of N-terminal glutamine and conversion of α-asparaginyl and aspartyl bonds to the β-linkage. Sequential degradation would cease when any of the last three reactions occurred [320]. In the Edman procedure, coupling is done at the milder 40° for 1 hr and cleavage at 40° for 15 min.

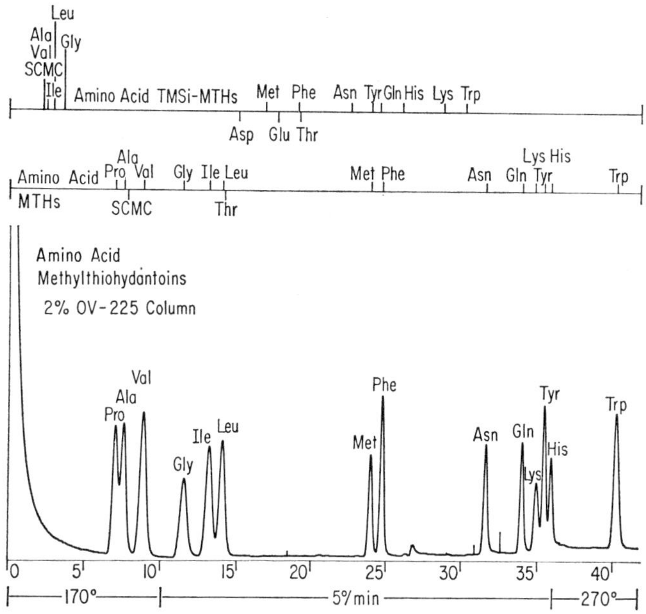

Fig. 9-6. Separation of amino acid methylthiohydantoins on 2% OV-225 column. Sample sizes: 1 μg each. Helium flow 60 ml/min. The initial isothermal temperature must be lowered in order to identify the trimethylsilyl derivatives of the earliest amino acid MTHs. (From Ref. [981])

I. Applications

Numerous applications of the GC method for amino acid PTH analysis have been reported. Two recent impressive applications are the amino acid sequence determinations of bovine parathyroid hormone, Fig. 9-7 [173], and the human high-density apolipoprotein, apolp-Gln-II, Fig. 9-8 [174]. The analyses were done with DC-560 and CFC columns.

Samples obtained from the sequencer may contain contaminants and/or buffer which could quickly foul a column. Such impurities are easily spotted at the time the thiazolinone solution is evaporated to dryness. They may be separated by extraction of the thiazolinone from an acid solution with ethyl acetate. The following procedure has been used successfully (E. Appella, personal communication). After evaporation of the chlorobutane with a nitrogen stream, 0.2 ml of 1 N ultrapure HCl is added and mixed thoroughly (Vortex). To the solution is added 1 ml of ethyl acetate and, after vortexing, the phases are separated by centrifugation. The ethyl acetate is removed and the residue extracted again. The ethyl acetate solutions are

Bovine PTH

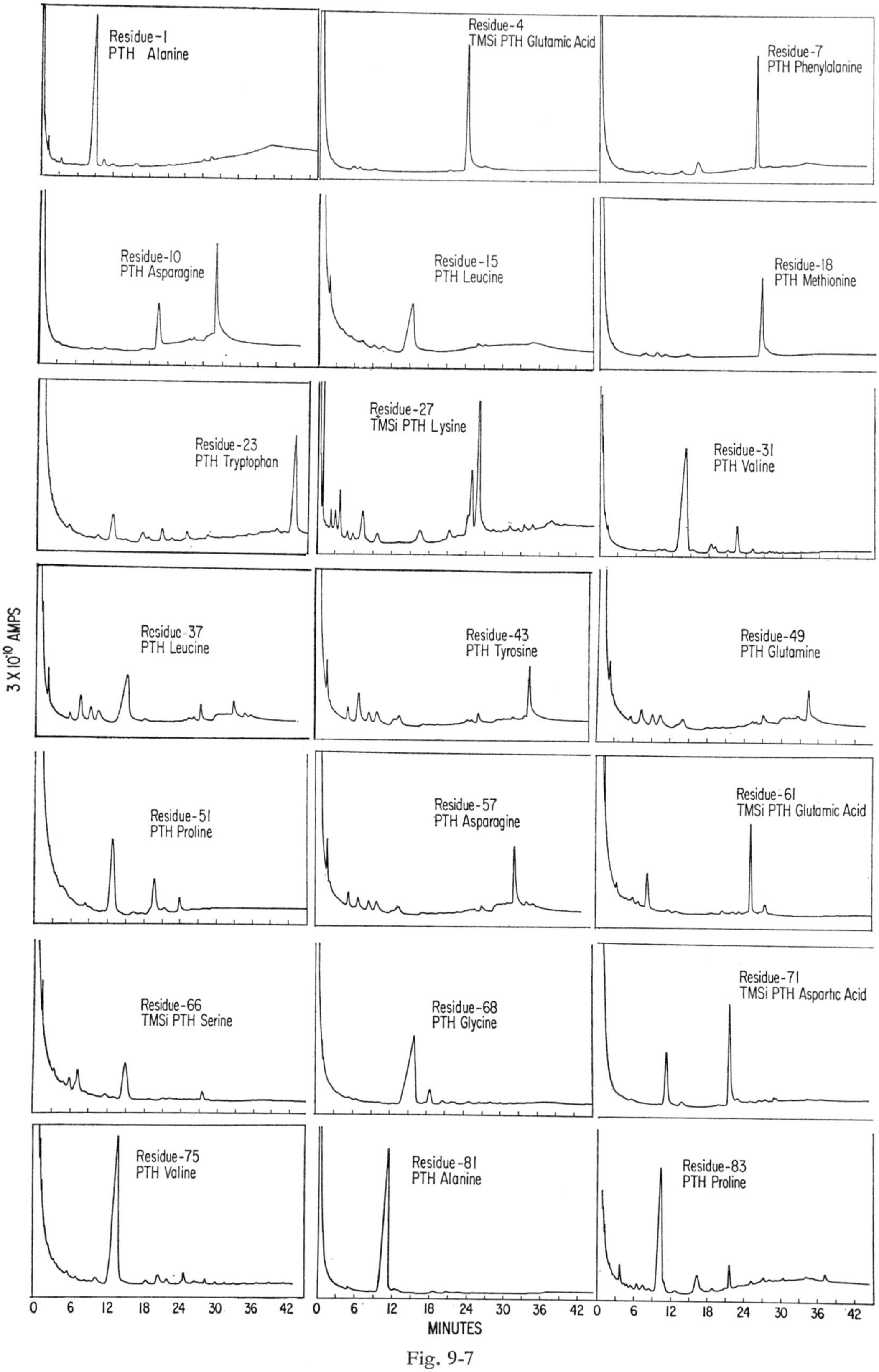

Fig. 9-7

combined and evaporated to dryness. The residue is now ready for conversion to the PTH, which is usually performed by the addition of 0.2 ml 1 N HCl and heating at 80° for 10 min. The solution is then extracted twice with ethyl acetate as above. PTHs in the extracts are identified after evaporation to dryness and resolution in a small volume of ethyl acetate.

III. High Performance Liquid Chromatography (HPLC)

The new technique of HPLC has been recently applied to the analysis of amino acid PTHs with encouraging results [1427]. Although resolution achieved so far is not as good as that attained by GC, the new technique has several meritorious features: (1) all the PTHs, including arginyl PTH, are eluted from a single column; (2) there is no need to prepare volatile derivatives of acidic PTHs; (3) all PTHs are stable during analysis and have similar UV absorbancies; (4) as little as 0.01 nmole of a PTH is easily detected and (5) less than 30 min is required for an analysis.

A. Instrumentation

Inasmuch as identification of PTHs is based upon retention times, it is important to select an instrument having a reliable pump and gradient maker capable of giving reproducible results. A UV monitor is the most convenient, sensitive and reliable detector for PTHs. One instrument meeting all these qualifications is the DuPont Model 830 liquid chromatograph equipped with their gradient maker and UV detector.

B. Columns and Reagents

Commercially available column packings which have been examined to date include Corasil II and C_{18}/Corasil purchased from Waters Associates, and Zorbax and ODS from the DuPont Company. Corasil II and Zorbax are absorbant silicas; C_{18}/Corasil and ODS are reverse phase packings having an octadecylsilane group covalently bonded to a silica core. Stainless steel columns 3 feet × 2 mm i.d. were dry packed [697]. Column pressures ranged from 700 to 1000 psi and flow rates 0.8 to 1.2 ml/min. Spectro-grade solvents distilled in glass were purchased from Burdick and Jackson Laboratories, Inc., Muskegun, Michigan.

Amino acid PTH standards were prepared in methanol, 1 mg/ml, and stored in the dark at −10°. All 20 of the common amino acids found in proteins are eluted from a C_{18}/Corasil column (Fig. 9-9). However, many derivatives are not adequately resolved. To some extent, the silica scolumn Corasil II and Zorbax complement the reserve phase columns but they are not advocated because of the need to use polar solvents, e.g., 0.5% acetic acid, to elute some derivatives. Re-equilibration of the column with less polar starting solvent is difficult and time-consuming. Retention

---

Fig. 9-7. Selected chromatograms of PTHs from the sequential degradation of bovine parathyroid hormone and its tryptic peptides by the automated (Steps 1 to 61) or manual Edman procedure. Columns: 10% DC-560 or CFC. (From Ref. [173])

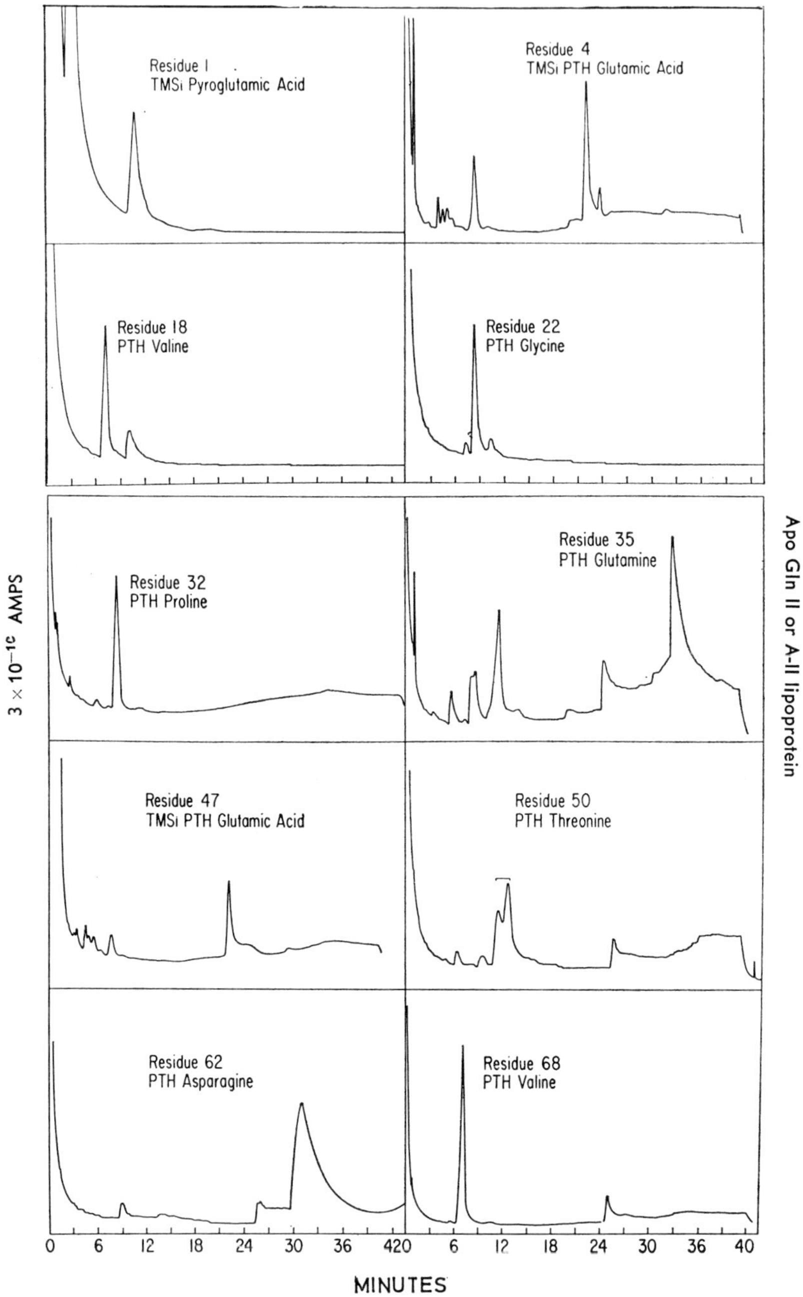

Fig. 9-8. Selected chromatograms of PTHs or their silyl derivatives from the automated or manual Edman degradation of tryptic and cyanogen bromide peptides of reduced and carboxymethylated apolp-Gln-II. Columns: 10% DC-560 or CFC. (From Ref. [174])

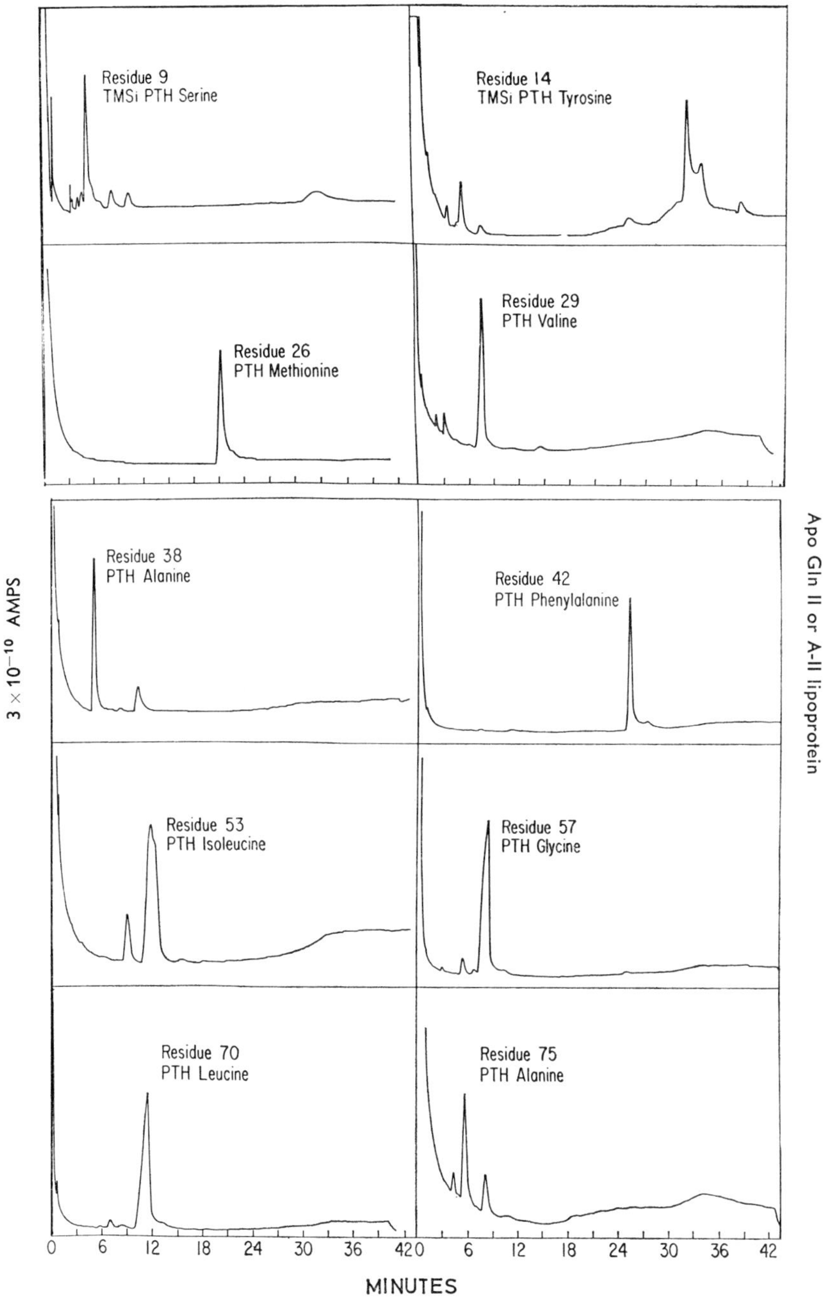
Residue 9
TMSi PTH Serine
Residue 14
TMSi PTH Tyrosine
Residue 26
PTH Methionine
Residue 29
PTH Valine
Residue 38
PTH Alanine
Residue 42
PTH Phenylalanine
Residue 53
PTH Isoleucine
Residue 57
PTH Glycine
Residue 70
PTH Leucine
Residue 75
PTH Alanine
3 × 10⁻¹⁰ AMPS
Apo Gln II or A-II lipoprotein
0 6 12 18 24 30 36 42
0 6 12 18 24 30 36 42
MINUTES

Fig. 9-8

times are not consistently reproduced even after one hour equilibration. Equilibration with reverse phase columns on the other hand is complete in less than 5 min.

A C_{18}/Corasil column has been used to identify amino acid PTHs obtained from automated Edman degradations (Fig. 9-10). Interference from U.V. absorbancy compounds is usually not a problem with the aqueous as well as the organic phases obtained at the conversion step. The extent of deamidation of glutamine and asparagine, often a difficult determination, is readily ascertained in HPLC because the amide and acid are seen in the same chromatogram.

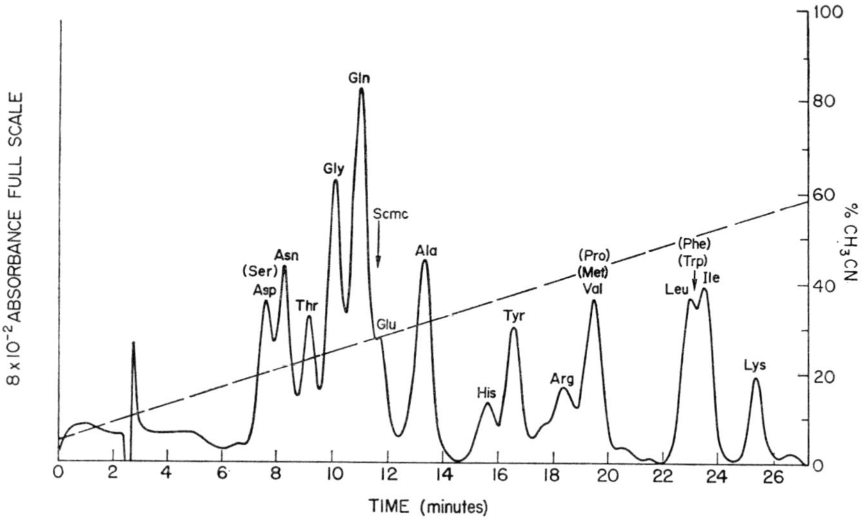

Fig. 9-9. Separation of amino acid phenylthiohydantoins on a 3 ft × 2 mm i.d. C_{18}/Corasil column. Solvent system: A, sodium acetate, 0.01 M, pH 3.80, acetonitrile, 95:5; B, acetonitrile. Solvent B was added at the rate of 2%/min. Pressure: 700 psi. Flow: 0.8 ml/min. Temperature: ambient. Sample size: ~1 μg of each derivative. (From Ref. [1471])

C. Comments on HPLC

The present deficiency in resolution seems to be a problem easy to be solved in the future. Advances in column technology and the preparation of packings of very small size particles (< 10 μ) have greatly improved speed and resolving power [804]. Tests with such new packings, e.g., Varian's MicroPak-CH, and the use of columns with different selectivities, e.g., DuPont's ETH, are very encouraging. With the combination of an ETH and C_{18}/Corasil or MicroPak-CH, and a dual photometer and recorder, it is possible to determine all the PTHs in a sample in under 25 min. The potential for optimizing HPLC for PTH analysis is high.

D. Quantitation

Quantitative end-group analysis provides valuable information about the native protein and the course of the Edman degradation. With the native protein it is

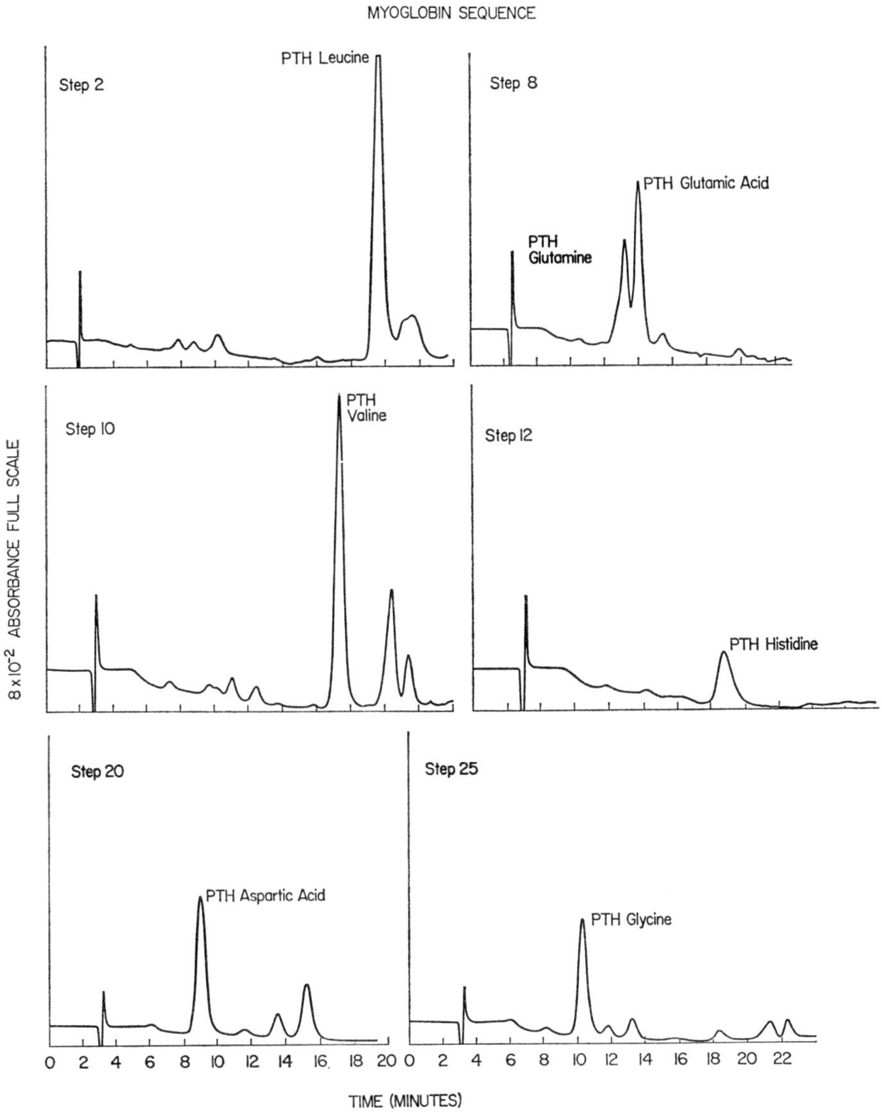

Fig. 9-10. Selected steps from the automated degradation of sperm whale myoglobin. Column, solvent and flow conditions same as Fig. 9-9. After conversion of the thiazolinones to the thiohydantoins, the latter were extracted into ethyl acetate, evaporated to dryness and the residue dissolved in 50 μl of methanol. Injected aliquot, 2 μl. PTH histidine in step 12 was recovered from the water layer in the conversion step of the Edman procedure. Sequenator conditions: 300 nmoles myoglobin, quadrol buffer, Beckman Instrument Co., Program number 072172. (From Ref. [1427])

possible (under ideal conditions) to determine the minimum molecular weight, the presence of non-identical subunits and contaminating polypeptides. Determination of the yield at each step of the degradation will reveal partial coupling or cleavage, non-specific cleavage, N-terminal blockage and the loss of peptide at the extraction steps. It is also possible to predict the number of degradations possible in a given run [320].

It should be remembered that there are two separate parts to the problem of quantitative analysis of PTHs: the Edman procedure and the GC or HPLC analysis. It is well-known that seryl, threonyl, asparaginyl, glutaminyl, histidyl, tryptophanyl, glycyl, and S-carboxymethylcysteinyl derivatives are not obtained in quantitative yields in the Edman procedure. In the GC analysis certain derivatives undergo thermal decomposition and others give reduced responses because they absorb to the column packing. However these effects occur equally with sample and standard and are remarkably reproducible. Derivatives which have been noted to give low responses on all GC columns tested, and their peak areas relative to alanine in a typical analysis are: threonyl, 0.9 (sum of anhydro- and native compound); seryl, 0.5 (anhydroserine); histidyl, 0.5; glutaminyl, asparaginyl and S-carboxymethylcysteinyl, 0.4; and lysyl, 0.1. S-carboxymethylcysteinyl PTH undergoes β-elimination upon injection and gives the same product as serine. Lysyl PTH undergoes considerable decomposition, and the peak obtained has not been characterized. The low responses for asparaginyl and glutaminyl PTHs are due to both decomposition and adsorption. With histidyl PTH the low yield is due primarily to adsorption. Yields may be improved by selecting the optimum injector temperature and the shortest possible analysis time. This is most readily seen with asparaginyl and glutaminyl PTHs. The limit of detection of Group II and some members of Group III is about 0.5 nmole but 0.1 nmole of the others can be readily identified. In HPLC as little as 0.01 of all derivatives is easily detected. Peak areas are proportional to concentration in both methods.

IV. Alternate Methods

A. Thin-Layer Chromatography (TLC)

A TLC method using polyamide-coated glass plates [735] has been modified by the use of 5 $\times$ 5 cm square doubly-faced polyamide sheets and the incorporation of a fluorescent indicator in the first solvent of a two dimensional system. The reported practical lower limit of detection is 0.05 to 0.2 nmole depending on the particular amino acid residue in question [1244]. With refinements in the technique [736] it has been claimed that 0.1 nmole quantities of 24 PTHs can be determined in 30 min. Also reported is a procedure which allows the simultaneous identification of 10 samples on a single 5 $\times$ 5 cm sheet.

Another TLC innovation involves the use of plastic-backed silica gel plates. Of special significance is the fact that many of the PTHs give characteristic colors when the plastic sheet is heated after spraying with ninhydrin [601]. This greatly facilitates identification.

B. Mass Spectrometry (MS)

Another valuable technique for PTH analysis is MS [347, 350, 472, 847, 1023, 1225]. With this technique it is possible unequivocally to identify about 0.05 nmole of any PTH in about 10 min. Some investigators advocate the use of MITC over

PITC claiming that the MTHs are more easily identified and quantitated. However, others have not seen any advantage. Ion formation by the chemical ionization method (CI) is preferred over the more common electron ionization (EI) mode because of the greater sensitivity and ease in interpretation of the spectra [350]. In most cases, the quasimolecular ion (protonated parent molecule) is the only ion in the spectrum. Leucyl and isoleucyl PTHs give identical spectra, and the instrument must be converted to the EI mode (a simple operation) where different fragmentation patterns are obtained. A unique feature of the MS methods is the capability of determining thiazolinones which convert to the thiohydantoins in the inlet of the spectrometer. Thus, it is possible to analyze the samples after cleavage and without prior conversion to the thiohydantoin. However, in practice, it is often useful to prepare the more stable thiohydantoin for use in confirmatory tests. Also, thiazolinone samples usually contain more impurities which can quickly contaminate the mass spectrometer. All the techniques mentioned, GC, HPLC, TLC and MS are about equally affected by contaminants.

Quantitation of PTH by MS is achieved with the aid of a computer or by using internal standards containing a heavy isotope. Standards made with ^{15}N amino acids [347] of pentadeuterophenylisothiocyanate [350] have been used. A special computer program has been written for the Finnegan quadrole MS (Finnegan Instruments, Inc., Sunnyvale, Calif.). The computer controls the mass spectrometer. As the sample evaporates off the probe it switches the MS between various preset masses while integrating the signal at each mass for a fixed time. The total ion current for any given mass is directly proportional to the amount of PTH in the sample. The method does not require internal standards and provides simultaneous identification and quantitation. The high cost and complexity (down time is often exasperatingly long) of this powerful technique are major deterrents to its more widespread use.

Chapter 10

Reconstruction of the Primary Sequence of a Protein from Peptides of Known Sequence

Alfred D. Goldstone and Saul B. Needleman

I. Introduction

The reconstruction of the complete amino acid sequence of a protein is not unlike assembling a jigsaw puzzle. The ease with which the task is accomplished depends upon the details known about the smaller units making up the total picture. Thus, in the jigsaw puzzle, one is guided by examining the shape, color, size and pictorial composition of each piece in the puzzle. Similarly, in reassembling the primary structure of a protein, one must have comparable information. This would include a description of the component parts of the protein — the total amino acid composition, the amino acid content of the subunits, i.e. the individual polypeptides and, preferably, the exact sequence of each of these peptides.

Generally, the protein is prepared for sequence determination by fragmenting the long amino acid chain, by chemical or enzymic means, into smaller units of less complexity. Following the structural analysis of each component peptide, the original protein structure is reconstructed by a logical and combinational examination of overlapping fragments obtained from more than a single method of breakdown. The merits of the various proteolytic enzymes and chemical hydrolytic procedures used in producing the peptide digests of the protein have been discussed elsewhere in this volume. Special emphasis has been placed, in each case, upon the limited range of reactivity, that is, the high degree of specificity, exhibited by each of these methods; this is particularly true for those techniques which make use of enzymic hydrolysis. Less specific methods for hydrolysis of the peptide bond may be applied to reducing the protein to smaller components. The loss in specificity greatly increases the complexity of the resulting mixture, which in turn increases the difficulty of resolving the primary structure of the protein in question. It should be noted, however, that computer methods have been suggested for constructing the amino acid sequence of a protein from less information than is required by the conventional manual operations. This has been accomplished with fragmentary peptides obtained through random acid hydrolysis of the protein. Under such conditions, no information regarding the specificity of the hydrolytic sites is obvious and reconstruction is effected only on the basis of the amino acid composition of the peptide and of the composition and determination of the N-terminal residues of the peptides. The use of computer techniques greatly facilitates the orientation of the large number of peptides obtained through any of the hydrolysis techniques. The difficulty in readily applying such methods to the random peptide procedure lies in the inability to obtain, in pure form, the diversity of small peptides resulting from random acid hydrolysis.

As the soft- and hardware essential for the computer oriented methods for the reassembly of peptides is not yet standard equipment in most laboratories, it will be best to examine first and in detail the operations involved in the manual reconstruction of the complete protein sequence before turning to the discussion of computer methods. To accomplish this, one must determine both the amino and carboxyl terminal peptides of the protein and correctly align the intervening peptides. Detailed discussions for the determination of the terminal residue are found in Chapter 3 but are summarized here in abbreviated form only to indicate how alignment of the terminal peptides is achieved.

II. Determination of the Amino Terminal Peptide

One identifies the amino terminal fragment of a protein by comparison of the amino terminal residue of the protein, previously shown to contain only a single polypeptide chain, with the amino terminal residues of the various peptides isolated after proteolytic digestion. This can be done only in proteins whose amino terminal residue has a free amino group. Several methods mentioned earlier have been utilized in identifying this peptide once the amino terminal residue of the protein has been determined by either the DNFB method of SANGER [1067], the EDMAN degradation [319] or, in certain cases, leucine aminopeptidase hydrolysis or other methods. In studies on lysozyme, FRAENKEL-CONRAT [385] showed that reduction and alkylation of the protein did not significantly alter the molecular weight from that of native lysozyme indicating it to be present as a single polypeptide chain. SCHROEDER [1107] and JOLLES [643], among others, showed by the DNFB method that the protein contained only one amino terminal residue, lysine. Upon partial hydrolysis of DNP-lysozyme with 6 N HCl, SCHROEDER isolated a DNP-tetrapeptide whose amino acid sequence was Lys–Val–Phe–Gly suggesting this to be the amino terminal segment of the protein. CANFIELD and coworkers [199], after tryptic hydrolysis of lysozyme, isolated 18 peptides. Rapid tryptic hydrolysis yielded a peptide with the sequence Lys–Val–Phe–Gly–Arg while slow tryptic hydrolysis gave a peptide whose sequence was Val–Phe–Gly–Arg together with free lysine. These observations together with the results obtained by SCHROEDER and the absence of any other tryptic peptide in which lysine was the amino terminal residue, characterized Lys–Val–Phe–Gly–Arg as the amino terminal segment of lysozyme. From the specificity of trypsin, the isolation of a peptide with lysine as the amino terminal residue is good evidence for its being the amino terminal tryptic peptide.

Another example of unequivocal identification of the amino terminal peptide by dinitrophenylation alone can be found in the structural work by TANIUCHI and ANFINSEN [1265], on an extracellular nuclease from *Staphylococcus aureus*. DNFB treatment of the pure protein produced DNP-alanine with a yield of 80 to 90%, thereby characterizing it as the amino terminal residue. The protein, which has four methionine residues, was treated with cyanogen bromide by a modification of the method of

Gross and Witkop [452]. DNFB treatment of the resulting five peptides, isolated by gel filtration on Sephadex G-50 gave only one peptide containing alanine as the amino terminal residue and it could therefore be designated the amino terminal peptide in the protein. In cases where more than one peptide has an amino terminal residue identical to that found in the protein, techniques such as peptide overlap, which will be discussed fully later, must be employed. It was shown by Heller and Smith [526], using the DNFB method, that glycine was the amino terminal residue in *Neurospora crassa* cytochrome *c*. From peptide overlap and homology all but 41 residues located at the amino terminus could be positioned. Of the five chymotryptic peptides which comprise this segment, two had glycine as their amino terminal residues. The sequences of these peptides were Gly–Phe [C-1] and Gly–Leu–Phe [C-5], respectively, leaving the amino terminus still in doubt. Tryptic digestion of the protein yielded six peptides which encompassed the amino terminal 41 residues. Of these, two had glycine as their amino terminal residues and their amino acid sequences were Gly–Phe–Ser–Ala–Gly–Asp–Ser–Lys [T-1] and Gly–Ala–Asn–Leu–Phe–Lys [T-3]. Since only C-1 and T-1 has a common sequence at the amino terminus, namely, Gly–Phe, the amino terminal segment of the protein could be established as being identical with the sequence of T-1.

In proteins such as cytochrome *c* where the free amino group of the amino terminal residue is blocked, the method of detecting the amino terminal peptide is somewhat different. None of the methods previously discussed are of use since all require a free amino group for their chemical reactivity. In fact the lack of reactivity towards these various reagents is used as a criterion for determining whether the amino terminal residue of a protein is blocked. The ε-amino group of lysine, the imidazole group of histidine and even the sulfhydryl function of cysteine and the tyrosyl hydroxyl will react with DNFB, dansyl chloride or some of the other N-terminal reagents, but comparison with proper standards will eliminate any ambiguity.

Lack of reactivity with the Edman reagent while usually a good indication of a blocked α-amino group may give an ambiguous answer if glutamine is the amino terminal residue. Formation of pyrrolidone carboxylic acid, unreactive with the Edman reagent, might ensue and lead to the erroneous conclusion that the α-amino group of the amino terminal residue is blocked. Leucine aminopeptidase hydrolysis in many instances can help circumvent this difficulty. If glutamine is the amino terminal residue of a protein it will be released upon proteolytic digestion with leucine aminopeptidase. The blocking of the α-amino group of the amino terminal residue on the other hand renders the peptide bond formed by the residues γ-carboxyl group nonsusceptible to proteolytic attack by leucine aminopeptidase. In this way the presence of glutamine as the amino terminal residue can be determined.

Once it has been determined that a protein has a blocked amino terminus, peptides isolated after proteolytic digestion are screened for the absence of a free α-amino group by the methods mentioned above. When the peptide which proved to be the amino terminal peptide of dogfish heart cytochrome *c* [423] was digested with leucine aminopeptidase for 24 hrs no release of any amino acid could be detected, indicating it had a blocked α-amino group. The problem of identifying the blocking group as well as the blocked amino terminal residue now arises. In both the studies on dogfish cytochrome *c* and wheat germ cytochrome *c* [1228], a similar procedure was employed. In the case of dogfish heart cytochrome *c*, the amino acid composition of the amino terminal peptide after chymotryptic digestion was [Gly_2,Asp,Val_2, Glu,Lys_3,Phe]. Hydrolysis of this peptide with trypsin yielded three peptides whose amino terminal residues had free α-amino groups and a pentapeptide whose α-amino group was blocked. This pentapeptide whose amino acid composition was [Gly,Asp, Val,Glu,Lys] could therefore be designated the amino terminal segment of the original chymotryptic decapeptide. From the specificity of trypsin, lysine could be designated the carboxyl-terminal residue of this pentapeptide. Partial acid hydrolysis with 0.3 N HCl for 16 hrs at 105° C hydrolyzes selectively peptide bonds formed by both the α-amino and β-carboxyl groups of aspartic acid. Such treatment resulted in the formation of three components which could be separated by electrophoresis at pH 1.9. Two of these components were ninhydrin positive indicating a free α-amino group. One of these components was free aspartic acid and the other component, which was neutral, had an amino acid composition of [Val,Glu,Lys]. This indicated that glutamic acid and not glutamine was present. Referring to the specificity of trypsin, lysine could be designated the carboxyl terminal residue of this tripeptide which had to be the carboxyl terminal segment of the pentapeptide. One step of the Edman degradation showed valine to be the amino terminus and this peptide could be written as Val–Glu–Lys. The third component was ninhydrin negative indicating no free amino group and characterizing it as the amino terminus of the peptide and of the protein. It was detected on paper by the method of Rydon and Smith [1051] and can be written as follows:

$$\text{—CO—NH-} + Cl_2 \rightarrow \underset{\text{N-chloro-peptide}}{\text{CO—NCl—}} \xrightarrow{\text{KI}} \text{—CO—NH—} + KCl + I_2$$

An aliquot of a peptide is chromatographed on paper, placed in an atmosphere of chlorine, and after thorough aeration, sprayed with a 1% starch-KI solution. The iodine released by the N-chloro-peptide is detected as blue-black spot on a faint blue background. This component migrated identically with authentic N-acetylglycine when chromatographed in butanol [4]:acetic acid [1]:water [5] for 16 hrs characterizing the amino terminal residue of dogfish cytochrome *c* as N-acetylglycine. The amino terminal pentapeptide could therefore be written N-acetyl–Gly–Asp–Val–Glu–Lys.

The approach used by Needleman and Margoliash [907] in establishing the N-terminus of rabbit heart cytochrome *c* involved enzymic and chemical operations different from those just described. The first 10 amino acids were obtained as the chymotryptic peptide. The specificity of chymotrypsin permitted assignment of phenylalanine to the C-terminal position of the decapeptide. Treatment of Peptide C1-10 with trypsin yielded five fragments including the N-terminal pentapeptide: (Ac,Gly,Asp,Val,Glu)–Lys, Gly–Lys, Gly–Lys–Lys, Ile–Phe and free lysine. In all but the one case, the lysyl residue was assigned to the C-terminal position because of the specificity of trypsin. The tryptic pentapeptide now was subjected to the action of elastase which, in spite of its usually broad range of specificity [809], hydrolyzed only the peptide bond which was C-terminal to valine in the present example. The order of residues in the terminal dipeptide was fixed by the specificity of trypsin. The blocked N-terminal tripeptide was insensitive to ninhydrin and was located on paper chromatograms with the Rydon-Smith procedure previously mentioned. A fourth enzymic operation, requiring papain at pH 4.25 [809], yielded the acylated N-terminal dipeptide and free valine. The identical dipeptide was obtained from the tryptic dipeptide T1-5 under similar conditions. The final identification was made on the basis of the reactivity of carboxamide bonds, but not the free carboxyl group, with anhydrous hydrazine [15] to form the acyl hydrazides. The reaction yielded acetyl hydrazide, glycyl hydrazide and free aspartic acid, confirming that aspartic acid rather than glycine was the C-terminal amino acid residue and had a free carboxylic acid group.

III. Determination of the Carboxyl Terminal Peptide

Kinetic studies with carboxypeptidase A and/or carboxypeptidase B is the procedure most widely used in the determination of the carboxyl terminal residue of proteins and of the carboxyl terminal peptide. When denatured cytochrome *c* [1 μmole]

Table 10-1. Residues released after carboxypeptidase A treatment of *Neurospora crassa* cytochrome *c* (Ref. [526])

Amino acid	μmoles liberated per mole cytochrome *c*		
	2 hrs	18 hrs	40 hrs
Alanine	0.07	0.37	0.48
Threonine	0.04	0.15	0.13
Glutamic acid	0.01	0.07	0.07
Methionine sulfoxide	0.02	0.14	0.16
Isoleucine	0.01	0.08	0.10
Phenylalanine	0.00	0.04	0.04

from *Neurospora crassa* [422] was incubated with 0.01 μmole DFP-treated carboxylpeptidase A for time intervals of 2 hrs, 18 hrs and 40 hrs, the various amino acids were released in the amounts shown in Table 10-1. It is evident from these results that alanine is the carboxyl terminal residue of the protein. The exact sequence of the remainder of the carboxyl terminal segment cannot be deduced from the above data, but the presence of threonine, glutamic acid, methionine, isoleucine and phenylalanine in this area is strongly suggested. In certain instances, more than the terminal amino acid can be ascertained. From the studies on α-corticotropin, a plot of the rate of release of amino acids after treatment with carboxypeptidase A, shown in Fig. 10-1, established Leu–Glu–Phe–COOH as the carboxyl terminal segment.

Hydrolysis of a protein with cyanogen bromide is another method by which one can quickly determine the carboxyl terminal segment of the protein. Cyanogen bromide attacks the sulfur atom of methionine causing cleavage at the carboxyl end of the methionyl residue with conversion to homoserine. All but one of the peptides isolated from a cyanogen bromide digest of the parent protein have a homoserine residue as their carboxyl terminal group. The peptide lacking the homoserine residue would comprise the carboxyl terminal peptide of the original protein. As mentioned above, TANIUCHI et al. [1265] found four methionine residues to be present in an extracellular nuclease of *S. aureus* that they had isolated. Cleavage of the protein with cyanogen bromide yielded five major peptides as would be expected from the specificity of cyanogen bromide. The amino acid composition of the cyanogen bromide segments obtained in this manner is shown in Table 10-2. From these results it can be concluded that cyanogen bromide fragment E which contains no homoserine residue is the carboxyl terminal segment of the protein. Evidence for the establisment of the carboxyl terminal peptide can also be obtained by examination of the amino acid composition of peptides obtained from digestion of the protein with proteases with relatively limited substrate specificity. Trypsin, for example, acts primarily on the

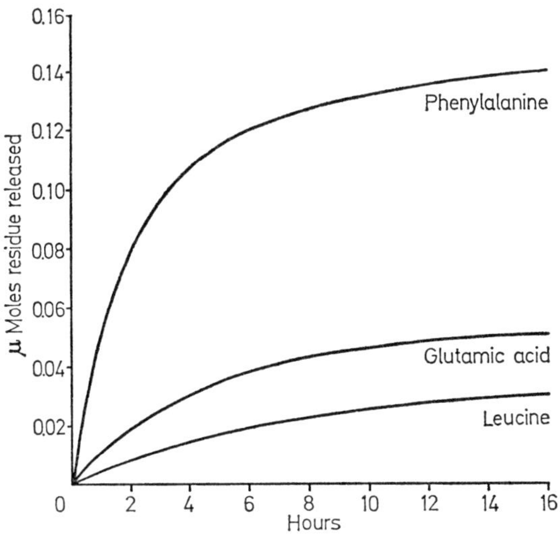

Fig. 10-1. Rate of release of amino acids from α-corticotropin by the action of carboxypeptidase A estimated as their DNP-derivatives (Ref. [512])

bond formed by the carboxyl group of arginine and lysine. Accordingly all peptides except the one found at the carboxyl terminus should contain at least one residue of arginine or lysine. The carboxyl terminal peptide, on the other hand, should contain neither of these residues except on the rate occasion when lysine or arginine happens to be the carboxyl terminal residue of the original protein. If the amino acid composition of five peptides which were isolated after tryptic digestion of a hypothetical protein were as found in Table 10-3, Peptide 3 could readily be designated the carboxyl terminal peptide since it lacks both lysine and arginine residues.

Several other chemical methods have been utilized for the determination of carboxyl terminal residues of a protein. Among these, hydrazinolysis and reduction of C-terminal carboxyl group to amino alcohols are two of the more extensively studied methods. Hydrazinolysis, based on the work of AKABORI [15], liberates the carboxyl terminal residue as a free amino acid while the remaining amino acids in peptide linkage are converted to the amino acid hydrazides. Further treatment with benzaldehyde transforms the amino acid hydrazides to water insoluble dibenzyl

Table 10-2. Amino acid content in cyanogen bromide fragments (Ref. [1265])

Amino acid	Fragments				
	A	B	C	D	E
Lysine	5	1	5—6	4—5	5—6
Histidine	1		1		1
Arginine			1	1—2	2
Aspartic acid	2		1	3—4	6—7
Threonine	3—4		3	1	1
Serine	1		1		2—3
Glutamic acid	1	1	3	4	9
Proline	1	1	3		1
Glycine	1	1	2	4	2
Alanine	2—3		2	3	6
Cysteine					
Valine	1		2	4	4
Methionine					
Isoleucine	2		1—2	1	1
Leucine	3		2—3	5	5
Tyrosine		1	1	2	2
Phenylalanine			1—2	1	
Tryptophan					1
Homoserine	1	1	1	1	

Table 10-3. Amino acid content of tryptic peptides from hypothetical protein

Amino acid	Peptides				
	1	2	3	4	5
Lysine	1			1	
Histidine		1	1	2	
Arginine		1			1
Aspartic acid	2		2		
Threonine	1	2	3	1	4
Serine	3	1	2	3	1
Glutamic acid		3		2	
Proline	1			1	
Glycine	3	2			
Alanine	1	3	1		
Cysteine				1	
Valine	2	1	2		1
Methionine	1	2			
Isoleucine	2	1	3		3
Leucine	1			3	2
Tyrosine		2			
Phenylalanine	2	1	2		
Tryptophan				1	

derivatives leaving the water soluble C-terminal amino acid in solution. The reactions are shown in Fig. 10-1. One serious disadvantage of this method is the low recovery of all C-terminal amino acids and the complete loss of the C-terminal amino acid if this residue is cystine, cysteine, glutamine or asparagine. The method of reduction of the C-terminal residue to an amino alcohol with agents such as lithium borohydride suffers from undesirable side reactions. These include partial reductive cleavage of some peptide bonds with subsequent production of amino alcohols making unequivocal determination of the C-terminal amino alcohol impossible in some cases.

IV. Alignment of Peptides by Analogy

In the case where there is a relatively large degree of homology between two proteins, and the amino acid sequence of one of the proteins is known, it is sometimes

Table 10-4. Comparison of homologous peptides from the pig and cow with the whale (Ref. [152])

Pig, Cow	N-Acetyl-Gly-Asp-Val-Glu-Lys-Gly-Lys-Lys-Ile-Phe^{10}
Whale	N-Acetyl-Gly-Asp-Val-Glu-Lys-Gly-Lys-Lys-Ile-Phe
	← C-1 →
Pig, Cow	Gly-Arg-Lys-Thr^{40}-Gly-Gln-Ala-Pro-Gly-Phe
Whale	Gly-Arg-Lys-Thr -Gly-Gln-Ala-*Val*-Gly-Phe
	← C-5 → ← C-6 →
Pig, Cow	Lys-Gly-Glu^{50}-Arg-Glu-Asp-Leu
Whale	Lys-Gly-Glu -Arg-*Ala*-Asp-Leu
	← C-15 →
Pig, Cow	Lys-Lys^{100}-Ala-Thr-Asn-Glu-COOH^{104}
Whale	Lys-Lys -Ala-Thr-Asn-Glu-COOH
	← C-17 →

possible to correctly align the peptides which arise from the digestion of the unknown protein with a single proteolytic enzyme by comparing these peptides with those similarly obtained from the known protein. Examples of this can be found in the studies done on cytochrome *c*.

Since the amino acid sequence of horse heart cytochrome *c* was determined by Margoliash and Smith [810], the sequences of cytochrome *c* from over 50 other species have been elucidated. The strong homology between some of these proteins made it possible properly to orient peptides of a chymotryptic digest of an unknown cytochrome *c* by comparison with the corresponding peptides of previously studied species. Immunological studies [152] indicated that *Cetacea* [whale] was more closely related to *Artiodactyla* [cow, pig] than to any other mammalian order. When the amino acid sequence of the 17 chymotryptic peptides of cytochrome *c* isolated from the California gray whale became known [422], only two peptides were found to differ from analogous peptides found in the pig and cow; it was therefore possible to deduce the complete sequence of the entire protein by analogy. Table 10-4 illustrates the extreme analogy between some of the chymotryptic peptides found in whale and in pig, sheep and cow. As can be seen, once the amino acid sequence of the porcine and bovine cytochromes *c* were known, one could unequivocally place the chymotryptic peptides isolated from the whale cytochrome *c* into their proper position along the polypeptide chain.

Another example of how comparison of analogous peptides can lead to the proper alignment of these peptides in the protein can be seen in the studies of primate cytochrome *c*. The amino acid sequence of human cytochrome *c* was completed by Matsubara et al. [822] in 1962. Needleman [906] has completed the sequences of peptides isolated after chymotryptic hydrolysis of several primate cytochromes *c*. The almost identical nature of these peptides when compared to chymotryptic peptides of human cytochrome *c*, permitted their proper alignment in the protein chain. In *Erythrocebus patas*, for example only one chymotryptic peptide shows any variance form the analogous peptide found in human cytochrome *c*, differing from it in one amino acid. An alanine residue in the human peptide corresponding to position number 50 in the protein is replaced by a glutamic acid residue in *E. patas*. Thus complete sequence of the protein often can be determined on the basis of a single proteolytic digest, saving both time and material, if the homology between the protein in question and a protein whose sequence is known is sufficiently close.

V. Alignment of Peptides by Peptide Overlap

A. Digestion of the Protein with Two Enzymes of Different Specificity

1. Order of Residues with Both Sets of Fragments Known

In the vast majority of cases involving determination of the primary sequence of a protein there is not enough similarity between the protein in question and proteins whose sequences are known to utilize the technique of peptide analogy. In these instances one must invoke the method of peptide overlap. Reconstruction of the primary sequence of a protein by this method involves the orientation of the set of peptides obtained from one enzymatic digest with respect to the set of peptides resulting from another method of breakdown in such a manner as to have peptides of either one set overlap the sites of hydrolysis in the other. A trivial example will illustrate the general procedure. In the simplest case, let us assume that one method of breakdown has produced the fragment peptides of known order as:

A–B–C D–E F–G H–I–J K   Set 1

where A, B, C, D, E, F, G, H, I, J, K represent amino acids and these occur only once in the protein. Let a second method of breakdown yield the peptides:

A–B C–D–E–F G–H I J–K   Set 2

It is then evident that peptide A–B–C– in Set 1 overlaps the peptides A–B and C–D–E–F of Set 2. Fragment F–G establishes the overlap between the protein thus far reconstructed and fragment G–H of Set 2. The reassembly continues with fragment H–I–J overlapping G–H and J–K thereby completing the sequence. The position of fragment I in Set 2 is fixed because of the known order of amino acids in peptide H–I–J in Set 1. The orientation of fragments from the two sets can be summarized as:

A–B–C D–E F–G H–I–J K	Set 1
A–B C–D–E–F G–H I J–K	Set 2

2. Order of Residues with One Set of Fragments Known

As a second example, consider the case more frequently encountered, where the amino acid composition of both sets of breakdown fragments is known but the order of the residues within the fragments is known only for one set. Let parenthesis indicate that the order of the residues they enclose is unknown. For example, (A,B,C) represents the six permutations of A,B and C; (D,E,) represents either D–E or E–D. Then (A,B,C) (D,E,) represents the $6 \times 2 = 12$ arrangements of amino acids resulting from the six permutations in (A,B,C) followed by D–E or E–D, etc. If the fragment peptides of Set 1 above represent the fragments of unknown order and the peptides of Set 2 those of known order, we can proceed with the reconstruction of the primary sequence.

Fragment (A,B,C) has possible the six permutations A–B–C, A–C–B, B–A–C, B–C–A, C–A–B and C–B–A. The known order of Set 2 fragment A–B reduces these possibilities to A–B–C and C–A–B both of which contain the residues A and B in the proper sequential order. Peptide fragment (D,E) may have the order D–E or E–D while fragment (F,G) may exist as F–G or G–F. Together with peptide A–B, the peptide of Set 2, C–D–E–F, reduces the permutations and combinations to the single possibility A–B–C–D–E–F–G. It should be noted that Set 1 fragment (D,E) is completely contained within Set 2 fragment C–D–E–F and only the known order of the amino acids in the Set 2 fragment establishes the order as D–E rather than E–D since no overlap peptides exist in the present example in which residue D is in one fragment and E in another.

3. Order of Residues with Neither Set of Fragments Known

In the final general example, let us consider the case where the amino acid components of the fragments are known but the order within the fragments is unknown. Again, let the parenthesis indicate that the order they enclose is unknown. Focusing our attention on the first two fragments of Set 1 and Set 2 we have (A,B,C) and (D,E) of Set 1 and (A,B) and (C,D) of Set 2. Clearly (C,D) of Set 2 overlaps (A,B,C) and (D,E) of Set 1, but (A,B) is contained within (A,B,C). Hence two possible reconstructions result, namely, (A,B) (C) (D) (E) and (E) (D) (C) (A,B) where, in each possibility, the order of (A,B) still remains unknown. Such partial reassemblies frequently occur but afford considerable advantage in that with only minimal expenditure of effort on sequencing (ordering) procedures one can pin-point those few areas of the molecule which will require further analytical manipulation.

B. Hydrolysis of the Protein with a Single Agent Having High Degree of Limited and Absolute Specificity

1. Chemical Modification of Enzymic Susceptible Sites

In the initial digestion of a protein into its peptide components, one of the primary characteristics one looks for in the choice of protease, as stated earlier, is that it have as limited and absolute specificity as possible. This reduces the number of peptides obtained after hydrolysis and thereby greatly facilitates the ease with which purified peptides can be isolated in good yield *via* chromatographic means. For this reason, trypsin is extremely popular, cleaving only peptide bonds in which lysyl or arginyl residues are the carboxyl donor partners in susceptible peptide bonds. By chemically modifying either lysine, arginine or cysteine residues of a protein, the versatility of trypsin can be further enhanced and, in some cases, the complete sequence of a protein can be elucidated with trypsin alone [1169b]. By selectively blocking the free ε-amino group of lysine with reversible reagents such as succinate, it no longer remains a suitable substrate for trypsin action. Proteins, whose lysine residues have been modified in this manner, will produce a set of peptides, upon tryptic digestion, derived solely from cleavage of arginyl residues. Arginine, on the other hand, can be rendered inactive towards tryptic digestion by treating it with cyclohexanedione [1171]. In this way a set of peptides is obtained through tryptic hydrolysis specifically due to cleavage at lysyl residues. LINDLEY [783] has shown that cysteine residues, normally not attacked by trypsin, can be modified by treatment with β-bromoethylamine to form S-aminoethylcysteine residues. These now become typically susceptible to cleavage with trypsin. Proteins whose cysteine residues are modified in this manner, and whose lysyl and arginyl residues are blocked by the methods described above, will produce tryptic peptides resulting from cleavage at S-aminoethylcysteinyl residues exclusively.

The application and utility of this selective modification of residues in order to limit the action of trypsin is illustrated below on the hypothetical peptide.

H_2N–Ala–Val–Lys–Gln–Leu–Arg–His–Val–Ala–Cys–Gly–Lys–Glu–Ser–COOH

Peptide A

Tryptic digestion of the above tetradecapeptide would yield the following set of smaller peptides:

1. (Ala,Val)–Lys,
2. (Gln,Leu)–Arg,
3. (His,Val,Ala,Cys,Gly)–Lys,
4. (Glu,Ser).

Analysis of the N-terminal residue of these peptides and of intact peptide A would establish peptide 1 as the N-terminal sequence. Although peptide 3 also contains alanine, histidine rather than alanine would show up as the N-terminal residue of this peptide. A single step of the Edman procedure, or the application of the dansyl or Sanger reagents followed by hydrolysis, would suffice to establish the sequence and location of peptide 4 as the carboxyl-terminus. It would be impossible to position peptides 2 and 3 from the available data.

If an aliquot of peptide A were treated with succinate, thereby modifying the lysyl residues, on tryptic digestion, the following peptides would be obtained:

5. Ala–(Val,Lys,Gln,Leu)–Arg,
6. His–(Val,Ala,Cys,Gly)–Lys–Glu–Ser.

From peptide 5, arginine is associated with the peptide containing the N-terminal alanyl residue. This places peptide 2 before peptide 3 in the larger peptide.

Alternately, if peptide A were treated with cyclohexanedione, modifying the arginine residue, the following peptides would result from trypsin action:

7. Ala–Val–Lys,
8. Gln–(Leu,Arg,His,Val,Ala,Cys,Gly)–Lys,
9. Glu–Ser.

Here, peptides 7 and 9 are identical with peptides 1 and 4 previously described. N-Terminal analysis of peptide 8 (see Chapter 6 for procedures involving glutamine) would in effect place peptide 2 prior to peptide 3. The position of the four peptides from the initial tryptic digestion are now unequivocally located by comparing with the peptides produced from the chemical modification studies.

The sequence of peptide 2 is established on the basis of N-terminal analysis and the known specificity of trypsin. The sequence of peptide 3 (or peptides 6 or 8) must depend on the application of other techniques. These can include progressive Edman degradations or the use of other enzymes as described in Section B2 below.

Chymotrypsin, the other endopeptidase frequently used in hydrolyzing a protein as the initial step in sequence determination, is not as amenable to the procedures described above as is trypsin. Although the specificity of chymotrypsin is reasonably limited, the variety of amino acid side chains subject to cleavage is so large as to make the method of chemical modification inapplicable with any degree of general usefulness.

2. Peptidase Specificity

Several proteins, other than trypsin, chymotrypsin, pepsin and papain recently have been found to be useful tools in the elucidation of the sequences of certain peptides. Among these is the enzyme, tripeptidase, which shows a marked predilection for removing the N-terminal residue from a tripeptide [221]. Aminotripeptidase from swine kidney has been isolated as three homogeneous proteins differing in properties such as chromatographic behavior and molecular weight [222]. These have been shown to be specific for tripeptides, being inactive against di-, tetra- or larger peptides. The enzyme requires a free α-amino group and a free carboxyl group for catalysis and will cleave off any NH_2-terminal residue from a tripeptide with the exception of those cases in which the terminal residue is glutamic or aspartic acid. Further, the enzyme does not act on tripeptides containing a residue having the D-configuration at the terminal or penultimate positions or on tripeptides not having an amide hydrogen at the susceptible peptide bond. In this regard, neither NH_2-X-Pro-Y nor NH_2-X-HyPro-Y will be attacked. There are several ways in which this tripeptidase can be rapidly and easily used in determining the sequence of a peptide:

1. Upon recovery of purified tripeptides obtained after digestion with an endoproteolytic enzyme of known specificity, individual treatment of these with the amino tripeptidase will release the amino-terminal amino acid in addition to the

residual dipeptide. The amino acid released can be identified by electrophoresis on paper at pH 1.9 and, from the known specificity of the endoprotease used in the initial phase of hydrolysis, the sequence of the tripeptide is obtained. For example, if after tryptic hydrolysis, a tripeptide is obtained having the amino acid composition (Gly,Ala,Arg) and if alanine is released after hydrolysis with aminotripeptidase, one can unequivocally write the sequence as H_2N–Ala–Gly–Arg. If tripeptides which are known to contain glutamic or aspartic acid residues, are not attacked by the tripeptidase, one can safely place the dicarboxylic acid residue at the amino-terminal end of the tripeptide.

2. Larger peptides such as tetra- or pentapeptides, after one or two steps of the Edman degradation procedure, respectively, can be subjected to aminotripeptidase hydrolysis. Only the free amino-terminal residue from the resultant tripeptide will be released and can be determined, as previously described, following paper electrophoresis at pH 1.9. Since for some peptides it becomes increasingly more difficult to determine the sequential removal of amino acids unequivocally after several steps of the Edman procedure, this method seems to have definite practical value.

3. Selective Chemical Cleavage

With the advent of automatic sequenator analysis methodology (Chapter 8) and subsequent modifications allowing reliable identification of 30 to 50 consecutive amino acid residues in a protein [434b], a new strategy has evolved in the area of sequence analysis. Instead of degrading a protein by breaking it down into a relatively large number of small peptides whose amino acid sequence generally is determined, after extensive purification, by methods such as the manual step-wise Edman procedure and aminopeptidase and carboxypeptidase kinetic digestion, the protein now can be cleaved selectively into a small number of relatively large peptides whose sequence can then be determined directly by sequenator analysis. Several means for accomplishing selective cleavage of a protein into a limited number of peptides have been proposed and some of these are discussed in greater detail elsewhere in the volume (Chapter 5). The most frequently used procedure for chemical modification makes use of cyanogen bromide which cleaves the protein on the carboxyl side of methionyl residues [452, 1386]. The numerous advantages of this strategy are discussed and illustrated below.

Treatment of a protein with cyanogen bromide will fragment the protein into a small number of large major peptides each of which usually can be isolated in excellent yield. Enzymatic hydrolysis, even with a protease whose specificity is as limited as trypsin, leads to a large number of minor peptides as well as several major peptides which cannot be isolated in nearly as good yield as can those obtained by the chemical method. The small number of peptides obtained upon chemical cleavage greatly simplifies the purification procedures necessary to isolate homogeneous peptides and usually results in obtaining these peptides in high yield. Enzymically derived peptides are more difficult to purify to any degree of homogeneity in good yield. Thus, the amount of pure protein needed to begin sequence determination of that protein is only a fraction of what was needed prior to the introduction of sequenator analysis techniques. As one consequence of this, it is now possible to undertake sequence analysis of proteins of much higher molecular weight than could previously have been attempted [530].

The advantages of selective chemical cleavage can best be illustrated by returning to the example of rabbit heart cytochrome *c* described previously [907].

With only two methionyl residues present in the protein, cleavage with cyanogen bromide will result in the formation of only three major peptides, one of 65 residues (A), one with 15 residues (B) and one with 24 residues (C). The 24 residue peptide is devoid of a homoserine residue after cleavage and can immediately be designated the carboxyl-terminal peptide and its sequence determined directly by sequenator analysis. From the absence of activity with amino-terminal reagents, peptide A (65 residues) can be placed at the amino-terminus of the original protein. The amino acid sequence of the internal 15 residue peptide can be determined quite readily by sequenator analysis. Further fragmentation of peptide A is necessary before its sequence can be elucidated unequivocally with the sequenator. One approach to further fragmentation is through trypsin hydrolysis of the large peptide after succinylation. Succinylation of all the lysyl residues in peptide A renders them immune to the action of trypsin and leaves only the arginine residue at position 38 susceptible to tryptic attack. The resulting two peptides, A-1 comprising 38 amino acids and A-2 with 27 residues, can then be subjected to analysis on the sequenator and their amino acid sequences determined. Again, lack of activity with the typical amino-terminal reagents permits locating peptide A-1 at the amino-end of parent peptide A and, hence, at the amino-terminal end of rabbit heart cytochrome *c* as well. The complete sequence of the protein can thus be reconstructed in the following manner:

```
   |--------------------A--------------------| |------B------| |------C------|
   |------ A–1 ------|------ A–2 ------|
N–Ac–Gly ------ Arg–Lys ------ Met Glu ------ Met Ile ------ Glu–COOH
    1            38   39         65  66         80  81          104
```

When the amino acid sequence of rabbit heart cytochrome *c* was determined by Needleman and Margiolash [907] employing the more classic procedure of peptide overlap following separate proteolytic digestion by trypsin and chymotrypsin, it was necessary to isolate in homogeneous form approximately 35 small peptides and then individually determine their amino acid sequences. It is quite apparent that the strategy employed in determining the amino acid sequence of rabbit heart cytochrome *c* by sequenator analysis would have greatly reduced the amount of purified rabbit protein needed for analysis and would have involved considerably less labor in achieving the same end.

C. Reconstruction of the Protein Sequence by Manual Operation

Unfortunately, the situation encountered in the laboratory as regards identifying the amino acid sequence of a protein rarely is as simple as the trivial examples just presented. More commonly, one might find between 100 and 250 amino acids linked through peptide bonds, and each of the different enzyme digests used to reduce the complexity of the protein chain may in itself contain 50 or more major peptide fragments which must be separated prior to their use in reconstructing the complete amino acid sequence of the original protein molecule. Since the number of combinations of n things taken r at a time ($n > r > 1$) increases more rapidly than does n

itself, it is to be expected that the difficulties in reassembling the fragments of a protein will increase proportionally faster than the number of amino acids in the protein. It is in the logic and combinational problems produced by the increasing complexity of protein structure subjected to analysis that the computer finds great application.

In turning to an actual example of the reconstruction of the peptide fragments of a protein, two new factors must be introduced; these are not encountered in the trivial examples above. The first which increases the complexity of the problem to a large extent relates to the fact that in natural proteins most of the component amino acids occur many times in the protein; hence it is to be expected that many breakdown

Table 10-5. Tryptic peptides of rabbit cytochrome *c* (Ref. [907])

Peptide	Amino acid composition
T 1– 5	Lys, Asp, Glu, Gly, Val
T 6– 8	Lys_2, Gly
T 9– 13	Lys, Glu, Val, Ile, Phe
T 14– 25	Lys_2, His, Thr, Glu_2, Gly_2, Ala, Val, Cys_2
T 23– 25	Lys, Gly_2
T 26– 27	Lys, His
T 28– 39	Lys, His, Arg, Asp, Thr, Pro, Gly_3, Leu_2, Phe
T 40– 53	Lys, Asp_2, Thr_2, Ser, Glu, Gly_2, Ala_2, Val, Phe, Tyr
T 54– 55	Lys, Asp
T 56– 72	Lys, Asp_2, Thr_2, Glu_3, Pro, Gly_2, Met, Ile, Leu_2, Tyr, Trp
T 56– 73	Lys_2, Asp_2, Thr_2, Glu_3, Pro, Gly_2, Met, Ile, Leu_2, Tyr, Trp
T 74– 79	Lys, Thr, Pro, Gly, Ile, Tyr
T 80– 86	Lys, Gly, Ala, Met, Ile_2, Phe
T 88– 91	Lys, Arg, Asp, Glu
T 92– 97	Asp, Ala_2, Ile, Leu, Tyr
T 98– 99	Lys, Leu
T 100–104	Lys, Asp, Thr, Glu, Ala
T 101–104	Asp, Thr, Glu, Ala

peptide fragments will contain the same amino acid residue or even combinations of several residues. Without the feature of unique occurrence given in the trivial examples above, the arrangement of fragments becomes considerably more involved.

It is perhaps fortunate that the second new feature works in the direction opposite to that just described. From the table giving the amino acid composition of rabbit [901] and dogfish [423] cytochrome *c*, it can be noted that several of the amino acid residues appear only one, two or three times in the protein. Thus in the rabbit protein, tryptophan (Trp) and serine (Ser) occur only once, arginine (Arg) and methionine (Met) twice and histidine (His) and proline (Pro) three times. Dogfish cytochrome *c* has one arginyl residue more and one methionyl residue less than the rabbit protein molecule. Such residues can serve as marker sites from which to begin the reassembly of the peptide fragments; this is of considerable aid in simplifying the reconstruction procedure. A second group of amino acid residues serve as marker in another fashion. The C-terminal residue of peptides released during enzymic or chemical hydrolysis generally are of a limited group relating to the specificity of the

Table 10-6. Amino acid sequence of chymotryptic peptides of rabbit heart cytochrome *c* (Ref. [907])

Peptide	Amino acid composition
C 1– 10	Acetyl-Gly-Asp-Val-Glu-Lys-Gly-Lys-Lys-Ile-Phe
C 11– 26	Val-Gln-Lys-Cys-Ala-Gln-Cys-His-Thr-Val-Glu-Lys-Gly-Gly-Lys-His
C 23– 26	Gly-Gly-Lys-His
C 27– 33	Lys-Thr-Gly-Pro-Asn-Leu-His
C 34– 36	Gly-Leu-Phe
C 37– 46	Gly-Arg-Lys-Thr-Gly-Gln-Ala-Val-Gly-Phe
C 47– 48	Ser-Tyr
C 49– 55	Thr-Asp-Ala-Asn-Lys-Asn-Lys
C 55– 59	Lys-Gly-Ile-Thr-Trp
C 60– 67	Gly-Glu-Asp-Thr-Leu-Met-Glu-Tyr
C 68– 74	Leu-Glu-Asn-Pro-Lys-Lys-Tyr
C 75– 80	Ile-Pro-Gly-Thr-Lys-Met
C 81– 82	Ile-Phe
C 83– 88	Ala-Gly-Ile-Lys-Lys-Lys
C 88– 94	Lys-Asp-Glu-Arg-Ala-Asp-Leu
C 95– 97	Ile-Ala-Tyr
C 98–104	Leu-Lys-Lys-Ala-Thr-Asn-Glu

Table 10-7. Comparison of differences in the amino acid sequence of peptides from dogfish and rabbit cytochromes *c* (Ref. [423 and 907])

Species	Peptide	Region	Difference
Rabbit	C 1– 10	8– 10	Lys-Ile-Phe
Dogfish			Lys-Val-Phe
Rabbit	C 11– 26	21– 23	Glu-Lys-Gly
Dogfish			Glu-Asn-Gly
Rabbit	C 27– 36	32– 34	Leu-His-Gly
Dogfish			Leu-Ser-Gly
Rabbit	C 27– 46	43– 45	Ala-Val-Gly
Dogfish			Ala-Gln-Gly
Rabbit	C 50– 56	50– 52	Lys-Asn-Lys
Dogfish			Lys-Ser-Lys
Rabbit	C 57– 64	57– 64	Gly-Glu-Asp-Thr-Leu-Met-Glu-Tyr
Dogfish			Gln-Gln-Glu-Thr-Leu-Arg-Ile-Tyr
Rabbit	C 83– 85	84– 85	Gly-Ile
Dogfish			Gly-Leu
Rabbit	C 86– 94	88– 93	Lys-Asp-Glu-Arg-Ala-Asp
Dogfish			Lys-Ser-Glu-Arg-Glu-Asp
Rabbit	C 98–104	100–104	Lys-Ala-Thr-Asn-Glu
Dogfish			Lys-Thr-Ala-Ala-Ser

enzyme or chemical procedure used in obtaining the peptides. The list of the major peptides obtained as the result of tryptic hydrolysis of rabbit heart cytochrome *c* are given in Table 10-5 and the peptides resulting from chymotrypsin action are in Table 10-6. For comparison, the peptides of the dogfish cytochrome *c* tryptic and chymotryptic digest are shown in Table 10-7.

One can begin the reconstruction from one end of the molecule and proceed in an orderly fashion across the length of the protein chain or one may prefer to make use of the marker information and begin the reconstruction at several loci in the chain permitting the segments of increasing length to "grow together". The consecutive approach will be illustrated later in the discussion of computer procedures. The present example will make use of the marker residues.

Since tryptophan and serine each occur only once in the total protein sequence, we will begin the reassembly of peptide fragments with these markers. From the list of tryptic peptides we note that only peptides T 56–72 and T 56–73 contain the tryptophanyl residue and from the list of chymotryptic peptides we need only C 55–59. Since T 56–72 obviously is completely contained in and is coincident with T 56–73 we may ignore this fragment in the present discussion. Since the order of residues in the chymotryptic peptide is known:

T 56–73 (Lys$_2$,Asp$_2$,Thr$_2$,Glu$_3$,Pro,Gly$_2$,Met,Ile,Leu$_2$,Tyr,Trp)
C 55–59 Lys-Gly-Ile-Thr-Trp

we may rewrite the T 56–73 fragment as:

Gly-Ile-Thr-Trp-(Asp$_2$,Thr,Glu$_3$,Pro,Gly,Met,Leu$_2$,Tyr)-Lys-Lys

The lysine in position 73 must be C-terminal in T 56–73 because of the enzyme specificity of trypsin while the position of the lysine in C 55–59 is known from sequencing studies on this peptide. The two lysine cannot be the same. Tryptophan, of course, must be C-terminal in C 55–59 because of chymotryptic specificity. We might now search for a peptide fragment containing both proline and methionine from the list of chymotryptic fragments. Only C 75–80 meets this requirement and it must be eliminated from consideration on the basis of the location of lysine in the peptide sequence. In T 56–73, both proline and methionine lie towards the N-terminal side of the lysine at position 73. If peptide C 75–80 were subjected to tryptic cleavage, the two fragments Ile-Pro-Gly-Thr-Lys and free Met would result, placing methionine C-terminal next to lysine contrary to what is thus far known. It is necessary then to find separate proline- and methionine-containing peptides which relate to this region of the sequence. These include:

C 27–33 Lys-Thr-Gly-Pro-Asn-Leu-His
C 68–74 Leu-Glu-Asn-Pro-Lys-Lys-Tyr
C 60–67 Gly-Glu-Asp-Thr-Leu-Met-Glu-Tyr

Having eliminated C 75–80 from consideration, C 60–67 remains as the only methionine-containing peptide available. Since no lysyl residue is contained within C 60–67 and methionine (and proline) must be N-terminal to Lys (73), the methionine residue must lie to the N-terminal side of proline and C 68–74 must be the proline-containing peptide needed to extend the sequence in this region. Thus by making use of the methionine peptide one may rewrite the sequence above as:

Lys-Gly-Ile-Thr-Trp-Gly-Glu-Asp-Thr-Leu-Met-Glu-Tyr-Leu-Glu-Asn-Pro-Lys-Lys-Tyr

Tyrosine is C-terminal in C 68–74 because of chymotrypsin specificity.

Only tyrosine remains without overlap in a tryptic peptide. It is highly undesirable to attempt to establish an overlap on the basis of a single residue or even of two amino acids unless these are of the unique types of residues which can serve as markers. Tyrosine occurs too frequently in cytochrome *c* to be considered a marker residue and it would be best to leave this sequence at this point and return to it later when other segments in the protein molecule have been completed.

Since serine is the only other residue occurring only once in the protein, we will list all of the tryptic and the chymotryptic peptides which contain this amino acid. Peptide C 47–48 is completely contained in T 40–53 leaving only a single peptide for consideration. At present we need not pursue this region further because too few marker residues are available here to aid in the reconstruction of the sequence.

Turning to methionine, as an example of a marker residue which occurs twice in the protein, it should be recalled that one of these already has been located in the reconstructed segment covering positions 55 to 74. Peptide C 75–80 contains the second methionine together with proline. In that T 80–86 is the sole methionine-containing tryptic peptide not already accounted for, it clearly must overlap C 75–80 by the single methionyl residue. Thus one obtains:

Ile-Pro-Gly-Thr-Lys-Met-(Gly, Ala, Ile_2, Phe)-Lys

Again lysine must be C-terminal because of enzyme specificity. Overlap is permissible on the basis of a single residue in this case because of the infrequent occurrence of methionine in the protein sequence. Peptide T 74–79 clearly overlaps the N-terminus of this peptide because it is the sole tryptic peptide containing both isoleucine and proline. This results in:

Tyr-Ile-Pro-Gly-Thr-Lys-Met-(Gly, Ala, Ile_2, Phe)-Lys

By similar logic, peptide C 81–82 must be completely contained within the unordered region of this segment, it being the only one of the two-Ile-Phe-containing chymotryptic peptides which would make sense at this point. The addition of this peptide leaves only three residues in unknown sequence, i.e. (Gly,Ile,Ala), but the arrangement of the two peptides, (Gly,Ile,Ala) and (Ile-Phe), also is unknown. Thus one

Tyr-Ile-Pro-Gly-Thr-Lys-Met-[(Gly,Ile,Phe), (Ile-Phe)]-Lys
-(Gly, Ile, Ala)-Ile-Phe-Lys or Ile-Phe-(Gly, Ile, Ala)-Lys

must find a chymotryptic peptide which contains part of this sequence. Peptide C 83–88 is coincident with the latter and establishes the overlap sequence uniquely from position 74 to position 88.

Arginine is another marker residue which may be used in the reconstruction of the cytochrome *c* sequence. Referring to the list of chymotryptic and tryptic digest peptides given in Tables 10-5 and 10-6, one finds two chymotryptic peptides, C 37–46 and C 88–94, and two tryptic peptides, T 28–39 and T 88–91, which contain arginine. From the order of amino acids, it is apparent that the two chymotryptic peptides are unique and occupy different parts of the sequence. Further, it is obvious that T 28–38, but not T 88–91, must overlap C 37–46. This is established by the absence of a glycyl residue in T 88–91 though one is found just prior to arginine in C 37–46. It is also noted that the composition of T 88–91 is compatible with that of C 88–94. Thus the orientation of these peptides gives:

I (His,Asp,Thr,Gly_2,Pro,Leu_2,Phe)-Gly-Arg-Lys-Thr-Gly-Gln-Ala-Val-Gly-Phe

and

II Lys-Asp-Glu-Arg-Ala-Asp-Leu.

Both satisfy the specificities of the enzymes used in obtaining these peptides as well as the requisite amino acid composition. The peptide fragment, C 27–33, contains both proline and histidine and clearly overlaps the unordered portion of I above to yield:

Lys-Thr-Gly-Pro-Asn-Leu-His-(Gly,Leu,Phe)-Gly-Arg-Lys-Thr-Gly-Gln-Ala-Val-Gly-Phe.

The unordered tripeptide is completely coincident with C 34–36 thereby establishing the order of amino acid residues over the entire region 27 to 46.

Of the tryptic peptides remaining unassigned, only T 92–97 has an aspartic acid and leucine occurring together. This establishes the overlap necessary to extend II giving:

Lys-Asp-Glu-Arg-Ala-Asp-Leu-(Ala,Ile,Tyr)

The order of the final three amino acid residues is fixed by the chymotryptic peptide fragment C 95–99.

Tryptic peptide T 101–104 is unusual in being completely free of any residue subject to hydrolysis either by trypsin or by chymotrypsin. It is the only such peptide listed in Tables 10-5 and 10-6 and must, thereby, hold the unique position of being the C-terminus of the protein. The sequence of C 98–104 contains these amino acids and provides the needed overlap for peptide fragments T 98–99, T 100–104 and T 101–104 thereby completing the C-terminal sequence. That this is in fact the C-terminal sequence can be established too by use of the techniques described in Chapter 5.

Heme-containing peptides have been obtained from cytochrome *c* by means of chymotrypsin, trypsin and pepsin digestion. The peptides obtained in the latter two cases are identical and are contained within the structure of C 11–26, the chymotryptic digestion heme peptide. Both T 1–5 and T 9–13 contain valine, glutamic acid and lysine necessary for establishing the overlap with the N-terminal end of the chymotryptic heme peptide. The selection of one of these is simplified by the fact that cytochrome *c* has been shown to have an acetyl blocked N-terminal glycyl residue. Peptide T 1–5, but not T 9–13, contains glycine and, accordingly, must constitute the N-terminal fragment while T 9–13 extends the sequence of C 11–26 to:

III (Ile,Phe)-Val-Gln-Lys-Cys-Ala-Gln-Cys-His-Thr-Val-Glu-Lys-Gly-Gly-Lys-His.

This selection can further be verified by examining both peptide fragments for the presence of a blocked N-terminal glycine.

It remains necessary now only to turn to peptide C 1–10 to provide the overlap through -(Ile-Phe)- with segment III above and with T 1–5 and T 6–8 to complete the N-terminal segment of the protein.

Reference to the table of amino acid composition shows that all the histidine residues have been accounted for in the long segments reassembled from the various peptide fragments. At this point, we may make use of peptide fragment T 26–27 to establish the overlap with segments extending from residues 1 to 26 and 27 to 46. Further, we may now use peptide fragment T 40–53 to fix the overlap with C 37–46,

Table 10-8. Amino acid composition of rabbit and dogfish heart cytochromes *c* (Refs. [423, 907])

Amino acid	Residues per molecule cytochrome *c*	
	Rabbit	Dogfish
Lysine	18	17
Histidine	3	2
Arginine	2	3
Aspartic acid	10	7
Threonine	8	8
Serine	1	5
Glutamic acid	10	12
Proline	3	3
Glycine	13	12
Alanine	7	7
Cysteine	2	2
Valine	4	4
Methionine	2	1
Isoleucine	6	5
Leucine	6	7
Tyrosine	4	4
Phenylalanine	4	4
Tryptophan	1	1

C 47–48 and C 49–55 in that no other fragments remain unused and the composition of T 40–53 can be accounted for by these chymotryptic peptides in the absence of any obvious marker residues.

The C-terminal -Lys-Lys-Tyr sequence of peptide C 68–74 overlaps the C-terminal part of T 56–73 and the N-terminal part of T 74–79 while peptide fragment C 83–88 is sufficient to fix the order of amino acids over the region 80–94 containing fragments C 83–88, C 88–94, T 80–86 and T 88–91. The C-terminal peptide is aligned on the basis of the absence of an amino acid residue susceptible to trypsin or chymotrypsin hydrolysis as well as on the basis of the composition of fragment T 98–99.

Now with reference to the composition in Table 10-8, it can be seen that all the amino acid residues have been accounted for in all of the segments and all of the reassembled segments contain the complete amino acid composition as given. Under these circumstances, chymotryptic fragment C 49–55 suffices to establish the overlap between the reconstructed segment spanning residues 1–55 and the segment including residues 55 to 104. The only residue in common to the two segments is the lysine in position 55. As stated earlier, an overlap established on the basis of a single amino acid residue must be viewed as being an equivocal overlap unless that residue is unique to the protein. In the case of cytochrome *c* where lysyl residues account for 15 to 20% of the total amino acid composition, other information must be included in order to consider the overlap assignment to be valid. Having accounted for all the amino acids and eliminating the possibility of two separate lysyl residues, it is safe to assume that the C-terminal lysine of segment 1–55 is identical with the N-terminal lysine of 55–104.

Dogfish cytochrome *c* with 14 amino acid differences from the rabbit sequence just presented illustrates the value in turning towards marker residues as aids in aligning peptide fragments for sequence reconstruction. In place of the single serine in the rabbit sequence, five residues must be located in the dogfish protein; only two histidine residues but three arginine and one methionine are to be accounted for. Of seven serine-containing peptides obtained through chymotryptic digestion, two, C 27–35 and C 99–104, may be eliminated from consideration because they are completely contained in the larger peptide fragments C 27–36 and C 98–104. Peptide fragment C 98–104 is quickly placed at the C-terminus of the protein because it has as its C-terminal residue an amino acid which is not recognized as a site of hydrolysis by either trypsin or chymotrypsin. Peptide C 47–48 may be treated as in the case of the identical fragment in the rabbit protein described above. Some difficulties might be anticipated in utilizing C 53–59 to establish an overlap with tryptic peptide fragments because of the alternate residue spacing of lysine residues. The tryptic peptide containing serine would contain only two residues and would give only an equivocal overlap without further information. The remaining two serine-containing chymotryptic fragments would afford no particular difficulties in orienting them into their proper location in the sequence. They form part of rather long tryptic fragments which can be readily differentiated on the basis of their amino acid compositions.

VI. Qualities of Computer Programs

It should be evident from the above discussion that the alignment of enzymically or chemically derived peptide fragments into the full protein structure is a complex operation and requires a degree of mechanical skill and logic as well as concrete chemical information regarding amino acid composition, sequence of at least one set of peptides, enzyme or chemical hydrolysis specificity, etc. The complexity of the operation increases rapidly as the number of amino acids in the protein increases. In addition, the number of errors in the fragments as well as other uncertainties increase. Hence, the logical and combinational problems become quite severe and the assistance of a computer becomes desirable or even necessary in the analysis. By exhaustively analyzing the possibilities of protein reconstructions, the computer may assist in determining the best next step to try in the chemical analysis processes eliminating much in the way of needless efforts. A number of programs have been made available for sequence analysis. These differ somewhat in the information required and in their approach to the problem [204b, 1054b].

A programming system for reconstructing the complete amino acid sequence of a protein consists of a number of independent sub-programs, each of which is designed to perform a specific type of operation which in turn becomes the input for one of the other sub-programs [99, 819, 1207].

The first program which need be mentioned may be called: Program ENLIST. From this program are obtained lists of peptide fragments resulting from various enzymic or chemical methods of hydrolysis. Thus KYLIST might represent the list of peptides obtained through chymotryptic digestion while TRLIST might include the trypsin hydrolysis peptide fragments. Depending upon the nature or flexibility of the programming system, the sequence of amino acids in each peptide or only their amino acid composition need be known.

Table 10-9. Composition of theoretical peptides obtained from two different methods of enzymic digestion

Breakdown C		Breakdown T	
c_1	[R] [A,B]	t_1	[A] [B,C,D]
c_2	[D] [B] [C,A]	t_2	[A] [C,A,C]
c_3	[A] [C] [D] [X,A] [C]	t_3	[X] [A,C,B]
c_4	[B] [D] [B] [D] [A,B,D]	t_4	[B] [A,A,C,D]
c_5	[C] [A] [Z]	t_5	[A] [B,C,D]

Table 10-10. Combinations of c fragments associated with each t fragment obtained from Table 10-9

t_1	t_2	t_3	t_4	t_5
$\mathbf{c_1c_2}$	$\mathbf{c_2c_3}$	$\mathbf{c_3c_4}$	c_1c_3	$\mathbf{c_1c_2}$
c_1c_4	c_2c_5		c_4c_3	c_2c_4
c_2c_4	c_3c_5		$\mathbf{c_4c_5}$	c_3c_2
c_4c_2				$\mathbf{c_3c_4}$
				c_4c_2
				$\mathbf{c_4c_5}$

Program MARKER searches all of the peptides generated by the ENLIST program and orients those peptides along the lines described for the manual operation above.

Program MAKLAP is an extension of the program just discussed in systematically examining all other peptide fragments for overlap of the segments already generated from peptides containing marker residues.

Program SEARCH supplements the two programs described above by systematically looking at each pair, or higher combination, of amino acids that occur in each peptide fragment. This information might supplement program MAKLAP and program MARKER in effecting the proper alignment of peptides. A further modification of this program could calculate the probability for the occurrence of each pair of amino acids from the amino acid frequency data as given in Table 10-8. From these probabilities, the program makes a list of those groups of amino acids known to occur together which are of improbable occurrence. Such information is included in the data utilized in orienting peptide fragments. Program MAKLAP finds the maximum possible overlap between any of the longer segments reconstructed from peptide fragments.

The variety of sub-programs possible is as great as the imagination of the programmer and can be as simple as or complex as the information available. Discussion of the programming methods will be clarified if a simple illustration is considered. Suppose that two lists of peptides obtained from different methods of enzymic digestion is made as in Table 10-9. Each peptide fragment c_i of breakdown method C must either overlap several fragments of Breakdown T or else be included within some fragment of T. One can prepare a list for each c_i of all possible such associated t fragments; Table 10-10 shows such a list for the present example. Entries in the list are made according to whether or not some arrangement of the amino acid residues

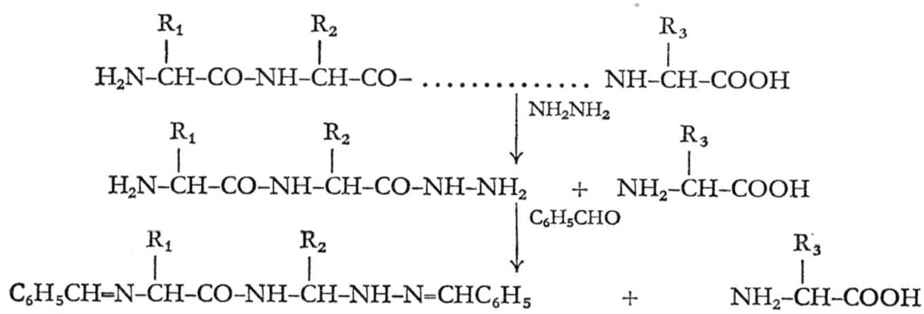

Fig. 10-2. Hydrazinolysis method of Akabori (Ref. [14])

in peptide fragments c_4 and c_5 are contained in fragment t_4. The maximum overlap of amino acids in t_4 and c_4 is (B,A,D). This leaves (A,C) of t_4 to be accounted for in peptide fragment c_5. This is clearly possible, resulting in the overlap:

c_4 c_5

[B] [D] [B] [D] [B] [A,D] [C] [A] [Z]

[B] [A,D] [C] [A]

t_4

In order to determine all the entries in the list, such empirical trials must be made by the computer for every pair of peptide fragments c_ic_j for each t_k.

The lists of all such possibilities can be reduced, however, if one leaves only those combinations which are consistent. For example, t_3 can arise only from c_3c_4. Accordingly c_3 must precede c_4 and c_4 must follow c_3. All other possibilities involving c_3 and c_4 must therefore be eliminated. This would include examples such c_1c_4 and c_2c_4 as in the t_1 list of Table 10-10 or c_3c_5 in the t_2 list, etc. This reduces the lists to those pairs of peptide fragments which are underlined in Table 10-10. This determines the structure as:

$$t_1t_2t_3t_4t_5$$

On proceeding similarly through the t_2, t_3, t_4 and t_5 lists the final structure can be deduced as:

(R) (A) (B) (D) (B) (A) (C) (A) (C) (D) (X) (A) (C) (B) (D) (B) (D) (B) (A,D) (C) (A) (Z).

The solutions of most problems, both scientific and otherwise, are accomplished by first breaking down the more complex complete entity into smaller, more fundamental segments, which can be more readily analyzed, and then by finding suitable interrelationships between the segments recombining them into a whole, which now becomes understandable. As has been described above, the method used in determining the amino acid sequence of proteins has followed this same pattern. The complex protein is first cleaved into smaller segments (peptides) either by chemical means such as cyanogen bromide hydrolysis or by means of proteolytic digestion. After the amino acid sequence of each smaller segment has been determined, the segments can be uniquely aligned properly to restore their orientation in the parent polypeptide chain by the method of homology or by peptide overlap. In this way the once formidable task of determining the complete amino acid sequence of a protein can be achieved.

Chapter 11

Peptide Synthesis

DUANE GISH

I. Introduction

The modern era of peptide synthesis began in 1932 with the introduction by BERGMANN and ZERVAS of the carbobenzoxy derivative for the protection of the α-amino group of amino acids and peptides [91]. Before that time some peptides had been synthesized by methods that were laborious and very limited in their application. The introduction of the carbobenzoxy group provided protection for the amino function that could be removed under very mild conditions (catalytic hydrogenolysis), conditions which left the peptide bond unaffected. Today we have numerous other very useful protecting agents, and new and widely applicable coupling methods have been developed, but peptide synthesis is essentially the same procedure it was in 1932. The carbobenzoxy group is still one of the most widely used groups for protection of the amino function.

Peptide synthesis in principle is a simple, straightforward process. Reactive groups of the amino acids or peptides which are not to participate in the coupling reaction are protected. The protected components are coupled, either by prior activation of the carboxyl group or by the use of various coupling reagents:

$$\mathrm{R^1NH\overset{R^2}{\overset{|}{C}}HCOX + H_2N\overset{R^3}{\overset{|}{C}}HCOOR^4 \xrightarrow{\text{-HX}} R^1NH\overset{R^2}{\overset{|}{C}}HCONH\overset{R^3}{\overset{|}{C}}HCOOR^4} \tag{1}$$

In Eq. (1), R^1 represents the protecting group on the α-amino function which is not to participate in the reaction, R^4 constitutes the protecting group on the non-participating α-carboxyl function, while R^2 and R^3 represent amino acid side chains. The carboxyl-activated amino acid derivative

$$\mathrm{R^2NH\overset{R^1}{\overset{|}{C}}HCOX} \text{ in (1)}$$

may be isolated in the form of a stable intermediate, as is the case with an activated ester, or it may be formed *in situ*, as with the azide procedure and with numerous coupling reagents. If a dipeptide is desired, all protecting groups are removed to complete the synthesis. If the peptide chain is to be lengthened further, the appropriate protecting group is selectively removed and another partially protected amino acid or peptide is coupled:

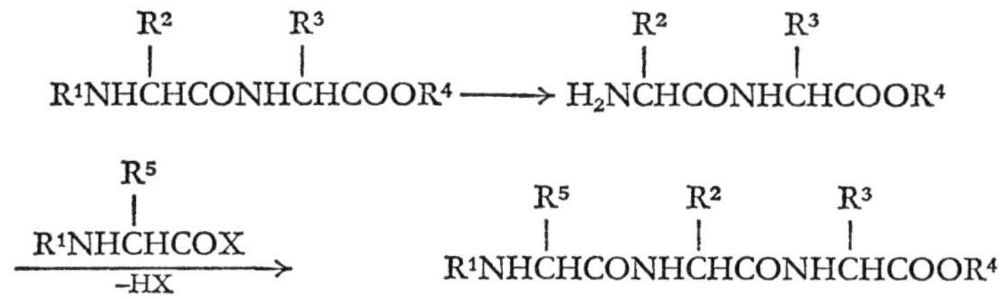

(2)

When the chain has reached the desired length, all protecting groups are removed.

While the basic scheme of peptide synthesis is simple, the actual synthesis of peptides is far from straightforward, as will be apparent from the ensuing discussion. The synthesis of a peptide is usually a research project. The wide choice of protecting groups and coupling methods tests the chemist's ingenuity in devising a synthetic scheme that will offer the best chance for success with reasonable effort.

In the above equations, R^2, R^3 and R^5 might represent reactive side-chains such as those bearing a mercapto group (cysteine), an amino group (lysine), or a carboxyl group (aspartic and glutamic acids). Protecting groups can be so chosen that reactive side chains of amino acids may remain covered while protecting groups on other amino and carboxyl groups may be selectively removed. Proper choice of the method for activating the carboxyl group in the coupling reaction may be the difference between a good yield and a poor yield, or perhaps no yield at all. The choice of solvent and other reaction conditions often determine the yield obtained and the nature and extent of undesirable side reactions.

Although some important innovations have been introduced, such as the Merrifield solid phase method [851], no procedure is universally applicable. Difficulties are commonly encountered. SCHROEDER reported that of the 29 amino acid residues in glucagon, 20 were found to cause difficulties in the synthesis of this polypeptide [1106]. Often the coupling reaction does not go to completion, resulting in reduced yield and an impure product which may be difficult to purify.

Side reactions are frequently encountered during the coupling step or during removal of the protecting groups. Intramolecular cyclization may take place during activation of the carboxyl group, as has been encountered with proline [427] and arginine [140]. Coupling reactions may take an undesired course, such as the formation of acylureas during activation of the carboxyl group with dicyclohexylcarbodiimide [1153], or the cleavage of mixed anhydrides in an unwanted manner during use of the mixed anhydride procedure [334, 1311]. Asparagine may undergo some dehydration to the nitrile compound when its carboxyl group is activated [414]. The use of the Merrifield solid phase technique [851], while usually shortening the time required for synthesis, introduces difficulties unique to this method. These will be discussed under the section devoted to that technique.

Special difficulties may be encountered during removal of protecting groups when certain amino acid residues are present. The catalytic hydrogenolysis of carbobenzoxy groups usually cannot be accomplished with sulfur-containing compounds except under special conditions [1366], and even in peptides which contain no sulfur, catalytic hydrogenolysis of carbobenzoxy groups becomes more difficult as the peptide chain lengthens. The use of sodium in liquid ammonia to remove protecting groups may result in certain side reactions, such as the cleavage of the proline ring in the peptide [1009]. The use of HBr in acetic acid or other solvents may cause difficulties with such amino acids as methionine [18] and tryptophan [146], unless special measures are taken.

Racemization may occur during peptide synthesis. This may always be expected to some extent when the carboxyl group of a peptide is activated, unless its C-terminal amino acid is glycine or proline. Since the azide method avoids racemization, it is almost always the method of choice when peptides terminating in other than glycine or proline are coupled. Racemization is avoided by use of the stepwise method in which appropriately protected amino acids are coupled one at a time to the growing peptide chain.

In spite of these many difficulties, such factors as the availability of a wide variety of protecting groups and coupling methods, and the innovation of techniques as the solid phase method have combined to make syntheses practical which would not have been attempted two decades ago. The most ambitious projects to date have been the syntheses of ACTH (39 amino acid residues) by SCHWYZER and co-workers [1136] and of insulin (two polypeptide chains, one of 21 residues and one of 30 residues, linked by disulfide bridges) by KATSOYANNIS and co-workers [670] and by a group of Chinese chemists [738]. While we may exceed these accomplishments with presently available techniques, it seems highly unlikely that practical syntheses for polypeptide chains approaching 100 residues will be devised using these techniques. To reach such a goal, a new technique, which would constitute a real breakthrough, must be devised. It is doubtful that true proteins of definite desired sequences will ever be synthesized without the use of the living cell, or apparatus derived therefrom.

Brief mention should be made here of two recently introduced procedures that are departures from the classical approach to peptide synthesis. In neither method are the usual protected derivatives required. HIRSCHMANN and co-workers have developed a controlled synthesis of peptides in aqueous medium using N-carboxy-α-amino acid anhydride [553]. In this synthesis, reaction between the anhydrides to form a peptide bond and subsequent release of the α-amino group from the carbamate formed during the reaction is accomplished by control of pH. This is a very convenient method and may prove to be of great value if it can be extended to the synthesis of long-chain polypeptides. A stepwise synthesis of peptides *via* an enzymatic method has been announced by SNELLMAN [1187]. Again, its extension to long-chain polypeptides remains to be demonstrated. These two procedures will be more fully discussed in later sections.

This review is necessarily limited in scope and discussion will generally be restricted to the more commonly used protecting groups and coupling methods. In the past few years, several fine books on peptide synthesis have been puplished [136, 721, 1104], and these should be consulted by those desiring a comprehensive review of peptide synthesis and application to specific sequences. Recent reviews include those of SCHWYZER [1117], LAW [755], ANDERSON [33], ZAHN [1414], and BODANSZKY and BODANSZKY [129].

II. Protecting Groups

A. Amino-Protecting Groups

1. Urethane-Type Protecting Groups

a) Benzyloxycarbonyl (Carbobenzoxy) and Substituted Benzyloxycarbonyl Groups

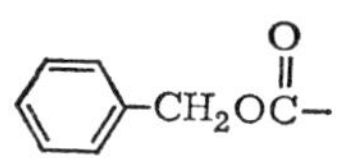

Carbobenzoxy Group

Carbobenzoxy-amino acids are easily prepared in aqueous alkaline solution from carbobenzoxy chloride and the amino acid [Eq. (3)]:

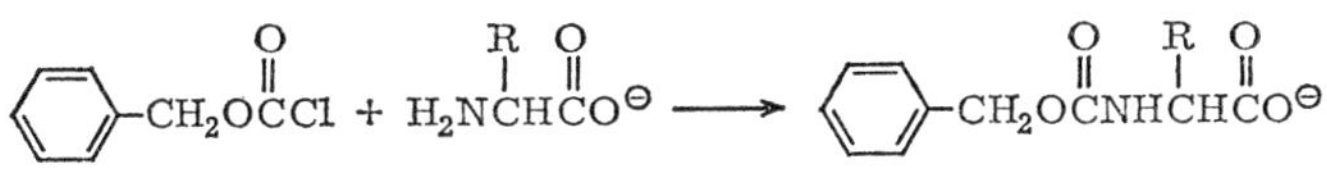

(3)

With certain amino acids, modification of the usual procedure may be necessary, such as careful control of pH. The monograph of Greenstein and Winitz [443b] describes the preparation of carbobenzoxy derivatives and lists the properties of a large number of the carbobenzoxy-amino acids. However, the commercial availability of all of the carbobenzoxy-amino acids at reasonable prices renders their preparation in the laboratory unnecessary.

The carbobenzoxy group is usually stable towards alkaline hydrolysis and ester groups in amino acid or peptide derivatives can ordinarily be readily hydrolyzed with an equivalent of alkali without disturbance of the carbobenzoxy group. Exceptions are noted when glycine is adjacent in the peptide chain to the carbobenzoxy-amid acid. Elimination of benzyl alcohol with formation of a hydantoin derivative may take place, and this can be followed by further hydrolysis of the hydantoin to a urea derivative [430] [Eq. (4)]:

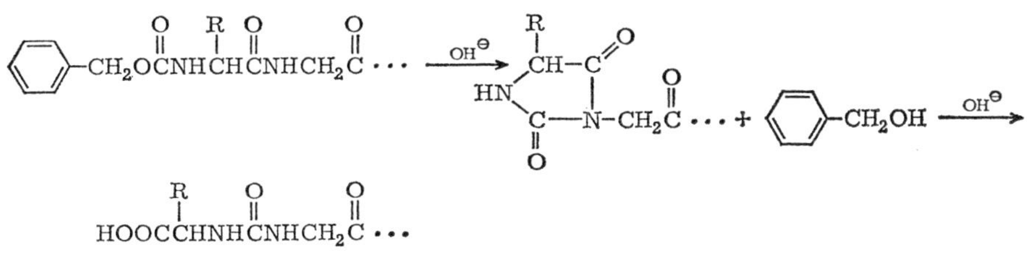

(4)

For the hydrolysis of this type of ester, excess alkali must be avoided. The same type of side reaction has been observed during amidation with ammonia of carbobenzoxy peptide esters containing glycine next to the carbobenzoxy-amino acid [113, 279, 952, 1254].

The carbobenzoxy group can be removed from non-sulfur containing compounds by catalytic hydrogenation. The hydrogenation can be performed at atmospheric pressure or at elevated pressures. As the peptide chain lengthens, the hydrogenation becomes more and more difficult until it may fail altogether. Catalytic hydrogenolysis usually fails with sulfur-containing amino acids and peptides. MEDZIHRADSZKY and SCHWERGER have recently reported that addition of cyclohexylamine makes possible decarbobenzoxylations of compounds containing methionine in other than the N-terminal position [843]. OKAMOTO and co-workers have now discovered that the use of boron trifluoride etherate along with a palladium catalyst is effective for the hydrogenolysis of methionine-containing peptides [937].

Fortunately, other methods for removing the carbobenzoxy group have been devised. The most commonly employed alternative to hydrogenation is treatment with hydrogen bromide in glacial acetic acid or other organic solvents [19, 79, 110, 620]. The concentration of hydrogen bromide may vary from 1 *N* to a saturated solution. The time required varies from a few minutes to up to about 1 hr at room temperature. Side reactions may occur during this treatment. When acetic acid is used, the hydroxy amino acids may be acetylated. To avoid this side reaction, trifluoroacetic acid or other organic solvents are used. In the case of methionine, treatment with hydrogen bromide and acetic acid may lead to the benzyl sulfonium bromide addition product [464], while treatment with hydrogen bromide in nitromethane may lead to the displacement of the S-methyl group to form S-benzylhomocysteine [19]. The addition of ethylmethyl sulfide prevents this side reaction [464]. Tryptophan-containing peptides may give rise to side products due to alkylation and to the presence of bromine in the reaction mixture. Addition of diethyl phosphite leads to satisfactory results [464].

Groups which are stable under conditions which permit complete decarbobenzoxylation with hydrogen bromide include the formyl, trityl, tosyl and phthalyl groups. The *tert*-butyloxycarbonyl group is readily cleaved by this treatment. The methyl and ethyl esters are stable to hydrogen bromide, while the benzyl ester is stable to limited treatment. The *p*-nitrobenzyl ester is considerably resistant to this treatment.

Other methods of cleaving the carbobenzoxy group include treatment with concentrated hydrochloric acid at 37° C, esters are simultaneously hydrolyzed [855], ethanolic hydrogen chloride [61], sodium in liquid ammonia [1158], hot trifluoroacetic acid [1361], liquid hydrogen fluoride [1061] and triethylsilane [111].

Modifications of the carbobenzoxy group that have found significant application in peptide synthesis include the *p*-nitrocarbobenzoxy group [208, 302, 414, 1048], the *p*-(phenylazo)-carbobenzoxy and *p*-(*p*'-methoxyphenylazo)-carbobenzoxy groups [1132, 1135, 1137], and the *p*-chlorocarbobenzoxy group [145, 699, 700, 701]. The derivatives of all of these groups show better crystallizability than the corresponding

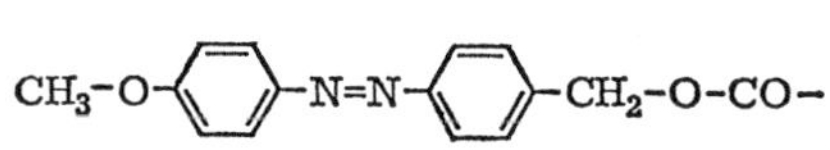

p-(*p*'-Methoxyphenylazo)-carbobenzoxy Group

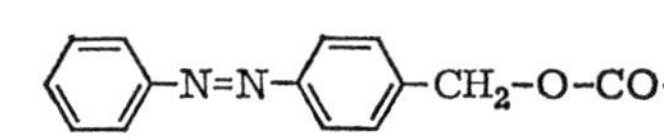

p-(Phenylazo)-carbobenzoxy Group

carbobenzoxy derivatives. Another advantage of the *p*-(phenylazo)-carbobenzoxy and *p*-(*p'*-methoxyphenylazo)-carbobenzoxy groups is the formation of colored derivatives, while derivatives bearing the *p*-nitrocarbobenzoxy group can be detected by UV absorption near 265 mμ. Their common disadvantages, compared to the corresponding carbobenzoxy derivatives, are their greater cost and less ready availability and the somewhat greater difficulty in removing these protecting groups by catalytic hydrogenation or by treatment with HBr-acetic acid.

b) tert-Butyloxycarbonyl Group

The *tert*-butyloxycarbonyl (*t*-BOC) group has found wide use since its introduction into peptide synthesis by McKay and Albertson [839] and by Anderson and McGregor [37] in 1957. Today it rivals the carbobenzoxy group in application.

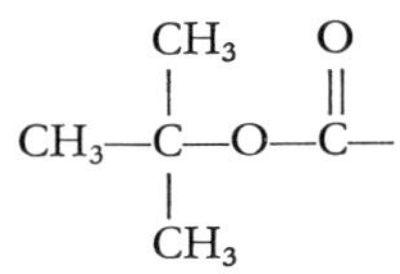

tert-Butyloxycarbonyl Group

The *t*-BOC amino acid derivatives can be conveniently prepared using *tert*-butyloxycarbonyl azide [209, 210, 1134]. Schnabel has published an improved method for the synthesis of *t*-BOC amino acids using the azide [1090]. Explosions have been reported during distillation of *t*-BOC azide, so this purification step should be omitted. The azide and the *t*-BOC amino acids and various active esters of the *t*-BOC amino acids are available from several commercial sources.

The great versatility of the *t*-BOC group is due to its ease of removal under mildly acidic conditions which leave the carbobenzoxy group intact, while, on the other hand, its stability to hydrogenation, sodium in liquid ammonia, and aqueous alkali leaves it intact under conditions that remove the carbobenzoxy and other groups. This has proven especially valuable when multifunctional amino acids, such as lysine and arginine, are present in the peptide being synthesized. The *t*-BOC group can be removed with anhydrous trifluoroacetic acid [724], formic acid [478], and HCl in various solvents such as benzene [37], diethylphosphite [37], acetic acid [839], nitromethane [839], and ethyl acetate [1135]. Mild conditions can be chosen so that the *t*-BOC group is readily split off while negligible splitting of the carbobenzoxy group occurs. When it is desired to selectively remove the carbobenzoxy group in the presence of the *t*-BOC group, catalytic hydrogenolysis is most often used, while sodium in liquid ammonia is used when catalytic hydrogenolysis cannot be used. Examples of this combination of protecting groups may be found in papers by Schwyzer and co-workers [1130, 1131, 1135]. The use of HBr in acetic acid or other solvents permits the simultaneous removal of both groups. In spite of the lability of the *t*-BOC group to acid, *t*-BOC amino acid and peptide hydrazides can be converted smoothly to the corresponding azides in the presence of HCl if the temperature is sufficiently lowered [619].

c) Other Aliphatic Urethane Groups

Numerous other aliphatic urethane groups have been tested for utility as masking groups in peptide synthesis. Almost all suffer from some serious disadvantage when compared to the *t*-BOC group. One group which may find important application in peptide synthesis is the 2-*p*-tolylsulfonylethoxycarbonyl group, which was suggested

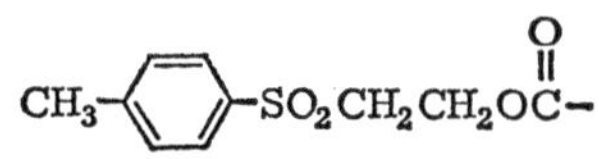

2-*p*-Tolylsulfonylethoxycarbonyl Group

for the synthesis of peptides by KADER and STIRLING [654]. This group is stable to acid and catalytic hydrogenolysis, but is cleaved by rather mildly basic conditions. It is sufficiently alkali stable to permit peptide synthesis to take place under usual conditions and to permit the formation of hydrazides with hydrazine. It nevertheless can be cleaved under alkaline conditions which leave an ester group intact. Treatment of a *p*-tolylsulfonylethoxycarbonyl dipeptide ethyl ester with 3 moles of *N* aqueous ethanolic sodium hydroxide for 1 hr at 20° C removed both the *p*-tolylsulfonylethoxycarbonyl group and the ester group, while treatment with 1 mole of ethanolic sodium ethoxide for 15 min at 20° C cleaved only the *p*-tolylsulfonylethoxycarbonyl group [654]. Treatment of the same protected dipeptide ester with concentrated HCl in acetic acid at reflux for 10 min yielded the *p*-tolylsulfonylethoxycarbonyl dipeptide acid in 92% yield.

2. Alkyl-Type Protecting Groups

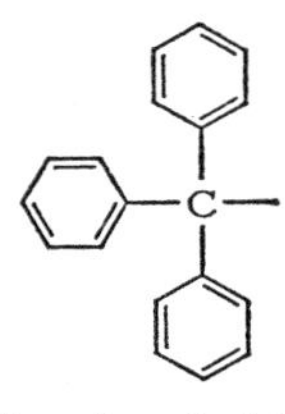

Triphenylmethyl Group

The triphenylmethyl (trityl) group is the only one of several alkyl type protecting groups investigated that has proven to be of fairly wide application in peptide synthesis [29, 538, 1222, 1425]. Ordinarily, monoalkylated amino groups can be acylated, but the bulky trityl group offers sufficient steric hindrance so that the amino group is completely protected. In fact, the steric hindrance offered by this group is great enough that some coupling methods do not succeed with certain of the trityl amino acids. Hydrolysis of esters of triphenylmethylated amino acids requires drastic conditions and the formation of the hydrazide from the ester fails with most amino acids. The exception is glycine, and in some cases, alanine. However, no difficulty is encountered when the carbodiimide method is employed with trityl amino acids, and all coupling methods applicable to peptides can be used when trityl peptides are to be coupled.

The trityl group can be removed by hydrolysis with aqueous acetic acid under conditions that do not affect the carbobenzoxy, *t*-butyloxycarbonyl, and most other groups [146, 1029, 1130]. An example of its use in conjunction with the *t*-BOC group is found in the paper by RITTEL [1029]. It is removed by other acidic conditions, such as hydrogen chloride in various solvents [1128, 1425] and trifluoroacetic acid [1133]. Due to greater stability of the trityl group when it is on other functional groups, the trityl group on the α-amino group can be selectively removed from α, ε-ditrityllysine [110], N,O-ditrityltyrosine [1222], N,S-ditritylcysteine [30], and α,*imid*-ditritylhistidine [1222].

Catalytic hydrogenolysis may also be used to remove the trityl group [1425]. Since the hydrogenation of this group proceeds slowly, such groups as the carbobenzoxy and benzyl ester groups can be removed selectively by stopping the hydrogenation after the appropriate amount of hydrogen has been taken up [1222].

3. Acyl-Type Protecting Groups

The use of acyl-type protecting groups in peptide synthesis has been fairly extensive. Certain of the acyl amino acids, such as the acetyl and benzoyl amino acids, tend to undergo racemization and have found little use. Until the development of more convenient groups, such groups as the formyl and tosyl groups were used extensively for protection of the ε-amino group of lysine. Other acyl-type groups, including the trifluoroacetyl and *o*-nitrophenylsulfenyl groups, because of their ease of removal under specialized conditions, have found important applications.

a) p-Toluenesulfonyl (Tosyl) Group

The tosyl group is stable to reagents commonly used to cleave other protecting

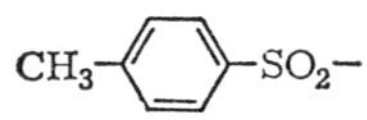

p-Toluenesulfonyl Group

groups. When it was discovered by DU VIGNEAUD and BEHRENS [300] that the tosyl group could be cleaved from amino acids and peptides by sodium in liquid ammonia, the applicability of this group to peptide synthesis became apparent. While the phthalyl, formyl, and *t*-BOC groups are stable to sodium in liquid ammonia, the carbobenzoxy and benzyl groups are cleaved. Some difficulties may be encountered during the treatment with sodium in liquid ammonia. The proline ring may be cleaved with resultant conversion of proline to δ-aminovaleric acid [1004]. When multiple tosyl groups are present, the compound may be difficultly soluble in liquid ammonia, or a product may precipitate before all tosyl groups have been removed. The product must be desalted if an ammonium salt or acetic acid has been used to neutralize the excess sodium. Desalting may be accomplished by direct addition of a resin [842].

Tosyl amino acids are prepared without difficulty [380, 665, 1273] and may be coupled by a variety of methods [63, 1096]. Before its displacement by the more convenient *t*-BOC group, the tosyl group was widely used as a protecting group for the ε-amino group of lysine. For an example, one is referred to its use in the ACTH series by LI and co-workers [774].

b) Phthalyl Group

The phthalyl group, introduced into peptide chemistry by SHEEHAN and FRANK [1152] and KING and KIDD [694], has found limited use in peptide synthesis. It has been used for syntheses in the glucagon series [107, 1101] and has found useful application as protection for the ε-amino group of lysine [1122]. The phthalyl group

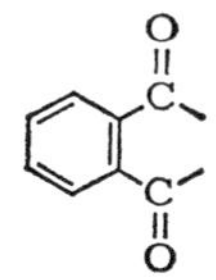

Phthalyl Group

can be removed only by hydrazinolysis. The use by SCHWYZER and co-workers of hydrazine hydrate in neutral solution for cleavage of the phthalyl group was an important modification, for it permits cleavage under very mild conditions [1122].

c) Formyl Group

Because of the tendency of formyl amino acids towards racemization and because of difficulties encountered with some coupling methods, the formyl group has not obtained a prominent place in peptide synthesis. Formyl amino acids are prepared

```
  O
  ||
HC—
```

Formyl Group

by the treatment of amino acids with a mixture of acetic anhydride and formic acid [301, 1157] and can be coupled *via* the carbodiimide method to form optically active peptides [1157]. The formyl group can be removed under mild acid conditions [301, 1157], or with acetyl chloride in benzyl alcohol [1325]. It is stable to alkali. YAJIMA and co-workers have found that the formyl group can be cleaved from the ε-amino group of lysine with hydrazine hydrate [1405]. This group was used by HOFMANN and co-workers as protection for the ε-amino group of lysine for the synthesis of peptides related to ACTH [568b], and, in like manner, by YAJIMA and co-workers for the synthesis of monkey β-MSH [1405].

d) Trifluoroacetyl Group

The trifluoroacetyl (TFA) group, first applied to peptide synthesis by WEYGAND

```
   F  O
   |  ||
F—C—C—
   |
   F
```

Trifluoroacetyl Group

and CSENDES [1350], shares in common with the *p*-tolylsulfonylethoxycarbonyl group the property of being readily cleaved with mild alkali while being stable to mild acid and catalytic hydrogenolysis, conditions which cleave other conveniently used protecting groups. The review by WEYGAND [1350] should be consulted for methods applicable to this group. Most peptide coupling methods can be employed. The stability to mild acid conditions does not apply in the case of the hydroxy amino acids, for TFA-serine and TFA-threonine are rapidly cleaved under very mildly acidic conditions [1352]. In contrast to the selective removal of the *p*-tolylsulfonylethoxycarbonyl group with alkali in the presence of an ester group [654], this is not possible except in special cases with the TFA group. Acid labile esters, such as the *t*-butyl ester, can be resorted to in order to circumvent this difficulty. This was demonstrated in the synthesis employed by MOORE and co-workers [875].

e) o-Nitrophenylsulfenyl (NPS) and the 2,4-Dinitrophenylsulfenyl (DNPS) Groups

The *o*-Nitrophenylsulfenyl [1420, 1421] and the 2,4-dinitrophenylsulfenyl [370] groups exhibit the same case of removal with mild acid as the trityl group, but offer significant advantages over the latter group. For example, these groups do not

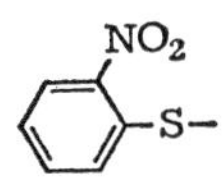

o-Nitrophenylsulfenyl Group

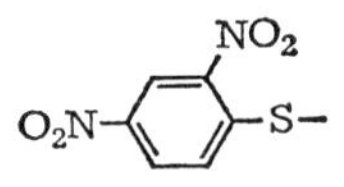

2,4-Dinitrophenylsulfenyl Group

exhibit the steric hindrance shown by the trityl group and the NPS and DNPS amino acids can be coupled in good yield *via* the diphenylphosphoryl or carbodiimide method, or as the active ester [370, 1420]. The NPS and DNPS groups are rapidly split by 2 moles of HCl in anhydrous solvents and can be split off selectively in the presence of such acid-sensitive groups as the *t*-BOC and *t*-butyl ester groups [1421]. The NPS and DNPS groups can also be removed with Raney nickel [845].

When the NPS and DNPS groups are cleaved with HCl, *o*-nitrophenylsulfenyl chloride and 2,4-dinitrophenylsulfenyl chloride are formed. These react with tryptophan [42] at the 2-position of the indole ring, as shown by FONTANA et al. [371]. These alkyl halides would also tend to react with methionine, and the Raney nickel desulfuration method is not applicable to such peptides. To circumvent these difficulties, thiol compounds and other strong nucleophiles can be employed in organic solvents to remove the NPS and DNPS groups [Eq. (5)] [159, 371, 648]:

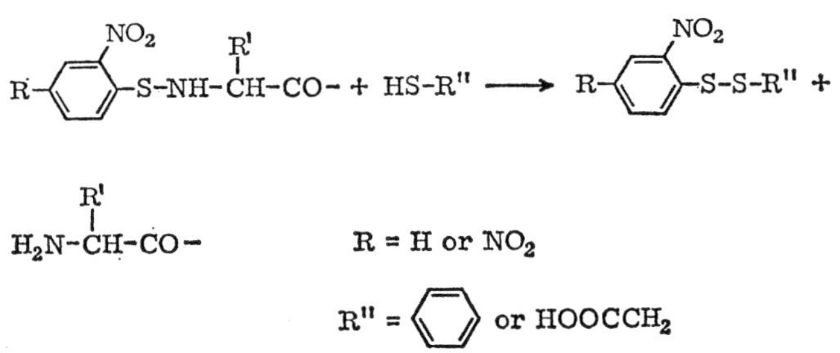

(5)

One hour or less at room temperature is sufficient to complete cleavage of the groups. Yields vary from 70 to 80%. WUENSCH and co-workers, in order to protect tryptophan residues from alkylation, carry out the cleavage in the presence of indoles, especially N-methylindole [1397]. SCOFFONE and co-workers have applied the sulfenyl chlorides as selective reagents for tryptophan in peptides and proteins [378, 1139, 1140]. The NPS and DNPS groups will find many important applications in peptide synthesis. A recent example is the use of the NPS group in the synthesis of human gastrin [885]. NPS-Amino acid derivatives are commercially available.

f) 5,5-Dimethyl-3-oxo-1-cyclohexen-1-yl Group

The application of the 5,5-dimethyl-3-oxo-1-cyclohexen-1-yl (dimedonyl or Dim)

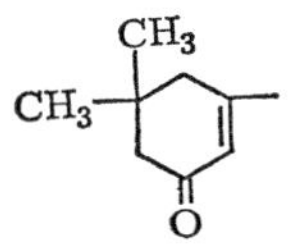

5,5-Dimethyl-3-oxo-1-cyclohexen-1-yl Group

group to peptide synthesis has been described by HALPERN and co-workers [270, 475, 476, 477]. This protecting group is stable to catalytic hydrogenolysis, sodium in liquid ammonia and to hydrolysis by acids or bases. It can be removed by treatment with bromine or nitrous acid. Nitrous acid appears to be an especially mild and useful reagent [477]. DEER [271] has recommended the dimedonyl group for use with the Merrifield solid phase method. In the final coupling step a dimedonyl amino acid was used. Hydrolysis of the resin ester bond followed by filtration and acidification gave dimedonyl peptides of high purity by direct crystallization. The protecting group can then be removed by treatment with nitrous acid. Because of its unique properties, the dimedonyl group will find application in special cases.

B. Carboxyl-Protecting Groups

As is the case with amino-protecting groups, a variety of carboxyl-protecting groups are available that can be selectively removed by various methods. It is thus possible to remove the carboxyl-protecting group without affecting amino-protecting groups, or *vice versa*, and to choose different protecting groups for α- and β- or γ-carboxyl groups so that one carboxyl-protecting group can be removed without affecting the others.

1. Methyl and Ethyl Esters

Methyl and ethyl esters, equally applicable to peptide synthesis, can be prepared with an excess of the alcohol in the presence of a strong acid, such as HCl, as catalyst [260]. This was the standard method of preparation until the more convenient method of BRENNER and HUBER [169] was introduced. In this method thionyl chloride is first added to the alcohol followed by the amino acid. Acetyl chloride may be used in place of thionyl chloride [488]. Peptides may be similarly esterified [358].

The β-ester of aspartic acid and the γ-ester of glutamic acid may be prepared by acid-catalyzed esterification if careful conditions are followed [92, 862]. The α-esters of aspartic acid and glutamic acid can be obtained from the intramolecular anhydrides of various N-protected derivatives by treatment with the alcohol [1355].

The removal of methyl and ethyl ester groups is usually accomplished by saponification. Ordinarily, from 20 to 60 min at room temperature with a small excess of alkali is sufficient. Various organic solvents in combination with aqueous alkali may be used. These include acetone [134], dioxane-acetone [664], dimethylsulfoxide [1243], pyridine [147], dioxane [147], methanol [147], and ethanol [157]. Caution must be used when peptides containing serine are saponified [464]. Difficulty has also been encountered during alkaline hydrolysis of α-N-*p*-nitrocarbobenzoxy-S-benzyl-*L*-cysteinyl-*L*-proline benzyl and methyl esters [415]. It is probable that β-elimination of the sulfur occurred due to the presence of the electrophilic nitro group, since no difficulty was encountered when the *p*-nitrocarbobenzoxy group was replaced with the carbobenzoxy group.

When difficulties occur with alkaline hydrolysis, as is sometimes the case with γ-esters of glutamic acid [69, 776], β-esters of aspartic acid [776] and with larger peptides, acid hydrolysis may be used. KATSOYANNIS used dioxane-aqueous HCl at 90° C for 1.5 hrs to hydrolyze a carbobenzoxy peptide ethyl ester [664]. The carbobenzoxy peptide was obtained in 94% yield. SHEEHAN and co-workers used aqueous HCl at 100° C for 2 hrs, or an acetone-aqueous HCl mixture at reflux for 2 hrs to hydrolyze phthalyl peptide esters to the phthalyl peptides [1151]. Yields were excellent. Use of concentrated HCl at about 40° C removes both the ester and the carbobenzoxy group [855].

2. *tert*-Butyl Esters

The introduction of the *t*-butyl ester group for peptide synthesis by TASCHNER and co-workers [1269], ANDERSON and CALLAHAN [36] and by ROESKE [1037] made available a carboxyl-protecting group that could be removed under the same mild acid conditions used for the cleavage of the *t*-BOC group. Its resistance to alkaline hydrolysis, ammonolysis and catalytic hydrogenolysis, and rapid cleavage under conditions that leave the carbobenzoxy and other groups intact, insured its wide use in peptide synthesis. These esters can be prepared by esterification of protected amino acids, followed by removal of the protecting group [36, 1105, 1269, 1270], or by direct esterification of free amino acids [1037, 1268]. They are commercially available from a number of sources.

The *t*-butyl ester group may be removed by such treatment as HCl in organic solvents [1105, 1271], aqueous HCl [1029], trifluoroacetic acid [1132, 1271] and HBr in acetic acid [36]. The utility of this group in peptide synthesis is illustrated by the synthesis of a decapeptide, described by SCHWYZER and KAPPELER [1129], and the synthesis of β-MSH by SCHWYZER and co-workers [1127].

Carbobenzoxy-aspartic acid α-benzyl-β-*t*-butyl ester can be hydrolyzed with alkali to the β-*t*-butyl ester in good yield [1123]. When the α-carboxyl group of aspartic acid is in peptide linkage, however, the β-*t*-butyl ester becomes labile to alkali [1127]. In this instance, the carboxyl group of the C-terminal residue may be protected by the *p*-nitrobenzyl ester. Subsequent removal of the *p*-nitrobenzyl ester group by catalytic hydrogenolysis leaves the β-*t*-butyl ester group unaffected.

3. Benzyl Esters

The benzyl ester group, the stability of which resembles very closely that of the carbobenzoxy group, was first used by BERGMANN and co-workers [96]. This group can be removed by catalytic hydrogenolysis, sodium in liquid ammonia, alkaline hydrolysis and by prolonged treatment with HBr in acetic acid. The carbobenzoxy group can be removed selectively from peptide benzyl esters by HBr in acetic acid or other solvents if treatment is sufficiently brief, although there will be some loss of the benzyl ester group [19, 79].

Benzyl esters of amino acids are most conveniently prepared from benzyl alcohol with organic acids such as *p*-toluene sulfonic acid and benzene sulfonic acids as catalyst and solvents which permit the azeotropic removal of the water formed [87, 110, 798, 1426]. SHIELDS and co-workers prepared benzyl and *p*-nitrobenzyl esters by passing the condensed azeotrope through a drying agent containing an indicator [1157a]. This permits the most convenient preparation, and completion of the esterification is easily detected.

The α-benzyl esters of aspartic and glutamic acid can be prepared by partial hydrolysis of the diesters or through the intramolecular anhydrides [94, 183, 1054]. The papers by SCHROEDER and KLIEGER [1102, 1103] illustrate the use of the α-esters for preparation of the ω-*t*-butyl esters by selective removal of the benzyl group. The γ-benzyl ester of glutamic acid can be prepared by selective esterification of glutamic acid [236], while the β-benzyl ester of aspartic acid may be obtained by partial acid hydrolysis of N-protected aspartic acid dibenzyl ester [87].

Examples of the application of benzyl esters in peptide synthesis include syntheses in the glucagon series [1100], the insulin series [668], and the synthesis of α-MSH [465].

4. Modified Benzyl Esters

a) p-Nitrobenzyl Esters

The introduction of the *p*-nitrobenzyl ester group into peptide synthesis [1134, 1150b] made available a carboxyl-protecting group with the same selectivity as the benzyl ester, with the added property of being much more stable to hydrogen bromide. This added stability permits the selective removal of the carbobenzoxy group with hydrogen bromide without significant effect on the *p*-nitrobenzyl ester group [1134, 1150b]. It was found that even the *p*-nitrocarbobenzoxy group could be selectively cleaved from a *p*-nitrobenzyl ester with hydrogen bromide [1157b]. Amino acid esters of *p*-nitrobenzyl alcohol are prepared similarly to esters of benzyl alcohol [1103, 1157b]. KATSOYANNIS has applied this ester in the insulin series [668].

b) 2,4,6-Trimethylbenzyl Esters

STEWART employed the 2,4,6-trimethylbenzyl (TMB) ester group [1229—1233, 1236] in order to provide a carboxyl-protecting group which would possess characteristics similar to the benzyl ester, yet could be cleaved by HBr under milder conditions, allowing the carbobenzoxy and TMB groups to be split simultaneously [1229]. Treatment of the TMB ester group with cold trifluoroacetic acid for a few minutes cleaves this group while leaving the carbobenzoxy group intact [1230, 1233].

Further, the TMB ester group is stable to methanolic HCl [1230], a reagent that rapidly splits off the amino-protecting nitrophenylsulfenyl group. The *t*-BOC group is readily split by HCl in organic solvents [31, 839, 1135], as is the *t*-butyl ester group. The TMB ester may be used with these and other groups to extend the versatility of peptide synthetic methods by permitting greater selectivity in removing various protecting groups.

Stewart has pointed out [1229] a possible application of methylated polystyrene resins in the Merrifield solid phase peptide synthesis technique. The procedure now used in the Merrifield technique utilizes an ester linkage between the peptide chain and benzyl groups on the resin. At the completion of the synthesis, the completed peptide chain must be cleaved from the resin with HBr. As mentioned in a previous section, the use of HBr can give rise to serious side reactions. The use of methylated polystyrene resins in the Merrifield technique would enable the completed peptide chain to be split off by brief treatment with trifluoroacetic acid, a very mild reagent which would eliminate side reactions encountered with the use of HBr.

Stewart has applied the TMB ester group in conjunction with the *o*-nitrophenylsulfenyl and carbobenzoxy groups in the synthesis of several peptides containing sequences found in ovine insulin [1235]. The TMB group gives promise of important applications in peptide synthesis in standard techniques as well as the solid phase method.

5. Miscellaneous Esters

Other esters which have been proposed as carboxyl-protecting groups include the *p*-methoxybenzyl [1356], the phthalimidomethyl [908] and the benzhydryl ester [554].

6. Amides

In a number of naturally occurring polypeptides such as α-MSH and the posterior pituitary hormones, vasopressin and oxytocin, the C-terminal amino acid is present as the amide. In many naturally occurring peptides, aspartic acid and glutamic acid occur as the ω-amides. Since the amide bond is about as inert as the peptide bond, the amide group can be introduced at the start of the synthesis of the chain and serve as the carboxyl-protecting group. ω-Amides are not as stable as α-amides, however, and this fact can give rise to certain side reactions. As previously mentioned, the amide group of asparagine may undergo dehydration to the nitrile when the carboxyl group of N-protected asparagine is activated [416]. ω-Amides may undergo partial hydrolysis when peptide esters are hydrolyzed with acid [1126]. Sondheimer and Holley have reported the conversion of the α-methyl esters of carbobenzoxy-asparagine and carbobenzoxy-glutamine to the imides when these compounds were treated with alkali [1193]. Nevertheless, when these special situations are taken into account, the amide group may serve conveniently as a carboxyl-protecting group when the amide is the final product desired.

Amides are usually prepared by saturating a methanol or ethanol solution of a methyl or ethyl ester with ammonia [94, 1176]. Chambers and Carpenter have described many of the amino acid amides [217]. Benzyl esters may also be amidated [1021]. As mentioned in an earlier section, hydantoin and urea derivatives tend to form when peptides containing glycine adjacent to an N-carbobenzoxy amino acid is amidated [266, 274, 397, 531].

7. Substituted Hydrazides

The phenylhydrazide group has been used as a carboxyl-protecting group [678, 867, 1323]. At the completion of synthesis of the peptide, the phenylhydrazide can be converted to the acid by oxidation in aqueous solution with copper acetate [1323], ferric chloride [867], or manganese dioxide in acetic acid [678]. The phenyl group cannot be removed from the hydrazide in order to permit further extension of the peptide chain *via* the azide method. MILNE and co-workers, however, have discovered that oxidation of phenylhydrazides of protected amino acids and peptides in non-aqueous solvents converts the phenylhydrazide into a phenyldiimide. The phenyldiimides couple with amino acid and peptide esters or amino acid and peptide phenylhydrazides to form peptide bonds [868, 869]. This will be discussed in more detail in a later section.

HOFMANN and co-workers [569, 570] used the *carbobenzoxyhydrazide* group for carboxyl protection in conjunction with the phthalyl, tosyl and carbobenzoxy groups for protection of the α-amino group. Selective removal of the amino-protecting group permitted extension of the peptide chain, the carbobenzoxyhydrazide group serving as carboxyl protection. Selective removal of the carbobenzoxy group from the carbobenzoxyhydrazide group of phthalyl and tosyl peptides permitted extension of the chains *via* the azide.

The *tritylhydrazide* group was used by WEYGAND and STEGLICH with the trifluoroacetyl group as the amino-protecting function [1361]. Since the trityl group could be removed selectively with acid, and the trifluoroacetyl group could be removed selectively with alkali, the peptide chain could be extended through either the amino group or *via* the azide through the carboxyl group.

The introduction of other protecting groups has increased the versatility of this method. PRESTON and WEINSTEIN have reversed the procedure of WEYGAND and STEGLICH. They have used an acid labile group, the *o*-nitrophenylsulfenyl group, for amino-protection, and the *trifluoroacetylhydrazide* group for carboxyl-protection [993]. Treatment with HCl permitted extension of the peptide chain through the α-amino group, while removal of the trifluoroacetyl group with alkali produced the free hydrazide for conversion to the azide and extension of the chain through the carboxyl group.

BOISSONNAS and co-workers [148] and SCHWYZER and co-workers [1138] have employed the *tert-butyloxycarbonylhydrazide group* for carboxyl protection, using it in conjunction with the carbobenzoxy group for amino protection. Further examples of this combination can be found in the papers by GUTTMANN [462] and SCHWYZER and AUNG TUN-KYI [1121].

By careful control of reaction conditions CHEUNG and BLOUT have succeeded in using the unprotected *hydrazide* group itself for carboxyl protection [225]. They took advantage of the difference in basicity of the α-amino group ($pK'_a = 7.4$) and the hydrazide function ($pK'_a = 2.8$) in α-amino acid hydrazides. Using Woodward's reagent [1393] for activation of the carboxyl group, they coupled carbobenzoxyglycine with glycine hydrazide to give carbobenzoxy-glycylglycine hydrazide in 62% yield. Under proper conditions, the hydrazide could be oxidized with N-bromosuccinimide to the carbobenzoxy peptide free acid. Conversion to the azide would have permitted extension of the chain *via* the carboxyl group, while catalytic hydrogenation would have permitted extension *via* the amino group. Where short chain peptide hydrazides are to be prepared, this simpler preparative method may be preferable, but for more complex peptide hydrazides the use of the more elegant and safer protected hydrazides may be advisable.

8. Protection by Salt Formation

A protected amino acid or peptide with an activated carboxyl group can be reacted with an amino acid or peptide when in the form of its salt [Eq. (6)]:

$$\text{R-NH}\overset{\overset{\text{R}^1}{|}}{\text{C}}\text{H}\overset{\overset{\text{O}}{\|}}{\text{C}}\text{X} + \text{H}_2\text{N}\overset{\overset{\text{R}^2}{|}}{\text{C}}\text{HCOO}^-\text{B}^+ \longrightarrow \text{RNH}\overset{\overset{\text{R}^1}{|}}{\text{C}}\text{H}\overset{\overset{\text{O}}{\|}}{\text{C}}\text{NH}\overset{\overset{\text{R}^2}{|}}{\text{C}}\text{HCOO}^-\text{B}^+ + \text{HX} \quad (6)$$

An aqueous or mixed aqueous-organic solvent mixture must usually be used since the salts of amino acids and lower peptides are generally insoluble in anhydrous solvents. The salts of some peptides, particularly higher peptides, with organic bases such as triethylamine or tributylamine, are soluble in dimethylformamide, and the use of this anhydrous solvent eliminates hydrolysis of the activated carboxyl component.

The advantage of this method is its simplicity and the avoidance of difficulties that sometimes occur during hydrolysis of peptide esters. On the other hand, purification of the product may be difficult due to the fact that the product and the starting carboxyl component (by hydrolysis) are both acid. When the carboxyl component is an activated amino acid derivative and the amino component is a long peptide chain, the product can often be purified without difficulty due to the considerable difference in solubility of the small carboxyl component and the long chain product.

The various methods used for activating the carboxyl components include the acid chloride [95, 665, 792], azide [689, 1092], mixed anhydride [336, 1417], and activated esters [466, 467, 531].

C. Sulfur-Protecting Groups

The sulfhydryl group in cysteine must always be protected during peptide synthesis because of its ease of oxidation to the disulfide, and because, as a nucleophile, it competes with the α-amino group for the acylating agent. The *S-benzyl group* [1158, 1330] was one of the earliest groups used and has been the most widely utilized, although today there is an increasing number of more convenient groups available for S-protection. The advantage of this group is that it is stable to most reagents used to remove a wide variety of N-protecting groups, while its disadvantage is that it must be removed by treatment with sodium in liquid ammonia, a procedure less convenient than other methods and which may lead to side reactions.

The S-benzyl group has been used in conjunction with many groups for protection of the α-amino function. These include the tosyl group [101, 303, 304, 579, 650, 800], the carbobenzoxy group [125, 245, 503, 504, 581, 616, 935, 1025], the trifluoroacetyl group [1236], the trityl group [232], the *tert*-butyloxycarbonyl group [37], the formyl group [581], the *p*-nitrocarbobenzoxy group [302, 464, 693], the *p*-chlorocarbobenzoxy group [699, 700], and the phthalyl group [58, 363, 693].

Modified benzyl groups used for S-protection include the *p-nitrobenzyl group* [101, 664], the *methoxybenzyl group* [17, 1060], and the *p-methoxy-α,α-diethylbenzyl group* [707].

All of these groups when used for S-protection are stable to HBr in acetic acid under conditions which cleave the N-carbobenzoxy group. Hot trifluoroacetic acid is used to remove S-*p*-methoxybenzyl and S-*p*-methoxy-α,α-diethylbenzyl groups [17, 707], while Berse and co-workers reported that the *p*-nitrobenzyl group could be removed by catalytic hydrogenation [101].

The *trityl group*, first applied by Amiard and co-workers [30], and by Zervas and Theodoropoulos [1425], has found wide use for S-protection. The elegant work of Zervas and co-workers [1422, 1423] and of Hiskey and co-workes [555, 557, 558, 559, 561], illustrate the use of the trityl group in combination with the carbobenzoxy, *t*-butyloxycarbonyl, *o*-nitrophenylsulfenyl, trityl, phthalyl, and benzhydryloxycarbonyl groups for N-protection, the methyl, ethyl, benzyl, *t*-butyl and benzhydryl groups for carboxyl protection, and the benzyl, benzhydryl, benzoyl, and diphenylmethyl groups for S-protection. The use of such a variety of protecting groups in conjunction with the S-trityl group, and the use of mild reagents for its selective removal, has resulted in the S-trityl group assuming an important role as an S-protecting group.

Veber and co-workers [1314] have recently introduced the *acetamidomethyl group* for S-protection. This new group is stable to trifluoroacetic acid, HBr in acetic

$$CH_3\overset{\overset{\displaystyle O}{\|}}{C}NHCH_2\text{—}$$

Acetamidomethyl Group

acid, HCl in ethanol, anhydrous HF at 0° C, and is stable in aqueous solution in the pH range 0 to 13. It is stable to hydrazine and thus applicable when the azide coupling method is used. The acetamidomethyl group is removed in high yield by mercuric ion at pH 4 at room temperature. It may be conveniently used in conjunction with the catbobenzoxy and *t*-butyloxycarbonyl groups.

Other groups used for S-protection include the carbobenzoxy [88, 664, 1422, 1424], diphenylmethyl [556, 1422], benzoyl [556, 1424], benzhydryl [555, 558, 560, 561], benzylthiomethyl [181, 562, 664, 1410], acetyl [1424], ethylcarbamoyl [463, 1041], ethylthio [614], and the 1-phenylcyclohexyl, 9-methylfluoren-9-yl, and piperonyl groups [707]. Where multiple disulfide bridges are to be synthesized, different S-protecting groups can be so chosen that particular –SH groups may be selectively uncovered. This assures formation of each disulfide bond in the manner desired.

D. Hydroxyl-Protecting Groups

Often it is possible to prepare peptides of serine, threonine, tyrosine, and hydroxyproline without protection of the hydroxyl groups. At times, however, acylation of the hydroxyl group, as well as other side reactions, occurs during the peptide bond-forming step. Paul reported acylation of the tyrosine hydroxyl group when *t*-BOC-valine was coupled with tyrosine ethyl ester [960]. Stewart and co-workers have reported 0-acylation of threonine when using the carbodiimide reagent in the Merrifield solid phase method [1237, 1238, 1412].

Groups finding use as protecting function for the hydroxyl group of tyrosine include the carbobenzoxy [621, 956], benzyl [1400], tosyl [667], and the *t*-butyl groups [195, 1099, 1401, 1402], all in conjunction with the carbobenzoxy group for protection of the α-amino group. Active esters of α-N-NPS-0-*t*-butyl-tyrosine have been reported [1398].

O-Benzyl-serine was first prepared by OKAWA [940] and GRASSMANN and co-workers [437] by rather difficult procedures. The recently announced procedure for the preparation of O-benzyl-threonine [873], if applicable to serine, would make preparation of O-benzyl-serine much more convenient. In spite of the difficulty of preparation, O-benzyl-serine has been used in the preparation of numerous peptides [379, 437, 612, 747, 938, 939].

The *tert*-butyl group has proven to be a very convenient and widely applicable group for the protection of the hydroxyl group of serine and threonine. These derivatives are prepared by acid-catalyzed esterification with isobutylene [105, 106, 195, 1099]. WÜNSCH and co-workers have prepared a variety of active esters of the α-N-carbobenzoxy-O-*t*-butyl and α-N-nitrophenylsulfenyl-O-*t*-butyl derivatives of serine and threonine [1398, 1403]. The O-*t*-butyl derivatives of serine and threonine have been applied to syntheses in the glucagon series [107, 1099, 1403].

A new hydroxyl-protecting group which may prove to be very useful is the 2,2,2-trifluoro-1-benzyloxycarbonylaminoethyl group, introduced by WEYGAND and co-workers [1363]. These derivatives of serine and threonine are prepared by treating

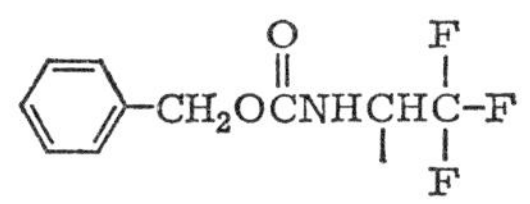

2,2,2-Trifluoro-1-benzyloxycarbonylaminoethyl group

the N-acylamino acid with the chloride. This group can be split off by catalytic hydrogenation, HBr in acetic acid, or with anhydrous HF.

Other groups that have been suggested for hydroxyl protection include the acetyl [1363] and the tetrahydropyranyl groups [618, 1062].

III. Peptide Bond Formation

Less than 20 years ago, only two general methods for forming the peptide bond were available, these being the acid chloride and azide procedures. During the past two decades several new general methods and a great variety of specific new reagents have been developed. Just during the past year, numerous new activated derivatives of N-protected amino acids have been suggested. It has become difficult, if not impossible, for each individual chemist to test all suggested methods.

Almost all methods that have assumed importance involve activation of the carboxyl group of an N-protected amino acid or peptide. Some derivatives, such as the acid chlorides, azides, and mixed anhydrides, are unstable and are prepared *in sito* or used immediately after preparation. On the other hand, most of the activated esters are practically indefinitely stable, and are isolated and stored in crystalline form. This is of considerable advantage, not only because they can be prepared in advance in pure form, but also because they are thus obtainable from commercial sources.

In other methods, the protected carboxyl and amino components are mixed and coupled by the addition of stable reagents, such as the ester phosphites, carbodiimides, Woodward's reagent, and carbonyldiimidazole.

All coupling methods involve the possibility of side reactions. Careful control of temperature, time, solvent, and pH may be used to minimize these side reactions. For instance, the use of mild [1059] and hindered bases [127] has been used to minimize racemization during coupling with active esters. The possibility of racemization is an ever present danger, and intermediates, solvents, and activating reagents must be so chosen to minimize or eliminate this side reaction.

A. Acid Chloride Method

Acid chlorides, first used by FISCHER for peptide synthesis [357], are rarely used today. The carbobenzoxy amino acid chlorides are unstable, splitting out benzyl chloride to form the N-carboxyanhydride [95]. Acyl peptide chlorides may give rise to oxazolones [948]. Tosyl amino acid chlorides are stable, crystallizable derivatives, but may undergo fragmentation in aqueous alkali [71]. The acid chloride procedure has been eclipsed by more convenient methods. Earlier, however, this procedure was used to prepare a variety of peptides a few of which may be cited [95, 414, 503, 792].

B. The Azide Procedure

Curtius introduced the azide method [259] at about the same time FISCHER began use of the acid chloride procedure. Unlike the acid chloride method, however, the azide procedure is still in wide use. As previously indicated, the azide method is the only method which is free of racemization when the carboxyl group of a protected peptide is activated. In addition, the azide procedure is still the method of choice in the preparation of histidyl [571b, 1092, 1345] and seryl peptides [8, 280, 359, 360, 507, 1092].

Hydrazides are most often prepared by the reaction of hydrazine hydrate with methyl or ethyl esters [273, 413, 464, 666, 1353, 1396], or, less often, with benzyl esters [666, 1077]. Alcohol is frequently used as solvent, although dimethylformamide may be used where solubility problems are encountered. The hydrazides are treated with nitrous acid to give the reactive azides.

$$\text{RNH}\overset{\overset{\displaystyle R^1}{|}}{\text{C}}\text{HCOOCH}_3 \xrightarrow{H_2NNH_2} \text{RNH}\overset{\overset{\displaystyle R^1}{|}}{\text{C}}\text{HCONHNH}_2 \xrightarrow{HNO_2} \text{RNH}\overset{\overset{\displaystyle R^1}{|}}{\text{C}}\text{HCON}_3 \qquad (7)$$

The azides react with the amino component with release of gaseous HN_3 [Eq. (8)].

$$\text{RNH}\overset{\overset{\displaystyle R^1}{|}}{\text{C}}\text{HCON}_3 + \text{H}_2\text{N}\overset{\overset{\displaystyle R^2}{|}}{\text{C}}\text{HCOOR}^3 \longrightarrow \text{RNH}\overset{\overset{\displaystyle R^1}{|}}{\text{C}}\text{HCONH}\overset{\overset{\displaystyle R^2}{|}}{\text{C}}\text{HCOOR}^3 + \text{NH}_3 \qquad (8)$$

The reaction may be carried out in anhydrous solvents or under Schotten-Bauman conditions.

Side reactions are sometimes encountered during preparation of hydrazides. It has been reported that benzyl mercaptan may be split off during formation of hydrazides of S-benzyl-cysteine derivatives [800], although hydrazides of such derivatives have been formed without difficulty, even at reflux temperature [666]. Threonine- and serine-containing peptides suffer slight racemization [1091]. Side reactions may be caused by the guanido group of arginine [396], although such side reactions will not occur when the guanido group is simply protonated. Phthalyl and trifluoroacetyl amino acid and peptide ersters cannot be converted to hydrazides by reaction with hydrazine since such treatment cleaves these groups. Hydrazides of these derivatives may be prepared by reacting the activated derivatives with carbobenzoxy or *t*-butyloxy-carbonyl hydrazides, as mentioned in an earlier section. Side reactions which occur under alkaline conditions must be considered a possibility during preparation of hydrazides.

Since azide formation takes place in strongly acidic media, mixtures of aqueous acetic acid and hydrochloric acid are often used. Where solubility difficulties are encountered, mixtures of aqueous hydrochloric acid with dimethylformamide may be used [464]. To suppress side reactions, the nitrite is added to the solution of the hydrazide at -5 to $-10°$ C. The azide, which forms within minutes, is extracted into organic solvents, dried briefly after washing, and used immediately for coupling. Sometimes the azide may be filtered off, washed, and used in moist form. When dimethylformamide is used as solvent, the solution may be neutralized with a tertiary base and the amino component added directly to this solution [464]. Azides of peptides containing acid-sensitive groups, such as the trityl or *t*-butyloxycarbonyl groups, may be prepared without difficulty if the temperature is sufficiently low and exposure to acid is brief [615, 619, 1129].

The substitution of *t*-butyl nitrite [580] or isoamyl nitrite [835] in the place of sodium nitrite permits the formation of the azide in mixtures of tetrahydrofuran and concentrated hydrochloric acid or in organic solvents with anhydrous HCl. Side reactions are minimized.

Side reactions encountered with azides include rearrangement to the isocyanate and hydrolysis to the amide. These side reactions have been reviewed and studied in detail by Schnabel [1090] and Honzl and Rudinger [580].

C. Mixed Anhydride Method

Mixed anhydrides between N-protected amino acids and a wide variety of organic and inorganic acids have been used in peptide synthesis. Those that have found the greatest application have been the anhydrides with monoesters of carbonic acid and with organic acids. The discussion here will be limited to these two types. Mixed anhydrides were introduced by Wieland and co-workers [1370, 1371]. A series of papers from several laboratories soon appeared, and the mixed anhydride procedure became permanently established in peptide chemistry.

1. Mixed Anhydrides with Monoesters of Carbonic Acid

This method was introduced independently by Wieland and Bernhard [1371], Boissonnas [143], and Vaughan [1310]. The anhydrides are formed by reaction of N-protected amino acids with various esters of chloroformic acid [Eq. (9)].

$$\mathrm{RNH\overset{R^1}{\overset{|}{C}}HCOOH + Cl\overset{O}{\overset{\|}{C}}OR^2 \xrightarrow{base} RNH\overset{R^1}{\overset{|}{C}}H\overset{O}{\overset{\|}{C}}{-}O{-}\overset{O}{\overset{\|}{C}}OR^2} \qquad (9)$$

Aminolytic cleavage of the anhydride yields only CO_2 and an alcohol, in addition to the peptide [Eq. (10)].

$$\mathrm{RNH\overset{R^1}{\overset{|}{C}}H\overset{O}{\overset{\|}{C}}{-}O{-}\overset{O}{\overset{\|}{C}}OR_2 + H_2N\overset{R^3}{\overset{|}{C}}HCOOR^4 \longrightarrow RNH\overset{R^1}{\overset{|}{C}}H\overset{O}{\overset{\|}{C}}NH\overset{R^3}{\overset{|}{C}}HCOOR^4 + R^2OH + CO_2} \qquad (10)$$

Wieland and Bernhard [1371] and Vaughan and Osato [1311] have carried out studies on the influence of the alkyl group on the course of the reaction. Other than methyl and phenyl esters which give unsatisfactory results, a wide variety of alkyl groups may be used. Ethyl chloroformate and isobutyl chloroformate have been most frequently used.

Normally the condensation of the acyl amino acid and chloroformate requires only a few minutes at -5 to $-10°$ C [1273, 1312]. Temperatures as high as $+10°$ C and reaction times up to 30 min have been used [338], although Wieland and Bernhard have reported that disproportionation of the anhydride occurs above 0° C [1372]. Tetrahydrofuran is favored as a solvent, although many other solvents have been used, including toluene [183, 1047, 1307], ethyl acetate [765], dimethylformamide [620, 668], dioxane [338, 567] and chloroform [29, 110, 183, 800]. Triethylamine is generally used to neutralize the HCl formed, although other tertiary amines have been used, including tri-*n*-butylamine [567, 1310], tri-*n*-propylamine [1086], and N-ethylpiperidine [1048].

The amino component is often added as a solution of its ester in an organic solvent. The reaction is generally allowed to proceed in the cold for a few hours and then overnight at room temperature. Sometimes brief heating is used to rapidly complete the reaction [1307, 1312]. Alternately, the amino component may be added as an aqueous solution of its salt [1418].

Few side reactions are encountered with this method of bond formation. Aminolytic cleavage of the anhydride at the wrong place may occur on rare occasions [413, 695, 1311, 1369]. No racemization occurs when protected amino acids are coupled, except for such N-acylated derivatives as the acetyl or benzoyl compounds. However, when the carboxyl group of a protected peptide is activated *via* the anhydride, some racemization must always be expected, although Vaughan and Eichler have reported the preparation of optically pure peptides by the mixed anhydride method [1313].

2. Mixed Anhydrides with Organic Acids

Mixed anhydrides of protected amino acids with carboxylic acids tend to cleave in a manner that does not lead to formation of the peptide bond [334, 1370]. The organic acids that have been used with good success have been very limited. Good results have been obtained with the sterically hindered isovaleric [399, 416, 666, 1024] and pivalic acid [762, 1419]. In the coupling of tosyl-glutaminyl-asparagine with S-benzyl-cysteine benzyl ester, where partial dehydration of the asparagine occurred with the use of tetraethyl pyrophosphite and dicyclohexylcarbodiimide, no dehydration product could be isolated when the mixed anhydride method was used [399]. The mixed anhydride with isovaleric acid gave the best yield.

3. Mixed Anhydrides with Esters of Phosphorous Acid

The ester chlorophosphites and pyrophosphites are used to form reactive amides with the amino component as well as reactive anhydrides with the carboxyl components [Eq. (11)]. For simplicity, however, these compounds will be considered in this section. The procedure involving prior formation of the reactive amide is known as

$$\mathrm{ROOC\overset{R^1}{\overset{|}{C}}HNH_2 + Cl\text{–}P(OC_2H_5)_2 \xrightarrow{base} ROOC\overset{R^1}{\overset{|}{C}}HNHP(OC_2H_5)_2} \qquad (11)$$

$$\mathrm{RNH\overset{R^1}{\overset{|}{C}}HCOOH + Cl\text{–}P(OC_2H_5)_2 \xrightarrow{base} RNH\overset{R^1}{\overset{|}{C}}HCOOP(OC_2H_5)_2}$$

the "amide" procedure and that involving formation of the anhydride as the "anhydride" procedure. More often, the "standard" procedure is used, involving addition of the chlorophosphite and base to a mixture of the amino and carboxyl components.

Anderson and co-workers introduced diethyl chlorophosphite for this purpose [39], and later, the more advantageous *o*-phenylene chlorophosphite [34, 40]. The most widely used of the phosphites has been tetraethylpyrophosphite [35, 36].

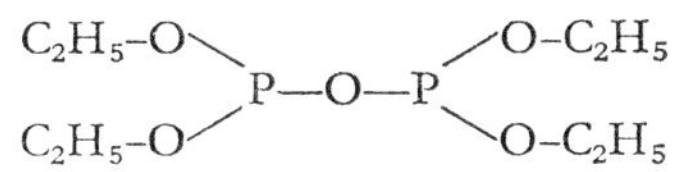

Tetraethyl Pyrophosphite

It has been used to prepare a number of naturally occurring polypeptides [304, 415, 579, 666, 1048]. In spite of the convenience of this reagent, and lack of side reactions, tetraethylpyrophosphite is in little, if any, use today. Perhaps this is because the reagent is not commercially available. The peptide chemist, like any other chemist, prefers to use what he can find on the shelf.

D. The Carbodiimide Method

Since its introduction into peptide synthesis by Sheehan and Hess in 1955 [1154], the carbodiimide method has become second to none. Its special applicability in the Merrifield solid phase procedure has added to its already wide use. The overall reaction may be represented as a dehydration [Eq. (12)]. The initial reaction involves

$$\mathrm{RNH\overset{R^1}{\overset{|}{C}}HCOOH + H_2N\overset{R^2}{\overset{|}{C}}HCOOR^3 + R^4\text{–}N\text{=}C\text{=}N\text{–}R^4 \longrightarrow RNH\overset{R^1}{\overset{|}{C}}HCONH\overset{R^2}{\overset{|}{C}}HCOOR^3 + R^4\text{–}NH\text{–}\overset{O}{\overset{\|}{C}}\text{–}NH\text{–}R^4} \qquad (12)$$

the formation of an intermediate [686, 1089], further reaction leading either to the desired peptide or to an acyl urea [Eq. (13)].

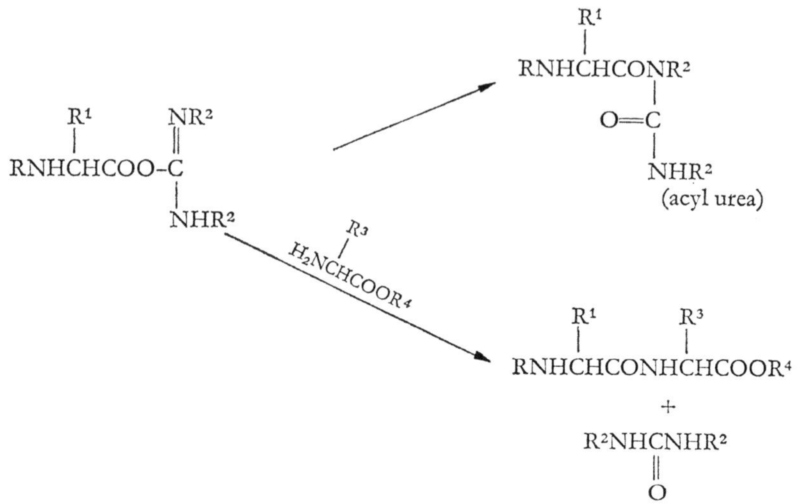

(13)

This tendency to form acyl ureas is the main side reaction encountered with the carbodiimide method [620, 1127, 1153, 1416]. Because low temperature reduces this tendency, a temperature of 0° C is generally used. It is possible to use a wide variety of solvents, including dimethylformamide [132, 146, 620, 1402], tetrahydrofuran [110, 855], aqueous tetrahydrofuran [108, 299], acetonitrile [464, 1127, 1153], chloroform [1063], pyridine [146, 525], and methylene chloride [855, 1153, 1402]. The formation of the acyl urea occurs more readily in dimethylformamide than in methylene chloride and acetonitrile. The use of strong bases also favors this side reaction [1089]. When the carboxyl group of asparagine is coupled *via* the carbodiimide method, dehydration of the amide group may occur [416].

The first carbodiimide used for peptide synthesis, and still most commonly used, was N,N'-dicyclohexylcarbodiimide. Dicyclohexylurea, the by-product of the reaction with this reagent, is only slightly soluble in most solvents, but difficulty is sometimes encountered in removing the last traces of the urea from the product [1155]. Sheehan and co-workers have prepared a number of water-soluble and acid-soluble carbodiimides by incorporating tertiary or quaternary bases in the structure [1151a, 1155]. The acylureas formed when these reagents are used are water-soluble and can be removed from the product by extraction with dilute acid.

Sheehan and associates have used the water-soluble carbodiimide 1-ethyl-3-(3-dimethylaminopropyl)carbodiimide for the synthesis of polypeptides containing up to seven amino acid residues without isolation of intermediates [1156]. A reaction time of 1 hr was allowed for each coupling reaction, and unreacted starting materials plus the urea and N-acylurea formed were washed out of the product with dilute acid and base. After removal of the aminoprotecting group, the product was used in the next step without further purification. The time required for synthesis of the peptides varied from 2.5 to 3.5 days.

Wolman and co-workers have reported the use of an insoluble carbodiimide, polyhexamethylenecarbodiimide, as a condensing agent in peptide synthesis [1387]. After the condensation was complete, the polymer was removed by filtration, and the product was then worked up in the usual manner. Yields of about 90% were reported for dipeptides.

E. Isoxazolium Salts

In 1961, WOODWARD and co-workers introduced the use of isoxazolium salts for activation of the carboxyl group [1394]. The N-ethyl-5-phenylisoxazolium-3'-sulfonate compound proved to be the most useful, and has become known as Woodward's reagent K, or simply, Woodward's reagent. A comprehensive report was published by these same workers in 1966 [1395].

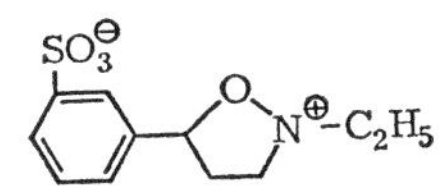

Woodward's Reagent

For activation, the N-protected amino acid or peptide is dissolved in solvent (preferably acetonitrile or nitromethane) containing triethylamine, and this solution is added to a suspension of Woodward's reagent in the same solvent. The suspension is stirred at 0° C until the reagent has dissolved. The amino acid ester or peptide ester (or the hydrochloride salt plus triethylamine) is added to this solution and the mixture is stirred overnight at room temperature. All reagents are used in equimolar amounts. Yields vary from about 80 to 95% [1395]. Even when the carboxyl group of carbobenzoxy-asparagine is activated, so that some dehydration of the amide group could be expected [416], the yields are 75 to 80% [1395]. Reference to the synthesis of a number of naturally occurring peptides using this reagent is given in the report by WOODWARD [1395].

Very little racemization is obtained when the carboxyl group of peptides is activated [1395]. The by-products of the reaction are water soluble, and therefore, easily removed from the product. Arginyl, seryl, asparaginyl, and glutaminyl peptides, ordinarily difficult to prepare in good yield, are obtained in excellent yield [1395]. Limitations on the choice of solvents is a disadvantage of the method.

KEMP and CHIEN have recently reported the activation of the carboxyl group by reaction with 2-ethyl-7-hydroxybenzisoxazolium fluoroborate [679]. The reaction product is an active ester, and this reagent will be considered in a later section devoted to active esters.

F. N,N-Carbonyldiimidazole

PAUL and ANDERSON, prompted by previous observations of WIELAND and SCHNEIDER [1374], and STAAB [1206], applied the reagent N,N-carbonyldiimidazole for the activation of the carboxyl group [38, 961]. Carbon dioxide and imidazole are

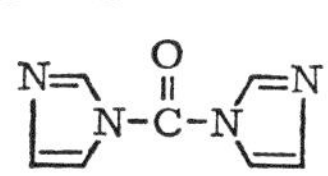

N,N-Carbonyldiimidazole

harmless by-products of the reaction, and carbon dioxide evolution drives the reaction to completion [Eq. (14)].

The formation of the activated derivative requires about 30 min, after which the

(14)

amino acid or peptide ester is added to the solution. Various solvents may be used, but for the activation step, the solvents must be dry since carbonyldiimidazole is highly sensitive to moisture [961]. The very convenient preparative procedure and the good yields obtained make this method highly useful. A disadvantage is the great sensitivity of the reagent to moisture. A number of naturally occurring peptides have been synthesized using this reagent [571, 657, 658, 962].

G. Active Ester Method

Active, or activated, esters have become, along with the carbodiimides, the most widely used reagents in peptide synthesis. Active esters were introduced by WIELAND and associates [1372].

A wide variety of active esters with great stability, enhanced reactivity and reduced tendency towards racemization make the method convenient and versatile. These active esters can be obtained in pure, crystalline form, and have a long shelf life. As a result, most are commercially available, greatly encouraging their use.

Ordinary alkyl esters are insufficiently activated for use as carboxyl components in peptide synthesis. An electrophilic alcoholic component, however, as, for example, in cyanomethyl esters [1124], thiophenyl esters [1124], and nitrophenyl esters [125, 126], correspondingly increases the electrophilic nature of the carbonyl carbon atom facilitating reaction with an amino group.

Systematic investigations by several groups have examined the influence of substituent groups on reactivity of esters [125, 126, 1118—1120, 1124, 1125, 1372]. The thiolesters and the cyanomethyl esters have such strong disadvantages in the synthesis of higher peptides that they will not be considered. The presence of sulfur in thiolesters tends to poison the catalyst in subsequent hydrogenations, while the cyanomethyl esters are more difficult to prepare than other equally active activated esters. JAKUBKE has published a recent review of the use of active esters in peptide synthesis [632].

1. O-Aryl Esters

a) p-Nitrophenyl Esters

Of the O-aryl esters, the *p*-nitrophenyl esters, first used by BODANSKY [125, 126], have assumed a leading position. A variety of methods have been used to prepare these esters [128, 329, 351, 417, 617, 1058, 1334], but the most convenient and widely used method is that employing dicyclohexylcarbodiimide [130, 131, 326, 1045]. The N-protected amino acid is treated with *p*-nitrophenol and dicyclohexylcarbodiimide. WOLMAN and co-workers have used a novel method for obtaining N-protected amino acid active esters directly from the amino acids [1388]. The amino acid sodium salt is treated with benzyl *p*-nitrophenyl carbonate, producing the carbobenzoxy amino acid and *p*-nitrophenol. Addition of an equivalent of dicyclohexylcarbodiimide then forms the *p*-nitrophenyl ester. Carbobenzoxy amino acid *p*-nitrophenyl esters were obtained in 65 to 80% yield. In a similar manner, using *t*-butyl *p*-nitrophenyl carbonate, the *t*-butyloxycarbonyl amino acid *p*-nitrophenyl esters were prepared in 70 to 75% yield. The paper by MEIENHOFER may be consulted for a list of *p*-nitrophenyl esters [844].

The *p*-nitrophenyl esters react with amino acid and peptide esters at room temperature in a variety of solvents. The reaction is complete within a few hours at the most, and the yields are generally excellent. The reaction is catalyzed by imidazole [834, 1375] and by certain bifunctional catalysts [109]. Imidazole was reported to give some racemization in a few cases, while 1,2,4-triazole gave no racemization and was a more efficient catalyst [109].

GOODMAN and STUEBEN prepared tripeptides by a novel method *via* the active esters [426]. Carbobenzoxy amino acid *p*-nitrophenyl esters were treated with HBr in acetic acid to yield crystalline amino acid *p*-nitrophenyl ester hydrobromide salts. These were coupled with carbobenzoxy amino acids *via* the mixed anhydride or dicyclohexylcarbodiimide method to give the carbobenzoxy dipeptide *p*-nitrophenyl esters. Addition of amino acid esters directly to the reaction mixture containing the active esters gave the carbobenzoxy tripeptide esters. KOVACS and RODIN have suggested a variation of this procedure [730]. They have found that carbobenzoxy amino acid and peptide *p*-nitrophenyl esters can be catalytically hydrogenated in the presence of an equivalent of hydrochloric acid to the amino acid and peptide *p*-nitrophenyl ester hydrochlorides, if the hydrogenation is stopped after uptake of one mole of hydrogen.

No side reactions are encountered during the coupling step with *p*-nitrophenyl esters, but intramolecular cyclization of some derivatives may occur. The *p*-nitrophenyl ester of carbobenzoxy-nitroarginine cyclizes to the lactam [963], and *p*-nitrophenyl esters of carbobenzoxy dipeptides with C-terminal cyclize to the carbobenzoxy diketopiperazine [427]. The amino acid *p*-nitrophenyl esters tend to racemize when in solution in the presence of strong bases [139]. Racemization is not a problem in the coupling reaction when no excess base is added, but some racemization may occur during the preparation of carbobenzoxy peptide *p*-nitrophenyl esters.

Preparation of the *p*-nitrophenyl esters of carbobenzoxy threonine and serine by the carbodiimide method yields mixtures due to O-acylation [137]. The 2,4-dinitrophenyl esters of carbobenzoxy threonine [1033] and serine [1136], however, can be prepared in good yield, due to the enhanced nucleophilicity of the phenolic group in dinitrophenol. BODANSZKY and ONDETTI also discovered that the 2,4-dinitrophenyl ester of carbobenzoxy-nitroarginine could be prepared without lactam formation [137].

A large number of naturally occurring peptides have been prepared *via* *p*-nitrophenyl esters. These include ACTH [1136] and its analogues [931, 1272], insulin [669, 847], secretin [138], oxytocin [133], lysine-vasopressin [135], partial sequences of glucagon [1397], and bradykinin [922, 1058].

b) Chlorophenyl Esters

KUPRYSZEWSKI [740—742] and PLESS and BOISSONNAS [987] have investigated the application of various chlorophenyl esters to peptide synthesis. PLESS and BOISSONNAS synthesized a great variety of chloro-, bromo-, and iodophenyl esters. Their results favored the use of the 2,4,5-trichlorophenyl ester, while good results were also obtained with the pentachlorophenyl ester. These active esters may be synthesized by the phosphorous oxychloride method [741], with tri-(2,4,5-trichlorophenyl)-phosphite [987], or with dicyclohexylcarbodiimide and the phenol [987]. Kovacs and associates have reported the synthesis of a series of *t*-butyloxycarbonyl amino acid pentachlorophenyl esters [729] and carbobenzoxy-amino acid pentachlorophenyl esters [728], including among the latter those of carbobenzoxy-nitro arginine and carbobenzoxy-serine. In a series of papers, they have reported on the polymerization of amino acid and peptides using pentachlorophenylesters (see reference [728]).

Sakakibara and Inukai have applied 2,4,5-trichlorophenyl trifluoroacetate and pentachlorophenyltrifluoroacetate to the synthesis of the carbobenzoxy-amino acid chlorophenyl esters [1058]. They also reported the direct synthesis of peptides by adding the trifluoroacetates to a mixture of the carboxyl and amino components.

Wolman and co-workers have synthesized 2,4,5-trichloro-, 2,4,6-trichloro-, and pentachlorophenyl esters of carbobenzoxy-amino acids directly from amino acids *via* the benzyl chlorophenyl carbonates and dicyclohexylcarbodiimide in the manner described for the *p*-nitrophenyl esters in the preceeding section [1388]. In an analogous manner using *t*-butyl 2,4,5-trichlorophenyl carbonate, Broadbent and associates have reported the preparation of *t*-butyloxycarbonyl amino acid 2,4,5-trichlorophenyl esters [177].

Stewart has prepared the 2,6-dichloro-4-nitrophenyl esters of several carbobenzoxy-amino acids but found them unsuitable for peptide synthesis [1234].

Kapoor has claimed the synthesis of carbobenzoxy-dipeptide and -tripeptide pentachlorophenyl esters by addition of a carbobenzoxy-amino acid or -dipeptide pentachlorophenyl ester, along with an equivalent of triethylamine and 2-hydroxypyridine, to a solution of the hydrochloride of an amino acid pentachlorophenyl ester [656].

2,4,5-Trichlorophenyl esters of carbobenzoxy-amino acids have been used in the preparation of several naturally occurring peptides [43, 250, 489].

c) 8-Hydroxyquinoline and 5-Chloro-8-hydroxyquinoline Esters

Jakubke and Voigt have prepared a number of the 8-hydroxyquinoline esters of amino acids with various protecting groups and have tested their utility in peptide synthesis [633]. They could be prepared by either the mixed anhydride or the carbodiimide method. They reported the preparation of peptides in good yield and free of racemization, even when coupling carbobenzoxy peptides. The 8-hydroxyquinoline and by-products could be removed by extraction with dilute acid.

Jakubke and Voigt have also reported the preparation of the 5-chloro-8-hydroxyquinoline esters of carbobenzoxy amino acids and the preparation of a number of peptides using these esters [634]. These esters are stable, crystalline derivatives and exhibited the same favorable characteristics as the 8-hydroxyquinoline esters. In contrast to the *p*-nitrophenyl esters, the 8-hydroxy- and 5-chloro-8-hydroxyquinoline esters are remarkably resistant to racemization in the presence of strong base.

2. N-Hydroxysuccinimide Esters

The N-hydroxysuccinimide active esters, introduced into peptide synthesis by Anderson and co-workers [41], have proven to be very useful. They are most conveniently synthesized by the carbodiimide method. A series of carbobenzoxy-

and *t*-butyloxycarbonyl-amino acid hydroxysuccinimide esters have been reported [41]. They are stable, crystalline compounds. Reaction with amino acid and peptide esters is usually complete within 1 hr. Hydroxysuccinimide is water-soluble and thus easily removed from the products. The hydroxysuccinimide esters may also be reacted with amino acids and peptides in the presence of bicarbonate in aqueous organic solvent mixtures in very good yield.

WEYGAND and associates have reported that addition of 2 moles of N-hydroxysuccinimide, when using the dicyclohexylcarbodiimide method to couple carbobenzoxy peptides, eliminates practically all racemization and formation of acyl urea [1354]. WÜNSCH and DREES increased the yield from 31% to 75% by the addition of N-hydroxysuccinimide when coupling a pentapeptide to a tripeptide *via* the dicyclohexylcarbodiimide method [1397].

H-Hydroxysuccinimide esters have been used in the synthesis of phyllokinin [97], in the glucagon series [1401, 1403, 1404], in the insulin series [1415], in the ACTH series [945, 841], and for the synthesis of an octapeptide including the "reactive" serine of chymotrypsin [747].

3. 1-Hydroxypiperidine Esters

HANDFORD and co-workers found that the 1-hydroxypiperidine esters of N-protected amino acids and peptides react readily at room temperature in high yield with amino acid and peptide esters [489]. They can be prepared by the carbodiimide method, the mixed anhydride method, the acid chloride method [489], and with WOODWARD's reagent [490]. The presence of acetic acid [489] or 1,2,4-triazole [490] considerably enhances the reaction rate during the peptide bond-forming step. These esters are optically stable in the presence of base and little or no racemization occurs when coupling carbobenzoxy-peptide 1-hydroxypiperidine esters [489, 490]. The synthesis of a series of dipeptides and tetrapeptides using 1-hydroxypiperidine esters has been described [490].

WEYGAND and co-workers have described a novel method for the preparation of dipeptides and tripeptides using 1-hydroxypiperidine esters [1357]. Amino acid N-carboxyanhydrides are condensed with the hydrochloride salt of 1-hydroxypiperidine to yield amino acid 1-hydroxypiperidine ester hydrochlorides. These are coupled as the amino components with carbobenzoxy-amino acids by a variety of methods to produce carbobenzoxy-dipeptide 1-hydroxypiperidine esters. These are further condensed with amino acid esters to yield carbobenzoxy-tripeptide esters. Yields were generally good with no detectable racemization.

4. Derivatives of N-Ethyl Benzisoxazolium Salts

Based on earlier studies of KEMP and WOODWARD [680], KEMP and CHIEN have prepared N-ethyl-7-hydroxybenzisoxazolium fluoroborate and have used it for peptide synthesis [679].

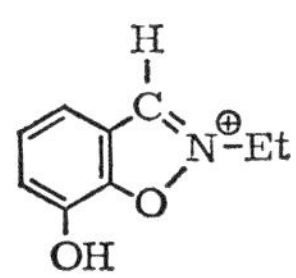

N-Ethyl-7-Hydroxyisoxazolium Cation

N-Protected amino acids in the presence of base react to form stable 3-acyloxy-2-hydroxy-N-ethylbenzamides. These compounds are optically stable in the presence of base and react with amino acid esters or salts in good yield.

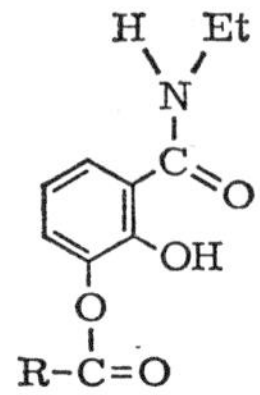

3-Acyloxy-2-hydroxy-N-ethylbenzamide

RAJAPPA and AKERKAR have synthesized 5,7-dichloro-N-ethylbenzisoxazolium fluoroborate and have prepared active esters with this compound, including a crystalline ester of carbobenzoxy-serine [1008]. They postulate that these active esters are 2-acyloxy-3,5-dichloro-N-ethylbenzamides:

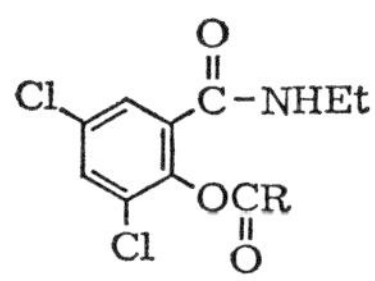

Carbobenzoxy-serylglycine *p*-nitrobenzyl ester was produced in about 80% yield using this active ester of carbobenzoxy-glycine.

5. Miscellaneous Active Esters

Other active esters that have been reported include those of N-hydroxyphthalimide [909, 910, 1212, 1334], N-hydroxyglutarimide [637b], 3-nitroacetophenonoxime [792b], and 1-phenyl-3-methylpyrazolone-(5) [792b].

6. Resin Active Esters

Several research groups have prepared derivatives of resins which yield active esters when coupled with N-protected amino acids. WIELAND and BIRR [1367] used a gel crosslinked with *p*-*p*'-dihydroxydiphenyl sulfone:

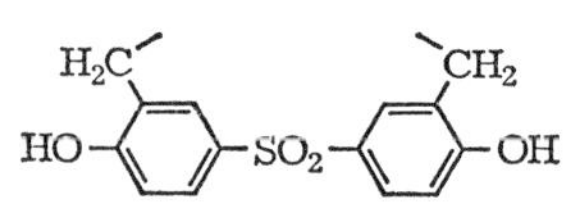

Active esters with carbobenzoxy and *t*-butyloxycarbonyl amino acids were formed *via* the mixed anhydride or carbodiimide method. The N-protected amino acid resin ester column was percolated with a solution of amino acid or peptide ester in dimethylformamide to yield di- and tripeptide esters in 56 to 71% yield.

Fridkin and co-workers [394] prepared active esters with carbobenzoxy amino acids and a poly-4-hydroxy-3-nitrostyrene crosslinked with divinylbenzene:

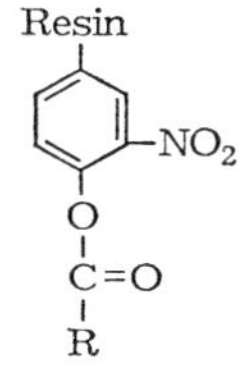

The resin active esters were prepared using the carbodiimide method. The carbobenzoxy amino acid resin active esters were stirred with a dimethylformamide solution of amino acid and peptide esters to yield polypeptides in good yield.

Laufer and associates [748] prepared copoly-(ethylene-N-hydroxymaleimide):

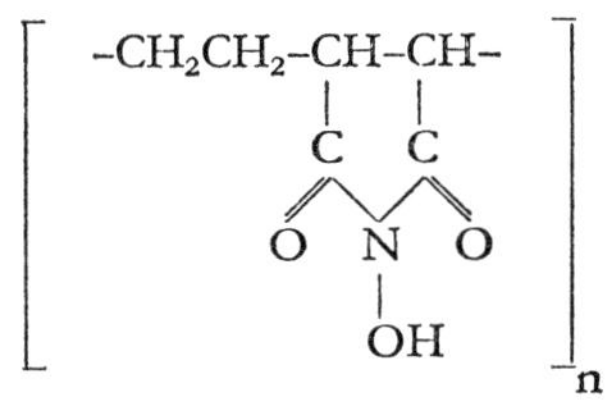

t-Butyloxycarbonyl amino acids were condensed with this resin to give active esters of the N-hydroxysuccinimide type. They were prepared *via* the mixed anhydride or carbodiimide method. Peptides were prepared by stirring the active ester resin with amino acid and peptide esters in dimethoxyethane or dimethylformamide. Insoluble by-products were filtered off and evaporation of the solvent gave chromatographically pure peptides in nearly quantitative yields. The preparation of a series of peptides by the method was described [748].

H. Coupling via Oxidation

1. Oxidation of N-Acyl Amino Acid Phenylhydrazides

Milne and Most have discovered that phenylhydrazides of carbobenzoxy amino acids can be oxidized in nonaqueous solutions with N-bromosuccinimide or lead tetraacetate to the crystalline carbobenzoxy amino acid phenyldiimides [868, 869]. These react with amino acid or peptide phenylhydrazides to yield carbobenzoxy peptide phenylhydrazides. They formulate these reactions as in Eq. (15):

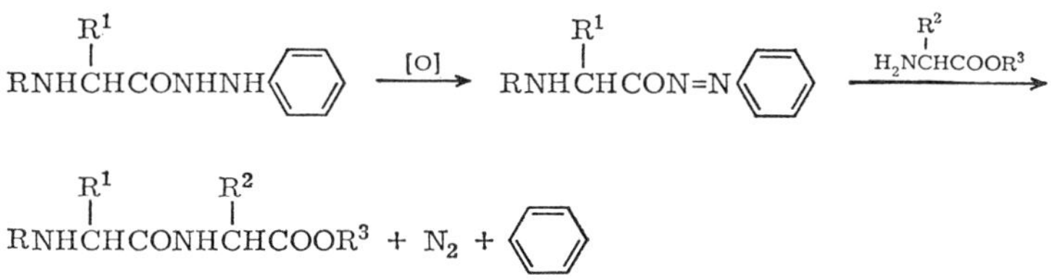

(15)

They also employed *o*-chlorophenylhydrazides for the same purpose [869]. Yields were good. The use of a carboxyl-protecting group that could participate directly in a coupling reaction would be very useful. However, disadvantages of these derivatives are the instability of the phenyldiimides and the significant amount of racemization encountered in some couplings. Furthermore, amino acids, which are sensitive to oxidation, such as tryptophan, could not be used.

2. Oxidation-Reduction Condensation

Mukaiyama and co-workers [891] have found that when a solution of the copper (II) or mercury (II) salt of a carboxylic acid and a sulfenamide is treated with triphenylphosphine, a carboxamide is formed [Eq. (16)]:

$$[(R^1CO_2)_2Cu] + 2R^2SNR^3R^4 + 2Ph_3P \longrightarrow [(R_2S)_2Cu] + 2R^1CONR^3R^4 + 2Ph_3P{=}O \quad (16)$$

When equimolar amounts of carbobenzoxy-phenylalanine, mercury (II) chloride, glycine ethyl ester, di-*o*-nitrophenyl disulfide, and triphenylphosphine in methylene chloride were stirred at room temperature for 3 hrs, the carbobenzoxy dipeptide ester was obtained in 89% yield.

o-Nitrophenylsulfenyl amino acid esters, being sulfenamides, could be used as the amino component. Thus, carbobenzoxy-phenylalanine was condensed with NPS-glycine ethyl ester to give carbobenzoxy-phenylalanylglycine ethyl ester in 68% yield.

I. Leuch's Anhydride Method

Hirschmann and co-workers have reported the controlled synthesis of peptides in aqueous medium using amino acid N-carboxyanhydrides (Leuch's anhydrides) and amino acid N-thiocarboxyanhydrides [278, 281, 553, 1314]. Previously, N-carboxyanhydrides had been used successfully only for the preparation of polyamino acids [662] [Eq. (17)].

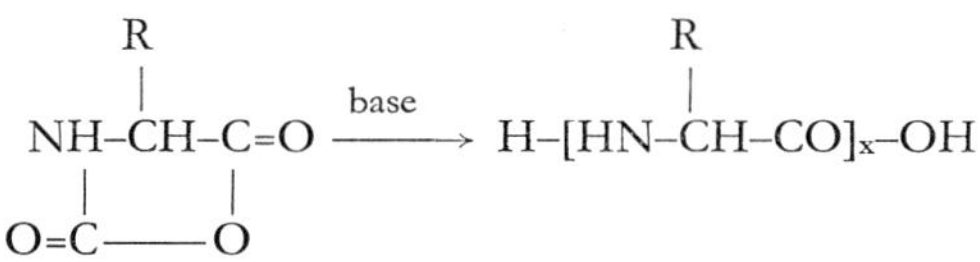

(17)

Polymers of various chain lengths are obtained. Now it has been found possible to use N-carboxyanhydrides in a stepwise synthesis by careful control of pH. A crystalline N-carboxyanhydride is added to a cold aqueous solution of an amino acid at pH 10.2 with high speed mixing, and after 2 min, the solution is acidified to yield the dipeptide [Eq. (18)].

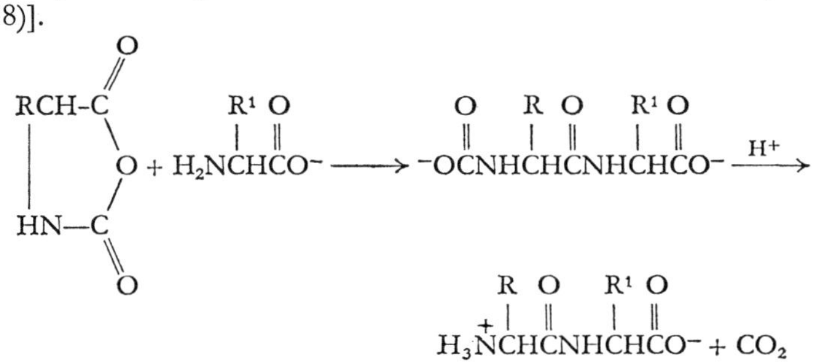

(18)

In the condensation step, the amino group of the resulting dipeptide is in the form of a carbamate ion, which protects the amino group from further reaction during the brief time required for completion of the coupling. Acidification decomposes the carbamate and releases the amino group for further coupling. After readjustment of the pH to 10.2 and addition of a second N-carboxyanhydride, the condensation may be repeated.

The advantages of this method is the lack of a requirement for blocking groups, and the speed of the reaction. Thus, free peptides are produced directly, and the time required for each step is but a few minutes. This would indeed be an ideal method for peptide synthesis, if it could be extended to the synthesis of higher peptides. The synthesis of many dipeptides, in yields of 90% or better, have been reported [554], as well as the synthesis of a tripeptide [1315].

The method is free of racemization when N-carboxyanhydrides are used [1315], but racemization occurs when N-thiocarboxyanhydrides are employed [281]. The use of the N-thiocarboxyanhydrides of glycine and histidine is advantageous. In a synthesis using the N-thiocarboxyanhydride of glycine, the yield obtained was increased from 37% to 95% as compared to the use of the N-carboxyanhydride. The use of the N-thiocarboxyanhydride of histidine was possible without protection for the imidazole group, which is not possible with the N-carboxyanhydride. Of the amino acids providing the amino component for the reaction, only lysine and cysteine require protecting groups on the side chains. It is reported that the N-carboxyanhydrides of amino acids, except those of proline and glycine, are stable for several months when stored at low temperature.

The successful use of a multi-step synthesis without isolation of intermediates requires nearly 100% reaction at each step without the occurrence of side reactions. Two side reactions occur during synthesis using the N-carboxyanhydride method. These are the formation of hydantoic acids and over-reaction [554]. These side reactions may impose serious limitations on the length of the peptide chain obtainable by this method.

J. Enzymatic Synthesis

SNELLMAN has reported the isolation of an enzyme from beef liver which can be used in a stepwise synthesis of polypeptides [1187]. In the presence of ascorbic acid and a nucleoside triphosphate, especially guanosine triphosphate (GTP), the enzyme couples the amino component to the carboxyl group of amino acid esters, alcohol being split out. The series of reactions which take place are formulated as follows [Eqs. (19) to (21)]:

$$H_2NR + GTP \longrightarrow GMP + (RNH\text{–}OP_2O_6^{--}) \qquad (19)$$

$$H_2N\underset{\displaystyle |}{\overset{\displaystyle R^1}{C}}HCOOR^2 + (RNH\text{–}OP_2O_6^{--}) \longrightarrow (H_2N\overset{R^1}{\overset{|}{C}}HCO\overset{R}{\overset{|}{N}}\text{–}P_2O_6^{--}) + R^2OH \qquad (20)$$

$$(H_2N\overset{R^1}{\overset{|}{C}}HCO\overset{R}{\overset{|}{N}}\text{–}P_2O_6)^{--} \longrightarrow H_2N\overset{R^1}{\overset{|}{C}}HCONHR + P_2O_6^{--} \qquad (21)$$

An amino component with an amino group pK of 7.8 to 8 is required. Since the pK's of the α-amino group of amino acids range from 9 to 10, and those of dipeptides are about 8.5, amino acids and dipeptides cannot be used as the amino component. Tripeptides, with pK's for the α-amino group of 7.8 to 8, are suitable.

The reaction is run as follows. To the enzyme solution is added an equivalent of tripeptide, followed by an equivalent of GTP. One minute later, one equivalent of the methyl ester of the first amino acid to be coupled is added. After 2 or 3 min, a second equivalent of GTP is added, followed in 1 min by the ester of the second amino acid. This is repeated until the synthesis of the peptide chain is complete. SNELLMAN has claimed by this method to have synthesized histidyl-phenylalanyl-arginyl-tryptophyl-glycyl-glycyl-glycine in 90% yield, and to have synthesized α-MSH and four of its analogs and the first 22 amino acid sequence of ACTH. This method apparently has great potential, if scale-up is possible. Further evaluation must await future publications.

IV. Merrifield Solid Phase Method

The introduction of the Merrifield solid phase method of peptide synthesis [851] has had a greater impact on this field than any other development since the introduction of the carbobenzoxy group. The inexperienced and the naive may feel that the Merrifield method has removed all problems, and that the automated synthesis of pure proteins is now assured. Others assume a more cautions attitude, realizing that every method involving the formation of chemical bonds is subject to limitations and side reactions. In a recent review [98], several workers in the field gave varying opinions. BODANSZKY characterized the solid phase method as a promising approach to the synthesis of long-chain polypeptides, while mentioning that several problems remain to be solved. ONDETTI stated that many people have experienced difficulty in applying the method. SCHWYZER characterized the Merrifield method as "the coming strategy".

The solid phase method has, in principle, greatly simplified the course required to synthesize a polypeptide, and has reduced the time to complete a synthesis to a fraction of that formerly required. Overall yields have been substantially increased. These results have been brought about by the fact that no intermediates are isolated, necessitating no time-consuming and yield-reducing purification procedures. As in other methods in which no intermediates are isolated, success depends upon nearly quantitative yield at each coupling step and the absence of side reactions. The occurrence of side reactions during coupling steps and during cleavage of protecting groups and of the completed peptide chain from the resin undoubtedly accounts for many of the problems that have been encountered. Nevertheless, outstanding success has been reported in the synthesis of a number of naturally occurring peptides by this method, including bradykinin and analogs [852, 853], angiotensin and analogs [688, 817], oxytocin [104], insulin [564, 808], and a decapeptide from tobacco mosaic virus protein [1237], to mention just a few. The method has now been automated [857], and the A chain of insulin was synthesized in 8 days using this apparatus [808].

In the solid phase method, an N-protected amino acid is bound to a resin by esterification [Eq. (22)]. The N-protecting group is removed [Eq. (23)], and a second

N-protected amino acid is coupled to the amino group of the resin-bound amino acid [Eq. (24)]. All side products and unreacted soluble materials are washed out of the resin, the N-protecting group is cleaved, and the coupling step is repeated with a third N-protected amino acid [Eq. (25)]. The process is continued until the synthesis is complete. The peptide is now cleaved from the resin [Eq. (26)], and purified by conventional techniques [in Eqs. (22) to (26), BOC = *t*-butyloxycarbonyl, and DCCI = dicyclohexylcarbodiimide].

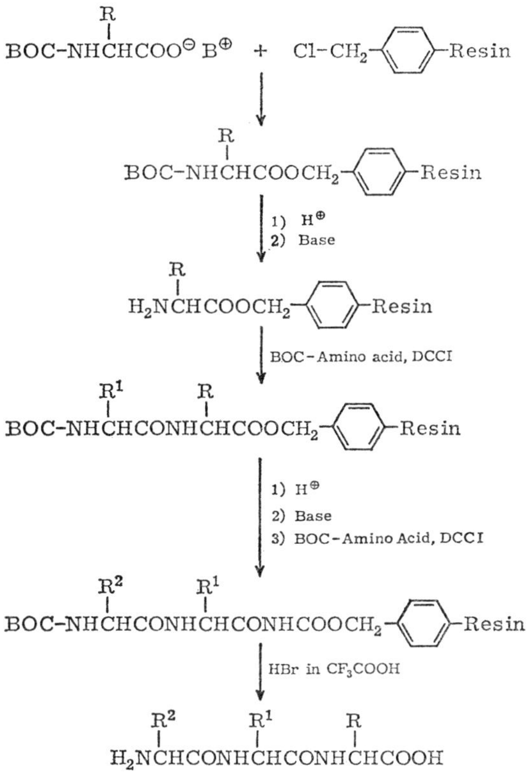

(22)

(23)

(24)

(25)

(26)

The resin used by MERRIFIELD [854] is a copolymer made from 98% styrene and 2% divinylbenzene. To provide a reactive group for attachment of the first amino acid, the benzene rings of the resin are partially chloromethylated. When this chloromethylated resin is treated with the triethylamine salt of an N-protected amino acid, a benzyl ester-type bond is formed. Such a bond is stable during the synthetic steps, but can be cleaved with HBr in acetic acid or trifluoroacetic acid, which simultaneously removes the N-protecting group and cleaves the peptide from the resin [Eq. (26)].

In earlier work [851], the carbobenzoxy group was used for N-protection, but the necessity to use HBr in acetic acid to remove this protecting group caused a number of problems. Use of the *t*-butyloxycarbonyl group, which is cleaved with HCl in acetic acid, a much milder reagent, alleviated these difficulties. The *t*-butyloxycarbonyl group is used almost exclusively, although the *o*-nitrophenylsulfenyl group can also be used with advantage [564].

The coupling reagent most widely used has been dicyclohexylcarbodiimide. Using methylene chloride as the solvent, and about a 50% excess of the *t*-butyloxycarbonyl amino acid and dicyclohexylcarbodiimide, the reaction is complete and quantitative in a few minutes [854]. Dimethylformamide is also used as solvent. Active esters can also be used in the coupling steps. The exclusive use of *p*-nitrophenyl esters have been reported by Hoernle for the synthesis of the A-chain of bovine insulin [564], and by Bodanszky and Sheehan for the synthesis of several polypeptides [141, 142]. Beyerman and co-workers used both *p*-nitrophenyl and N-hydroxysuccinimide esters, with the addition of 1,2,4-triazole to catalyze the condensation [104]. Bodanszky and Sheehan [142] used a hydroxymethylated resin rather than a chloromethylated resin. This permitted the esterification to be accomplished using dicyclohexylcarbodiimide, a much milder method than is necessary for esterification to a chloromethylated resin.

Difficulties encountered in this method may be due to various factors. These include incomplete coupling, creating chains of various lengths and amino acid sequence, side reactions during coupling, such as might occur between side chains, and side reactions induced by the treatment with anhydrous acid during the removal of protective groups and cleaving the peptide from the resin. These problems which no doubt increase with chain length, may finally determine the ultimate limit to which the chain may be lengthened. The synthesis of true proteins is not yet in sight.

References

1. Abderhalden, E., Brockmann, H.: Biochem. Z. **225**, 386 (1930)
2. Abderhalden, E., Komm, E.: Z. physiol. Chem., Hoppe-Seyler's **139**, 147 (1924)
3. Acher, R., Crocker, C.: Biochim. Biophys. Acta **9**, 704 (1952)
4. Acher, R., Laurila, U. R.: Bull. soc. chim. biol. **35**, 413 (1953)
5. Ackers, G. K.: Biochemistry (U.S.A.) **3**, 723 (1964)
6. Adams, J. M.: J. Mol. Biol. **33**, 571 (1968)
7. Adams, J. M., Capecchi, M. R.: Proc. Natl. Acad. Sci. (US) **55**, 147 (1966)
8. Adkins, H., Roos, R. M., Schroeder, D. C., Maloney, C. L., Gilbert, W. W.: J. Am. Chem. Soc. **76**, 147 (1954)
9. Africa, B., Carpenter, F. H.: Biochem. Biophys. Res. Commun. **24**, 113 (1966)
10. Akabori, S., Ikenaka, T.: J. Biochem. (Tokyo) **42**, 603 (1956)
11. Akabori, S., Ikenaka, T., Okada, Y., Kohno, K.: Proc. Japan Acad. **29**, 509 (1953)
12. Akabori, S., Narita, K.: In: Protein chemistry, vol. 2 (eds. S. Akabori, S. Mizushima), p. 170. Tokyo: Kyoritsu Shuppan 1954 (in Japanese)
13. Akabori, S., Ohno, K., Ikenaka, T., Nagata, A., Haruna, I.: Proc. Japan Acad. **29**, 561 (1953)
14. Akabori, S., Ikenaka, T., Nagata, A., Okada, Y., Hanafusa, H., Haruna, H., Tsugita, A., Sugae, K., Matsushima, T.: Bull. Chem. Soc. Japan **29**, 507 (1956)
15. Akabori, S., Ikenaka, T., Narita, K.: Bull. Chem. Soc. Japan **25**, 214 (1952)
16. Akabori, S., Narita, K., Hanabusa, H., Tsugita, A.: Nippon Kagaku Zasshi **76**, 225 (1955).
17. Akabori, S., Sakakibara, S., Shimonishi, Y., Nobuhara, Y.: Bull. Chem. Soc. Japan **37**, 433 (1964)
18. Albertson, N. F., McKay, F. C.: J. Am. Chem. Soc. **75**, 5323 (1953)
19. Albertson, N. F., McKay, F. C.: J. Am. Chem. Soc. **75**, 5325 (1953)
20. Alberty, R. A.: In: The proteins, vol. I, part A (1st ed.) (eds. N. Neurath, K. Bailey). New York: Academic Press 1956
21. Alexander, N. M.: Anal. Chem. **30**, 1292 (1958)
22. Allison, W. S., Admiraal, J., Kaplan, N. O.: J. Biol. Chem. **244**, 4743 (1969)
23. Alving, R. E., Laki, K.: Biochemistry (U.S.A.) **5**, 2597 (1966)
24. Alving, R. E., Moczar, E., Laki, K.: Biochem. Biophys. Res. Commun. **23**, 540 (1966)
25. Ambler, R. P.: Biochem. J. **96**, 32p (1965)
26. Ambler, R. P.: In: Methods in enzymology, vol. XI (ed. C. H. W. Hirs), p. 155, 436. New York: Academic Press 1967
27. Ambler, R. P., Brown, L. H.: Biochem. J. **104**, 784 (1967)
28. Ambler, R. P.: In: Methods in enzymology, vol. XXV (eds. C. H. W. Hirs, S. N. Timasheff), p. 143, 262. New York: Academic Press 1972
29. Amiard, G., Heymes, R., Velluz, L.: Bull. soc. chim. France **1955**, 191
30. Amiard, G., Heymes, R., Velluz, L.: Bull. soc. chim. France **1956**, 698
31. Amirkhanyan, M. M., Stephanov, V. M.: Biokhimiya **35**, 825 (1970)
32. Anderer, F. A., Uhlig, H., Weber, E., Schramm, G.: Nature **186**, 922 (1960)
33. Anderson, G. W.: Ann. Rep. Med. Chem. **1966**, 296
34. Anderson, G. W., Blodinger, J., Young, R. W., Welcher, A. D.: J. Am. Chem. Soc. **74**, 5304 (1952)
35. Anderson, G. W., Blodinger, J., Welcher, A. D.: J. Am. Chem. Soc. **74**, 5309 (1952)
36. Anderson, G. W., Callahan, F. M.: J. Am. Chem. Soc. **82**, 3359 (1960)
37. Anderson, G. W., McGregor, A. C.: J. Am. Chem. Soc. **79**, 6180 (1957)
38. Anderson, G. W., Paul, R.: J. Am. Chem. Soc. **80**, 4423 (1958)
39. Anderson, G. W., Welcher, A. D., Young, R. W.: J. Am. Chem. Soc. **73**, 501 (1951)

40. ANDERSON, G. W., YOUNG, R. W.: J. Am. Chem. Soc. **74**, 5307 (1952)
41. ANDERSON, G. W., ZIMMERMAN, J. E., CALLAHAN, F. M.: J. Am. Chem. Soc. **86**, 1839 (1964)
42. ANDERSON, J. C., BARTON, M. A., HARDY, P. M., KENNER, G. W., MACLEOD, J. K., PRESTON, J., SHEPPARD, R. C.: Acta Chim. Acad. Sci. Hung. **44**, 187 (1965)
43. ANDERSON, J. C., KENNER, G. W., MACLEOD, J. K., SHEPPARD, R. C.: Tetrahedron Letrs. **39** (Suppl. 8) (1966)
44. ANDO, T., TOBITA, T., YAMASAKI, M.: J. Biochem. (Tokyo) **45**, 285 (1958)
45. ANFINSEN, C. B., HABER, E.: J. Biol. Chem. **236**, 1361 (1961)
46. ANFINSEN, C. B., SELA, M., COOKE, J. P.: J. Biol. Chem. **237**, 1825 (1962)
47. ANFINSEN, C. B., SELA, M., TRITSCH, H.: Arch. Biochem. Biophys. **65**, 156 (1956)
48. ANSON, M. L.: J. Gen. Physiol. **20**, 663 (1937)
49. ATFIELD, G. N., MORRIS, C. J. D. R.: Biochem. J. **81**, 606 (1961)
50. ATTRILL, J. E., BUTTS, W. C., RAINEY, W. T., Jr., HOLLEMAN, J. W.: Anal. Letters **3**, 59 (1970)
51. AWE, W., REINECKE, I., THUM, J.: Naturwissenschaften **41**, 528 (1954)
52. BABA, T., SUGIYAMA, H., SETO, S.: J. Biochem. (Tokyo) **72**, 1571 (1972)
53. BAHL, O. P., SMITH, E. L.: J. Biol. Chem. **240**, 3585 (1965)
54. BAILEY, J. L.: In: Techniques in protein chemistry, p. 190. Amsterdam, London, New York: Elsevier Publ. Co. 1967
55. BAILEY, J. L.: Biochem. J. **60**, 173 (1955)
56. BAILEY, J. L., MOORE, S., STEIN, W. H.: J. Biol. Chem. **221**, 143 (1956)
57. BAKER, L. E.: J. Biol. Chem. **193**, 809 (1951)
58. BALENOVIC, K., GAŠPERT, B.: Chem. & Ind. (London) **1957**, 115
59. BANGHAM, A. D., HILL, M. W.: Nature **237**, 408 (1972)
60. BAPTIST, W. H., BULL, H. B.: J. Am. Chem. Soc. **75**, 1727 (1953)
61. BARKDOLL, A. E., ROSS, W. F.: J. Am. Chem. Soc. **66**, 951 (1944)
62. BARMAN, T. E., KOSHLAND, D. E., Jr.: J. Biol. Chem. **242**, 5771 (1967)
63. BARRAS, B. C., ELMORE, D. T.: J. Chem. Soc. **1957**, 3134
64. BARRETT, G. C.: Chem. Commun. **1967**, 487
65. BARRETT, G. C.: J. Chem. Soc. (C) **1967**, 1
66. BARRETT, G. C.: Nature **194**, 1171 (1962)
67. BARRETT, G. C., KHOKHAR, A. R.: J. Chromatog. **39**, 47 (1969)
68. BARRON. E. S. G.: Advances in Enzymol. **11**, 201 (1951)
69. BATTERSBY, A. R., ROBINSON, J. C.: J. Chem. Soc. **1955**, 259
70. BEAVEN, G. H., HOLIDAY, R. E.: Advances in Protein Chem. **7**, 319 (1952)
71. BEECHAM, A. F.: J. Am. Chem. Soc. **79**, 3257 (1957)
72. BELLO, C. D., SIGNOR, A.: J. Chromatog. **17**, 506 (1965)
73. BENCZE, W. L., SCHMID, K.: Anal. Chem. **25**, 1193 (1957)
73a. BENEDIT, R. C., STEDMAN, R. L.: Analyst **95**, 296 (1970)
74. BENESCH, R., BENESCH, R. E.: In: Methods of biochemical analysis, vol. 10, p. 43. New York: Interscience Publ. Co. 1962
75. BENESCH, R., BENESCH, R. E.:, GUTCHO, M., LAUFER, L.: Science **123**, 981 (1956)
76. BENESCH, R., BENESCH, R. E.: Biochim. Biophys. Acta **23**, 658 (1957)
77. BENESCH, R., STEDMAN, R. L.: Analyst **95**, 296 (1970)
78. BENESCH, R. E., LARDY, H. A., BENESCH, R.: J. Biol. Chem. **216**, 663 (1955)
79. BEN-ISHAI, D., BERGER, A.: J. Org. Chem. **17**, 1564 (1952)
80. BENISEK, W. F., COLE, R. D.: Biochem. Biophys. Res. Commun. **20**, 655 (1965)
81. BENISEK, W. F., RAFERTY, M. A., COLE, R. D.: Biochemistry (U.S.A.) **6**, 3780 (1967)
82. BENNETT, H. S., YPHANTIS, P. A.: J. Am. Chem. Soc. **70**, 3522 (1948)
83. BENNETT, J. C.: In: Methods in enzymology, vol. 11 (ed. C. H. W. HIRS), p. 330. New York: Academic Press 1967
84. BENSON, J. V., GORDON, M. J., PATTERSON, J. A.: Anal. Biochem. **18**, 228 (1967)
85. BENSON, J. V., JONES, R. T., CORMICK, J., PATTERSON, J. A.: Anal. Biochem. **16**, 91 (1966)

86. BENSON, J. V., PATTERSON, J. A.: In: New techniques in amino acid, peptide, and protein analysis (eds. A. NIEDERWIESER, G. PATAKI), p. 1. Ann Arbor (Mich.): Ann Arbor Science Publishers 1971
87. BERGER, A., KATCHALSKI, E.: J. Am. Chem. Soc. **73**, 4084 (1951)
88. BERGER, A., NOGUCHI, J., KATCHALSKI, E.: J. Am. Chem. Soc. **78**, 4483 (1956)
89. BERGMANN, E. D., BENTOV, M.: J. Org. Chem. **26**, 1480 (1961)
90. BERGMANN, M., FRUTON, J. S.: J. Biol. Chem. **145**, 247 (1942)
91. BERGMANN, M., ZERVAS, L.: Chem. Ber. **65**, 1192 (1932)
92. BERGMANN, M., ZERVAS, L.: Z. physiol. Chem., Hoppe-Seyler's **221**, 51 (1933)
93. BERGMANN, M., ZERVAS, L.: J. Biol. Chem. **113**, 341 (1936)
94. BERGMANN, M., ZERVAS, L., ROSS, W. F.: J. Biol. Chem. **111**, 245 (1935)
95. BERGMANN, M., ZERVAS, L., SCHLEICH, H., LEINERT, F.: Z. physiol. Chem., Hoppe-Seyler's **212**, 72 (1932)
96. BERGMANN, M., ZERVAS, L., SALZMANN, L.: Ber. deut. chem. Ges. **66**, 1288 (1933)
97. BERNARDI, L., BOSISIO, G., DE CASTIGLIONE, R., GOFFREDO, O.: Experientia **22**, 425 (1966)
98. BERNHARD, R.: Scientific Research, June 24, 1968
99. BERNHARD, S. A., DUDE, W. L., BRADLEY, D. F.: In: Data acquisition and processing in biology and medicine, vol. 2 (ed. K. ENSLEIN). New York: The Macmillan Co. 1964
100. BERNIER, I., JOLLES, P.: Compt. rend. **253**, 745 (1961)
101. BERSE, C., MASSIAH, T., PICHE, L.: J. Org. Chem. **26**, 4514 (1961)
102. BETHELL, D., METCALFE, G. E., SHEPPARD, R. C.: Chem. Commun. **1965**, 189
103. BETTELHEIM, F. R.: J. Biol. Chem. **212**, 235 (1955)
104. BEYERMAN, H. C., BOERS-BOONEKAMP, C. A. M., VAN DEN BRINK-ZIMMERMANNOVA, H. M.: Rec. trav. chim. **87**, 257 (1968)
105. BEYERMAN, H. C., BONTEKOE, J. S.: Proc. Chem. Soc. 249 (1961)
106. BEYERMAN, H. C., BONTEKOE, J. S.: Rec. trav. chim. **81**, 691 (1962)
107. BEYERMAN, H. C., BONTEKOE, J. S.: Rec. trav. chim. **81**, 699 (1962)
108. BEYERMAN, H. C., BONTEKOE, J. S.:, KOCH, A. C.: Rec. trav. chim. **78**, 935 (1959)
109. BEYERMAN, H. C., VAN DEN BRINK, W. M., WEYGAND, F., PROX, A., KÖNIG, W., SCHMIDHAMMER, L., NINTZ, E.: Rec. trav. chim. **84**, 213 (1965)
110. BEZAS, B., ZERVAS, L.: J. Am. Chem. Soc. **83**, 719 (1961)
111. BIRKHOFER, L., BIERWIRTH, E., RITTER, A.: Chem. Ber. **94**, 821 (1961)
112. BISERTE, G., OSTEUX, R.: Bull. soc. chim. biol. **33**, 50 (1951)
113. BLACKBURN, S., LEE, G. R.: Biochem. J. **58**, 227 (1954)
114. BLACKBURN, S.: Amino acid determination: methods and techniques, p. 24, 86. New York: Marcel Dekker 1968
114a. BLACKBURN, S., LOWTHER, A. G.: Biochem. J. **48**, 126 (1951)
115. BLOCK, R. J.: In: A laboratory manual of analytical methods of protein chemistry, vol. 2 (eds. P. ALEXANDER, R. J. BLOCK), p. 1. New York: Pergamon Press 1960
116. BLOCK, R. J.: Arch. Biochem. Biophys. **31**, 266 (1951)
117. BLOCK, R. J., BOLLING, D.: In: The amino acid composition of proteins and foods. Springfield (Ill.): Charles C. Thomas 1951
118. BLOCK, R. J., DURRUM, E. L., ZWEIG, G.: In: Paper chromatography and paper electrophoresis. New York: Academic Press 1958
119. BLOMBÄCK, B., BLOMBÄCK, M., EDMAN, P.: Acta Chem. Scand. **17**, 1184 (1963)
120. BLOMBÄCK, B., BLOMBÄCK, M., EDMAN, P., HESSEL, B.: Biochim. Biophys. Acta **115**, 371 (1966)
121. BLOMBÄCK, B., BLOMBÄCK, M., HESSEL, B., IWANAGA, S.: Nature **215**, 1445 (1967)
122. BLOMBÄCK, B., DOOLITTLE, R. F.: Acta Chem. Scand. **17**, 1816 (1963)
123. BOCCÙ, E., VERONESE, F. M., FONTANA, A., BENASSI, C. A.: Europ. J. Biochem. **13**, 188 (1970)
124. BOCK, R. M., LING, N.-S.: Anal. Chem. **26**, 1543 (1954)
125. BODANSZKY, M.: Acta Chim. Acad. Sci. Hung. **10**, 335 (1957)
126. BODANSZKY, M.: Nature **175**, 685 (1955)
127. BODANSZKY, M., BATH, R. J.: Chem. Commun. **1968**, 766
128. BODANSZKY, M., BIRKHIMER, C. A.: Chem. & Ind. (London) **1962**, 1620

129. Bodanszky, M., Bodanszky, A.: Am. Scientist **55**, 185 (1967)
130. Bodanszky, M., Denning, G. S., du Vigneaud, V.: In: Biochemical preparations, vol. 10 (ed. G. B. Brown), p. 122. New York: Wiley 1963
131. Bodanszky, M., du Vigneaud, V.: In: Biochemical preparations, vol. 9 (ed. M. J. Coon), p. 110. New York: Wiley 1962
132. Bodanszky, M., du Vigneaud, V.: J. Am. Chem. Soc. **81**, 2504 (1959)
133. Bodanszky, M., du Vigneaud, V.: J. Am. Chem. Soc. **81**, 5688 (1959)
134. Bodanszky, M., du Vigneaud, V.: J. Am. Chem. Soc. **81**, 6072 (1959)
135. Bodanszky, M., Meienhofer, J., du Vigneaud, V.: J. Am. Chem. Soc. **82**, 3195 (1960)
136. Bodanszky, M., Ondetti, M. A.: In: Peptide synthesis. New York: Interscience Publishers 1966
137. Bodanszky, M., Ondetti, M. A.: Chem. & Ind. (London) **1966**, 26
138. Bodanszky, M., Ondetti, M. A., Levine, S. D., Williams, N. J.: J. Am. Chem. Soc. **89**, 6753 (1967)
139. Bodanszky, M., Sheehan, J. T.: Ann. N.Y. Acad. Sci. **88**, 655 (1960)
140. Bodanszky, M., Sheehan, J. T.: Chem. & Ind. (London) **1960**, 1268
141. Bodanszky, M., Sheehan, J. T.: Chem. & Ind. (London) **1964**, 1423
142. Bodanszky, M., Sheehan, J. T.: Chem. & Ind. (London) **1966**, 1597
143. Boissonnas, R. A.: Helv. Chim. Acta **34**, 874 (1951)
144. Boissonnas, R. A., Haselbach, C. H.: Helv. Chim. Acta **36**, 576 (1953)
145. Boissonnas, R. A., Preitner, G.: Helv. Chim. Acta **36**, 875 (1953)
146. Boissonnas, R. A., Guttmann, St., Huguenin, R. L., Jaquenod, P.-A., Sandrin, E.: Helv. Chim. Acta **41**, 1867 (1958)
147. Boissonnas, R. A., Guttmann, St., Jacquenod, P.-A., Waller, J. P.: Helv. Chim. Acta **38**, 1491 (1955)
148. Boissonnas, R. A., Guttmann, St., Jacquenod, P.-A.: Helv. Chim. Acta **43**, 1349 (1960)
149. Bolling, D., Sober, H. A., Block, R. J.: Federation Proc. **8**, 185 (1949)
150. Boulton, A. A., Bush, I. E.: Biochem. J. **92**, 11p (1964)
151. Bovarnick, M., Clarke, H. T.: J. Am. Chem. Soc. **60**, 2426 (1938)
152. Boyden, A., Gemeroy, D.: Zoologica **35**, 145 (1950)
153. Boyer, P. D.: J. Am. Chem. Soc. **76**, 4331 (1954)
154. Boyer, S. H., Hathaway, P., Pascasio, F., Orton, C., Bordley, J.: Science **153**, 1539 (1966)
155. Bradbury, J. H.: Biochem. J. **68**, 482 (1958)
156. Brand, E., Edsall, J. T.: Ann. Rev. Biochem. **16**, 724 (1947)
157. Brand, E., Erlanger, B. F., Sachs, H., Polatnick, J.: J. Am. Chem. Soc. **73**, 3510 (1951)
158. Brand, E., Kassel, B.: J. Biol. Chem. **145**, 365 (1942)
159. Brandenberg, D.: Tetrahedron **49**, 6201 (1966)
159a. Braun, V., Schroeder, W. H.: Arch. Biochem. Biophys. **118**, 241 (1967)
159b. Brattin, W. J., Jr., Smith, E. L.: J. Biol. Chem. **246**, 2400 (1971)
160. Braunitzer, G.: Chem. Ber. **88**, 2025 (1955)
161. Braunitzer, G.: Biochim. Biophys. Acta **19**, 574 (1956)
162. Braunitzer, G.: In: Recent developments in the chemical study of protein structures, p. 3. Meeting in Montpellier, 28—29 Sept. 1971
163. Braunitzer, G., Schrank, B., Ruhfus, A.: Z. physiol. Chem., Hoppe-Seyler's **351**, 1589 (1970)
164. Braunitzer, G., Schrank, B., Ruhfus, A., Petersen, S., Petersen, U.: Z. physiol. Chem., Hoppe-Seyler's **352**, 1730 (1971)
165. Braunitzer, G., Chén, R., Schrank, B., Stancl, A.: Z. physiol. Chem., Hoppe-Seyler's **353**, 832 (1972)
166. Braunitzer, G., Chén, R., Schrank, B., Stancl, A.: Z. physiol. Chem., Hoppe-Seyler's **354**, 867 (1973)
167. Brautlecht, C. A.: J. Biol. Chem. **10**, 139 (1911)
168. Bray, H. G., Thorpe, W. V., White, K.: Biochem. J. **46**, 271 (1950)

169. BRENNER, M., HUBER, W.: Helv. Chim. Acta **36**, 1109 (1953)
170. BRENNER, M., NIEDERWIESER, A., PATAKI, G.: Experientia **17**, 145 (1961)
171. BRENNER, M., NIEDERWIESER, A., PATAKI, G.: In: Thin layer chromatography (ed. E. STAHL), p. 427. New York: Academic Press 1965
172. BREWER, C. F., RIEHM, J. P.: Anal. Biochem. **18**, 248 (1967)
173. BREWER, H. B., Jr., RONAN, R.: Proc. Natl. Acad. Sci. (US) **67**, 1862 (1970)
174. BREWER, H. B., Jr., LUX, S. E., RONAN, R., JOHN, K. M.: Proc. Natl. Acad. Sci. (US) **69**, 1304 (1972)
175. BREWER, H. B., Jr., FAIRWELL, T., RONAN, R., SIZEMORE, G. W., ARNAUD, C. D.: Proc. Natl. Acad. Sci. (US) **69**, 3585 (1972)
176. BRITTEN, R. J., ROBERTS, R. B.: Science **131**, 32 (1960)
177. BROADBENT, W., MORLEY, J. S., STONE, B. E.: J. Chem. Soc. (C) **1967**, 2632
178. BROWN, F., HALL, L. P.: Nature **166**, 66 (1950)
179. BROWN, J. R., HARTLEY, B. S.: Biochem. J. **101**, 214 (1966)
180. BROWN, J. R., KAUFFMAN, D. L., HARTLEY, B. S.: Biochem. J. **103**, 497 (1967)
181. BROWNLEE, P. J. E., COX, M. E., HANDFORD, B. O., MARSDEN, J. C., YOUNG, G. T.: J. Chem. Soc. **1964**, 3832
182. BRUTON, C. J., HARTLEY, B. S.: J. Mol. Biol. **52**, 165 (1970)
183. BRYANT, P. M., MOORE, R. H., PIMLOTT, P. J., YOUNG, G. T.: J. Chem. Soc. **1959**, 3868
184. BUCKINGHAM, D. A., COLLMAN, J. P.: Inorg. Chem. **6**, 1803 (1967)
185. BUCKINGHAM, D. A., COLLMAN, J. P., HAPPER, D. A. R., MARZILLI, L. G.: J. Am. Chem. Soc. **89**, 1082 (1967)
186. BUCKLEY, III, C. E., WHITNEY, P. L., TANFORD, C.: Proc. Natl. Acad. Sci. (US) **50**, 827 (1963)
187. BULL, H. B., HAHN, J. W.: J. Am. Chem. Soc. **70**, 2132 (1948)
188. BURKHARDT, III, W. A., WILCOX, P. E.: Biochem. Biophys. Res. Commun. **28**, 803 (1967)
189. BURR, M. E., KOSHLAND, D. E., Jr.: Proc. Natl. Acad. Sci. (US) **52**, 1017 (1964)
189a. BURSTEIN, Y., WALSH, K. A., NEURATH, H.: Biochemistry (U.S.A.) **13**, 205 (1974)
190. BURTON, H.: Biochim. Biophys. Acta **29**, 193 (1958)
191. BURTON, R. B., ZAFFARONI, A., KENTMANN, E. H.: J. Biol. Chem. **188**, 763 (1951)
192. BUTLER, P. J. G., HARRIS, J. I., HARTLEY, B. S., LEBERMAN, R.: Biochem. J. **103**, 78p (1967)
192a. BUTLER, P. J. G., HARRIS, J. I., HARTLEY, B. S., LEBERMAN, R.: Biochem. J. **112**, 679 (1969)
193. CAHNMANN, H. J., ARNON, R., SELA, M.: J. Biol. Chem. **241**, 3247 (1966)
194. CALDIN, D. J. Mc.: Chem. Rev. **60**, 39 (1960)
195. CALLAHAN, F. M., ANDERSON, G. W., PAUL, R., ZIMMERMAN, J. E.: J. Am. Chem. Soc. **85**, 201 (1963)
196. CALLAHAN, F. M., MCDONALD, J. K., ELLIS, S.: Federation Proc. **28**, 661 (1969)
197. CALLAHAN, F. M., MCDONALD, J. K., ELLIS, S.: In: Methods in enzymology, vol. 25 (eds. S. P. COLOWICK, N. O. KAPLAN), p. 282. New York, London: Academic Press 1972
198. CALLAHAN, F. M., MCDONALD, J. K., ELLIS, S.: Federation Proc. **31**, 1105 (1972)
199. CANFIELD, R. E.: J. Biol. Chem. **238**, 2691 (1963)
200. CANFIELD, R. E.: J. Biol. Chem. **238**, 2698 (1963)
201. CANFIELD, R. E., ANFINSEN, C. B.: J. Biol. Chem. **238**, 2684 (1963)
202. CANFIELD, R. E., ANFINSEN, C. B.: In: The proteins, 2nd ed., vol. 1 (ed. H. NEURATH), p. 311. New York: Academic Press 1963
203. CANN, J. R., GOAD, W. B.: J. Biol. Chem. **240**, 148 (1965)
204. CANN, J. R.: In: Interacting macromolecules. New York: Academic Press 1970
204a. CANNON, E. L.: Anal. Biochem. **46**, 32 (1972)
205. CAPPUGI, G., NASSI, P., TREVES, C., REMPONI, G.: Experientia **27**, 237 (1971)
206. CAREY, W. F., WELLS, J. R. E.: J. Biol. Chem. **247**, 5573 (1972)
207. CARPENTER, D., VAN HOLDE, K. E.: Biochemistry (U.S.A.) **12**, 2231 (1973)
208. CARPENTER, F. H., GISH, D. T.: J. Am. Chem. Soc. **74**, 3818 (1952)
209. CARPINO, L. A.: J. Am. Chem. Soc. **79**, 4427 (1957)
210. CARPINO, L. A., GIZA, C. A., CARPINO, B. A.: J. Am. Chem. Soc. **81**, 955 (1959)

211. Casassa, E. F., Eisenberg, H.: Advances in Protein Chem. **19**, 287 (1964)
212. Cecil, R.: Biochim. Biophys. Acta **18**, 154 (1955)
213. Cecil, R.: In: The proteins, 2nd ed., vol. 1 (ed. H. Neurath), p. 380. New York: Academic Press 1963
214. Cecil, R., McPhee, J. R.: Biochem. J. **59**, 234 (1955)
215. Cecil, R., McPhee, J. R.: Biochem. J. **60**, 496 (1955)
216. Cecil, R., McPhee, J. R.: Advances in Protein Chem. **14**, 255 (1959)
217. Chambers, R. W., Carpenter, F. H.: J. Am. Chem. Soc. **77**, 1522 (1955)
218. Chargaff, E., Levine, C., Green, C.: J. Biol. Chem. **175**, 67 (1948)
219. Chauvet, J. P., Acher, R.: Febs Letters **1**, 305 (1968)
220. Chan, S. K., Margoliash, E.: J. Biol. Chem. **241**, 507 (1966)
221. Chenoweth, D., Mitchel, R. E. J., Smith, E. L.: J. Biol. Chem. **248**, 1672 (1973)
222. Cherbuliez, E., Baehler, B., Rabinowitz, J.: Helv. Chim. Acta **43**, 1871 (1960)
223. Cherbuliez, E., Sussmann, A. R., Rabinowitz, J.: Helv. Chim. Acta **44**, 319 (1961)
224. Chervenka, C. H., Wilcox, P. E.: J. Biol. Chem. **222**, 612 (1956)
225. Cheung, H. T., Blout, E. R.: J. Org. Chem. **30**, 315 (1965)
226. Chibnall, A. C.: J. Intern. Soc. Leather Trades' Chemists **30**, 1 (1946)
227. Chibnall, A. C., Mangan, J. L., Rees, M. W.: Biochem. J. **68**, 111 (1958)
228. Chibnall, A. C., Mangan, J. L., Rees, M. W.: Biochem. J. **68**, 114 (1958)
229. Chibnall, A. C., Rees, M. W.: Biochem. J. **68**, 105 (1958)
230. Chibnall, A. C., Rees, M. W., Williams, E. F.: Biochem. J. **37**, 354 (1943)
231. Chibnall, A. C., Spahr, P. F.: Biochem. J. **68**, 135 (1958)
232. Chillemi, D., Scarso, L., Scoffone, E.: Gazz. chim. ital. **87**, 1356 (1957)
233. Chu, R. C. L., Yasunobu, K. T.: Biochim. Biophys. Acta **89**, 148 (1964)
234. Clark-Walker, G. D., Robinson, H. C.: J. Chem. Soc. **1961**, 2810
235. Clarke, H. T.: J. Biol. Chem. **97**, 235 (1932)
236. Clayton, D. W., Kenner, G. W., Sheppard, R. C.: J. Chem. Soc. **1956**, 371
237. Cleland, W. W.: Biochemistry (U.S.A.) **3**, 480 (1964)
238. Clerc-Bory, M., Pacheco, H., Mentzer, Ch.: Compt. rend. **238**, 525 (1954)
239. Cohn, E. J., Edsall, J. T.: In: Proteins, amino acids, and peptides. New York: Reinhold 1943
240. Cole, R. D.: In: Methods in enzymology, vol. 11 (ed. C. H. W. Hirs), p. 206. New York: Academic Press 1967
241. Cole, R. D., Stein, W. H., Moore, S.: J. Biol. Chem. **233**, 1359 (1958)
242. Collman, J. P., Buckingham, D. A.: J. Am. Chem. Soc. **85**, 3039 (1963)
243. Collman, J. P., Kimura, R.: J. Am. Chem. Soc. **89**, 6096 (1967)
244. Consden, R.: Nature **162**, 1007 (1948)
245. Consden, R., Gordon, A. H.: Biochem. J. **46**, 8 (1950)
246. Consden, R., Gordon, A. H., Martin, A. J. P., Synge, R. L. M.: Biochem. J. **41**, 596 (1947)
247. Conway, E. J.: Microdiffusion analysis and volumetric error, 4th ed., p. 90. London: Crosby-Lockwood 1957
248. Cope, A. C., Towe, P. H.: J. Am. Chem. Soc. **71**, 3423 (1949)
249. Cory, R. P., Becker, R. R., Rosenbluth, R., Isenberg, I.: J. Am. Chem. Soc. **90**, 1643 (1968)
250. Costopanagiotis, A. A., Hanford, B. O., Weinstein, B.: J. Org. Chem. **33**, 1261 (1968)
251. Crammer, J. L., Neuberger, A.: Biochem. J. **37**, 302 (1943)
252. Creeth, J. R., Pain, H. H.: Progr. Biophys. Molec. Biol. **17**, 217 (1967)
253. Crestfield, A. M.: Anal. Chem. **28**, 117 (1956)
254. Crestfield, A. M., Moore, S., Stein, W. H.: J. Biol. Chem. **238**, 622 (1963)
255. Crestfield, A. M., Stein, W. H., Moore, S.: Arch. Biochem. Biophys. **1** (Suppl.), 217 (1962)
256. Criddle, R. S., Edwards, D. L., Petersen, T. G.: Biochemistry (U.S.A.) **5**, 578 (1966)
257. Crombie, G., Foster, J. A., Franzblau, C.: Biochem. Biophys. Res. Commun. **52**, 1228 (1973)
258. Cromwell, L. D., Stark, G. R.: Biochemistry (U.S.A.) **8**, 4735 (1969)

259. Curtius, T.: Ber. deut. chem. Ges. **35**, 3226 (1902)
260. Curtius, T., Goebel, F.: J. prakt. Chem. **37**, 150 (1888)
261. Dakin, H. D.: J. Chem. Soc. **107**, 434 (1915)
262. Dalgliesh, C. E.: Biochem. J. **52**, 3 (1952)
263. DasGupta, B. R.: Anal. Biochem. **30**, 288 (1969)
264. Dautrevau, M., Biserte, G.: Compt. rend. **240**, 1153 (1955)
265. Davidson, B. E., Sajgo, M., Noller, H. F., Harris, J. I.: Nature **216**, 1181 (1967)
266. Davis, N. C., Smith, E. L.: J. Biol. Chem. **223**, 935 (1956)
267. Davis, N. C., Smith, E. L.: J. Biol. Chem. **224**, 261 (1957)
268. Dayhoff, M. O.: Atlas of protein sequence and structure, vol. 5. Washington (D.C.): National Biomedical Research Foundation 1972
269. Deal, W. C., Rutter, W. J., van Holde, K. E.: Biochemistry (U.S.A.) **2**, 246 (1963)
270. Deer, A., Fried, J. H., Halpern, B.: Australian J. Chem. **20**, 797 (1967)
271. Deer, A.: Angew. Chem. Int. Ed. Engl. **5**, 1041 (1966)
272. Delia-Llosa, P., Jatisz, M., Scoffone, E.: Bull. soc. chim. France **1960**, 1621
273. Dekker, C. A., Fruton, J. S.: In: Methods in medical research, vol. 3 (ed. R. W. Gerard), p. 272. Chicago: Year Book Pub., Inc. 1950
274. Dekker, C. A., Taylor, S. P., Fruton, J. S.: J. Biol. Chem. **180**, 155 (1949)
275. deLange, R. J., Smith, E. L., Fambrough, D. M., Bonner, J.: Proc. Natl. Acad. Sci. (US) **61**, 1145 (1968)
276. Delpierre, G. R., Fruton, J. S.: Proc. Natl. Acad. Sci. (US) **56**, 1817 (1966)
277. deMarco, C., Graziani, M. T., Mosti, R.: Anal. Biochem. **15**, 40 (1966)
278. Denkewalter, R. G., Schwam, H., Strachan, R. G., Beesley, T. E., Veber, D. F., Schoenewaldt, E. F., Barkemeyer, H., Paleveda, W. J., Jr., Jacob, T. A., Hirschmann, R.: J. Am. Chem. Soc. **88**, 3163 (1966)
279. Desnuelle, P., Casal, A.: Biochim. Biophys. Acta **2**, 64 (1948)
280. Determann, H., Torff, H.-J., Zipp, O.: Ann. Chem., Justus Liebigs **670**, 141 (1963)
281. Dewey, R. S., Schoenewaldt, E. F., Joshua, H., Paleveda, W. J., Jr., Swam, H., Barkemeyer, H., Arison, B. H., Veber, D. F., Denkewalter, R. G., Hirschmann, R.: J. Am. Chem. Soc. **90**, 3254 (1968)
282. Deyl, Z., Rosmus, J.: J. Chromatog. **20**, 514 (1965)
283. Dickman, S. R., Crockett, A. L.: J. Biol. Chem. **220**, 957 (1956)
284. Dickman, S. R., Westcott, W. L.: J. Biol. Chem. **210**, 481 (1954)
285. Dijkstra, A., Billiet, H. A., van Doninck, A. H., van Velthuysen, H., Maat, L., Beyerman, H. C.: Rec. trav. chim. **86**, 65 (1967)
286. Djerassi, C., Undheim, K., Sheppard, R. C., Terry, W. G., Sjoberg, B.: Acta Chem. Scand. **15**, 903 (1961)
287. Dohan, J. S., Woodward, G. E.: J. Biol. Chem. **129**, 393 (1939)
288. Doolittle, R. F.: Biochem. J. **94**, 742 (1965)
289. Doolittle, R. F., Armentrout, R. W.: Biochemistry (U.S.A.) **7**, 516 (1968)
290. Doolittle, R. F.: In: Methods in enzymology, vol. 25 (eds. C. W. H. Hirs, S. N. Timasheff), p. 231. New York: Academic Press 1972
291. Doolittle, R. F.: In: Methods in enzymology, vol. 19 (eds. G. E. Perlmann, L. Loland), p. 555. New York: Academic Press 1970
292. Dopheide, T. A. A., Jones, W. M.: J. Biol. Chem. **243**, 3906 (1968)
292a. Dopheide, T. A. A., Trikojus, V. M.: Nature **201**, 1128 (1964)
293. Dowling, L. M., Stark, G. R.: Biochemistry (U.S.A.) **8**, 4728 (1969)
294. Downs, F., Agman, W.: J. Protein Res. **1**, 181 (1969)
295. Dreyer, W. J., Bynum, E.: In: Methods in enzymology, vol. 11 (ed. C. H. W. Hirs), p. 33. New York: Academic Press 1967
296. Dunker, A. K., Ruechert, R. R.: J. Biol. Chem. **244**, 5074 (1969)
297. Durant, J. A.: Nature **169**, 1062 (1952)
298. Dus, K., Lindroth, S., Pabst, R., Smith, R. M.: Anal. Biochem. **18**, 532 (1967)
299. du Vigneaud, V., Bartlett, M. F., Jöhl, A.: J. Am. Chem. Soc. **79**, 5572 (1957)
300. du Vigneaud, V., Behrens, O. K.: J. Biol. Chem. **117**, 27 (1937)
301. du Vigneaud, V., Dorfmann, R., Loring, H. S.: J. Biol. Chem. **98**, 577 (1932)
302. du Vigneaud, V., Gish, D. T., Katsoyannis, P. G.: J. Am. Chem. Soc. **76**, 4751 (1954)

303. du Vigneaud, V., Hunt, M.: J. Biol. Chem. **125**, 269 (1938)
304. du Vigneaud, V., Ressler, C., Swan, J. M., Roberts, C. W., Katsoyannis, P. G.: J. Am. Chem. Soc. **76**, 3115 (1954)
305. du Vigneaud, V., Ressler, C., Trippitt, S.: J. Biol. Chem. **205**, 949 (1953)
306. Easley, C. W.: Biochim. Biophys. Acta **107**, 386 (1965)
307. Edelhoch, H.: Biochemistry (U.S.A.) **6**, 1948 (1967)
308. Edlbacher, S., Baur, H., Staehelin, H. R., Zeller, A.: Z. Physiol. Chem., Hoppe-Seyler's **270**, 158 (1941)
309. Edman, P.: Arch. Biochem. Biophys. **22**, 475 (1949)
310. Edman, P.: Acta Chem. Scand. **4**, 277 (1950)
311. Edman, P.: Acta Chem. Scand. **4**, 283 (1950)
312. Edman, P.: Nature **177**, 667 (1956)
313. Edman, P.: Acta Chem. Scand. **10**, 761 (1956)
314. Edman, P.: Proc. Roy. Austral. Chem. Inst. **1957**, 434
315. Edman, P.: Acta Chem. Scand. **7**, 700 (1953)
316. Edman, P.: Ann. N.Y. Acad. Sci. **88**, 602 (1960)
317. Edman, P., Begg, G.: Europ. J. Biochem. **1**, 80 (1967)
318. Edman, P., Lauber, K.: Acta Chem Scand. **10**, 466 (1956)
319. Edman, P., Sjöquist, J.: Acta Chem. Scand. **10**, 1507 (1956)
320. Edman, P.: In: Protein sequence determination (ed. S. B. Needleman), p. 211. Berlin-Heidelberg-New York: Springer 1970
321. Edmundson, A. B.: Nature **198**, 354 (1963)
322. Edmundson, A. B.: Nature **205**, 883 (1965)
323. Edward, J. T., Nielsen, S.: Chem. & Ind. (London) **1953**, 197
324. Edwards, B. G.: Arch. Biochem. Biophys. **21**, 103 (1949)
325. Eisen, H. M., Belman, S., Carsten, M. E.: J. Am. Chem. Soc. **75**, 4583 (1953)
326. Elliot, D. F., Russel, D. W.: Biochem. J. **66**, 49p (1957)
327. Ellis, S., Fairwell, T., Lovins, R. E.: Biochem. Biophys. Res. Commun. **49**, 1407 (1972)
328. Ellman, G. L.: Arch. Biochem. Biophys. **82**, 70 (1959)
329. Elmore, D. T., Smyth, J.: Proc. Chem. Soc. **1963**, 18
330. Elmore, D. T., Toseland, P. A.: J. Chem. Soc. **1954**, 4533
331. Elmore, D. T., Toseland, P. A.: J. Chem. Soc. **1957**, 2460
332. Elmore, D. T., Toseland, P. A.: J. Chem. Soc. **1965**, 188
333. Elving, P. J., Rosenthal, I., Cramer, M. K.: J. Am. Chem. Soc. **73**, 1717 (1951)
334. Emery, A. R., Gold, V.: J. Chem. Soc. **1950**, 1443
335. Emery, A. R., Gold, V.: J. Chem. Soc. **1950**, 1447
336. Emery, A. R., Gold, V.: J. Chem. Soc. **1950**, 1455
337. Erlanger, B. F., Cohen, W.: J. Am. Chem. Soc. **85**, 348 (1963)
338. Erlanger, B. F., Sachs, H., Brand, E. J.: J. Am. Chem. Soc. **76**, 1806 (1954)
339. Erspamer, V., Anastasi, A.: Experientia **18**, 58 (1962)
340. Erspamer, V., Boretti, G.: Arch. intern. pharmacodynamie **88**, 296 (1951)
341. Ertingshausen, G., Adler, H. J., Reichler, A. S.: J. Chromatog. **42**, 355 (1969)
342. Eveleigh, J. W., Winter, G. D.: In: Protein sequence determination (ed. S. B. Needleman), p. 91. Berlin-Heidelberg-New York: Springer 1970
343. Eyem, J., Sjöquist, J.: Anal. Biochem. **52**, 255 (1973)
344. Eylar, E. H.: Proc. Natl. Acad. Sci. (US) **67**, 1425 (1970)
345. Fahmy, R. A., Niederwieser, A., Pataki, G., Brenner, M.: Helv. Chim. Acta **44**, 2022 (1961)
346. Fahrney, D., Gold, A. M.: J. Am. Chem. Soc. **85**, 997 (1963)
347. Fairwell, T., Barnes, W. T., Richards, F. F., Lovins, R. E.: Biochemistry (U.S.A.) **9**, 2260 (1970)
348. Fairwell, T., Ellis, S., Lovins, R. E.: Anal. Biochem. **53**, 115 (1973)
349. Fairwell, T., Lovins, R. E.: Biochem. Biophys. Res. Commun. **43**, 1280 (1971)
350. Fales, H. M., Nagai, Y., Milne, G. W. A., Brewer, H. B., Jr., Bronzert, T. J., Pisano, J. J.: Anal. Biochem. **43**, 288 (1971)
351. Farrington, J. A., Kenner, G. W., Turner, J. M.: Chem. & Ind. (London) **1955**, 601

352. Fava, A., Iliceto, A., Camera, E.: J. Am. Chem. Soc. **79**, 833 (1957)
352a. Feigl, F.: In: Spot tests in organic analysis, p. 228. Amsterdam, London, New York: Elsevier Publ. Co. 1956
353. Fenn, M. D., Bradbury, J. H.: Anal. Biochem. **49**, 498 (1972)
354. Felix, A. M., Terkelsen, G.: Anal. Biochem. **56**, 610 (1973)
355. Felix, A. M., Terkelsen, G.: Arch. Biochem. Biophys. **157**, 177 (1973)
356. Fields, R.: In: Methods in enzymology, vol. 25 (eds. C. W. H. Hirs, S. N. Timasheff), p. 464. New York: Academic Press 1972
357. Fischer, E.: Ber. deut. chem. Ges. **36**, 2982 (1903)
358. Fischer, E.: Ber. deut. chem. Ges. **39**, 2893 (1906)
359. Fischer, R. F., Whetstone, R. R.: J. Am. Chem. Soc. **76**, 5076 (1954)
360. Fischer, R. F., Whetstone, R. R.: J. Am. Chem. Soc. **77**, 750 (1955)
361. Fish, W. W., Mann, K. G., Tanford, C.: J. Biol. Chem. **244**, 3989 (1969)
362. Flecher, J. C.: Biochem. J. **102**, 815 (1967)
363. Fleš, D., Markovac-Prpić, A., Tomašić, V.: J. Am. Chem. Soc. **80**, 4654 (1958)
363a. Fletcher, A. P., Marshall, R. D., Neuberger, A.: Biochem. J. **88**, 37p (1963)
364. Folin, O., Ciocalteu, V.: J. Biol. Chem. **73**, 627 (1927)
365. Folk, J. E., Gladner, J. A.: J. Biol. Chem. **231**, 379 (1958)
366. Folk, J. E., Piez, K. A., Carroll, W. R., Gladner, J. A.: J. Biol. Chem. **235**, 2272 (1960)
367. Folk, J. E., Schirmer, E. W.: J. Biol. Chem. **238**, 3884 (1963)
368. Foltman, B., Hartley, B. S.: Biochem. J. **104**, 1064 (1967)
369. Fontana, A.: Unpublished results
370. Fontana, A., Marchiori, F., Moroder, L.: Ricerca sci. **36**, 261 (1966)
371. Fontana, A., Marchiori, F., Moroder, L., Scoffone, E.: Tetrahedron **26**, 2985 (1966)
372. Fontana, A., Scoffone, E., Benassi, C. A.: Biochemistry (U.S.A.) **7**, 980 (1968)
373. Fontana, A., Scoffone, E.: In: Mechanism of reactions of sulfur compounds, vol. 4 (ed. N. Kharasch), p. 15. Santa Monica (Calif.): Intra-Science Res. Found. 1969
374. Fontana, A., Scoffone, E.: In: Methods in enzymology, vol. 25 (eds. C. H. W. Hirs, S. N. Timasheff), p. 482. New York: Academic Press 1972
375. Fontana, A., Marchiori, F., Rocchi, R., Pajetta, P.: Gazz. chim. ital. **96**, 1301 (1966)
376. Fontana, A., Toniolo, C.: In: The chemistry of the thiol group (ed. S. Patai). London: J. Wiley and Sons 1974 (in press)
377. Foster, J. A., Bruenger, E., Hu, C. L., Albertson, K., Franzblau, C.: Biochem. Biophys. Res. Commun. **53**, 70 (1973)
378. Fowden, L.: Nature **167**, 1030 (1951)
379. Fox, S. W., Hayakawa, T., Harada, K.: Bull. Chem. Soc. Japan **36**, 1050 (1963)
380. Fox, S. W., Hurst, T. L., Griffith, J. F., Underwood, O.: J. Biol. Chem. **77**, 3118 (1955)
381. Fraenkel-Conrat, H., Harris, J. I., Levy, A. L.: In: Methods of biochemical analysis, vol. 2 (ed. D. Glick), p. 359. New York: Interscience 1955
382. Fraenkel-Conrat, H., Simpson, M. E., Evans, H. M.: J. Biol. Chem. **142**, 107 (1942)
383. Fraenkel-Conrat, H., Singer, B.: J. Am. Chem. Soc. **76**, 180 (1954)
384. Fraenkel-Conrat, H., Tsung, C. M.: In: Methods in enzymology, vol. XI (ed. C. H. W. Hirs), p. 151. New York: Academic Press 1967
385. Fraenkel-Conrat, H., Mahammad, A., Ducay, E. D., Mecham, D. K.: J. Am. Chem. Soc. **73**, 625 (1951)
386. Frank, H., Petersen, H.: Z. physiol. Chem., Hoppe-Seyler's **299**, 1 (1955)
387. Frantz, J. D., Hare, P. E.: Carnegie Inst. Wash. Publ., Year Book **72**, 704 (1973)
388. Frater, R., Light, A., Smith, E. L.: J. Biol. Chem. **240**, 253 (1965)
389. Friedmann, E.: Biochim. Biophys. Acta **9**, 65 (1952)
390. Friedmann, E., Marrian, D. H., Simon-Reuss, I.: Brit. J. Pharmacol. **4**, 105 (1949)
391. Friedman, M., Noma, A.: Textile Research J. **40**, 1073 (1970)
392. Friedman, M., Krull, L. H., Cavins, J. F.: J. Biol. Chem. **245**, 3868 (1970)
393. Friedman, M., Finley, J. W.: J. Agr. Food Chem. **19**, 626 (1971)
394. Fridkin, M., Patchornik, A., Katchalski, E.: J. Am. Chem. Soc. **88**, 3164 (1966)

395. Fromageot, C., Jutisz, M., Meyer, D., Penasse, L.: Biochim. Biophys. Acta **6**, 283 (1950)
396. Fruton, J. S.: Advances in Protein Chem. **5**, 1 (1949)
397. Fruton, J. S., Bergmann, M.: J. Biol. Chem. **145**, 253 (1942)
398. Fruton, J. S., Bergmann, M., Anslow, W. P., Jr.: J. Biol. Chem. **127**, 627 (1939)
399. Fruton, J. S., Hearn, W. R., Ingram, V. M., Wiggans, D. S., Winitz, M.: J. Biol. Chem. **204**, 891 (1953)
400. Fujita, H.: In: Mathematical theory of sedimentation analysis. New York: Academic Press 1962
400a. Fujita, H., Williams, J. W.: J. Phys. Chem. **70**, 309 (1966)
401. Fujika, H., Braunitzer, G., Rudloff, V.: Z. physiol. Chem., Hoppe-Seyler's **351**, 901 (1970)
402. Gaetjens, E., Barany, M.: Biochim. Biophys. Acta **117**, 176 (1966)
403. Galiazzo, G., Jori, G., Scoffone, E.: Biochem. Biophys. Res. Commun. **31**, 158 (1968)
404. Gawron, O., Fernando, J.: J. Am. Chem. Soc. **83**, 2906 (1961)
405. Gaitonde, M. K.: Biochem. J. **104**, 627 (1967)
406. Gaitonde, M. K., Dovey, T.: Biochem. J. **117**, 907 (1970)
407. Gehrke, C. W., Takeda, H.: J. Chromatog. **76**, 63 (1973)
408. Gehrke, C. W., Zumwalt, R. W., Kuo, K., Rash, J. J., Aue, W. A., Stalling, D. L., Kvenvolden, K., Ponnamperuma, C.: Space Life Sci. **3**, 439 (1972)
409. Georgiadis, A. G., Coffey, J. W.: Anal. Biochem. **56**, 121 (1973)
410. Gerngross, O., Voss, K., Herfeld, H.: Ber. deut. chem. Ges. **66**, 435 (1933)
411. Gerwin, B. I., Stein, W. H., Moore, S.: J. Biol. Chem. **241**, 3331 (1966)
412. Ghosh, P. B., Whitehouse, M. W.: Biochem. J. **108**, 155 (1968)
413. Gillessen, D., Schnabel, E., Meienhofer, J.: Ann. Chem., Justus Liebigs **667**, 164 (1963)
414. Gish, D. T., Carpenter, F. H.: J. Am. Chem. Soc. **75**, 5872 (1953)
415. Gish, D. T., du Vigneaud, V.: J. Am. Chem. Soc. **79**, 3579 (1957)
416. Gish, D. T., Katsoyannis, P. G., Hess, G. P., Stedman, R. J.: J. Am. Chem. Soc. **78**, 5954 (1956)
416a. Gladner, J. A., Folk, J. E.: J. Biol. Chem. **231**, 393 (1958)
417. Glatthard, T., Matter, M.: Helv. Chim. Acta **46**, 795 (1963)
418. Glozer, A. N.: Ann. Rev. Biochem. **39**, 108 (1970)
419. Goddard, D. R., Michaelis, L.: J. Biol. Chem. **112**, 361 (1935)
420. Goldberger, R. F., Anfinsen, C. B.: Biochemistry (U.S.A.) **1**, 401 (1962)
421. Goldstein, J., Konigsberg, M., Hill, R. J.: J. Biol. Chem. **238**, 2016 (1963)
422. Goldstone, A. D., Smith, E. L.: J. Biol. Chem. **241**, 4480 (1966)
423. Goldstone, A. D., Smith, E. L.: J. Biol. Chem. **242**, 4702 (1967)
424. Goodman, L., Ross, L. O., Baker, B. R.: J. Org. Chem. **23**, 1954 (1958)
425. Goodman, M., Levine, L.: J. Am. Chem. Soc. **86**, 2918 (1964)
426. Goodman, M., Stueben, K. C.: J. Am. Chem. Soc. **81**, 3980 (1959)
427. Goodman, M., Stueben, K. C.: J. Am. Chem. Soc. **84**, 1279 (1962)
428. Goodwin, T. W., Morton, R. A.: Biochem. J. **40**, 628 (1946)
429. Gordon, A. H., Keil, B., Sebesta, K., Knesse, O., Šorm, F.: Collection Czechoslov. Chem. Commun. **15**, 1 (1950)
430. Gordon, A. H., Martin, A. J. P., Synge, R. L. M.: Biochem. J. **35**, 1369 (1941)
431. Gornall, A. G., Bordowill, C. J., David, M. M.: J. Biol. Chem. **177**, 751 (1949)
432. Gottlieb, A. J., Sussman, H. H.: Biochim. Biophys. Acta **194**, 170 (1969)
433. Graffeo, A. P., Haag, A., Karger, B. L.: Anal. Letters **6**, 505 (1973)
434. Grannis, G. F.: Arch. Biochem. Biophys. **91**, 255 (1960)
435. Grassetti, D. R., Murray, J. F., Jr.: Arch. Biochem. Biophys. **119**, 41 (1967)
436. Grassmann, W., Hörmann, H.: Z. physiol. Chem., Hoppe-Seyler's **292**, 24 (1953)
437. Grassmann, W., Wünsch, E., Deufel, P., Zwick, A.: Chem. Ber. **91**, 538 (1958)
438. Gray, W. R.: In: Methods in enzymology, vol. 25, part B (eds. C. H. W. Hirs, S. N. Timasheff), p. 121, 333. New York: Academic Press 1972
439. Gray, W. R., Hartley, B. S.: Biochem. J. **89**, 379 (1963)

440. GREEN, F. C., KAY, L. M.: Anal. Chem. **24**, 726 (1952)
441. GREEN, L. C., HARRAP, B. S.: Anal. Biochem. **42**, 377 (1971)
442. GREEN, M., STAHMANN, M. A.: J. Biol. Chem. **197**, 771 (1952)
443. GREENE, R. D., BLACK, A.: Proc. Soc. Exp. Biol. Med. **54**, 322 (1943)
443a. GREENSTEIN, J. P., WINITZ, M.: In: Chemistry of amino acids, vol. 2, p. 887. New York: Wiley 1961
444. GREGORY, J. D.: J. Am. Chem. Soc. **77**, 3922 (1955)
445. GREGORY, H., HARDY, P. M., JONES, D. S., KENNER, G. W., SHEPPARD, R. C.: Nature **204**, 931 (1964)
446. GREIG, C. G., LEABACK, D. H.: Nature **188**, 310 (1960)
447. GRESSNER, A. M.: Anal. Biochem. **56**, 532 (1973)
448. GROSKOPF, W. R., HOLLEMAN, J. W., MARGOLIASH, E., KLOTZ, I. M.: Biochemistry (U.S.A.) **5**, 3783 (1966)
449. GROSS, D.: J. Chromatog. **5**, 194 (1961)
450. GROSS, E., MORELL, J. L.: J. Biol. Chem. **241**, 3638 (1966)
451. GROSS, E., WITKOP, B.: J. Am. Chem. Soc. **83**, 1510 (1961)
452. GROSS, E., WITKOP, B.: J. Biol. Chem. **237**, 1856 (1962)
453. GUERIN, M. R., SHULTS, W. D.: J. Chromatog. Sci. **7**, 701 (1969)
454. GUEST, J. R., YANOFSKY, C.: J. Biol. Chem. **241**, 1 (1966)
455. GUIDOTTI, G.: Biochim. Biophys. Acta **42**, 177 (1960)
456. GUIDOTTI, G., HILL, R. J., KONIGSBERG, W.: J. Biol. Chem. **237**, 2184 (1962)
457. GUIDOTTI, G., KONIGSBERG, W.: J. Biol. Chem. **239**, 1474 (1964)
458. GURD, F. R. N., WILCOX, P. E.: Advances in Protein Chem. **11**, 311 (1956)
459. GURD, F. R. N.: In: Methods in enzymology, vol. 25 (eds. C. H. W. HIRS, S. N. TIMASHEFF), p. 424. New York: Academic Press 1972
460. GURIN, S., CLARKE, H. T.: J. Biol. Chem. **107**, 395 (1934)
461. GUSSIN, G. N., CAPECCHI, M. R., ADAMS, J. M., ARGETSINGER, J. E., WEBER, K., WATSON, J. D.: Cold Spring Harbor Symposia Quant. Biol. **36**, 257 (1966)
462. GUTTMANN, ST.: Helv. Chim. Acta **44**, 721 (1961)
463. GUTTMANN, ST.: Helv. Chim. Acta **49**, 83 (1966)
464. GUTTMANN, ST., BOISSONNAS, R. A.: Helv. Chim. Acta **41**, 1852 (1958)
465. GUTTMANN, ST., BOISSONNAS, R. A.: Helv. Chim. Acta **42**, 1257 (1959)
466. GUTTMANN, ST., BOISSONNAS, R. A.: Helv. Chim. Acta **44**, 1713 (1961)
467. GUTTMANN, ST., PLESS, J., BOISSONNAS, R. A.: Helv. Chim. Acta **45**, 170 (1962)
468. HAARMAN, W.: Biochem. Z. **314**, 1 (1943)
469. HABEEB, A. F. S. A., CASSIDY, A. H., SINGER, S. J.: Biochim. Biophys. Acta **29**, 587 (1958)
470. HAGEL, D., GERDING, J. J. T.: Anal. Biochem. **28**, 47 (1969)
471. HAGEN, P., BLACK, W.: Federation Proc. **23**, 371 (1960)
472. HAGENMAIER, H., EBBIGHAUSEN, W., NICHOLSON, G., VÖTSCH, W.: Z. Naturforsch. **256**, 681 (1970)
473. HAGLUND, H.: Sci. Tools **14**, No. 2 (1967)
474. HALEY, E., CORCORAN, B. J.: Biochemistry (U.S.A.) **6**, 2668 (1967)
475. HALPERN, B., JAMES, L. B.: Nature **202**, 592 (1964)
476. HALPERN, B.: Australian J. Chem. **18**, 417 (1965)
477. HALPERN, B., CROSS, A. D.: Chem. & Ind. (London) **1965**, 1183
478. HALPERN, B., NITECKI, D. E.: Tetrahedron Letrs. **1967**, 3031
479. HALSEY, Y. D., NEURATH, H.: J. Biol. Chem. **217**, 247 (1955)
480. HAMILTON, L. D. G.: In: Analytical methods of protein chemistry, vol. 2 (eds. P. ALEXANDER, R. J. BLOCK), p. 61. New York: Pergamon Press 1960
481. HAMILTON, P. B.: Anal. Chem. **30**, 914 (1948)
482. HAMILTON, P. B.: Anal. Chem. **35**, 2055 (1963)
483. HAMILTON, P. B.: Ann. N.Y. Acad. Sci. **102, 55** (1962)
484. HAMILTON, P. B.: Nature **205**, 284 (1965)
485. HAMILTON, P. B.: In: Advances in chromatography, vol. 2 (eds. J. C. GIDDINGS, R. A. KELLER), p. 3. New York: Marcel Dekker 1966
486. HAMILTON, P. B., NAGY, B.: Space Life Sci. **3**, 432 (1972)

487. Hamilton, P. B., Myoda, T. T.: Clin. Chem. **20**, 687 (1974)
488. Hanby, W. E., Waley, S. G., Watson, J.: J. Chem. Soc. **1950**, 3239
489. Hanford, B. O., Hylton, T. A., Preston, J., Weinstein, B.: J. Org. Chem. **32**, 1243 (1967)
490. Hanford, B. O., Jones, J. H., Young, G. T., Johnson, T. F. N.: J. Chem. Soc. **1965**, 6814
491. Hangaard, N.: J. Biol. Chem. **164**, 265 (1946)
492. Hare, P. E.: Carnegie Inst. Wash. Publ., Year Book **64**, 232 (1965)
493. Hare, P. E.: Federation Proc. **25**, 709 (1966)
494. Hare, P. E., Abelson, P. H.: Carnegie Inst. of Wash. Publ., Year Book **66**, 526 (1968)
495. Hare, P. E.: In: Organic geochemistry: methods and results (eds. G. Eglinton, M. T. J. Murphy), p. 438. Berlin-Heidelberg-New York: Springer 1969
496. Hare, P. E., Harada, K., Fox, S. W.: Proc. Apollo 11 Lunar Sci. Conf., Geochim. Cosmochim. Acta **2** (Suppl. 1) 1799 (1970)
497. Hare, P. E.: Carnegie Inst. Wash. Publ., Year Book **70**, 268 (1971)
498. Hare, P. E.: Space Life Sci. **3**, 354 (1972)
499. Hare, P. E.: Carnegie Inst. Wash. Publ., Year Book **72**, 701 (1973)
500. Harley-Mason, J., Archer, A. A. P. G.: Biochem. J. **69**, 60 (1958)
501. Harman, R. E., Patterson, J. L., van den Heuvel, W. J. A.: Anal. Biochem. **25**, 452 (1968)
502. Harrap, B. S., Green, L. C.: Anal. Biochem. **42**, 398 (1971)
503. Harrington, C. R., Mead, T. H.: Biochem. J. **30**, 1598 (1936)
504. Harrington, C. R., Pitt-Rivers, R. V.: Biochem. J. **38**, 417 (1944)
505. Harris, J. I.: Nature **177**, 471 (1956)
506. Harris, J. I.: Biochem. J. **71**, 451 (1959)
507. Harris, J. I., Fruton, J. S.: J. Biol. Chem. **191**, 143 (1951)
508. Harris, J. I., Hindley, J.: J. Mol. Biol. **3**, 117 (1961)
509. Harris, J. I., Knight, C. A.: Nature **170**, 613 (1952)
510. Harris, J. I., Lerner, A. B.: Nature **179**, 1346 (1957)
511. Harris, J. I., Li, C. H.: J. Am. Chem. Soc. **76**, 2607 (1954)
512. Harris, J. I., Li, C. H.: J. Biol. Chem. **213**, 499 (1955)
513. Harrison, P. M., Hoffman, T.: Biochem. J. **80**, 389 (1961)
514. Harshman, S., Najjar, V. A.: Biochemistry (U.S.A.) **4**, 2526 (1965)
515. Hartley, B. S., Kauffman, D. L.: Biochem. J. **101**, 229 (1966)
516. Hartley, B. S., Massey, V.: Biochim. Biophys. Acta **21**, 58 (1956)
517. Hartley, B. S.: Biochem. J. **119**, 805 (1970)
518. Harvey, D. G., Miller, E. J., Robson, W.: J. Chem. Soc. **1941**, 153
519. Hauptschein, M., Stokes, C. S., Nodiff, E. A.: J. Am. Chem. Soc. **74**, 4005 (1952)
520. Hayashi, R., Aibara, S., Hata, T.: Biochim. Biophys. Acta **212**, 359 (1970)
521. Hayashi, R., Moore, S., Stein, W. H.: J. Biol. Chem. **248**, 2296 (1973)
522. Heathcote, J. G., Haworth, C.: J. Chromatog. **43**, 84 (1969)
523. Heffter, B.: Med. Naturw. Arch. **1**, 81 (1907)
524. Heffter, B.: Chem. Zentr. **1907**, 822
525. Helferich, B., Böshagen, H.: Chem. Ber. **92**, 2813 (1959)
526. Heller, J., Smith, E. L.: J. Biol. Chem. **241**, 3165 (1966)
527. Hellerman, L.: Physiol. Revs. **17**, 454 (1937)
528. Henning, U., Helinski, D. R., Chao, F. C., Yanofsky, C.: J. Biol. Chem. **237**, 1523 (1962)
529. Henry, R. J., Smith, E. C.: Science **104**, 426 (1946)
530. Hermodson, M. A., Ericsson, L. H., Titani, K., Neurath, H., Walsh, K. A.: Biochemistry (U.S.A.) **11**, 4493 (1972)
531. Heynes, K., Legler, G.: Z. physiol. Chem., Hoppe-Seyler's **321**, 161 (1960)
532. Hickman, K., White, I., Stark, E.: Science **180**, 15 (1973)
533. Hill, R. L.: Advances in Protein Chem. **20**, 37 (1965)
534. Hill, R. L., Konigsberg, W.: J. Biol. Chem. **237**, 3151 (1962)
535. Hill, R. L., Schmidt, W. R.: J. Biol. Chem. **237**, 389 (1962)
536. Hill, R. L., Smith, E. L.: J. Biol. Chem. **228**, 577 (1957)

537. Hill, R. L., Spackman, D. H., Brown, D. M., Smith, E. L.: Biochem. Preparations **6**, 35 (1958)
538. Hillmann-Elies, Hillmann, G., Jatzkewitz, H.: Z. Naturforsch. **8**, 445 (1953)
539. Hilschmann, N., Craig, L. C.: Proc. Natl. Acad. Sci. (US) **53**, 1403 (1965)
540. Hilse, K., Sorger, V., Braunitzer, G.: Z. physiol. Chem., Hoppe-Seyler's **344**, 466 (1966)
541. Himmelhoch, S. R., Peterson, E. A.: Biochemistry (U.S.A.) **7**, 2085 (1968)
542. Himmelhoch, S. R.: In: Methods in enzymology, vol. XIX (eds. G. E. Perlmann, L. Lorand), p. 508. New York: Academic Press 1970
543. Hirs, C. H. W.: J. Biol. Chem. **219**, 611 (1956)
544. Hirs, C. H. W.: Ann. Rev. Biochem. **33**, 597 (1964)
545. Hirs, C. H. W.: In: Methods in enzymology, vol. XI (ed. C. H. W. Hirs). New York: Academic Press 1967
546. Hirs, C. H. W., Moore, S., Stein, W. H.: J. Biol. Chem. **195**, 669 (1952)
547. Hirs, C. H. W., Moore, S., Stein, W. H.: J. Biol. Chem. **219**, 623 (1956)
548. Hirs, C. H. W., Stein, W. H., Moore, S.: J. Biol. Chem. **221**, 151 (1956)
549. Hirs, C. H. W., Moore, S., Stein, W. H.: J. Biol. Chem. **235**, 633 (1960)
550. Hirs, C. H. W., Stein, W. H., Moore, S.: J. Biol. Chem. **211**, 941 (1954)
551. Hirs, C. H. W., Timasheff, S. N.: Methods in enzymology, vol. XXV. New York: Academic Press 1972
552. Hirs, C. H. W., Moore, S., Stein, W. H.: J. Biol. Chem. **235**, 633 (1960)
553. Hirschmann, R., Strachan, R. G., Schwan, H., Schoenewaldt, E. F., Joshua, H., Barkemeyer, B., Veber, D. F., Paleveda, W. J., Jacob, T. A., Beesley, T. E., Denkewalter, R. G.: J. Org. Chem. **32**, 3415 (1967)
554. Hiskey, R. G., Adams, J. B., Jr.: J. Am. Chem. Soc. **87**, 3969 (1965)
555. Hiskey, R. G., Adams, J. B., Jr.: J. Org. Chem. **31**, 2178 (1966)
556. Hiskey, R. G., Mizoguchi, T., Inui, T.: J. Org. Chem. **31**, 1192 (1966)
557. Hiskey, R. G., Mizoguchi, T., Smithwick, E. L., Jr.: J. Org. Chem. **32**, 97 (1967)
558. Hiskey, R. G., Staples, J. T., Smith, R. L.: J. Org. Chem. **32**, 2772 (1967)
559. Hiskey, R. G., Harpold, M. A.: J. Org. Chem. **33**, 559 (1968)
560. Hiskey, R. G., Harpp, D. N.: J. Am. Chem. Soc. **87**, 3965 (1965)
561. Hiskey, R. G., Smith, R. L.: J. Am. Chem. Soc. **90**, 2677 (1968)
562. Hiskey, R. G., Tucker, W. P.: J. Am. Chem. Soc. **84**, 4789 (1962)
563. Hoenders, H. J., Tol, J. V., Bloemendal: Biochim. Biophys. Acta **160**, 238 (1968)
564. Hoernle, S.: Z. physiol. Chem., Hoppe-Seyler's **348**, 1355 (1967)
565. Hofmann, K., Fimm, F. M., Limetti, M., Montibeller, J., Zanetti, G.: J. Am. Chem. Soc. **88**, 3633 (1966)
566. Hofmann, K., Haas, W., Smithers, M. J., Wells, R. D., Wolman, Y., Yanaihara, N., Zanetti, G.: J. Am. Chem. Soc. **87**, 620 (1965)
567. Hofmann, K., Pechham, W. D., Rheiner, A.: J. Am. Chem. Soc. **78**, 238 (1956)
568. Hofmann, K., Woolner, M. E., Spühler, G., Schwartz, E. T.: J. Am. Chem. Soc. **80**, 1486 (1958)
568a. Hofmann, K., Yajima, H., Yanaihara, N., Lui, T., Lande, S.: J. Am. Chem. Soc. **78**, 238 (1956)
569. Hofmann, K., Lindenmann, A., Magee, M. Z., Kahn, N. H.: J. Am. Chem. Soc. **74**, 470 (1952)
570. Hofmann, K., Magee, M. Z., Lindenmann, A.: J. Am. Chem. Soc. **72**, 7814 (1950)
571. Hofmann, K., Yanaihara, N., Lande, S., Yajima, H.: J. Am. Chem. Soc. **84**, 4470 (1962)
571a. Hofmann, K., Kappeler, H., Furlenmeier, A. E., Woolner, M. E., Schwartz, E. T., Thompson, T. A.: J. Am. Chem. Soc. **79**, 1641 (1957)
572. Hofmann, T.: Biochemistry (U.S.A.) **3**, 356 (1964)
573. Holcomb, G. N., James, S. A., Ward, D. N.: Biochemistry (U.S.A.) **7**, 1291 (1968)
574. Holiday, E. R.: Biochem. J. **30**, 1795 (1936)
575. Holley, R. W., Holley, A. D.: J. Am. Chem. Soc. **74**, 5445 (1952)
576. Holmquist, R.: J. Mol. Biol. **64**, 145 (1972)
577. Holmquist, W. R., Schroeder, W. A.: Biochemistry (U.S.A.) **5**, 2489 (1966)

578. HONEGGER, C. G.: Helv. Chim. Acta **44**, 173 (1961)
579. HONZL, J., RUDINGER, I.: Collection Czechoslov. Chem. Commun. **20**, 1190 (1955)
580. HONZL, J., RUDINGER, I.: Collection Czechoslov. Chem. Commun. **26**, 2333 (1961)
581. HOOPER, K. C., RYDON, H. N., SCHOFIELD, J. A., HEATON, G. S.: J. Chem. Soc. **1956**, 3148
582. HORINISHI, H., HACHIMORI, Y., KURIHARA, K., SHIBATA, K.: Biochim. Biophys. Acta **86**, 477 (1964)
583. HORMANN, H., JOSEPH, K. TH.: Biochim. Biophys. Acta **100**, 598 (1965)
584. HORNING, E. C., VAN DEN HEUVEL, W. J. A., CREECH, B. G.: Methods Biochem. Anal. **11**, 69 (1963)
585. HOROWITZ, M. G., KLOTZ, I. M.: Arch. Biochem. Biophys. **63**, 77 (1956)
586. HORTON, H. R., KOSHLAND, D. E., Jr.: J. Am. Chem. Soc. **87**, 1126 (1965)
587. HORTON, H. R., KOSHLAND, D. E., Jr.: In: Methods in enzymology, vol. 25 (eds. C. H. W. HIRS, S. N. TIMASHEFF), p. 468. New York: Academic Press 1972
588. HORTON, H. R., TUCKER, W. P.: J. Biol. Chem. **245**, 3397 (1970)
589. HSIEH, W. T., GUNDERSEN, L. E., VESTLING, C. S.: Biochem. Biophys. Res. Commun. **43**, 69 (1971)
590. HUBBARD, R. W.: Biochem. Biophys. Res. Commun. **19**, 679 (1965)
591. HUGLI, T. E., MOORE, S.: J. Biol. Chem. **247**, 2828 (1972)
592. HUNT, J. A., INGRAM, V. M.: Biochim. Biophys. Acta **49**, 520 (1961)
593. HUNTER, D., BAMFORD, R. R., RUSSELL, D. S.: Quart. J. Med. **33**, 193 (1940)
594. ICHISHIMA, E.: Tanpakushitsu, Kakusan, Koso (Protein, Nucleic acid, Enzyme) **17**, 58 (1972) (in Japanese)
595. ICHISHIMA, E.: Biochim. Biophys. Acta **258**, 274 (1972)
596. ICHISHIMA, E.: Abstracts, IVth Intern. Fermentation Symp. (Kyoto), p. 45 (1972)
597. IKAWA, M., SNELL, E. E.: J. Biol. Chem. **236**, 1955 (1961)
598. IKENAKA, T., BAMMERLIN, H., KAUFMANN, H., SCHMID, K.: J. Biol. Chem. **241**, 5560 (1966)
599. IKENAKA, T., SCHMID, K.: Nature **215**, 66 (1967)
600. ILSE, D., EDMAN, P.: Australian J. Chem. **16**, 411 (1963)
601. INAGAMI, T., MURAKAMI, K.: Anal. Biochem. **47**, 501 (1972)
602. INAGAMI, T.: Anal. Biochem. **52**, 318 (1973)
603. INGLIS, A. S., LIU, T. Y.: J. Biol. Chem. **245**, 112 (1970)
604. INGLIS, A. S., NICHOLLS, P. W., ROXBURGH, C. M.: Australian J. Biol. Sci. **24**, 1247 (1971)
605. INGRAM, V. M.: J. Chem. Soc. **1953**, 3717
606. INGRAM, V. M.: Nature **166**, 1038 (1950)
607. INGRAM, V. M.: J. Biol. Chem. **202**, 193 (1953)
608. INGRAM, V. M.: Biochim. Biophys. Acta **20**, 577 (1956)
609. INGRAM, V. M.: Biochim. Biophys. Acta **28**, 539 (1958)
610. INMAN, J. K., HANNON, J. E., APPELLA, E.: Biochem. Biophys. Res. Commun. **46**, 2075 (1972)
611. INOUYE, K., FRUTON, J. S.: Biochemistry (U.S.A.) **6**, 1765 (1967)
612. INOUYE, K., OTSUKA, H.: Bull. Chem. Soc. Japan **34**, 1 (1961)
613. INOUYE, K., VOYNICK, I. M., DELPIERRE, G. R., FRUTON, J. S.: Biochemistry (U.S.A.) **5**, 2473 (1966)
614. INUKAI, N., NAKANO, K., MURAKAMI, M.: Bull. Chem. Soc. Japan **40**, 2913 (1967)
615. ISELIN, B.: Arch. Biochem. Biophys. **78**, 532 (1958)
616. ISELIN, B., FEURER, M., SCHWYZER, R.: Helv. Chim. Acta **38**, 1508 (1955)
617. ISELIN, B., RITTEL, W., SIEBER, P., SCHWYZER, R.: Helv. Chim. Acta **40**, 373 (1957)
618. ISELIN, B., SCHWYZER, R.: Helv. Chim. Acta **39**, 57 (1956)
619. ISELIN, B., SCHWYZER, R.: Helv. Chim. Acta **44**, 169 (1961)
620. ISELIN, B., SCHWYZER, R.: Helv. Chim. Acta **45**, 1499 (1962)
621. ISHIDA, Y., ONISHI, M.: Tokushima Daigaku Yakugokubu Kenkyû Nempo **14**, 5 (1965) (Engl.); — Chem. Abs. **68**, 13371 a (1968)
622. ITANO, H. A., GOTTLIEB, A. J.: Biochem. Biophys. Res. Commun. **12**, 405 (1963)
623. IVANOV, CH. P., MANCHEVA, I. N., GAITANDJIEV, S.: Compt. rend. acad. bulgare sci. **20**, 799 (1967)

624. Ivanov, Ch. P., Mancheva, I. N.: Compt. rend. acad. bulgare sci. **21**, 785 (1968)
625. Ivanov, Ch. P., Mancheva, I. N.: Compt. rend. acad. bulgare sci. **25**, 349 (1972)
626. Ivanov, Ch. P., Mancheva, I. N.: J. Chromatog. **75**, 129 (1973)
627. Ivanov, Ch. P., Mancheva, I. N.: Anal. Biochem. **53**, 420 (1973)
628. Ivanov, Ch. P., Vladovska-Yukhnovska, Y.: Biochim. Biophys. Acta **194**, 345 (1969)
629. Izumi, Y.: Anal. Biochem. **12**, 1 (1965)
630. Jacobs, W. A., Graig, L. C.: J. Biol. Chem. **113**, 759 (1936)
631. Jacobsen, C. J., Leonis, J., Linderstrom-Lang, K., Ottessen, M.: In: Methods of biochemical analysis, vol. IV (ed. D. Glick), p. 171. New York: Wiley-Interscience 1955
632. Jakubke, H. D.: Z. Chem. **6**, 52 (1966)
633. Jakubke, H. D., Voigt, A.: Chem. Ber. **99**, 2419 (1966)
634. Jakubke, H. D., Voigt, A.: Chem. Ber. **99**, 2944 (1966)
635. James, L. B.: J. Chromatog. **59**, 178 (1971)
636. Jeppsson, J., Sjöquist, J.: Anal. Biochem. **18**, 264 (1967)
637. Jeppsson, J. B., Smith, I.: Nature **172**, 1100 (1953)
637a. Jeschkeit, H.: Z. Chem. **8**, 20 (1968)
638. Jerchel, D., Müller, R.: Naturwissenschaften **38**, 561 (1951)
639. John, W. D., Young, G. T.: J. Chem. Soc. **1954**, 2870
640. Johnson, T. B., Nicolet, B. H.: J. Am. Chem. Soc. **33**, 1973 (1911)
641. Jolles, P., Fromageot, C.: Biochim. Biophys. Acta **9**, 287 (1952)
642. Jolles, P., Fromageot, C.: Biochim. Biophys. Acta **9**, 416 (1952)
643. Jolles, P., Thaureaux, J., Fromageot, C.: Arch. Biochem. Biophys. **69**, 290 (1957)
644. Jolles, J., Jolles, P.: Helv. Chim. Acta **54**, 2668 (1971)
645. Jones, J. H., Young, G. T.: J. Chem. Soc. **1968**, 53
646. Jones, J. H., Liberek, B., Young, G. T.: J. Chem. Soc. **1967**, 2371
647. Jones, R. T., Weiss, G.: Anal. Biochem. **9**, 377 (1964)
648. Jones, S. R., Hofmann, T.: Can. J. Biochem. **50**, 1297 (1972)
649. Jori, G., Galiazzo, G., Scoffone, E.: Intern. J. Protein Res. **1**, 289 (1969)
650. Jost, K., Rudinger, J.: Collection Czechoslov. Chem. Commun. **26**, 2345 (1961)
651. Julian, P. L., Meyer, E. W., Printy, H. C.: In: Heterocyclic compounds, vol. 3 (ed. R. C. Elderfield), p. 84. New York: John Wiley and Sons, Inc. 1952
652. Juneja, K. K., Sule, A. D., Chipalkatti, V. B.: Textile Research J. **38**, 461 (1968)
653. Jutisz, M., Ritschard, W.: Biochim. Biophys. Acta **17**, 548 (1955)
654. Kader, A. T., Stirling, C. J. M.: J. Chem. Soc. **1964**, 258
655. Kagamiyama, H., Watanabe, T., Wada, H.: Biochem. Biophys. Res. Commun. **32**, 678 (1968)
656. Kapoor, A.: J. Pharm. Sci. **56**, 1532 (1967)
657. Kappeler, H.: Helv. Chim. Acta **44**, 476 (1961)
658. Kappeler, H., Schwyzer, R.: Helv. Chim. Acta **43**, 1453 (1960)
659. Karush, F., Klinman, N. R., Marks, R.: Anal. Biochem. **9**, 100 (1964)
660. Kasper, C. B., Matsubara, H., Smith, E. L.: J. Biol. Chem. **240**, 1131 (1965)
661. Kasper, C. B., Smith, E. L.: J. Biol. Chem. **241**, 3754 (1966)
662. Katchalski, E.: Advances in Protein Chem. **6**, 123 (1951)
663. Katchalski, E., Benjamin, G. S., Gross, Y.: J. Am. Chem. Soc. **79**, 4069 (1957)
664. Katsoyannis, P. G.: J. Am. Chem. Soc. **83**, 4053 (1961)
665. Katsoyannis, P. G., du Vigneaud, V.: J. Am. Chem. Soc. **76**, 3113 (1954)
666. Katsoyannis, P. G., Gish, D. T., Hess, G. P., du Vigneaud, V.: J. Am. Chem. Soc. **80**, 2558 (1958)
667. Katsoyannis, P. G., Gish, D. T., du Vigneaud, V.: J. Am. Chem. Soc. **79**, 4516 (1957)
668. Katsoyannis, P. G., Suzuki, K.: J. Am. Chem. Soc. **83**, 4057 (1961)
669. Katsoyannis, P. G., Tometsko, A. M., Ginos, J. Z., Tilak, M. A.: J. Am. Chem. Soc. **88**, 164 (1966)
670. Katsoyannis, P. G., Tometsko, A. M., Zalut, C.: J. Am. Chem. Soc. **88**, 166 (1966)
671. Katsuki, S., Scott, J. E., Yamashina, J.: Biochem. J. **97**, 25c (1965)
672. Katz, A. M., Dreyer, W. J., Anfinsen, C. B.: J. Biol. Chem. **234**, 2897 (1959)
673. Katz, J. J.: Nature **174**, 509 (1954)

674. Kawahara, K., Tanford, C.: Biochimistry (U.S.A.) **5**. 1578 (1966)
675. Kawanishi, Y., Iwai, K., Ando, T.: J. Biochem. (Tokyo) **56**, 314 (1964)
676. Kaweran, E., Wieland, Th.: Nature **168**, 77 (1951)
677. Keil, B., Knesslova, V., Šorm, F.: Chem. listy **46**, 167 (1952)
678. Kelley, R. B.: J. Org. Chem. **28**, 453 (1963)
679. Kempf, D. S., Chien, S. W.: J. Am. Chem. Soc. **89**, 2745 (1967)
680. Kempf, D. S., Woodward, R. B.: Tetrahedron **21**, 3019 (1965)
681. Kenner, G. W., Khorana, H. G., Stedman, R. J.: J. Chem. Soc. **1953**, 673
682. Kenner, G. W., Khorana, H. G.: J. Chem. Soc. **1952**, 2076
683. Kenner, G. W., Khorana, H. G., Stedman, R. J.: J. Chem. Soc. **1953**, 673
684. Kessler, W., Iselin, B.: Helv. Chim. Acta **49**, 1330 (1966)
685. Keyser, J. W.: Biochem. J. **43**, 488 (1948)
686. Khorana, H. G.: Chem. Rev. **53**, 145 (1953)
687. Khorana, H. G.: J. Chem. Soc. **1952**, 2081
688. Khosla, M. C., Smeby, R. R., Bumpus, F. M.: Biochemistry (U.S.A.) **6**, 754 (1967)
689. Kienhuis, H., van de Linde, A., van der Holst, J. P. J., Verweij, A.: Rec. trav. chim. **80**, 1278 (1961)
690. Kimmel, J. R., Kato, G. K., Paiva, A. C. M., Smith, E. L.: J. Biol. Chem. **237**, 2525 (1962)
691. Kimmel, J. R., Smith, E. L.: J. Biol. Chem. **215**, 55 (1955)
692. Kimmel, J. R., Thompson, E. O. P., Smith, E. L.: J. Biol. Chem. **217**, 151 (1955)
693. King, F. E., Clark-Lewis, J. W., Wade, R.: J. Chem. Soc. **1957**, 886
694. King, F. E., Kidd, D. A. A.: J. Chem. Soc. **1949**, 3315
695. King, F. E., Clark-Lewis, J. W., Kidd, D. A. A., Smith, G. R.: J. Chem. Soc. **1954**, 1039
696. Kirk, K. L., Cohen, L. A.: J. Org. Chem. **34**, 395 (1969)
697. Kirkland, J. J.: J. Chromatog. Sci. **10**, 129 (1972)
698. Kirsten, E., Kirsten, R.: Biochem. Biophys. Res. Commun. **7**, 76 (1962)
699. Kisfaludy, L., Dualszky, S.: Acta Chim. Acad. Sci. Hung. **24**, 301 (1960)
700. Kisfaludy, L.: Acta Chim. Acad. Sci. Hung. **24**, 309 (1960)
701. Kisfaludy, L., Dualszky, S., Medzihradszky, K., Bajusz, S., Bruckner, V.: Acta Chim. Acad. Sci. Hung. **30**, 473 (1962)
702. Kjaer, A., Eriksen, P.: Acta Chem. Scand. **6**, 448 (1952)
703. Klebe, J. F., Finkbeiner, H., White, D. M.: J. Am. Chem. Soc. **88**, 3390 (1966)
704. Klotz, I. M., Ayers, J.: J. Am. Chem. Soc. **79**, 4078 (1957)
705. Knox, R., Kohler, G. O., Palter, R., Walker, H. G.: Anal. Biochem. **36**, 136 (1970)
706. Koch, G., Weidel, W.: Z. physiol. Chem., Hoppe-Seyler's **303**, 213 (1956)
707. Koenig, W., Geiger, R., Siedel, W.: Chem. Ber. **101**, 681 (1968)
708. Kohn, J.: Clin. Chim. Acta **2**, 297 (1957)
709. Kolthoff, I. M., Lingane, J. J.: In: Polarography, 2nd ed. New York: Interscience 1952
710. Kolthoff, I. M., Stricks, W.: Anal. Chem. **23**, 763 (1951)
711. Kolthoff, I. M., Stricks, W.: J. Am. Chem. Soc. **72**, 1952 (1950)
712. Kolthoff, I. M., Harris, W. E.: Ind. Eng. Chem., Anal. Ed. **18**, 161 (1946)
713. Kolthoff, I. M., Furman, N. H.: Potentiometric titrations, 2nd ed. New York: J. Wiley and Sons 1949
714. Komatsu, S.: Chem. Zentralblatt II, 537 (1911)
715. Konigsberg, W.: In: Methods in enzymology, vol. XI (eds. S. P. Colowick, N. O. Kaplan), p. 461. New York, London: Academic Press 1967
716. Konigsberg, W., Goldstein, J., Hill, R. J.: J. Biol. Chem. **238**, 2028 (1963)
717. Konigsberg, W., Hill, R. J.: J. Biol. Chem. **237**, 2547 (1962)
718. Konigsberg, W., Hill, R. J.: J. Biol. Chem. **237**, 3157 (1962)
719. Konigsberg, W., Weber, K., Notani, G., Zinder, N.: J. Biol. Chem. **241**, 2579 (1966)
720. Konigsberg, W.: In: Methods in enzymology, vol. 25, part B (eds. C. H. W. Hirs, S. N. Timasheff), p. 185, 326. New York: Academic Press 1972
721. Kopple, K. D.: In: Peptides and amino acids. New York: W. A. Benjamin, Inc. 1966
722. Kopple, K. D., Bächli, E.: J. Org. Chem. **24**, 2053 (1959)

723. KORMAN, S., CLARKE, H. T.: J. Biol. Chem. **221**, 113, 133 (1956)
724. KOSHLAND, D. E., Jr., KARKHANIS, Y. D., LATHAM, H. G.: J. Am. Chem. Soc. **86**, 1448 (1964)
725. KOSHLAND, D. E., Jr., STRUMEYER, D. H., RAY, W. J., Jr.: Brookhaven Symposia in Biol. **15**, 101 (1962)
726. KOSTKA, V., CARPENTER, F. H.: J. Biol. Chem. **239**, 1799 (1964)
727. KOTAKI, A., HARODA, M., YAGI, K.: J. Biochem. (Tokyo) **55**, 553 (1964)
728. KOVACS, J., CEPRINI, M. Q., DUPRAZ, C. A., SCHMITT, G. N.: J. Org. Chem. **32**, 3696 (1967)
729. KOVACS, J., GIANOTTI, R., KAPOOR, A.: J. Am. Chem. Soc. **88**, 2282 (1966)
730. KOVACS, J., RODIN, R. L.: J. Org. Chem. **33**, 2418 (1968)
731. KREIL, G., KREIL-KISS, G.: Biochem. Biophys. Res. Commun. **27**, 275 (1967)
732. KREIL, G., TUPPY, H.: Nature **192**, 1121 (1961)
733. KRIVTSOV, V. F., STEPHANOV, V. M.: J. Gen. Chem. (U.S.S.R.) **35**, 554 (1965)
734. KUBO, H., NAKAJIMA, T., TAMURA, Z.: Proceedings of the 8th symposium on peptide chemistry (ed. T. KANEKO), p. 171. Osaka: Protein Research Foundation 1970
735. KULBE, K. D.: Anal. Biochem. **44**, 548 (1971)
736. KULBE, K. D.: Anal. Biochem. **59**, 564 (1974)
737. KUNDZIN, S. F. R. M., DE BUN, M., NORD, F. F.: J. Am. Chem. Soc. **73**, 4615 (1951)
738. KUNG YUEH-TING: Sci. Sinica (Peking) **15**, 544 (1966)
739. KUNKEL, H. G., TISELIUS, A.: J. Gen. Physiol. **35**, 89 (1951)
740. KUPRYSZEWSKI, G.: Roczniki Chem. **35**, 595 (1961)
741. KUPRYSZEWSKI, G.: FORMELA, M.: Roczniki Chem. **35**, 1533 (1961)
742. KUPRYSZEWSKI, G.: Roczniki Chem. **37**, 593 (1963)
743. KURIHARA, K., HORINISHI, H., SHIBATA, K.: Biochim. Biophys. Acta **74**, 678 (1963)
744. KUSAMA, K.: J. Biochem. (Tokyo) **44**, 375 (1957)
745. LAMKIN, W. M., WEATHERFORD, J. W., JONES, N. S., PAN, T., WARD, D. N.: Anal. Biochem. **58**, 422 (1974)
746. LANDMANN, W. A., DRAKE, M. P., DILLAHA, J.: J. Amer. Chem. Soc. **75**, 3638 (1953)
747. LAUFER, D. A., BLOUT, E. R.: J. Am. Chem. Soc. **89**, 1246 (1967)
748. LAUFER, D. A., CHAPMAN, T. H., MARLBOROUGH, D. I., VAIDYA, V. M., BLOUT, E. R.: J. Am. Chem. Soc. **90**, 2696 (1968)
749. LAURSEN, R. A.: J. Am. Chem. Soc. **88**, 5344 (1966)
750. LAURSEN, R. A.: Biochem. Biophys. Res. Commun. **37**, 663 (1969)
751. LAURSEN, R. A.: Europ. J. Biochem. **20**, 89 (1971)
752. LAURSEN, R. A.: In: Recent developments in the chemical study of protein structures, p. 11. Meeting in Montpellier, 28—29 Sept. 1971
753. LAURSEN, R. A., HORN, M. J., BONNER, A. G.: Febs Letters **21**, 67 (1972)
754. LAVER, W. G.: Biochim. Biophys. Acta **53**, 469 (1961)
755. LAW, H. D.: Ann. Repts Progr. Chem. (Chem. Soc. London) **63**, 517 (1966)
756. LEACH, S. J., PARKHILL, E. M. J.: Proc. Intern. Wool Textiles Res. Conf. Australia, Vol. C, 92 (1955)
757. LEACH, S. J.: Australian J. Chem. **13**, 520 (1960)
758. LEACH, S. J.: In: Analytical methods of protein chemistry, vol. 4 (eds. P. ALEXANDER, M. P. LUNDGREN), p. 3. New York: Pergamon Press 1966
759. LEONIS, J., LEVY, A. L.: Compt. rend. trav. lab. Carlsberg **29**, 57 (1954)
760. LEONIS, J., LEVY, A. L.: Compt. rend. trav. lab. Carlsberg **29**, 87 (1954)
761. LEONIS, J., LI, C. H., CHUNG, D.: J. Am. Chem. Soc. **81**, 419 (1959)
762. LEPLAWY, M. T., JONES, D. S., KENNER, G. W., SHEPPARD, R. C.: Tetrahedron Letrs. **11**, 39 (1960)
763. LEQUIN, R. M., NIALL, H. D.: Biochim. Biophys. Acta **257**, 76 (1972)
764. LESLIE, J., WILLIAMS, D. L., GORIN, G.: Anal. Biochem. **3**, 257 (1962)
765. LEVIN, Y., BERGER, A., KATCHALSKI, R.: Biochem. J. **63**, 308 (1956)
766. LEVY, A. L.: J. Chem. Soc. **1950**, 404
767. LEVY, A. L.: Biochim. Biophys. Acta **15**, 589 (1954)
768. LEVY, A. L.: Nature **174**, 126 (1954)
769. LEVY, A. L., CHUNG, D.: Biochim. Biophys. Acta **17**, 454 (1955)

770. LEVY, M.: In: Methods in enzymology, vol. IV (eds. S. P. COLOWICK, N. O. KAPLAN), p. 238. New York: Academic Press 1957
771. LI, C. H., BERTSCH, L.: J. Biol. Chem. **235**, 2638 (1960)
772. LI, C. H., CUMMINGS, J. T.: J. Biol. Chem. **233**, 73 (1958)
773. LI, C. H., GESCHWIND, I. I., COLE, R. D., RAACKE, I. D., HARRIS, J. I., DIXON, J. S.: Nature **176**, 687 (1955)
774. LI, C. H., RAMACHANDRAN, J., CHUNG, D., GORUP, B.: J. Am. Chem. Soc. **86**, 2703 (1964)
775. LIAO, T., ROBINSON, G., SALNIKOW, J.: Anal. Chem. **45**, 2286 (1973)
776. LIEFLÄNDER, M.: Z. physiol. Chem., Hoppe-Seyler's **320**, 35 (1960)
777. LIENER, L. E.: J. Biol. Chem. **225**, 1061 (1957)
778. LIGHT, A. L.: In: Methods in enzymology, vol. XI (ed. C. H. W. HIRS), p. 417. New York: Academic Press 1967
779. LIGHT, A. L., GREENBERG, J.: J. Biol. Chem. **240**, 258 (1965)
780. LIGHT, A. L.: In: Methods in enzymology, vol. 25 (eds. C. H. W. HIRS, S. N. TIMASHEFF), p. 253. New York: Academic Press 1972
781. LIGHT, A. L., SMITH, E. L.: J. Biol. Chem. **237**, 2537 (1962)
782. LIGHT, T. S., SWARTZ, J. L.: Anal. Letters **1**, 825 (1968)
783. LINDLEY, H.: Nature **178**, 647 (1956)
784. LINDLEY, H.: Biochem. J. **126**, 683 (1972)
785. LIU, T-Y., STEIN, W. H., MOORE, S., ELLIOT, S. D.: J. Biol. Chem. **240**, 1143 (1965)
786. LIU, T-Y., CHANG, Y. H.: J. Biol. Chem. **246**, 2842 (1971)
787. LIU, T-Y.: In: Methods in enzymology, vol. 25 (eds. C. H. W. HIRS, S. N. TIMASHEFF), p. 44. New York: Academic Press 1972
788. LIU, T-Y., INGLIS, A. S.: In: Methods in enzymology, vol. 25 (eds. C. H. W. HIRS, S. N. TIMASHEFF), p. 55. New York: Academic Press 1972
789. LOEHR, J. S., MASON, H. S.: Biochem. Biophys. Res. Commun. **51**, 741 (1973)
790. LONDON, G. M., PORTSMOUTH, D., LUKTON, A., KOSHLAND, D. E., Jr.: J. Am. Chem. Soc. **91**, 2792 (1969)
791. LONG, C. L., GEIGER, J. W.: Anal. Biochem. **29**, 265 (1969)
792. LOSSE, G., MÜLLER, G.: Chem. Ber. **94**, 2768 (1961)
792a. LOSSE, G., HOFFMANN, K.-H., HETZER, G.: Ann. Chem., Justus Liebigs **684**, 236 (1965)
793. LOVINS, R. E., CRAIG, J., FAIRWELL, T., MCKINNEY, C.: Anal. Biochem. **47**, 539 (1972)
794. LOWRY, D. H., ROSEBROUGH, N. A., FARR, A. L., RANDALL, R. J.: J. Biol. Chem. **193**, 265 (1951)
795. LOWTHER, A. G.: Nature **167**, 767 (1951)
796. LUDOWIEG, J., DORFMAN, A.: Biochim. Biophys. Acta **38**, 212 (1960)
797. LUGG, J. W. H.: Biochem. J. **26**, 2144 (1932)
798. LUTZ, W. B., RESSLER, C., NETTLETON, D. E., Jr., DU VIGNEAUD, V.: J. Am. Chem. Soc. **81**, 167 (1959)
799. MACKWORTH, J. F.: Biochem. J. **42**, 82 (1948)
800. MACLAREN, J. A., SAVIGE, W. E., SWAN, J. M.: Australian J. Chem. **11**, 345 (1958)
801. MACLAREN, J. A., SWEETMAN, B. J.: Australian J. Chem. **19**, 2355 (1966)
802. MAEDA, M., TSUJI, A., GANNO, S., ONISHI, Y.: J. Chromatog. **77**, 434 (1973)
803. MAHOWALD, T.: Biochemistry (U.S.A.) **4**, 732 (1965)
804. MAJORS, R. E.: J. Chromatog. Sci. **11**, 88 (1973)
805. MALKIN, R., RABINOWITZ, J. C.: Biochemistry (U.S.A.) **6**, 3880 (1967)
806. MANNING, J. M., MOORE, S.: J. Biol. Chem. **243**, 5591 (1968)
807. MANNING, J. M.: J. Am. Chem. Soc. **92**, 7449 (1970)
808. MARGLIN, A., MERRIFIELD, R. B.: J. Am. Chem. Soc. **88**, 5051 (1966)
809. MARGOLIASH, E.: J. Biol. Chem. **237**, 2161 (1962)
810. MARGOLIASH, E., SMITH, E. L.: J. Biol. Chem. **237**, 2151 (1962)
810a. MARGOLIASH, E., SMITH, E. L., KREIL, G., TUPPY, H.: Nature **192**, 1125 (1961)
811. MARKLAND, F. S., KREIL, G., RIBADEAU-DUMAS, B., SMITH, E. L.: J. Biol. Chem. **241**, 4642 (1966)

812. MARKLAND, F. S., RIBADEAU-DUMAS, B., SMITH, E. L.: J. Biol. Chem. **242**, 5174 (1967)
813. MARKLAND, F. S., SMITH, E. L.: J. Biol. Chem. **242**, 5198 (1967)
814. MARKUS, G., KARUSH, F.: J. Am. Chem. Soc. **79**, 134 (1957)
815. MARRIAN, D. H.: J. Chem. Soc. **1949**, 1515
816. MARRIG, C., LUCCIONI, F., LAURENT, B.: Biochim. Biophys. Acta **105**, 606 (1965)
817. MARSHALL, G. R., MERRIFIELD, R. B.: Biochemistry (U.S.A.) **4**, 2394 (1965)
818. MARTIN, R. B., AMES, B.: J. Biol. Chem. **236**, 1372 (1961)
819. MASON, E. E.: In: Computer applications in medicine. Springfield (Ill.): C. C. Thomas Co. 1966
820. MATSUBARA, H., SASAKI, R. M., CHAIN, R. K.: Proc. Natl. Acad. Sci. (US) **57**, 439 (1967)
821. MATSUBARA, H., SASAKI, R. M.: Biochem. Biophys. Res. Commun. **35**, 175 (1969)
822. MATSUBARA, H., SMITH, E. L.: J. Biol. Chem. **241**, 2732 (1963)
822a. MATSUBARA, H.: In: Methods in enzymology, vol. XIX (eds. G. E. PERLMANN, L. LORAND), p. 642. New York: Academic Press 1970
822b. MATSUBARA, H., FEDER, J.: In: The enzymes **3**, 721 (1971)
823. MATSUO, H.: Protein Nucleic Acid Enzyme **14**, 523, 589 (1969)
824. MATSUO, H., FUJIMOTO, Y., TATSUNO, T.: Tetrahedron Letrs. **39**, 3465 (1965)
825. MATSUO, H., FUJIMOTO, Y., TATSUNO, T.: Biochem. Biophys. Res. Commun. **22**, 69 (1966)
826. MATSUO, H., FUJIMOTO, Y., TATSUNO, T.: Chem. Pharm. Bull. (Tokyo) **15**, 716 (1967)
827. MATSUO, H., MATSUBARA, H.: Proc. Soc. Exp. Biol. Med. **129**, 564 (1968)
828. MATSUO, H.: Tanpakushitsu, Kakusan, Koso (Protein, Nucleic acid, Enzyme) **14**, 523, 589 (1969) (in Japanese)
829. MATSUO, H., BABA, Y., NAIR, R. M. G., ARIMURA, A., SCHALLY, A. V.: Biochem. Biophys. Res. Commun. **43**, 1334 (1971)
830. MATSUO, H., NARITA, K.: Proc. 21st Symp. on Protein Structure (Japan), p. 73, 1970 (in Japanese)
831. MATSUO, H., TODA, H., NARITA, K.: Seikagaku **44**, 446 (1972) (in Japanese)
831a. MATTHEWS, B. W., JANSONIUS, J. N., COLMAN, P. M., SCHOENBOEN, B. P., DUPOURQUE, D.: Nature (New Biology) **238**, 37 (1972)
832. MAYNARD, J. L.: J. Am. Chem. Soc. **54**, 2108 (1932)
833. MAYO, F. R., WALLING, C.: Chem. Revs. **27**, 351 (1940)
834. MAZUR, R. H.: J. Org. Chem. **28**, 2498 (1963)
835. MAZUR, R. H., SCHLATTER, J. M.: U.S. Patent 3,280,099 (1966)
836. MCDONALD, J. K., CALLAHAN, P. X., ELLIS, S.: In: Methods in enzymology, vol. 25 (eds. C. H. W. HIRS, S. N. TIMASHEFF), p. 272, 282. New York: Academic Press 1972
837. MCDOWALL, M. A., SMITH, E. L.: J. Biol. Chem. **240**, 281 (1965)
838. MCFARLAND, B. G., INQUE, Y., NAKANISHI, K.: Tetrahedron Ltrs. **11**, 857 (1969)
839. MCKAY, F. C., ALBERTSON, N. F.: J. Am. Chem. Soc. **79**, 4686 (1957)
840. MCMURRAY, C. H., TRENTHAM, D. R.: Biochem. J. **115**, 913 (1969)
841. MCPHEE, J. R.: Biochem. J. **64**, 22 (1956)
842. MECHANIC, G. L., LEVY, M.: J. Am. Chem. Soc. **81**, 1889 (1959)
843. MEDZIHRADSZKY, K., SCHWEIGER, H. M.: Acta Chim. Acad. Sci. Hung. **30**, 105 (1962)
844. MEIENHOFER, J.: Chimia (Switz.) **16**, 385 (1962)
845. MEIENHOFER, J.: Nature **205**, 73 (1965)
846. MEIENHOFER, J., SCHNABEL, E., BREMER, H., BRINKHOFF, O., ZABEL, R., SROKA, W., KLOSTERMEYER, H., BRANDENBURG, D., OKUDA, T., ZAHN, H.: Z. Naturforscher **18**, 1120 (1963)
847. MELVAS, B. W.: Acta Chem. Scand. **23**, 1679 (1969)
848. MELVILLE, J.: Biochem. J. **29**, 179 (1935)
849. MERIGAN, T. C., DREYER, W. J., BERGER, A.: Biochim. Biophys. Acta **62**, 122 (1962)
850. MERRIFIELD, R. B.: J. Am. Chem. Soc. **85**, 2194 (1963)
851. MERRIFIELD, R. B.: J. Org. Chem. **29**, 3100 (1964)
852. MERRIFIELD, R. B.: Biochemistry (U.S.A.) **3**, 1385 (1964)
853. MERRIFIELD, R. B.: Science **150**, 178 (1965)
854. MERRIFIELD, R. B., WOOLEY, D. W.: J. Am. Chem. Soc. **78**, 4646 (1956)

855. Merrifield, R. B., Wooley, D. W.: J. Am. Chem. Soc. **80**, 6635 (1958)
856. Merrifield, R. B., Stewart, J. M., Jernberg, N.: Anal. Chem. **38**, 1905 (1966)
857. Meites, L.: In: Polarographic techniques. New York: Interscience 1955
858. Meselson, M., Stahl, F. W., Vinograd, J.: Proc. Natl. Acad. Sci. (US) **41**, 581 (1959)
859. Michl, H.: Monatsh. Chem. **82**, 489 (1951)
860. Michl, H.: J. Chromatog. **1**, 93 (1958)
861. Mikeš, O., Holeyšovsky, V., Tomášek, V., Šorm, F.: Biochem. Biophys. Res. Commun. **24**, 346 (1966)
862. Miller, H. K., Waelsch, H.: Arch. Biochem. Biophys. **35**, 176 (1952)
863. Millon, E.: Compt. rend. **28**, 40 (1949)
864. Mills, G. L.: In: Chromatographic and electrophoretic techniques, vol. I (ed. I. Smith), p. 143. London: William Heinemann 1960
865. Mills, G. L.: Nature **165**, 403 (1950)
866. Mills, G. L.: Biochem. J. **50**, 707 (1952)
867. Milne, H. B., Halver, J. E., Ho, D. S., Mason, M. S.: J. Am. Chem. Soc. **79**, 637 (1957)
868. Milne, H. B., Kilday, W. D.: J. Org. Chem. **30**, 64 (1965)
869. Milne, H. B., Most, C. F., Jr.: J. Org. Chem. **33**, 169 (1968)
870. Milstein, C.: Biochem. J. **100**, 40c (1966)
871. Milstein, C., Sanger, F.: Biochem. J. **79**, 456 (1961)
872. Mirsky, A. E., Anson, M. L.: J. Gen. Physiol. **18**, 307 (1935)
873. Mizoguchi, T., Levin, G., Woolley, D. W., Stewart, J. M.: J. Org. Chem. **33**, 903 (1968)
874. Moffat, E. D., Lytle, R. I.: Anal. Chem. **31**, 926 (1959)
875. Moore, A. T., Rydon, H. N., Smithers, M. J.: J. Chem. Soc. **1966**, 2349
876. Moore, A. T., Cole, R. D., Gundlach, H. G., Stein, W. H.: Proc. 4th Int. Congr. Biochem., Vienna 1958, Symposium **8**, 51 (1959)
877. Moore, A. T., Stein, W. H.: J. Biol. Chem. **176**, 367 (1948)
878. Moore, A. T., Stein, W. H.: J. Biol. Chem. **178**, 53, 79 (1949)
879. Moore, A. T., Stein, W. H.: J. Biol. Chem. **192**, 663 (1951)
880. Moore, A. T., Stein, W. H.: J. Biol. Chem. **211**, 893, 907 (1954)
881. Moore, A. T.: J. Biol. Chem. **243**, 6281 (1968)
882. Moore, A. T., Spackman, D. H., Stein, W. H.: Anal. Chem. **30**, 1185 (1958)
883. Moore, A. T., Stein, W. H.: In: Methods in enzymology, vol. 6 (eds. S. P. Colowick, N. O. Kaplan), p. 819. New York: Academic Press 1963
884. Moore, A. T.: In: Chemistry and biology of peptides (ed. J. Meienhofer), p. 629. Ann Arbor: Ann Arbor Science Publishers 1972
885. Morley, J. S.: J. Chem. Soc. **1967**, 2410
886. Morris, A. J.: Biochem. J. **81**, 606 (1961)
887. Morse, D., Horecker, B. L.: Anal. Biochem. **14**, 429 (1966)
888. Mross, G. A., Doolittle, R. F.: Federation Proc. **30**, 1241 (1971)
889. Mross, G. A., Doolittle, R. F.: Arch. Biochem. Biophys. **122**, 674 (1967)
890. Mross, G. H., Ohms, J. I.: Abstr. Commun. Mtg. Fed. Europ. Biochem. Soc. **8**, 819 (1972)
891. Mukaiyama, T., Ueki, M., Maruyama, H., Matsueda, R.: J. Am. Chem. Soc. **90**, 4490 (1968)
892. Munier, R. L., Sarrazin, G.: J. Chromatog. **22**, 347 (1966)
893. Muraoka, S., Takata, J.: Yakugaku Zasshi **77**, 1155 (1957) (in Japanese)
894. Narita, K.: Biochim. Biophys. Acta **28**, 184 (1958)
895. Narita, K.: Biochim. Biophys. Acta **30**, 352 (1958)
896. Narita, K.: Biochim. Biophys. Acta **31**, 372 (1959)
897. Narita, K.: Biochem. Biophys. Res. Commun. **5**, 160 (1961)
898. Narita, K., Ishii, J.: J. Biochem. (Tokyo) **52**, 367 (1962)
899. Narita, K., Murakami, H., Ikenaka, T.: J. Biochem. (Tokyo) **59**, 170 (1966)
900. Narita, K., Ohta, Y.: Bull. Chem. Soc. Japan **32**, 1023 (1959)
901. Narita, K.: In: Protein sequence determination (ed. S. B. Needleman), p. 25. Berlin-Heidelberg-New York: Springer 1970

902. NARITA, K., LEE, C. Y.: Biochem. Biophys. Res. Commun. **41**, 339 (1970)
903. NAUGHTON, M. A., SANGER, F., HARTLEY, B. S., SHAW, D. C.: Biochem. J. **77**, 149 (1960)
904. NEDKOF, P., GENOV, N.: Biochim. Biophys. Acta **127**, 541 (1966)
905. NEEDLE, D. J., POLLITT, R. J.: Biochem. J. **97**, 607 (1965)
906. NEEDLEMAN, S. B.: Atlas of protein sequences and structure. Natl. Biomed. Res. Fdn. **1968**, 86
907. NEEDLEMAN, S. B., MARGOLIASH, E.: J. Biol. Chem. **241**, 853 (1966)
908. NEFKENS, G. H. L.: Nature **193**, 974 (1962)
909. NEFKENS, G. H. L., TESSER, G. I.: J. Am. Chem. Soc. **83**, 1263 (1961)
910. NEFKENS, G. H. L., TESSER, G. I., NIVARD, R. J. F.: Rec. trav. chim. **81**, 683 (1962)
911. NEUBERGER, A.: Advances in Protein Chem. **4**, 339 (1948)
911a. NEUBERGER, A., GOTTSCHALK, A., MARSHALL, A. D.: In: Glycoproteins (ed. A. GOTTSCHALK), p. 273. Amsterdam, London, New York: Elsevier Publ. Co. 1966
912. NEUBERGER, A., SANGER, F.: Biochem. J. **36**, 662 (1942)
913. NEUMAN, H., GOLDBERGER, R. F., SELA, M.: J. Biol. Chem. **239**, 1536 (1964)
914. NEUMAN, H., SMITH, R. A.: Arch. Biochem. Biophys. **122**, 354 (1967)
915. NEUMAN, H., STEINBERG, J. Z., BROWN, J. R., GOLDBERGER, R. F., SELA, M.: Europ. J. Biochem. **3**, 171 (1967)
916. NEUMAN, H., STEINBERG, J. Z., KATCHALSKI, E.: J. Am. Chem. Soc. **87**, 3841 (1965)
917. NEUMAN, N. P., LAMPEN, J. O.: Biochemistry (U.S.A.) **6**, 468 (1968)
918. NEURATH, H.: In: Methods in enzymology, vol. II (eds. S. P. COLOWICK, N. O. KAPLAN), p. 77. New York: Academic Press 1955
919. NEURATH, H.: In: The enzymes, 2nd ed., vol. IV (eds. P. D. BOYER, H. LARDY, K. MYRBACK), p. 11. New York: Academic Press 1960
920. NEWMAN, M. S.: J. Am. Chem. Soc. **72**, 4783 (1950)
921. NIALL, H. D.: Personal communication
921a. NIALL, H., EDMAN, P.: J. Gen. Physiol. **45** (Suppl.), 185 (1962)
922. NICOLAIDES, E. D., DE WALD, H. A.: J. Org. Chem. **26**, 2872 (1961)
923. NIEDERWIESER, A.: In: Methods in enzymology, vol. 25 (eds. C. H. W. HIRS, S. N. TIMASHEFF), p. 60. New York: Academic Press 1972
924. NIU, C. I., FRAENKEL-CONRAT, H.: J. Am. Chem. Soc. **77**, 5882 (1955)
925. NOCKOLDS, C. E., KRETSINGER, R. H., COFFEE, C. J., BRADSHAW, R. A.: Proc. Natl. Acad. Sci. (US) **69**, 581 (1972)
926. NOLAN, C., MARGOLIASH, E.: J. Biol. Chem. **241**, 1049 (1966)
927. NOMOTO, M., NARAHASHI. Y., MURAKAMI, M.: J. Biochem. (Tokyo) **48**, 593, 906 (1960)
928. NORTHROP, J. H.: J. Gen. Physiol. **9**, 1217 (1927)
929. NYGAARD, A. P.: Acta Chem. Scand. **10**, 397 (1956)
930. O'DONNELL, I. J., THOMPSON, E. O. P.: Australian J. Biol. Sci. **17**, 271 (1964)
931. OELOFSEN, W., LI, C. H.: J. Am. Chem. Soc. **88**, 4254 (1966)
932. OELSHLEGEL, F. J., Jr., SCHROEDER, J. R., STAHMAN, M. A.: Anal. Biochem. **34**, 331 (1970)
933. OFFER, G. W.: Biochim. Biophys. Acta **90**, 193 (1964)
934. OHNO, K.: J. Biochem. (Tokyo) **41**, 345 (1954)
935. OHNO, K., KATO, T., MITSUYASU, N., WAKI, M., MAKISUMI, S., IZUMIYA, N.: Bull. Chem. Soc. Japan **40**, 204 (1967)
936. OHNO, K.: J. Pharm. Soc. Japan **76**, 713 (1956)
937. OKAMOTO, M., KIMOTO, S., OSKIMA, T., KINOMURA, Y., KAWASAKI, K., YAJIMA, H.: Chem. Pharm. Bull. **15**, 1618 (1967)
938. OKAWA, K.: Bull. Chem. Soc. Japan **29**, 486 (1956)
939. OKAWA, K.: Bull. Chem. Soc. Japan **30**, 976 (1957)
940. OKAWA, K.: Bull. Chem. Soc. Japan **30**, 110 (1957)
941. OKUNUKI, K., HAGIHARA, B., MATSUBARA, H., NAKAYAMA, T.: J. Biochem. (Tokyo) **43**, 453 (1956)
942. OKUYAMA, T., SATAKE, K.: J. Biochem. (Tokyo) **47**, 454 (1960)
943. OLSON, O. E., KUBY, S. A.: J. Biol. Chem. **239**, 460 (1964)
944. ORO, J., SKEWES, H. B.: Nature **207**, 1042 (1965)

945. Otsuka, H., Inouye, K., Shinozaki, F., Kanayama, M.: J. Biochem. (Tokyo) **58**, 512 (1965)
946. Overell, B. G., Petrow, J.: J. Chem. Soc. **1955**, 232
947. Pacheco, H.: Bull. soc. chim. biol. **33**, 1915 (1951)
948. Pacsu, E., Wilson, E. J.: J. Org. Chem. **7**, 117 (1942)
949. Palmer, G., Massey, V.: Biochim. Biophys. Acta **58**, 349 (1962)
950. Papastamatis, S. C., Williams, J. F.: Nature **167**, 724 (1951)
951. Parker, A. J., Kharasch, N.: J. Am. Chem. Soc. **82**, 3071 (1960)
952. Partridge, S. M., Davis, H. F.: Nature **165**, 62 (1950)
953. Pataki, G.: Dünnschicht-Chromatographie in der Aminosäure und Peptid-Chemie. Berlin: Walther De Gruyter 1966
954. Pataki, G.: Techniques of thin layer chromatography in amino acid and peptide chemistry. Ann Arbor (Mich.): Ann Arbor-Humphry Science Publ. 1969
954a. Pataki, G.: Chimia (Switz.) **18**, 24 (1964)
955. Patchornik, A., Lawson, W. B., Gross, E., Wittkop, B.: J. Am. Chem. Soc. **82**, 5923 (1960)
956. Patchornik, A., Lawson, W. B., Witkop, B.: J. Am. Chem. Soc. **80**, 4747 (1958)
957. Patchornik, A., Wilchek, M., Sarid, S.: J. Am. Chem. Soc. **86**, 1457 (1964)
958. Patton, A. R., Foreman, E. M.: Science **109**, 339 (1943)
960. Paul, R.: J. Org. Chem. **28**, 236 (1963)
961. Paul, R., Anderson, G. W.: J. Am. Chem. Soc. **82**, 4596 (1960)
962. Paul, R., Anderson, G. W.: J. Org. Chem. **27**, 2094 (1962)
963. Paul, R., Anderson, G. W., Callahan, F. M.: J. Org. Chem. **26**, 3347 (1961)
964. Pauly, H.: Z. physiol. Chem., Hoppe-Seyler's **42**, 517 (1904)
965. Pauly, H.: Z. physiol. Chem., Hoppe-Seyler's **94**, 288 (1915)
966. Pechere, J. F., Dixon, G. H., Maybury, R. H., Neurath, H.: J. Biol. Chem. **233**, 1364 (1958)
967. Penhoet, E., Kochman, M., Valentine, R., Rutter, W. J.: Biochemistry (U.S.A.) **6**, 2940 (1967)
968. Perham, R., Appella, E., Potter, M.: Science **154**, 391 (1966)
969. Perham, R., Jones, G. M. T.: Europ. J. Biochem. **2**, 84 (1967)
970. Perrin, C. H.: Anal. Chem. **25**, 968 (1953)
971. Peterson, E. A., Sober, H. A.: Anal. Chem. **31**, 857 (1959)
972. Peterson, E. A., Sober, H. A.: J. Am. Chem. Soc. **78**, 751 (1956)
973. Peterson, J. D., Nehrlich, S., Oyer, P. E., Steiner, D. F.: J. Biol. Chem. **247**, 4866 (1972)
974. Pfleiderer, G.: In: Methods in enzymology, vol. 19 (eds. G. E. Perlmann, L. Lorand), p. 514. New York: Academic Press 1970
975. Phillips, D. M. F.: Biochem. J. **87**, 258 (1963)
976. Phillips, D. M. F.: Biochem. J. **86**, 397 (1963)
977. Phillips, D. M. F., Stephen, J. M. L.: Nature **162**, 152 (1948)
978. Piez, K. A., Morris, L.: Anal. Biochem. **1**, 187 (1960)
979. Pisano, J. J., Bronzert, T. J.: Federation Proc. **28**, 661 (1969)
980. Pisano, J. J., Bronzert, T. J.: J. Biol. Chem. **224**, 5597 (1969)
981. Pisano, J. J., Bronzert, T. J.:, Brewer, H. B., Jr.: Anal. Biochem. **45**, 43 (1972)
982. Pisano, J. J.: In: Theory and applications of gas chromatography in industry and medicine (eds. H. S. Kroman, S. P. Bender), p. 147. New York: Grune 1968
983. Pisano, J. J., Zimmerman, C., Appella, E., Morris, S.: Federation Proc. **33**, 1502 (1974)
984. Pisano, J. J.: In: Methods in enzymology, vol. 25, part B (eds. C. H. W. Hirs, S. N. Timasheff), p. 27. New York: Academic Press 1972
985. Pisano, J. J., van den Heuvel, W. J. A., Horning, E. C.: Biochem. Biophys. Res. Commun. **7**, 82 (1962)
986. Plapp, B. V., Raftery, M. A., Cole, R. D.: J. Biol. Chem. **242**, 265 (1967)
987. Pless, J., Boissonnas, R. A.: Helv. Chim. Acta **46**, 1609 (1963)
988. Popenoe, E. A., Lawler, H. C., du Vigneaud, V.: J. Am. Chem. Soc. **74**, 3713 (1952)
989. Porath, J.: Biochim. Biophys. Acta **39**, 193 (1960)

990. Porter, R. R.: Biochem. J. **46**, 304 (1950)
991. Porter, R. R.: Sanger, F.: Biochem. J. **42**, 287 (1948)
992. Prager, E. M., Arnheim, N., Mross, G. A., Wilson, A. C.: J. Biol. Chem. **247**, 2905 (1972)
993. Preston, J., Weinstein, B.: Experientia **23**, 265 (1967)
994. Previero, A., Coletti-Previero, M. A.: Bull. soc. chim. biol. **49**, 1059 (1967)
995. Previero, A., Coletti, M. A., Cavadore, J. C.: Biochim. Biophys. Acta **147**, 453 (1967)
996. Previero, A., Scoffone, E., Pajetta, P., Benassi, C. A.: Gazz. chim. ital. **93**, 841 (1963)
997. Previero, A., Coletti-Previero, M. A., Galzigna, L.: Biochem. Biophys. Res. Commun. **16**, 195 (1964)
998. Previero, A., Pechere, J. F.: Biochem. Biophys. Res. Commun. **40**, 549 (1970)
999. Previero, A., Derancourt, J., Coletti-Previero, M.-A., Laursen, R. A.: Febs Letters **33**, 135 (1973)
1000. Previero, A., Prota, G., Coletti-Previero, M.-A.: Biochem. Biophys. Acta **285**, 267 (1972)
1001. Price, C. A., Campbell, C. W.: Biochem. J. **65**, 512 (1957)
1002. Prochazka, Z.: Chem. listy **47**, 1643 (1953)
1003. Putnam, F. W.: In: The proteins, vol. 2, part 1 (ed. H. Neurath), p. 893. New York: Academic Press 1953
1004. Raftery, M. A., Cole, R. D.: J. Biol. Chem. **241**, 3457 (1966)
1005. Raftery, M. A., Cole, R. D.: Biochem. Biophys. Res. Commun. **10**, 467 (1963)
1006. Rajagopalan, T. G., Moore, S., Stein, W. H.: J. Biol. Chem. **241**, 4940 (1966)
1007. Rajagopalan, T. G., Stein, W. H., Moore, S.: J. Biol. Chem. **241**, 4295 (1966)
1008. Rajappa, S., Akerkar, A. S.: Chem. Commun. **22**, 826 (1966)
1009. Ramachandran, J.: Nature **206**, 927 (1965)
1010. Ramachandran, J., Witkop, B.: J. Am. Chem. Soc. **81**, 4028 (1959)
1011. Ramachandran, J., Epp, A., McConnell, W. B.: Anal. Chem. **27**, 1734 (1955)
1011a. Ramachandran, J., McConnell, W. B.: J. Am. Chem. Soc. **78**, 1255 (1956)
1012. Ramshaw, J. A. M., Thompson, E. W., Boulter, D.: Biochem. J. **119**, 535 (1970)
1013. Rash, J. J., Gehrke, C. W., Zumwalt, R. W., Kuo, K. C., Kvenvolden, K. A., Stalling, D. L.: J. Chromatog. Sci. **10**, 444 (1972)
1014. Ratney, R. S., Godshalk, M. F., Joice, J., James, K. W.: Anal. Biochem. **19**, 357 (1967)
1015. Ray, W. J., Jr., Koshland, D. E., Jr.: Brookhaven Symposia in Biol. **13**, 135 (1960)
1016. Raymond, S., Weintraub, L. S.: Science **130**, 711 (1959)
1017. Rechnitz, G. A., Hseu, J. M.: Anal. Chem. **40**, 1054 (1968)
1018. Redfield, R. R., Anfinsen, C. B.: J. Biol. Chem. **221**, 385 (1956)
1019. Rees, M. W.: Biochem. J. **40**, 632 (1946)
1019a. Reith, W. S., Waldron, N. M.: Biochem. J. **56**, 116 (1954)
1020. Reindel, F., Hoppe, W.: Chem. Ber. **87**, 1103 (1954)
1021. Ressler, C., du Vigneaud, V.: J. Am. Chem. Soc. **76**, 3107 (1954)
1022. Reynolds, J. A., Tanford, C.: J. Biol. Chem. **245**, 5161 (1970)
1023. Richards, F. F., Lovins, R. E.: In: Methods in enzymology, vol. 25, part B (eds. C. H. W. Hirs, S. N. Timasheff), p. 314. New York: Academic Press 1972
1024. Richards, F. F., Barnes, W. T., Lovins, R. E., Salomone, R., Waterfield, M. D.: Nature **221**, 1241 (1969)
1025. Ried, W., Franz, G.: Ann. Chem., Justus Liebigs **644**, 141 (1961)
1026. Riordan, J. F., Sokolovsky, M., Vallee, B. L.: J. Am. Chem. Soc. **88**, 4104 (1966)
1027. Riordan, J. F., Sokolovsky, M., Vallee, B. L.: Biochemistry (U.S.A.) **6**, 358 (1967)
1028. Riordan, J. F., Wacker, W. E. C., Vallee, B. L.: Biochemistry (U.S.A.) **4**, 1758 (1965)
1029. Rittell, W.: Helv. Chim. Acta **45**, 2465 (1962)
1030. Roach, D., Gehrke, C. W.: J. Chromatog. **52**, 393 (1970)
1031. Robel, E. J.: Anal. Biochem. **18**, 406 (1967)
1032. Roberts, E., Rouser, G.: Anal. Chem. **30**, 1291 (1958)
1033. Rocchi, R., Marchiori, F., Scoffone, E.: Gazz. chim. ital. **93**, 823 (1963)

1034. ROCHAT, C., ROCHAT, H., EDMAN, P.: Anal. Biochem. **37**, 259 (1970)
1035. ROCHE, J. N., THOAI, V., HATT, J. L.: Biochim. Biophys. Acta **14**, 71 (1954)
1036. RODA, G., ZAMORANI, A.: J. Chromatog. **46**, 315 (1970)
1037. ROESKE, R. W.: Chem. & Ind. **1959**, 1121
1038. ROSEAU, G., PANTEL, P.: J. Chromatog. **44**, 392 (1969)
1039. ROSMUS, J., DEYL, Z.: Chromatog. Rev. **13**, 163 (1971)
1040. ROSMUS, J., DEYL, Z.: J. Chromatog. **70**, 221 (1972)
1041. ROSS, D. L., SKINNER, C. G., SHIVE, W.: J. Med. Pharm. Chem. **3**, 519 (1961)
1042. ROSS, A. F., STANLEY, W. M.: J. Gen. Physiol. **22**, 165 (1938)
1043. ROTH, M.: Anal. Chem. **43**, 880 (1971)
1044. ROTH, M., HAMPAII, A.: J. Chromatog. **83**, 353 (1973)
1045. ROTHE, M., KUNITZ, F.-W.: Ann. Chem., Justus Liebigs **609**, 88 (1957)
1046. ROVERY, M., FABRE, C., DESNUELLE, P.: Biochim. Biophys. Acta **12**, 547 (1953)
1047. ROWLANDS, D. A., YOUNG, G. T.: Biochem. J. **65**, 516 (1957)
1048. RUDINGER, J., HONZL, M., ZAORAL: Collection Czechoslov. Chem. Commun. **21**, 202 (1956)
1049. RUSSELL, D. W.: Biochem. J. **87**, 1 (1963)
1050. RUTTENBERG, M. A., KING, T. P., CRAIG, L. C.: Biochemistry (U.S.A.) **3**, 758 (1964)
1051. RYDON, H. N., SMITH, P. W. G.: Nature **169**, 922 (1952)
1052. RYLE, A. P., SANGER, F.: Biochem. J. **60**, 535 (1955)
1053. RYLE, A. P., SANGER, F., SMITH, L. F., KITAI, R.: Biochem. J. **60**, 541 (1955)
1054. SACHS, H., BRAND, E.: J. Am. Chem. Soc. **75**, 4610 (1953)
1054a. SACKIN, M. J.: Biochem. Genetics **5**, 287 (1971)
1055. SAKAGUCHI, S.: J. Biochem. (Tokyo) **5**, 133 (1925)
1056. SAKAGUCHI, S.: J. Biochem. (Tokyo) **5**, 25 (1925)
1057. SAKAGUCHI, S.: J. Biochem. (Tokyo) **37**, 231 (1950)
1058. SAKAKIBARA, S., INUKAI, N.: Bull. Chem. Soc. Japan **38**, 1979 (1965)
1059. SAKAKIBARA, S., ITOH, M.: Bull. Chem. Soc. Japan **40**, 656 (1967)
1060. SAKAKIBARA, S., NOBUHARA, Y.: Japanese Patent 2710 (1967); cf. Chem. Abs. **68**, 13402 (1968)
1061. SAKAKIBARA, S., SHIMONISHI, Y.: Bull. Chem. Soc. Japan **38**, 1412 (1965)
1062. SAKAMI, W., TOENNIS, G.: J. Biol. Chem. **144**, 203 (1942)
1063. SAKIYAMA, F., OKAWA, K., YAMAKAWA, T., AKABORI, S.: Bull. Chem. Soc. Japan **31**, 926 (1958)
1064. SANBORN, B. M., BRUMMEL, M. C., STEGINK, L. D., VESTLING, C. S.: Biochim. Biophys. Acta **221**, 125 (1970)
1065. SANGER, F.: Biochem. J. **39**, 507 (1945)
1066. SANGER, F.: Advances in Protein Chem. **7**, 1 (1952)
1067. SANGER, F.: Biochem. J. **44**, 126 (1949)
1068. SANGER, F., THOMPSON, E. O. P.: Biochim. Biophys. Acta **9**, 225 (1952)
1069. SANGER, F., THOMPSON, E. O. P.: Biochem. J. **53**, 353 (1953)
1070. SANGER, F., THOMPSON, E. O. P.: Biochem. J. **53**, 366 (1955)
1071. SANGER, F., TUPPY, H.: Biochem. J. **49**, 463 (1951)
1072. SANTARIUS, K., BELITZ, H. D.: Z. Lebensm.-Untersuch. u. Forsch. **150**, 180 (1970)
1073. SARGES, R., WITKOP, B.: J. Am. Chem. Soc. **86**, 1862 (1963)
1074. SATAKE, K., OKUYAMA, T., OHASHI, M., SHINODA, T.: J. Biochem. (Tokyo) **47**, 654 (1960)
1075. SATAKE, K., SASAKAWA, S., MARUYAMA, T.: J. Biochem. (Tokyo) **53**, 516 (1963)
1076. SATAKE, K., TANAKA, M., SHINO, H.: J. Biochem. (Tokyo) **50**, 6 (1961)
1077. SAVIGE, W. E.: Australian J. Chem. **14**, 664 (1961)
1078. SCANES, F. S., TOZER, B. T.: Biochem. J. **63**, 282, 565 (1956)
1079. SCHACHMAN, H. K.: In: Ultracentrifugation in biochemistry. New York: Academic Press 1959
1080. SCHACHMAN, H. K., EDELSTEIN, S. J.: Biochemistry (U.S.A.) **5**, 2681 (1966)
1081. SCHAFFER, N. K., LANG, R. P., SIMET, L., DRISKO, R. W.: J. Biol. Chem. **230**, 185 (1958)
1082. SCHAFFER, N. K., SIMET, L., HARSHMAN, S., ENGLE, R. R., DRISKO, R. W.: J. Biol. Chem. **225**, 197 (1957)

1083. Schellenberger, A., Jeschkeit, H., Graubaum, H., Mech, C., Sternkopf, G.: Z. Chem. **12**, 62 (1972)
1083a. Schellenberger, A., Jeschkeit, H., Henkel, R., Lehmann, H.: Z. Chem. **7**, 191 (1967)
1084. Schelling, V.: J. Biol. Chem. **96**, 17 (1932)
1085. Schlack, P., Kumpf, W.: Z. physiol. Chem., Hoppe Seyler's **151**, 125 (1926)
1086. Schlögl, K., Wessely, F., Woidich, H.: Monatsh. Chem. **87**, 425 (1956)
1087. Schlack, P., Kumpf, W.: Chem. Abst. **20**, 3298 (1926)
1088. Schlack, P., Schmer, G., Kreil, G.: Anal. Biochem. **29**, 186 (1969)
1089. Schlüssler, H., Zahn, H.: Chem. Ber. **95**, 1076 (1962)
1090. Schnabel, E.: Ann. Chem., Justus Liebigs **659**, 168 (1962)
1091. Schnabel, E., Zahn, H.: Ann. Chem., Justus Liebigs **614**, 141 (1958)
1092. Schneider, F.: Z. physiol. Chem., Hoppe Seyler's **321**, 38 (1960)
1093. Schöberl, A., Hamm, R.: Chem. Ber. **81**, 210 (1948)
1094. Schöberl, A., Kawohl, M.: Chem. Ber. **90**, 2077 (1957)
1095. Schoellman, N. G., Shaw, E.: Biochemistry (U.S.A.) **2**, 252 (1963)
1096. Schönheimer, R.: Z. physiol. Chem., Hoppe-Seyler's **154**, 203 (1926)
1097. Schramm, M.: Biochemistry (U.S.A.) **3**, 1231 (1964)
1098. Schramm, M., Moore, S., Bigwood, E. J.: Biochem. J. **57**, 33 (1954)
1099. Schröder, E.: Ann. Chem., Justus Liebigs **670**, 127 (1963)
1100. Schröder, E.: Ann. Chem., Justus Liebigs **681**, 241 (1965)
1101. Schröder, E.: Ann. Chem., Justus Liebigs **688**, 250 (1965)
1102. Schröder, E., Klieger, E.: Ann. Chem., Justus Liebigs **673**, 196 (1964)
1103. Schröder, E., Klieger, E.: Ann. Chem., Justus Liebigs **673**, 208 (1964)
1104. Schröder, E., Lübke, K.: In: The peptides, vol. I and II. New York: Academic Press 1965
1105. Schröder, E., Lübke, K.: Ann. Chem., Justus Liebigs **655**, 211 (1962)
1106. Schröder, E.: In: Peptides. Proc. European Peptide Symp., 8th, p. 245. Noordwijk, Neth. 1966
1107. Schroeder, W. A.: J. Am. Chem. Soc. **74**, 5118 (1952)
1108. Schroeder, W. A.: In: Methods in enzymology, vol. XI (eds. S. P. Colowick, N. O. Kaplan), p. 445. New York, London: Academic Press 1967
1109. Schroeder, W. A., Cua, J. T., Matsuda, G., Fenninger, W. D.: Biochim. Biophys. Acta **63**, 532 (1963)
1110. Schroeder, W. A., Kay, L. M., le Gette, J., Honnen, L., Green, F. C.: J. Am. Chem. Soc. **76**, 3556 (1954)
1111. Schroeder, W. A., Jones, R. T., Cormick, J., McCalla, K.: Anal. Chem. **34**, 1570 (1962)
1112. Schroeder, W. A., Robberson, B.: Anal. Chem. **37**, 1583 (1965)
1113. Schroeder, W. A., Shelton, J. R., Robberson, B.: Biochim. Biophys. Acta **147**, 590 (1967)
1114. Schroeder, W. A., Shelton, J. R., Shelton, J. B., Cormick, J., Jones, R. T.: Biochemistry (U.S.A.) **2**, 992 (1963)
1115. Schultz, J., Allison, H., Grice, M.: Biochemistry (U.S.A.) **1**, 694 (1962)
1116. Schwarz, H., Arakawa, K.: J. Am. Chem. Soc. **81**, 5691 (1959)
1117. Schwyzer, R.: In: Peptides. Proc. European Peptide Symp., 8th, p. 1. Noordwijk, Neth. 1966
1118. Schwyzer, R.: Helv. Chim. Acta **36**, 414 (1953)
1119. Schwyzer, R.: Helv. Chim. Acta **37**, 647 (1954)
1120. Schwyzer, R.: Helv. Chim. Acta **37**, 674 (1954)
1121. Schwyzer, R., Aung, Tun-Kyi: Helv. Chim. Acta **45**, 859 (1962)
1122. Schwyzer, R., Costopanagiotes, A., Sieber, P.: Chimia (Switz.) **16**, 295 (1962)
1123. Schwyzer, R., Dietrich, H.: Helv. Chim. Acta **44**, 2003 (1961)
1124. Schwyzer, R., Feurer, M., Iselin, B., Kagi, H.: Helv. Chim. Acta **38**, 80 (1955)
1125. Schwyzer, R., Iselin, B., Feurer, M.: Helv. Chim. Acta **38**, 69 (1955)
1126. Schwyzer, R., Iselin, B., Kappeler, H., Riniker, B., Rittel, W., Zuber, H.: Helv. Chim. Acta **41**, 1273 (1958)

1127. Schwyzer, R., Iselin, B., Kappeler, H., Riniker, B., Rittel, W., Zuber, H.: Helv. Chim. Acta **46**, 1975 (1963)
1128. Schwyzer, R., Iselin, B., Rittel, W., Sieber, P.: Helv. Chim. Acta **39**, 872 (1956)
1129. Schwyzer, R., Kappeler, H.: Helv. Chim. Acta **44**, 1991 (1961)
1130. Schwyzer, R., Rittel, W.: Helv. Chim. Acta **44**, 159 (1961)
1131. Schwyzer, R., Rittel, W., Costopanagiotis, A.: Helv. Chim. Acta **45**, 2473 (1962)
1132. Schwyzer, R., Rittel, W., Kappeler, H., Iselin, B.: Angew. Chem. **72**, 915 (1960)
1133. Schwyzer, R., Sieber, P.: Helv. Chim. Acta **40**, 624 (1957)
1134. Schwyzer, R., Sieber, P.: Helv. Chim. Acta **42**, 972 (1959)
1135. Schwyzer, R., Sieber, P.: Helv. Chim. Acta **43**, 1910 (1960)
1136. Schwyzer, R., Sieber, P.: Helv. Chim. Acta **49**, 134 (1966)
1137. Schwyzer, R., Sieber, P., Zatsko, K.: Helv. Chim. Acta **41**, 491 (1958)
1138. Schwyzer, R., Surbeck-Wegmann, E., Dietrich, H.: Chimia (Switz.) **14**, 366 (1960)
1139. Scoffone, E., Fontana, A., Rocchi, R.: Biochem. Biophys. Res. Commun. **25**, 170 (1966)
1140. Scoffone, E., Fontana, A., Rocchi, R.: Biochemistry (U.S.A.) **7**, 971 (1968)
1141. Scoffone, E., Vianello, E., Lorenzini, A.: Gazz. chim. ital. **87**, 354 (1957)
1142. Scoffone, E., DeLa Llosa, P., Jutisz, M.: Bull. soc. chim. France **1959**, 1553
1143. Seibles, Th. S., Weil, L.: In: Methods in enzymology, vol. 11 (ed. C. H. W. Hirs), p. 204. New York: Academic Press 1967
1144. Seiler, N.: In: Methods of biochemical analysis, vol. 18 (ed. D. Glick), p. 259. New York: Interscience Publishers 1970
1145. Seiler, N., Wiechman, J.: Experientia **20**, 559 (1964)
1146. Seiler, N., Wiechman, J.: Experientia **21**, 203 (1965)
1147. Seiler, N., Wiechman, J.: J. Chromatog. **28**, 351 (1967)
1148. Seki, T.: J. Biochem. (Tokyo) **47**, 253 (1960)
1149. Shaw, J. L. D., McFarlane, W. D.: Can. J. Research **16B**, 361 (1938)
1150. Shaw, R.: Biochim. Biophys. Acta **92**, 558 (1964)
1151. Sheehan, J. C., Chapman, D. W., Roth, R. W.: J. Am. Chem. Soc. **74**, 3822 (1952)
1151a. Sheehan, J. C., Cruickshank, P. A., Boshart, G. L.: J. Org. Chem. **26**, 2525 (1961)
1152. Sheehan, J. C., Frank, V. S.: J. Am. Chem. Soc. **71**, 1856 (1949)
1153. Sheehan, J. C., Goodman, M., Hess, G. P.: J. Am. Chem. Soc. **78**, 1367 (1956)
1154. Sheehan, J. C., Hess, G. P.: J. Am. Chem. Soc. **77**, 1067 (1955)
1155. Sheehan, J. C., Hlavka, J. J.: J. Org. Chem. **21**, 439 (1956)
1156. Sheehan, J. C., Preston, J., Cruickshank, P. A.: J. Am. Chem. Soc. **87**, 2492 (1965)
1157. Sheehan, J. C., Yang, D. D. H.: J. Am. Chem. Soc. **80**, 1154 (1958)
1157a. Shields, J. E., McGregor, W. H., Carpenter, F. H.: J. Org. Chem. **26**, 1491 (1961)
1158. Sifferd, R. H., du Vigneaud, V.: J. Biol. Chem. **108**, 753 (1953)
1159. Signor, A., Biondi, L., Tamburro, A. M., Bordignon, E.: Europ. J. Biochem. **7**, 328 (1969)
1160. Signor, A., Bordignon, E.: Tetrahedron **24**, 6995 (1968)
1161. Signor, A., Bordignon, Vidali, G.: J. Org. Chem. **32**, 1135 (1967)
1162. Signor, A., Previero, A., Terbojevich, M.: Nature **205**, 596 (1965)
1163. Skrabka-Blotnicka, T.: Chem. Anal. (Warsaw) **10**, 199 (1965)
1164. Slobin, L. I., Singer, S. J.: J. Biol. Chem. **243**, 1777 (1968)
1165. Sjöquist, J.: Acta Chem. Scand. **7**, 447 (1953)
1166. Sjöquist, J.: Arkiv Kemi **11**, 151 (1957)
1167. Sjöquist, J.: Biochim. Biophys. Acta **41**, 20 (1960)
1168. Sjöquist, J.: Arkiv Kemi **11**, 129 (1957)
1169. Slobin, L. I., Carpenter, F. H.: Biochemistry (U.S.A.) **5**, 499 (1966)
1169a. Slobin, L. I., Singer, S. J.: J. Biol. Chem. **243**, 1777 (1968)
1170. Slump, P., Schreuder, K. A. W.: Anal. Biochem. **27**, 182 (1969)
1171. Sluyterman, L. A.: Biochim. Biophys. Acta **25**, 402 (1957)
1172. Smillie, L. B., Hartley, B. S.: Biochem. J. **101**, 232 (1966)
1173. Smillie, L. B., Hartley, B. S.: Biochem. J. **105**, 1125 (1967)
1174. Smith, E. L.: In: The enzymes, vol. I, part 2 (eds. J. B. Summer, K. Myrbäck), p. 793. New York: Academic Press 1951

1175. SMITH, E. L., HILL, R. L.: In: The enzymes, vol. IV (eds. P. D. BOYER, H. LARDY, K. MYRBÄCK), p. 37. New York: Academic Press 1960

1176. SMITH, E. L., POLGLASE, W. J.: J. Biol. Chem. **180**, 1209 (1949)

1176a. SMITH, E. L., SPACHMAN, D. H.: J. Biol. Chem. **212**, 271 (1955)

1177. SMITH, I.: Nature **169**, 60 (1952)

1178. SMITH, I.: Nature **171**, 43 (1953)

1179. SMITH, M. M.: J. Theor. Biol. **13**, 261 (1966)

1180. SMITHIES, O.: Advances in Protein Chem. **14**, 65 (1959)

1181. SMITHIES, O., GIBSON, D. M., FANNING, E. M., GOODFLIESH, R. M., GILLMAN, J. G., BALLANTYNE, D. L.: Biochemistry (U.S.A.) **10**, 4912 (1971)

1182. SMYTH, D. G., BATTAGLIA, F. C., MESCHIA, G.: J. Gen. Physiol. **44**, 889 (1961)

1183. SMYTH, D. G., NAGAMATSU, A., FRUTON, J. S.: J. Am. Chem. Soc. **82**, 4600 (1960)

1184. SMYTH, D. G., STEIN, W. H., MOORE, S.: J. Biol. Chem. **237**, 1845 (1962)

1185. SMYTH, D. G., STEIN, W. H., MOORE, S.: J. Biol. Chem. **238**, 227 (1963)

1186. SMYTH, D. S., UTSUMI, S.: Nature **216**, 332 (1967)

1187. SNELLMAN, K. O.: Netherlands Patent **6**, 508, 512 (1965)

1188. SODEK, J., HOFFMANN, T.: In: Methods in enzymology, vol. 19 (eds. G. E. PERLMAN, L. LORAND), p. 460. Academic Press 1970

1189. SOGAMI, M., FOSTER, J. F.: J. Biol. Chem. **237**, 2514 (1962)

1190. SOKOLOVSKY, M., RIORDAN, J. F., VALLEE, B. L.: Biochemistry (U.S.A.) **5**, 3582 (1966)

1191. SOKOLOVSKY, M., SADEH, T., PATCHORNIK, A.: J. Am. Chem. Soc. **86**, 1212 (1964)

1192. SOKOLOVSKY, M., VALLEE, B. L.: Biochemistry (U.S.A.) **5**, 3574 (1966)

1193. SONDHEIMER, E., HOLLEY, R. W.: J. Am. Chem. Soc. **76**, 2476 (1954)

1194. SORU, E., ZAHARIA, O.: Arch. Biochem. Biophys. **119**, 498 (1967)

1195. SPACKMAN, D. H.: Federation Proc. **22**, 244 (1963)

1196. SPACKMAN, D. H., SMITH, E. L., BROWN, D. M.: J. Biol. Chem. **212**, 255 (1955)

1197. SPACKMAN, D. H.: In: Methods in enzymology, vol. 11 (ed. C. H. W. HIRS), p. 3. New York: Academic Press 1967

1198. SPACKMAN, D. H., STEIN, W. H., MOORE, S.: Anal. Chem. **30**, 1190 (1958)

1199. SPANDE, T. F., WILCHEK, M., WITKOP, B.: J. Am. Chem. Soc. **90**, 3256 (1968)

1200. SPANDE, T. F., WITKOP, B.: In: Methods in enzymology, vol. 11 (ed. C. H. W. HIRS), p. 498. New York: Academic Press 1967

1201. SPECK, J. C., RIEHM, J. P., EVARD, R.: Federation Proc. **20**, 220 (1961)

1202. SPIES, J. R., CHAMBERS, D. C.: Anal. Chem. **21**, 1249 (1949)

1203. SPIES, J. R.: Anal. Biochem. **39**, 1412 (1967)

1204. SPIES, J. R.: J. Agr. Food Chem. **16**, 514 (1968)

1205. SPRÖSSLER, B., HELLMANN, H.-D., GRAMPP, E., UHLIG, H.: Z. physiol. Chem., Hoppe-Seyler's **352**, 1524 (1971)

1206. STAAB, H. A.: J. Am. Chem. Soc. **609**, 75 (1957)

1207. STACEY, R. W., WAXMAN, B. D.: In: Computers in biochemical research. New York: Academic Press 1965

1208. STAHL, E.: Thin layer chromatography. Berlin-Heidelberg-New York: Springer 1970

1209. STAHL, E., KALDEWEY, H.: Z. physiol. Chem., Hoppe-Seyler's **323**, 182 (1961)

1210. STALLING, D. L., GEHRKE, C. W., ZUMWALT, R. W.: Biochem. Biophys. Res. Commun. **31**, 616 (1968)

1211. STARK, G. R.: Biochemistry (U.S.A.) **7**, 1796 (1968)

1212. STARK, G. R.: Federation Proc. **24**, 225 (1965)

1213. STARK, G. R.: In: Methods in enzymology, vol. 25 (eds. C. H. W. HIRS, S. N. TIMASHEFF), p. 103, 369. New York: Academic Press 1972

1214. STARK, G. R., STEIN, W. H., MOORE, S.: J. Biol. Chem. **235**, 3177 (1960)

1215. STARK, G. R., SMYTH, D. G.: J. Biol. Chem. **238**, 214 (1963)

1216. STEERS, E., Jr., CRAVEN, G. R., ANFINSEN, C. B., BETHUNE, J. L.: J. Biol. Chem. **240**, 2478 (1965)

1217. STEGEMANN, H.: Z. physiol. Chem., Hoppe-Seyler's **213**, 255 (1968)

1218. STEGNIK, L. D., VESTLING, C. S.: J. Biol. Chem. **241**, 4923 (1966)

1219. STEGNIK, L. D., MYER, P. D., BRUMMEL, M. C.: J. Biol. Chem. **246**, 3001 (1971)

1220. STEIN, S., BOHLEN, P., STONE, J., DAIRMAN, W., UDENFRIEND, S.: Arch. Biochem. Biophys. **155**, 203 (1973)
1221. STEIN, W. H., MOORE, S., BERGMANN, M.: J. Biol. Chem. **154**, 191 (1944)
1222. STELAKATOS, G. C., THEODOROPOULOS, D. M., ZERVAS, L.: J. Am. Chem. Soc. **81**, 2884 (1959)
1223. STEPHANOV, V. M., KRIVTSOV, V. F.: J. Gen. Chem. (U.S.S.R.) **35**, 49 (1965)
1224. STEPHANOV, V. M., KRIVTSOV, V. F.: J. Gen. Chem. (U.S.S.R.) **35**, 988 (1965)
1225. STEPHANOV, V. M., VULLFSON, N. S., PUCHOW, V. A., ZYAKUN, A. M.: J. Gen. Chem. (U.S.S.R.) **34**, 3771 (1964)
1226. STEPHANOV, V. M., VULLFSON, N. S., PUCHOW, V. A., ZYAKUN, A. M.: J. Gen. Chem. (U.S.S.R.) **34**, 3822 (1964)
1227. STERN, R., DE LUCA, M., MEHLER, A. H., MCELROY, W. D.: Biochemistry (U.S.A.) **5**, 126 (1966)
1228. STEVENS, F., GLAZER, A. N., SMITH, E. L.: J. Biol. Chem. **242**, 2764 (1967)
1229. STEWART, F. H. C.: Australian J. Chem. **19**, 1067 (1966)
1230. STEWART, F. H. C.: Australian J. Chem. **19**, 1511 (1966)
1231. STEWART, F. H. C.: Australian J. Chem. **19**, 2373 (1966)
1232. STEWART, F. H. C.: Australian J. Chem. **20**, 365 (1967)
1233. STEWART, F. H. C.: Australian J. Chem. **20**, 2243 (1967)
1234. STEWART, F. H. C.: Australian J. Chem. **21**, 477 (1968)
1235. STEWART, F. H. C.: Australian J. Chem. **20**, 1991 (1967)
1236. STEWART, F. H. C.: Australian J. Chem. **21**, 1101 (1968)
1237. STEWART, J. M., YOUNG, J. D., BENJAMINI, E., SHIMIZU, M., LEUNG, C. Y.: Biochemistry (U.S.A.) **5**, 3396 (1966)
1238. WOOLLEY, D. W., STEWART, J. M.: J. Med. Chem. **6**, 599 (1963)
1239. STRICKS, W., KOLTHOFF, I. M.: J. Am. Chem. Soc. **73**, 4569 (1951)
1240. STRICKS, W., KOLTHOFF, I. M., KAPOOR, R. C.: J. Am. Chem. Soc. **77**, 2057 (1955)
1241. STRICKS, W., KOLTHOFF, I. M., TANAKA, N.: Anal. Chem. **26**, 299 (1954)
1242. STRICKS, W., KOLTHOFF, I. M., HEYNDRICKX, A.: J. Am. Chem. Soc. **76**, 1515 (1954)
1243. STUDER, R. O., LERGIER, W., VOGLER, K.: Helv. Chim. Acta **46**, 612 (1963)
1244. SUMMERS, M. R., SMYTHERS, G. W., OROSZLAN, S.: Anal. Biochem. **53**, 624 (1973)
1245. SUN, T., LOVINS, R. E.: Anal. Biochem. **45**, 176 (1972)
1246. SUN, T., LOVINS, R. E.: Org. Mass Spectrom. **6**, 39 (1972)
1247. SURAN, A. A.: Arch. Biochem. Biophys. **113**, 1 (1966)
1248. SVEDBERG, T., PEDERSEN, K. O.: In: The ultracentrifuge. Oxford: University Press 1940
1249. SWALLOW, D. L., ABRAHAM, E. P.: Biochem. J. **70**, 364 (1958)
1250. SWAN, J. M.: Australian J. Sci. Research **A 5**, 711 (1952)
1250a. SWAN, J. M.: Nature **180**, 643 (1957)
1251. SWEETMAN, B. J., MACLAREN, J. A.: Australian J. Chem. **19**, 2347 (1966)
1252. SWENSEN, A. D., BOYER, P. D.: J. Am. Chem. Soc. **79**, 2174 (1957)
1253. SYNGE, R. L. M.: Chem. Revs. **32**, 135 (1943)
1254. SYNGE, R. L. M.: Biochem. J. **39**, 351 (1945)
1255. SZENT-GYÖRGYI, A.: Arch. Biochem. Biophys. **42**, 305 (1953)
1256. SZEWZUK, A., MULCZYK, M.: Europ. J. Biochem. **8**, 63 (1969)
1257. TAKAHASHI, K.: J. Biol. Chem. **240**, PC4117 (1965)
1258. TAMURA, Z., NAKAJIMA, T.: Tanpakushitsu, Kakusan, Koso (Protein, Nucleic acid, Enzyme) **12**, 729 (1967) (in Japanese)
1259. TAMURA, Z., NAKAJIMA, T., NAKAYAMA, T., PISANO, J. J., UDENFRIEND, S.: Anal. Biochem. **52**, 595 (1973)
1260. TANAKA, M., NAKASHIMA, T., BENSON, A., MOWER, H., YASUNOBU, K. T.: Biochemistry (U.S.A.) **5**, 1666 (1966)
1261. TANFORD, C.: In: Physical chemistry of macromolecules. New York: Wiley Publishing Co. 1961
1262. TANFORD, C., KAWAHARA, K., LAPANJE, S.: J. Am. Chem. Soc. **89**, 729 (1967)
1263. TANG, J., HARTLEY, B. S.: Biochem. J. **102**, 593 (1967)
1264. TANIUCHI, H., ANFINSEN, C.: J. Biol. Chem. **241**, 4366 (1966)

1265. Taniuchi, H., Anfinsen, C., Sodja, A.: J. Biol. Chem. **242**, 4736 (1967)
1266. Taniuchi, H., Anfinsen, C., Sodja, A.: J. Biol. Chem. **242**, 4752 (1967)
1267. Tarnowiski, G. S., Barclay, R. K., Mountain, I. M., Nakamura, M., Satterwhite, H. G., Solney, E. M.: Arch. Biochem. Biophys. **110**, 210 (1965)
1268. Taschner, E., Chimiak, A., Bator, B., Sokolowska, T.: Ann. Chem., Justus Liebigs **646**, 134 (1961)
1269. Taschner, E., Liberek, B., Wasielewski, Cz., Biernat, J.: Angew. Chem. **71**, 743 (1959)
1270. Taschner, E., Wasielewski, Cz., Biernat, J. F.: Ann. Chem., Justus Liebigs **646**, 119 (1961)
1271. Taschner, E., Wasielewski, Cz., Sokolowska, T., Biernat, J. F.: Ann. Chem., Justus Liebigs **646**, 127 (1961)
1272. Tesser, G. I., Schwyzer, R.: Helv. Chim. Acta **49**, 1013 (1966)
1273. Theodoropoulos, D., Craig, L. C.: J. Org. Chem. **21**, 1379 (1956)
1274. Thompson, A. R.: Nature **169**, 495 (1952)
1275. Thompson, A. R.: Biochem. J. **61**, 253 (1955)
1276. Thompson, E. O. P.: Biochim. Biophys. Acta **15**, 440 (1954)
1277. Thompson, E. O. P., O'Donnell, I. J.: Biochim. Biophys. Acta **53**, 447 (1961)
1278. Thompson, E. W., Laycock, M. V., Ramshav, J. A. M., Boulter, D.: Biochem. J. **117**, 183 (1970)
1279. Thomsen, J., Kristiansen, K., Brunfeldt, K., Sundby, F.: Febs Letters **21**, 315 (1972)
1280. Tibbs, J.: Nature **168**, 910 (1951)
1281. Timasheff, S. N., Gorbunoff, M. J.: Ann. Rev. Biochem. **36**, 13 (1967)
1282. Tiselius, A.: Biochem. J. **31**, 1464 (1937)
1283. Titani, K., Narita, K., Okunuki, K.: J. Biochem. (Tokyo) **51**, 350 (1962)
1283a. Titani, K., Narita, K., Okunuki, K., Hermodson, M. A., Ericsson, L. H., Walsh, K. A., Neurath, H.: Nature (New Biology) **238**, 35 (1972)
1284. Toennies, G., Homiller, R. P.: J. Am. Chem. Soc. **64**, 3054 (1942)
1285. Toennies, G., Kolb, J. J.: Anal. Chem. **23**, 823 (1951)
1286. Toennies, G., Kolb, J. J.: Nature **177**, 281 (1956)
1287. Toi, K., Bynum, E., Norris, E., Itano, H. A.: J. Biol. Chem. **242**, 1036 (1967)
1288. Tower, D. B., Peters, E. L., Wherrett, J. R.: J. Biol. Chem. **237**, 1861 (1962)
1289. Tower, D. B.: In: Methods in enzymology, vol. 11 (ed. C. H. W. Hirs), p. 76. New York: Academic Press 1967
1290. Trautman, R.: In: Instrumental methods of experimental biology (ed. D. W. Newman), p. 211. New York: Mac Millan Co. 1964
1291. Travis, J., McElroy, W. D.: Biochemistry (U.S.A.) **5**, 2170 (1966)
1292. Tristram, G. R.: Advances in Protein Chem. **5**, 83 (1949)
1293. Tschesche, H., Kupfer, S.: Europ. J. Biochem. **26**, 33 (1972)
1294. Tschesche, H., Schneider, M., Wachter, E.: Febs Letters **23**, 367 (1972)
1295. Tschesche, H., Wachter, E.: Europ. J. Biochem. **16**, 187 (1970)
1296. Tsung, C. M., Fraenkel-Conrat, H.: Biochemistry (U.S.A.) **4**, 793 (1965)
1297. Tsung, C. M., Fraenkel-Conrat, H.: Biochemistry (U.S.A.) **5**, 2061 (1966)
1298. Turner, R. A., Schmerzler, G.: Biochim. Biophys. Acta **13**, 553 (1954)
1299. Udenfriend, S., Velick, S. F.: J. Biol. Chem. **190**, 733 (1951)
1300. Udenfriend, S., Stein, S., Bohlen, P., Dairman, W., Leimgruber, W., Weigele, M.: Science **178**, 871 (1972)
1301. Udenfriend, S., Stein, S., Bohlen, P., Dairman, W.: In: Chemistry and biology of peptides (ed. J. Meienhofer), p. 655. Ann Arbor (Mich.): Ann Arbor Science Publishers 1972
1301a. Vallee, B., Neurath, H.: J. Biol. Chem. **217**, 253 (1955)
1302. Vance, D. E., Feingold, D. E.: Anal. Biochem. **36**, 30 (1970)
1303. van Eys, J., Judd, J., Ford, J., Womak, W. B.: Biochemistry (U.S.A.) **3**, 1755 (1964)
1304. Van Holde, K. E.: Fractions **1** (1967)
1305. Van Holde, K. E., Baldwin, R. L.: J. Phys. Chem. **62**, 734 (1958)
1306. Van Holde, K. E.: In: Physical biochemistry. Englewood Cliffs (N.J.): Prentice-Hall 1971

1307. van Orden, H. O., Smith, E. L.: J. Biol. Chem. **208**, 751 (1954)
1308. van Slyke, D. D.: J. Biol. Chem. **73**, 99 (1927)
1309. van Slyke, D. D.: J. Biol. Chem. **83**, 425 (1929)
1310. Vaughan, J. R.: J. Am. Chem. Soc. **73**, 3547 (1951)
1311. Vaughan, J. R., Jr., Osato, R. L.: J. Am. Chem. Soc. **74**, 676 (1952)
1312. Vaughan, J. R., Jr.: J. Am. Chem. Soc. **74**, 6137 (1952)
1313. Vaughan, J. R., Jr., Eichler, J. A.: J. Am. Chem. Soc. **75**, 5556 (1953)
1314. Veber, D. F., Milkowski, J. D., Denkewalter, R. G., Hirschmann, R.: Tetrahedron Letrs. **26**, 3057 (1968)
1315. Veber, D. F., Pfister, K., Hirschmann, R.: J. Med. Chem. **10**, 968 (1967)
1316. Velick, S. F., Udenfriend, S.: J. Biol. Chem. **191**, 233 (1951)
1317. Veronese, F. M., Boccu, E., Fontana, A.: Ann. Chim. (Rome) **58**, 1309 (1968)
1318. Veronese, F. M., Boccu, E., Fontana, A.: Intern. J. Protein Research **2**, 67 (1970)
1319. Visuri, K., Mikola, J., Enari, T. M.: Europ. J. Biochem. **7**, 193 (1969)
1320. Voisetnet, M. E.: Bull. soc. chim. **33**, 1198 (1905)
1321. Wachsmuth, E. D., Fritze, I., Pfleiderer, G.: Biochemistry (U.S.A.) **5**, 169 (1966)
1322. Wachsmuth, E. D., Fritze, I., Pfleiderer, G.: Biochemistry (U.S.A.) **5**, 175 (1966)
1323. Waldschmidt, E., Kuhn, K.: Chem. Ber. **84**, 381 (1951)
1324. Waley, S. G., Watson, J.: J. Chem. Soc. **1951**, 2394
1325. Waley, S. G., Watson, J.: Biochem. J. **57**, 529 (1954)
1326. Wang, K. T.: J. Chinese Chem. Soc. (Series II) **8**, 241 (1961)
1327. Wang, K. T., Huang, J. M. K., Wang, I. S. Y.: J. Chromatog. **22**, 362 (1966)
1328. Wang, K. T., Wang, I. S. Y.: J. Chromatog. **24**, 460 (1966)
1329. Wang, K. T., Wang, I. S. Y., Lin, A. L., Wang, C. S.: J. Chromatog. **26**, 323 (1967)
1330. Wang, K. T., Wang, I. S. Y.: J. Chromatog. **27**, 318 (1967)
1331. Wang, K. T., Wu, P. H.: J. Chromatog. **37**, 353 (1968)
1332. Wang, S., Carpenter, F. H.: J. Biol. Chem. **240**, 1619 (1965)
1333. Ward, D. N., Coffey, J. A., Ray, D. B., Lamkin, W. M.: Anal. Biochem. **14**, 243 (1966)
1334. Wasielewski, C.: Roczniki Chem. **41**, 1915 (1967); cf. Chem. Abs. **68**, 87571n (1958)
1335. Waterfield, M. D., Haber, E.: Biochemistry (U.S.A.) **9**, 832 (1970)
1336. Weber, C. J.: J. Biol. Chem. **86**, 217 (1930)
1337. Weber, K. L., Konigsberg, W.: J. Biol. Chem. **242**, 3563 (1967)
1338. Weber, K. L., Osborn, M.: J. Biol. Chem. **244**, 4406 (1969)
1339. Weber, R. E., Engelhardt, D. L., Zinder, N. D.: Proc. Natl. Acad. Sci. (US) **55**, 155 (1966)
1340. Wehmiller, J., Hare, P. E.: Science **173**, 907 (1971)
1341. Weigele, M., DeBernardo, S., Leimgruber, W.: Biochem. Biophys. Res. Commun. **50**, 352 (1973)
1342. Weil, L., Seibles, T. S.: Arch. Biochem. Biophys. **95**, 470 (1961)
1343. Weil, L., Telka, M.: Arch. Biochem. Biophys. **71**, 473 (1957)
1344. Weiner, A. M., Platt, T., Weber, K.: J. Biol. Chem. **247**, 3242 (1972)
1345. Weiss, M., Szobolew, N.: Biochem. Z. **58**, 119 (1913—14)
1346. Wells, J. R. E.: Biochem. J. **97**, 228 (1965)
1347. Wells, J. R. E.: Anal. Biochem. **19**, 448 (1967)
1348. Wells, J. R. E.: Biochim. Biophys. Acta **167**, 388 (1967)
1349. Wenck, H., Schwabe, F., Schneider, F., Flohe, L.: Z. anal. Chem. **258**, 267 (1972)
1350. Weygand, F., Csendes, E.: Angew. Chem. **64**, 136 (1952)
1351. Weygand, F.: Bull. soc. chim. biol. **43**, 1269 (1961)
1352. Weygand, F., Reiher, M.: Chem. Ber. **88**, 26 (1955)
1353. Weygand, F., Geiger, R.: Chem. Ber. **89**, 647 (1956)
1354. Weygand, F., Hoffman, D., Wunsch, E.: Z. Naturforsch. **21b**, 426 (1966)
1355. Weygand, F., Hunger, K.: Z. Naturforsch. **13b**, 50 (1958)
1356. Weygand, F., Hunger, K.: Chem. Ber. **95**, 1 (1962)
1357. Weygand, F., Konig, W., Nintz, E., Hoffmann, D., Huber, P., Kahn, N., Prinz, W.: Z. Naturforsch. **21b**, 325 (1966)

1358. WEYGAND, F.: Z. anal. Chem. **243**, 2 (1968)
1359. WEYGAND, F., OBERMEIER, R.: Europ. J. Biochem. **20**, 72 (1971)
1360. WEYGAND, F., RINNO, H.: Chem. Ber. **92**, 517 (1959)
1361. WEYGAND, F., STEGLICH, W.: Z. Naturforsch. **14b**, 472 (1959)
1362. WEYGAND, F., STEGLICH, W.: Chem. Ber. **92**, 313 (1959)
1363. WEYGAND, F., STEGLICH, W., PIETTA, P.: Tetrahedron Letrs. **1966**, 3751
1364. WHITAKER, J. R.: Anal. Chem. **35**, 1950 (1963)
1365. WHITE, F. H., Jr.: J. Biol. Chem. **235**, 383 (1960)
1366. WHITE, J.: J. Biol. Chem. **106**, 141 (1934)
1367. WIELAND, T., BIRR, CH.: Angew. Chem. (Int.) **5**, 310 (1966)
1368. WIELAND, T., HEINKE, B., VOGLER, K., MORIMOTO, H.: Ann. Chem., Justus Liebigs **655**, 189 (1962)
1369. WIELAND, T., KERN, W., SCHRING, R.: Ann. Chem., Justus Liebigs **569**, 117 (1950)
1370. WIELAND, T., SCHRING, R.: Ann. Chem., Justus Liebigs **569**, 122 (1950)
1371. WIELAND, T., BERNHARD, H.: Ann. Chem., Justus Liebigs **572**, 190 (1951)
1392. WIELAND, T., SCHÄFER, W., BOKELMANN, E.: Ann. Chem., Justus Liebigs **573**, 99 (1951)
1373. WIELAND, T., BAUER, L.: Angew. Chem. **63**, 511 (1951)
1374. WIELAND, T., SCHNEIDER, G.: Ann. Chem., Justus Liebigs **580**, 159 (1953)
1375. WIELAND, T., VOGELER, K.: Angew. Chem. **74**, 904 (1962)
1376. WILCOX, H. G., HEIMBERG, M.: Biochem. Biophys. Acta **152**, 424 (1968)
1377. WILCOX, P. E.: In: Methods in enzymology, vol. 11 (ed. C. H. W. HIRS), p. 63. New York: Academic Press 1967
1378. WILLIAMS, K. T., BEVENUE, A.: Science **113**, 582 (1951)
1379. WILLIAMS, J. W., VAN HOLDE, K. E., BALDWIN, R. L., FUJITA, H.: Chem. Revs. **58**, 715 (1958)
1380. WILLIAMS, R. C., Jr., RAJAGOPALAN, T. G.: J. Biol. Chem. **241**, 4951 (1966)
1381. WINEGARD, H. M., TOENNIES, G., BLOCK, R. J.: Science **108**, 506 (1948)
1382. WINKLER, S.: Z. physiol. Chem., Hoppe-Seyler's **228**, 50 (1934)
1383. WINSTEAD, J., WOLD, F.: Biochemistry (U.S.A.) **3**, 791 (1964)
1384. WINTERSBERGER, E., COX, D. J., NEURATH, H.: Biochemistry (U.S.A.) **1**, 1069 (1962)
1385. WINZOR, P. J.: In: Physical principles and techniques of protein chemistry, part A (ed. S. LEACH). New York: Academic Press 1969
1386. WITKOP, B.: Advances in Protein Chem. **16**, 221 (1961)
1387. WOLMAN, Y., KIVITY, S., FRANKEL, M.: Chem. Commun. **1967**, 629
1388. WOLMAN, Y., LADKANY, D., FRANKEL, M.: J. Chem. Soc. (C) **1967**, 689
1389. WOLMAN, Y., MILLER, S. L.: Nature **234**, 548 (1971)
1390. WOOD, J. L., DU VIGNEAUD, V.: J. Biol. Chem. **130**, 109 (1939)
1391. WOODS, K. R., WANG, K. T.: Biochim. Biophys. Acta **133**, 369 (1967)
1392. WOODWARD, G. E., FRY, E. G.: J. Biol. Chem. **97**, 465 (1932)
1393. WOODWARD, R. B., OLOFSON, R. A.: J. Am. Chem. Soc. **83**, 1107 (1961)
1394. WOODWARD, R. B., OLOFSON, R. A., MAYER, H.: J. Am. Chem. Soc. **83**, 1010 (1961)
1395. WOODWARD, R. B., OLOFSON, R. A., MAYER, H.: Tetrahedron Letrs. **8** (Suppl.), 321 (1966)
1396. WOOLLEY, D. W.: J. Biol. Chem. **172**, 71 (1948)
1397. WUENSH, E., DREES, F.: Chem. Ber. **99**, 110 (1966)
1398. WUENSH, E., FONTANA, A.: Chem. Ber. **101**, 323 (1968)
1399. WUENSH, E., FONTANA, A., DREES, F.: Z. Naturforsch. **22b**, 607 (1967)
1400. WUENSH, E., FRIES, G., ZURCK, A.: Chem. Ber. **91**, 542 (1958)
1401. WUENSH, E., ZWICK, A.: Chem. Ber. **99**, 105 (1966)
1402. WUENSH, E., ZWICK, A., WENDLBERGER, G.: Chem. Ber. **100**, 173 (1967)
1403. WUENSH, E., ZWICK, A., FONTANA, A.: Chem. Ber. **101**, 326 (1968)
1404. WUENSH, E., ZWICK, A., JAEGER, E.: Chem. Ber. **101**, 336 (1968)
1405. YAJIMA, H., OKADA, Y., KINOMURA, Y., MINAMI, H.: J. Am. Chem. Soc. **90**, 527 (1968)
1406. YAMASHIRO, D.: Nature **201**, 76 (1964)
1407. YAMASHITA, S.: Biochim. Biophys. Acta **229**, 301 (1971)

1408. YAMASHITA, S., ISHIKAWA, N.: In: Chemistry and biology of peptides: Proc. of the 3rd American peptide symposium (ed. J. MEIENHOFER), p. 701. Ann Arbor (Mich.): Ann Arbor Science Publishers 1972
1409. YANG, C. A.: Biochim. Biophys. Acta **133**, 346 (1967)
1410. YOUNG, G. T.: Angew. Chem. **71**, 742 (1959)
1411. YOUNG, G. T.: J. Chem. Soc. **1963**, 1105
1412. YOUNG, J. D., BENJAMINI, E., STEWART, J. M., LEUNG, C. Y.: Biochemistry (U.S.A.) **6**, 1455 (1967)
1413. YPHANTIS, D. A.: Biochemistry (U.S.A.) **3**, 297 (1964)
1414. ZAHN, H.: Naturwissenschaften **54**, 396 (1967)
1415. ZAHN, H.: DANHO, W., GUTTE, B.: Z. Naturforsch. **21b**, 763 (1966)
1416. ZAHN, H., DIEHL, J. F.: Angew. Chem. **69**, 135 (1957)
1417. ZAHN, H., LAFRANCE, N. H.: Ann. Chem., Justus Liebigs **630**, 37 (1960)
1418. ZAHN, H., LAFRANCE, N. H.: Z. physiol. Chem., Hoppe-Seyler's **319**, 143 (1960)
1419. ZAORAL, M.: Collection Czechoslov. Chem. Commun. **27**, 1273 (1962)
1420. ZERVAS, L., BOROVAS, D., GAZIS, E.: J. Am. Chem. Soc. **85**, 3660 (1963)
1421. ZERVAS, L., HAMALIDIS, C.: J. Am. Chem. Soc. **87**, 99 (1965)
1422. ZERVAS, L., PHOTAKI, I.: J. Am. Chem. Soc. **84**, 3887 (1962)
1423. ZERVAS, L., PHOTAKI, I., COSMATOS, A., BOROVAS, D.: J. Am. Chem. Soc. **87**, 4922 (1965)
1424. ZERVAS, L., PHOTAKI, I., GHELIS, N.: J. Am. Chem. Soc. **85**, 1337 (1963)
1425. ZERVAS, L., THEODOROPOULOS, D. M.: J. Am. Chem. Soc. **78**, 1359 (1956)
1426. ZERVAS, L., WINITZ, M., GREENSTEIN, J. P.: J. Org. Chem. **22**, 1515 (1957)
1427. ZIMMERMAN, C. L., PISANO, J. J., APPELLA, E.: Biochem. Biophys. Res. Commun. **55**, 1220 (1973)
1428. ZUBER, H.: Nature **201**, 613 (1964)
1429. ZUBER, H.: Z. physiol. Chem., Hoppe-Seyler's **349**, 1337 (1968)
1430. ZUBER, H., MATILE, PH.: Z. Naturforsch. **23b**, 663 (1968)
1431. ZUBER, H.: Abstr. 7th Intern. Congr. Biochem., Tokyo, p. 541, 1967

Subject Index

Molecular Biology, Biochemistry and Biophysics

Editors:
A. Kleinzeller
G. F. Springer
H. G. Wittmann

Vol. 1: J.H.van't Hoff
Imagination in Science
Translated into English with notes and a general introduction by G.F.Springer
1 portrait. VI, 18 pp. 1967
ISBN 3-540-03933-3 DM 9,80
ISBN 0-387-03933-3 (North America) $2.70

Vol. 2: K.Freudenberg, A.C.Neish
Constitution and Biosynthesis of Lignin
10 figs. IX, 129 pp. 1968
ISBN 3-540-04274-1
Cloth DM 38,–
ISBN 0-387-04274-1 (North America) $11.40

Vol. 3: T.Robinson
The Biochemistry of Alcaloids
37 figs. X, 149 pp. 1968
ISBN 3-540-04275-X
Cloth DM 48,–
ISBN 0-387-04275-X (North America) $14.60

Vol. 5: B.Jirgensons
Optical Activity of Proteins and Other Macromolecules
2nd revised and enlarged edition
71 figs. IX, 199 pp. 1973
ISBN 3-540-06340-4
Cloth DM 64,–
ISBN 0-387-06340-4 (North America) $24.60

Vol. 6: F.Egami, K.Nakamura
Microbial Ribonucleases
5 figs. IX, 90 pp. 1969
ISBN 3-540-04657-7
Cloth DM 38,–
ISBN 0-387-04657-7 (North America) $11.40

Vol. 8: **Protein Sequence Determination**
A Sourcebook of Methods and Techniques
Edited by S.B.Needleman
2nd revised and enlarged edition
79 figs. Approx. 400 pp. 1975
ISBN 3-540-07256-X
In preparation

Vol. 9: R.Grubb
The Genetic Markers of Human Immunoglobulins
8 figs. XI, 152 pp. 1970
ISBN 3-540-05211-9
Cloth DM 48,–
ISBN 0-387-05211-9 (North America) $17.30

Vol. 10: R.J.Lukens
Chemistry of Fungicidal Action
8 figs. XIII, 136 pp. 1971
ISBN 3-540-05405-7
Cloth DM 48,–
ISBN 0-387-05405-7 (North America) $14.90

Vol. 11: P.Reeves
The Bacteriocins
9 figs. XI, 142 pp. 1972
ISBN 3-540-05735-8
Cloth DM 58,–
ISBN 0-387-05735-8 (North America) $19.60

Vol. 12: T.Ando, M.Yamasaki, K.Suzuki
Protamines
Isolation, Characterization, Structure and Function
24 figs. 17 tables. IX, 114 pp. 1973. ISBN 3-540-06221-1
Cloth DM 58,–
ISBN 0-387-06221-1 (North America) $19.60

Vol. 13: P.Jollès, A.Paraf,
Chemical and Biological Basis of Adjuvants
24 figs. 41 tables. VIII, 153 pp. 1973. ISBN 3-540-06308-0
Cloth DM 58,–
ISBN 0-387-06308-0 (North America) $19.60

Vol. 14: **Micromethods in Molecular Biology**
Edited by V.Neuhoff
275 figs. (2 in color). 23 tables XV, 428 pp. 1973
ISBN 3-540-06319-6
Cloth DM 98,–
ISBN 0-387-06319-6 (North America) $40.20

Vol. 15: M.Weissbluth
Hemoglobin
Cooperativity and Electronic Properties
50 figs. VIII, 175 pp. 1974
ISBN 3-540-06582-2
Cloth DM 72,–
ISBN 0-387-06582-2 (North America) $29.60

Vol. 16: S.Shulman
Tissue Specificity and Autoimmunity
32 figs. XI, 196 pp. 1974
ISBN 3-540-06563-6
Cloth DM 60,–
ISBN 0-387-06563-6 (North America) $24.60

Vol. 17: Y.A.Vinnikov
Sensory Reception
Cytology, Molecular Mechanisms and Evolution
Translated from the Russian by W.L.Gray and B.M.Crook
124 figs. (173 separate ill.) IX, 392 pp. 1974
ISBN 3-540-06674-8
Cloth DM 118,–
ISBN 0-387-06674-8 (North America) $48.40

Vol. 18: H.Kersten, W.Kersten
Inhibitors of Nucleic Acid Synthesis
Biophysical and Biochemical Aspects
73 figs. IX, 184 pp. 1974
ISBN 3-540-06825-2
Cloth DM 68,–
ISBN 0-387-06825-2 (North America) $27.80

Vol. 19: M.B.Mathews
Connective Tissue
Macromolecular Structure and Evolution
31 figs. XII, 318 pp. 1975
ISBN 3-540-07068-0
Cloth DM 78,–
ISBN 0-387-07068-0 (North America) $31.90

Vol. 20: M.A.Lauffer
Entropy-Driven Processes in Biology
Polymerization of Tobacco Mosaic Virus Protein and Similar Reactions
90 figs. X, 264 pp. 1975
ISBN 3-540-06933-X
Cloth DM 73,–
ISBN 0-387-06933-X (North America) $29.80

Prices are subject to change without notice

Distribution rights for U.K., Commonwealth and the Traditional British Market (excluding Canada): Chapman & Hall Ltd., London

Springer-Verlag
Berlin
Heidelberg
New York